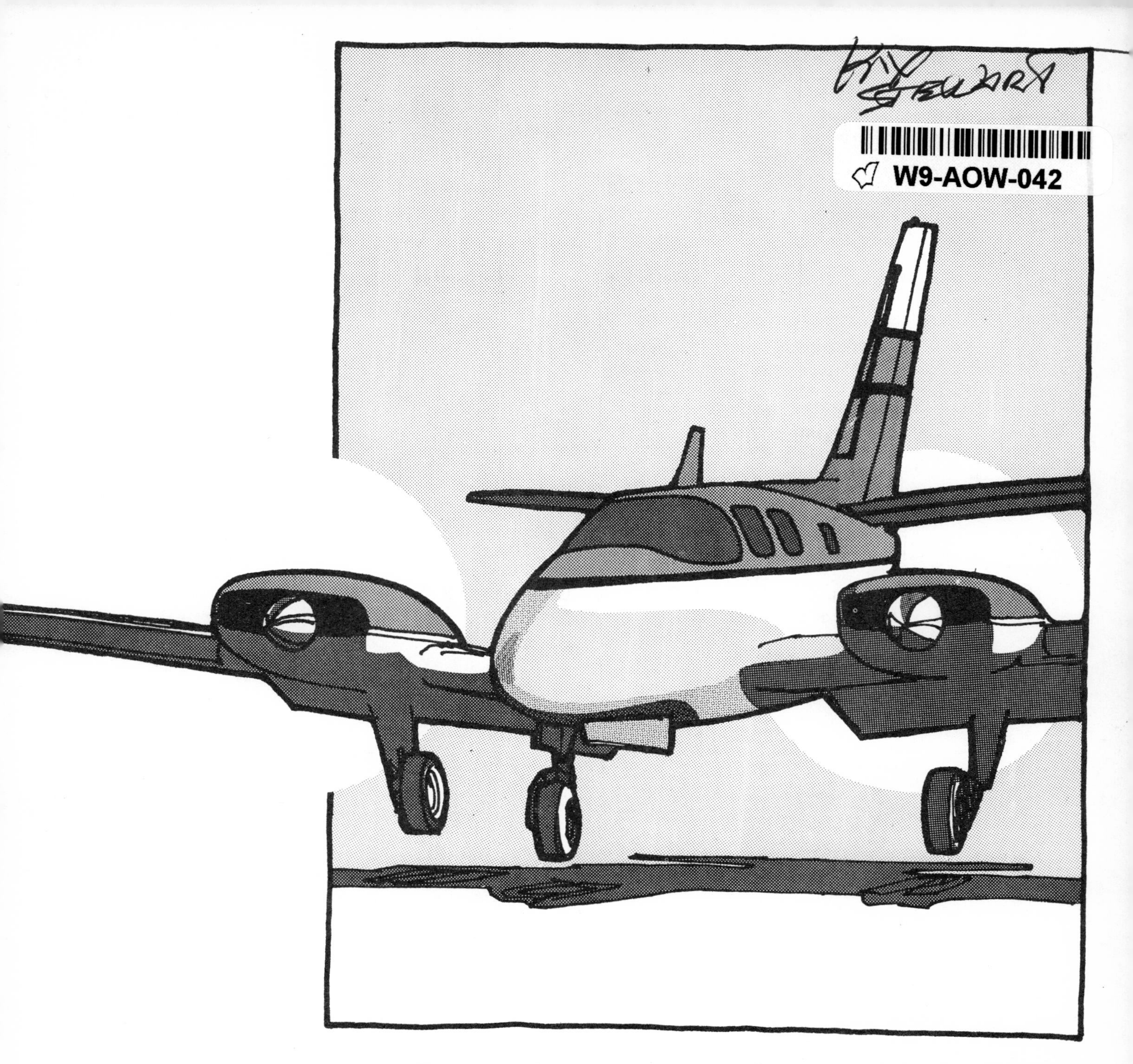

THE ADVANCED PILOT'S FLIGHT MANUAL

- ☐ AIRPLANE PERFORMANCE AND STABILITY FOR PILOTS
- ☐ CHECKING OUT IN ADVANCED MODELS AND TYPES
- ☐ EMERGENCIES AND UNUSUAL SITUATIONS
- ☐ CROSS-COUNTRY AND NIGHT FLYING
- ☐ PREPARING FOR THE COMMERCIAL FLIGHT TEST

WILLIAM K. KERSHNER

IOWA STATE UNIVERSITY PRESS, AMES, 50010

Third edition, 1970
Second printing, 1971
Third printing, 1972
Fourth printing, 1973
Fifth printing, 1974

International Standard Book Number: 0-8138-1300-X
Library of Congress Catalog Card Number: 70-126167

ILLUSTRATED BY THE AUTHOR

WILLIAM KERSHNER has two decades of experience in flying over 50 types and models of airplanes from light trainers to jets. As a ground instructor and flight instructor he has taught many students—using the principles which he stresses in his books. He also holds the commercial and airline transport pilot certificate, and is the author of THE STUDENT PILOT'S FLIGHT MANUAL and THE INSTRUMENT FLIGHT MANUAL (Iowa State University Press).

TO MAC AND SARAH LISENBEE

INTRODUCTION

IT HAS LONG BEEN the writer's opinion that the average pilot could learn the basics of airplane performance very easily if the involved mathematics were bypassed. One of the purposes of *The Advanced Pilot's Flight Manual* is to bridge the gap between theory and practical application. If the pilot knows the principles of performance he can readily understand the effects of altitude, temperature and other variables on the operation of his airplane.

Throughout this book there is reference to the "Airplane Flight Manual." However, airplanes having a gross weight of 6000 pounds or less are not required to have an Airplane Flight Manual as such, but are required to have certain limitations information available to the operator in the form of clearly stated placards, markings, or manuals. The Owner's Manuals (or Owner's Handbooks) for some makes of airplanes of less than 6000 pounds gross weight contain all this information required by the FAA. Other Owner's Manuals contain only descriptions and suggestions for operating the airplane and information such as airspeed limitations, weight and balance limits, etc., is furnished in the form of a separate sheet or sheets (Operations Limitations or Airplane Flight Manual) which are kept in the airplane. An airplane with the "Complete Airplane Flight Manual" type of Owner's Handbook must still have the information in the airplane at any time it's being operated.

When it is said in this book for instance, "You can find the center of gravity limits in the Airplane Flight Manual," the term is meant to cover all sources of information available to the operator. You may be able to open the Owner's Handbook and get this information *or* you may have to go out to the airplane and look on the sheets of paper termed "Operations Limitations" or "Airplane Flight Manual."

So, rather than repeat each time, "You can find this information in either the Owner's Handbook (or Owner's Manual) or Pilot's Handbook, or Flight Handbook, or Airplane Flight Manual, or Operations Limitations or in the form of placards or markings," the writer has used one term: "Airplane Flight Manual."

Thumb rules are used throughout as a means of presenting a clearer picture of the recommended speeds for various performance requirements such as maximum range, maximum endurance, maximum angle of climb, etc. Being rules of thumb, they are not intended in any way to replace the figures as given by the Airplane Flight Manual or comparable information sources, if available. However, the knowledge even of the approximate speed ranges for various maximum performance requirements will enable the pilot to obtain better performance than if he had no idea at all of the required airspeeds. Naturally, this practice must be tempered with judgment. If a pilot flies at rich mixture and high power settings until he has only a couple of gallons of fuel left, setting up either the rule of thumb or the manufacturer's recommended airspeed for maximum range still won't allow him to then make an airport 75 miles farther on. The same would apply to maximum endurance. Waiting to the last minutes of fuel to set up the maximum endurance speed will have no perceptible effect on increasing endurance.

Many of the rules of thumb are based on the use of airspeed indicator markings, which are in turn based on the *gross weight* (which unless otherwise noted will be the maximum certificated weight of the airplane). Airplane weight variation effects on recommended airspeeds are to be ignored unless specifically mentioned.

The reader of the earlier *Student Pilot's Flight Manual* will notice that certain parts of that book, such as instrument theory and emergency flying by reference to instruments, have been repeated here, though in somewhat more detail. This was done to make sure that this book is complete in itself. With the new emphasis by the FAA that pilots have the ability to fly on instruments in an emergency situation, and the safety gained by having this ability, it behooves the private pilot to learn as much as possible in this area. Pilots receiving their certificates before May, 1960, were not required to demonstrate this ability, but the reader who falls into that category may want to obtain a Blue Seal on his private certificate, or continue on for the commercial certificate. He is definitely encouraged to do so.

The material in this book includes what is felt by the writer to be of most interest to the private pilot. For instance, the chapters on checking out in advanced models and types are intended to cover the questions most often asked by private pilots checking out in those airplanes.

It is hoped that the material whets the reader's desire for more information. If so, then the mission of this book will have been accomplished. The books listed in the bibliography are recommended for further study.

The writer was fortunate in having knowledgeable people available to comment on the manuscript and would like to thank the following people for their help (these acknowledgments are not meant to imply endorsement of *The Advanced Pilot's Flight Manual* by any of the organizations mentioned, and should not be construed as such):

John Paul Jones, Chief of the Engineering and Manufacturing Branch of the FAA Training Center, and H. E. Smith, Jr., engineering pilot of the Center, who reviewed the complete manuscript and made valuable comments and suggestions.

Dr. Mervin Strickler, of the FAA Public Affairs Office, who aided in getting the manuscript to the right people in the FAA.

Leighton Collins, Editor of *Air Facts*, who reviewed the first half of the manuscript and whose suggestions on methods of presenting some of the more technical parts were of particular value.

Charles Maggart, of the Army Aviation School at Fort Rucker, Ala., whose comments on this manuscript as in that of the *Student Pilot's Flight Manual* were greatly appreciated.

Jack LeBarron, of the Navy's *Approach* magazine, who always managed to find and furnish performance information when it was most needed.

Ken Landis, Chief Controller at the Williamsport, Pennsylvania, airport, who kindly reviewed that part of the manuscript dealing with communications procedures.

People at Piper Aircraft who reviewed parts of the manuscript as it covered their special areas and who made valuable suggestions and comments included: Clyde R. Smith, Flight Test Supervisor; Calvin Wilson, Aerodynamicist; and Elliot Nichols, Design Engineer, Power Plants Installation.

Harold Andrews of the Stability and Control Section of the Bureau of Naval Weapons, Washington, D.C., for his pertinent comments on the chapter on Stability and Control and other parts.

Bernard Carson, instructor in the Aeronautical Engineering Department at Pennsylvania State University, who reviewed the first half of the manuscript and made comments on flight limitations which were an aid in writing that part.

W. D. Thompson, Chief of Flight Test and Aerodynamics of Cessna Aircraft for furnishing earlier information written by himself concerning the use of thumb rules for performance, in addition to furnishing answers to specific questions concerning Cessna aircraft.

I appreciate information furnished by Richard Wible and others at Piper as well as by people at Aero Design, Beech and Mooney and other companies, who helped me in quests for sometimes obscure reports and information.

Special acknowledgment must be given to my editor, Merritt Bailey of the Iowa State University Press, who was instrumental in assuring that this book, like the *Student Pilot's Flight Manual*, was published in a manner more readable than that evidenced by my sometimes rambling discussions in the manuscript.

To my wife, whose encouragement never faltered and who furnished practical help in the form of typing the manuscripts, I am most indebted.

INTRODUCTION TO THE LATER EDITIONS

This manual has been revised and updated in an attempt to give the reader the most up-to-date information. Although the basics of flight do not change, the method of presentation of services available to the pilot in the way of teletype weather information, NOTAMS and other sources of important flight data are in a constant state of flux. As a major revision may take up to six months from completion to printing it is suggested that the reader use Chapters 19 and 20 as basic information and obtain the latest detailed information from the *Airman's Information Manual* for such things as frequencies used by the various FAA facilities, etc. This also goes for Weather Bureau services available.

The biggest problem for the writer of an aviation text is that of knowing several months ahead of a proposed new pilot service that is slated to come just before or shortly after the book would be off the press; proposed services or changes in service sometimes are dropped at the last minute and this book is not written with the hope of getting a "scoop" that may not materialize. There are several proposed changes in pilot services that may (or may not) materialize before this book gets to the reader's hands but they have not been mentioned in the text.

A new chapter on flight mechanics has been added in an attempt to shoot down old theories that Lift makes the airplane climb or that Thrust merely acts to make the airplane go faster. The subject of flight mechanics has long been available in aeronautical engineering texts but all too often the word has not been passed to the pilot, and that is the purpose of Chapter 2.

The Drag and power curves have all been changed to give representative values for typical high performance single or light twin engine general aviation airplanes as noted. In most cases the curves have been modified slightly to obtain even values for certain important airspeeds, but have been designed to be within reason for the types of airplanes discussed.

The primary purpose of this book is that of examining airplane performance and the effects of the Four Forces (Lift, Weight, Drag and Thrust) from a pilot's viewpoint. The Four Forces have been capitalized throughout to bring to the reader's attention the importance of these forces in the performance of any maneuver.

The reader with an engineering background may note that "illegitimate" scales were used on some of the curves, particularly for airspeed values. This was done because of space limitations in trying to show the entire range of airspeeds, and with the realization that they were not "working" curves.

Again, I was fortunate in having knowledgeable individuals furnishing information in preparing this second edition; however, any errors in this book are mine. I would like to thank the following people:

W. D. Thompson, Chief of Flight Test and Aerodynamics of Cessna Aircraft Company, for furnishing actual curves of Drag and Thrust horsepower required and available, and other information on Cessna airplanes.

Allen W. Hayes, veteran flight instructor of Ithaca, N.Y., whose cogent arguments on the subject of flight mechanics were the greatest factor in researching and inserting a chapter on that subject.

Delbert W. Robertson, W. R. Wright and Hugh Pritchard of the Chattanooga Weather Bureau Airport Station and Jack Merryman of the Nashville WBAS who answered questions and furnished actual sequence reports and other weather data for publication.

Calvin Wilson, Aerodynamicist, and Richard Wible of that department of Piper Aircraft Corporation, for furnishing data on Piper airplanes.

Fred C. Stashak, Chief Stress Engineer of Piper Aircraft Corporation, for furnishing information on Piper airplanes.

William Schmedel, Director of Training, National Aviation Academy, St. Petersburg, Fla., who pointed out errors in earlier printings.

Appreciation is expressed to *Business/Commercial Aviation* magazine for permission to use material from some of my articles on airplane performance.

Thanks must also go to American Aviation Publications for permission to take parts of articles I wrote for *Skyways* magazine.

As was the case for the first edition, these could not have been completed without the help of my wife whose typing and pasting talents were invaluable.

William K. Kershner
Sewanee, Tennessee

TABLE OF CONTENTS

PART I / AIRPLANE PERFORMANCE AND STABILITY FOR PILOTS

1. THE FOUR FORCES

BACKGROUND

Lift, Drag, Thrust and Weight are the Four Forces acting on an airplane in flight. The actions of the airplane are affected by the balance (or imbalance) of these forces and while each will be discussed separately in this chapter don't get the idea that each works completely separately.

Sure, you can fly an airplane without knowing how Lift or the other of the Four Forces work, *but* a good idea of the factors affecting each of them can lead to analyzing and predicting the performance of your airplane under different conditions of Weight, altitude, etc. The job of this chapter will be to take a look at how each Force is developed and the next chapter will show how it *acts* in flight.

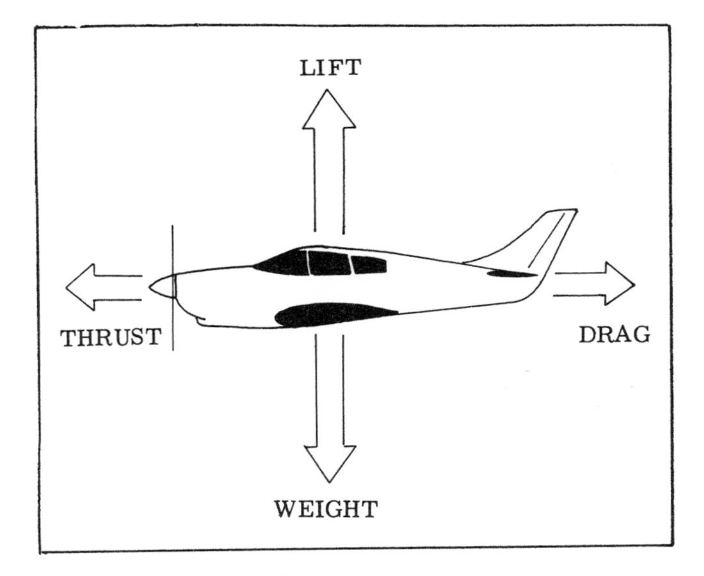

Fig. 1-1. The Four Forces.

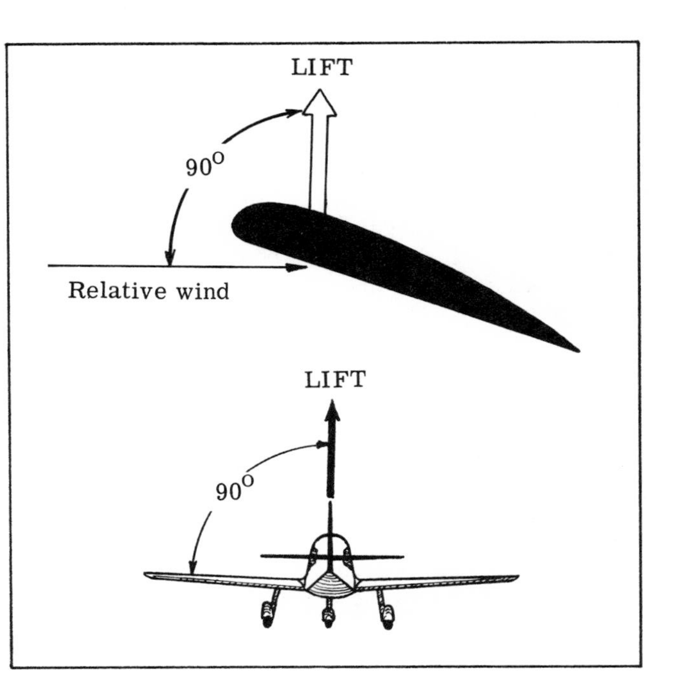

Fig. 1-2. Lift acts perpendicular to the relative wind and to the wingspan.

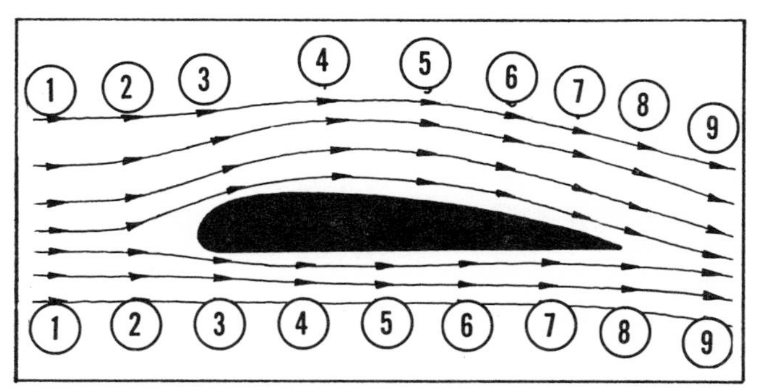

Fig. 1-3.

LIFT

Lift is the force exerted primarily by the wings (although the fuselage contributes, and even the tail helps under certain special conditions) and is created by the action of the air moving past the airfoil (cross sectional shape of the wing). *Lift is considered to act perpendicularly to the relative wind and to the wingspan.* (Fig. 1-2)

The action of the airfoil can best be explained by Bernoulli's Theorem which basically states, "The faster a fluid moves past a body, the less sidewise pressure is exerted on the body by the fluid." The fluid is air in this case. The body is an airfoil. Shown in Fig. 1-3 is the typical reaction of an airflow past an airfoil such as is normally used on light planes.

The distance that the air must travel is greater over the top of the airfoil than under the bottom. The air passing over the top surface must travel faster in an attempt to re-establish equilibrium at the trailing edge of the airfoil. Thus, less sidewise force is exerted on top; a difference in pressure now exists, and the force known as Lift is created. Fig. 1-4 gives some of the nomenclature of the airfoil. Note that the

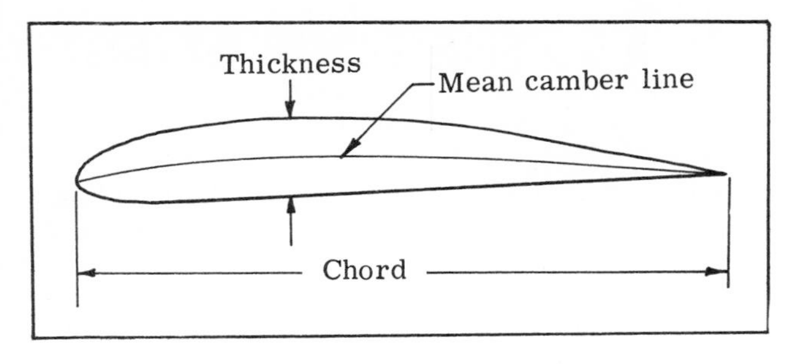

Fig. 1-4. Airfoil Nomenclature.

mean camber (average curve) line is equidistant from each surface.

The theory of airfoils is a lot more complicated than just described, but at this point there isn't much need of going into such things as circulation, bound vortices and the like. If you are interested in more information in this regard you might check the recommended texts at the end of this book.

Standard sea level pressure is 2116 pounds per square foot (psf). The airfoil on an airplane sitting stationary at sea level has this force acting on every square foot of the surface. There's no pressure difference and therefore no forces at work. (Fig. 1-5)

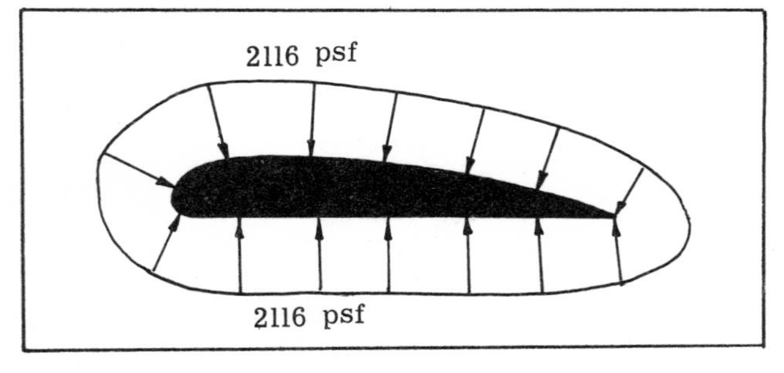

Fig. 1-5.

As the airplane starts moving, the pressure on top of the wing drops; the pressure on the bottom is slightly greater than normal and Lift begins to become a force with which to be reckoned. At the beginning of the take-off run Lift is weak, but as the speed increases it soon becomes enough to support the airplane in flight.

About 75 per cent of the wing's Lift is caused by the drop in pressure on the top surface, with the increase in pressure on the bottom contributing the other 25 per cent; these percentages vary slightly with angle of attack. There is certainly not a vacuum on top of the wing as many people mistakenly believe. Assume that an airplane equipped with the airfoil just discussed weighs 3000 pounds and has a wing area of 150 square feet. (Fig. 1-6)

For level cruising flight, Lift is considered to equal Weight and therefore each square foot of the wing must carry a load of 20 pounds. This is the wing loading, 20 pounds per square foot. Different parts of the wing carry different loads but this is the average for the example airplane. Assuming that 75 per cent of the Lift is due to the decreased pressure on top and 25 per cent to the increased pressure

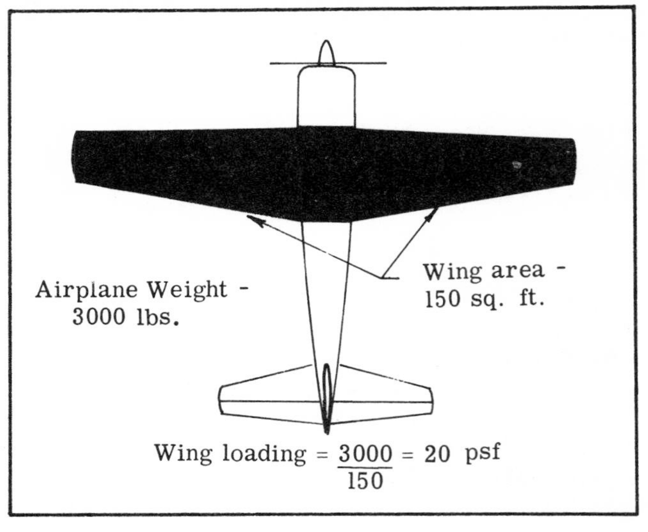

Fig. 1-6.

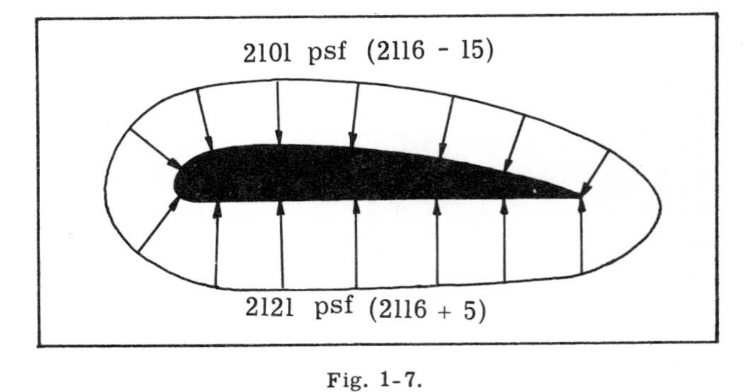

Fig. 1-7.

underneath, this means a pressure drop of 15 psf on top and a pressure increase of 5 psf underneath. The pressures acting on the airfoil are now as shown in Fig. 1-7.

Lift is actually a result of several factors and the equation for Lift is $L = C_L S \frac{\rho}{2} V^2$ or $L = C_L \times S \times \frac{\rho}{2} \times V^2$.

Where L - Lift, in pounds

C_L - Coefficient of Lift. (Varies with the type of airfoil used and angle of attack.)

S - Wing area, in square feet.

$\frac{\rho}{2}$ - The air density (ρ) divided by 2. Rho (ρ) is air density, which for standard sea level conditions is 0.002378 slugs per cubic foot. If you want to know the mass of an object in slugs divide the Weight (in pounds) by the acceleration of gravity, or 32.2. (The acceleration caused by gravity is 32.2 feet per second, per second at the earth's surface.)

and V^2 - True velocity (true airspeed) of the air particles in feet per second (squared).

Coefficient of Lift is a relative measure of an airfoil's lifting capabilities. Comparatively high-lift airfoils such as the Clark Y type (the airfoil used in the drawings earlier), with its curved or cambered upper surface and flat lower surface, may have a maximum coefficient of Lift of 1.8. A thin airfoil, such as might be used on a jet, may have a maximum coefficient of Lift of only 0.9. The airplane having the higher maximum coefficient of Lift, or $C_{L_{max}}$, will use less runway on landing (assuming two airplanes of equal wing loading and operating in the same air density). Having available the Coefficient of Lift versus Angle of Attack curves of various airfoils, the engineer can decide which airfoil would be better to use for a particular airplane. Your contact with the term is only through control of the angle of attack while flying.

A plot of C_L versus Angle of Attack for a NACA 23012, a typical general aviation airplane type of airfoil, shows that the C_L increases in a straight line with an increase in angle of attack until the stalling angle is reached, at which point C_L drops off rapidly. (Fig. 1-8)

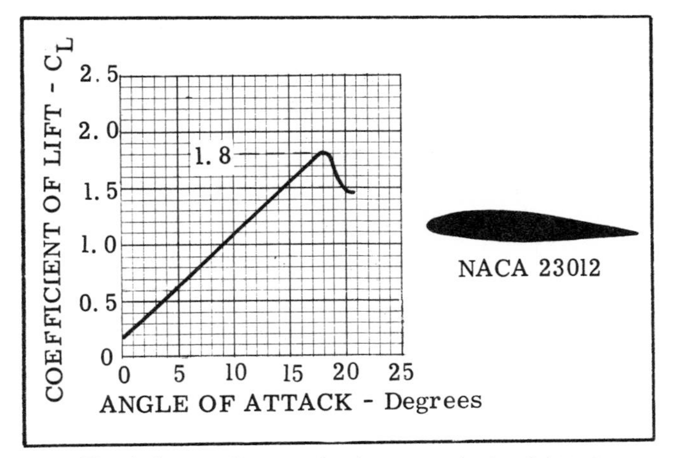

Fig. 1-8. Coefficient of Lift versus Angle of Attack — NACA 23012 airfoil.

Now look at Fig. 1-9 which shows the C_L vs. Angle of Attack for an NACA 0006, a high speed symmetrical airfoil such as may be used on jets. Its maximum coefficient of Lift is only 0.9 which means that the airplane would have a high landing speed.

These airfoil designations describe the airfoil properties and shape. The 23012 is an unsymmetrical airfoil. The 12 indicates that the airfoil maximum thickness is 12 per cent of its chord. The 0006 airfoil is a symmetrical airfoil (the first two zeros tell this) with a thickness ratio of 0.06 or 6 per cent.

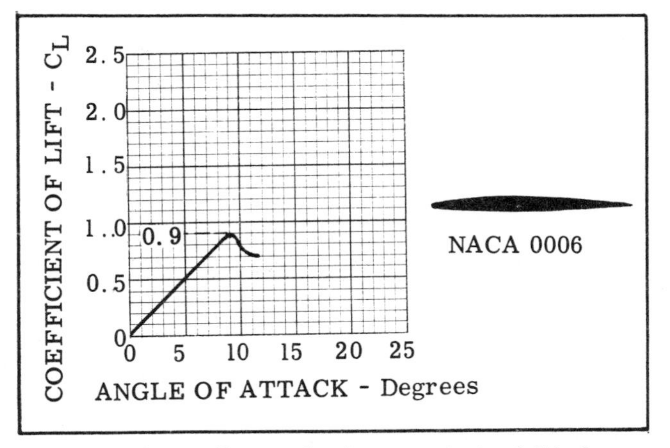

Fig. 1-9. Coefficient of Lift versus Angle of Attack — NACA 0006 airfoil

Take the 3000 pound plane with the 150 square foot wing area of the earlier example. Assume that there are two airplanes exactly alike with this Weight and wing area; one has a 23012 and the other a 0006 airfoil. The plane with the 0006 high speed airfoil will cruise faster because of less Drag, but will also land faster. In fact, it may land so fast as to be useless for many smaller airports.

It has been found that a "birdlike" airfoil (Fig. 1-10) has a comparatively high $C_{L_{max}}$. Earlier planes used this type and had low landing speeds — but also had low cruise speeds because of the higher Drag at all angles of attack. A good setup would be to have a 0006 airfoil for cruising and a birdlike airfoil for landing. Flaps accomplish just that; when you lower the flaps you raise the $C_{L_{max}}$ and have a lower landing speed.

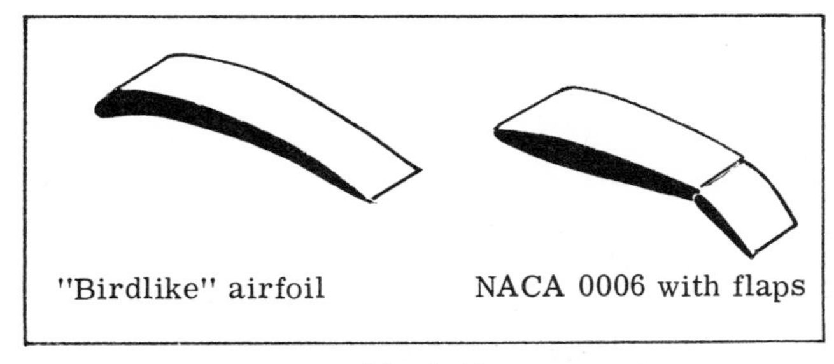

Fig. 1-10.

Fig. 1-11 shows a comparison of the Coefficient of Lift versus Angle of Attack for the NACA 23012 airfoil without flaps, and an NACA 0006 airfoil with flaps and without flaps. Notice the similarity of the curve of the 0006 with flaps with the 23012 curve. It can be readily seen that although the flaps installation adds Weight, flaps make it possible for fast airplanes to land at low speeds. The two airplanes of the same Weight and configuration in the example would probably use the same amount of runway for landing, assuming the plane with the 0006 airfoil used 60° of flaps and the one with the 23012 airfoil did not have flaps and, of course, the air density was the same for both airplanes.

By using the Lift equation and solving for velocity an actual comparison of the two airplane landing speeds can be found. ($L = C_L S \frac{\rho}{2} V^2$.) For convenience, we'll assume that the two airplanes are landing at sea level and that Lift just equals Weight at the touchdown and that the C_L in the equation is $C_{L_{max}}$ (both will touch down at the maximum angle of attack or minimum speed).

First, solving for the landing speed of the airplane with the 23012 airfoil and doing a little algebraic shuffling of the Lift equation:

$V = \sqrt{\frac{2L}{C_{L_{max}} S \rho}}$. Everything inside of the square root enclosure is known:

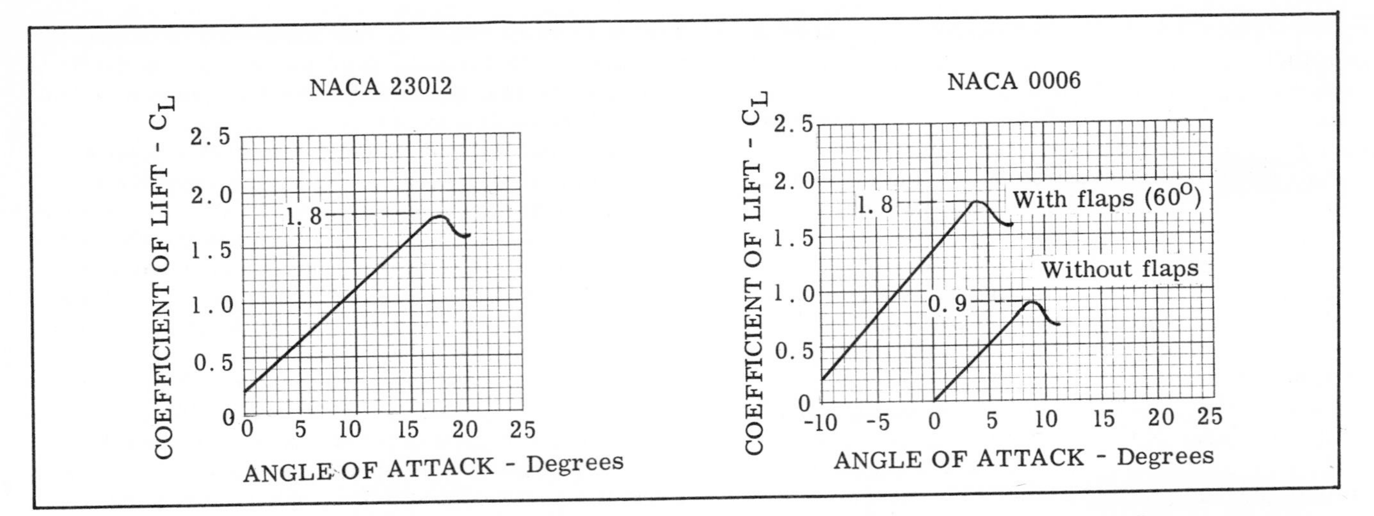

Fig. 1-11. Comparison of plain NACA 23012 airfoil and NACA 0006 airfoil with and without flaps.

L - (Weight) = 3000 pounds

$C_{L_{max}}$ - (for this airfoil) 1.8

S - 150 square feet

ρ - 0.002378 slugs per cubic foot

$$V = \sqrt{\frac{6000}{1.8 \times 150 \times 0.002378}} = \sqrt{\frac{6000}{0.64}} = \sqrt{9380} = 97 \text{(fps)}$$

Thinking in terms of miles per hour, V = *66 mph* (true airspeed).

Doing the same thing for the airplane with the 0006 airfoil (no flaps) which means that $C_{L_{max}}$ is 0.9 instead of 1.8 (everything else is the same):

$$V = \sqrt{\frac{6000}{0.9 \times 150 \times 0.002378}} = \sqrt{\frac{6000}{0.32}} = \sqrt{18{,}750} = 137$$

(fps) or *93.5 mph* (true airspeed).

(At sea level the indicated and true airspeeds will be the same, assuming no airspeed instrument or position error.)

A pretty "hot" airplane but it could be cooled down by adding the flaps mentioned earlier.

Maybe you're not interested in working out the algebra of the landing but the problem just covered was intended to show that things are not so mysterious about the workings of airplanes — if you have the right equation. Of course, the answers were rounded off and a few assumptions were made, and square roots are hard to hack out without a slide rule or logarithm table, so you're not apt to go through these calculations every time you land.

Getting back to Lift in general, the funny thing about it is that you really don't worry about how much Lift you have in normal flying — you fly the airplane and Lift takes care of itself. Sure, you can be flying along at cruise and increase Lift by pulling back on the wheel but you can *feel* that Lift is greater than it should be by the way you are being pressed down in the seat. As Drag is increased by the increase of Lift, you'll find that the airplane slows — and Lift will tend to regain its old value again.

One time when you *are* interested in watching Lift increase is on take-off. If your 3000 pound airplane with the 23012 airfoil (or any airfoil) is taking off at a high elevation (where the density is low) it must have a higher V^2 (true airspeed, squared) in order to make up for the lower density to get the required Lift for lift-off. This is *one* of the reasons why a longer take-off run is required at airports at higher elevations. (The big reason is that the engine is producing less power in the less dense air, but this will be covered in later chapters.)

HIGH LIFT DEVICES

The effects of flaps have been discussed and, as noted, can increase the maximum coefficient of Lift as compared to the wing without flaps. While the term "high lift device" is commonly used, actually *the purpose of flaps, slots or slats is to provide the same Lift as before (say, 3000 pounds) at a lower airspeed and they are actually not used to increase the Lift over that required.* If you are flying at a high speed and suddenly put the flaps down it is true that Lift will be increased suddenly and the airplane will *accelerate* upward — and you will again be pressed down sharply in the seat. You could find that by a simple matter of "increasing the Lift" in such a fashion certain problems could arise (such as leaving a trail of flaps and other parts of the airplane fluttering behind). *So the high lift devices are used at low speeds.*

FLAPS

Flaps are the most widely used high lift device and there are many types in use, a few of which will be covered here.

1. Plain Flap — A simple means of changing the camber of the airfoil for use at low speeds. (Fig. 1-12)
2. Split Flap — You can see in Fig. 1-13 that there is a low pressure region between the flap and the wing so that for equal flap areas and settings, the split flap tends to cause greater Drag than the plain flap, particularly at lesser angles of deflection.
3. Fowler Flap — This flap combines a camber

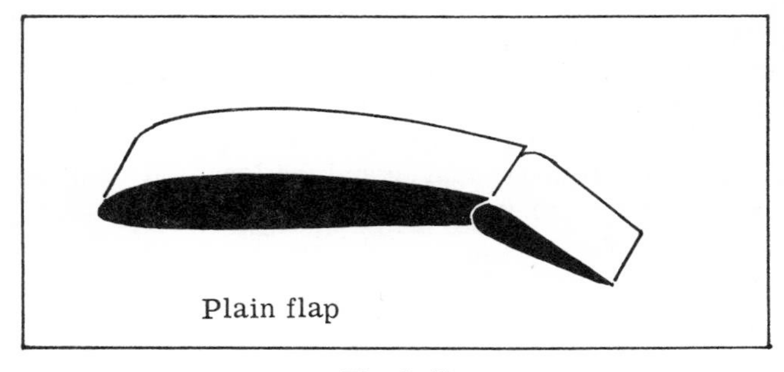

Fig. 1-12.

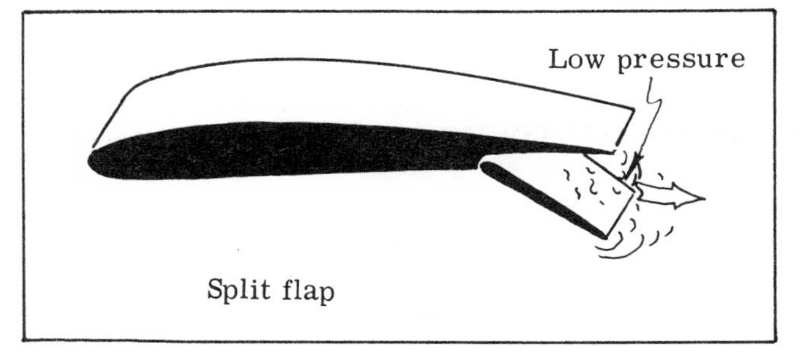

Fig. 1-13.

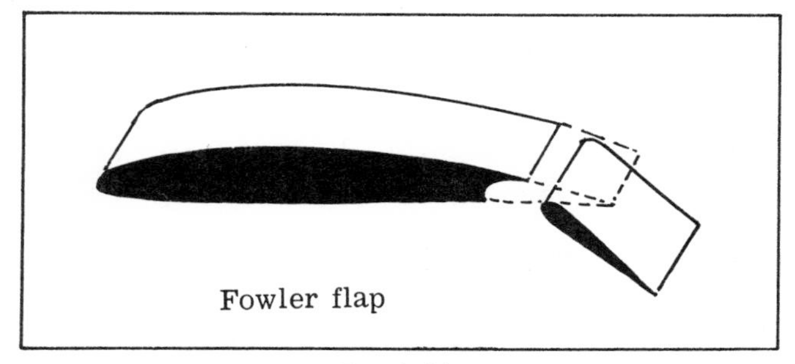

Fig. 1-14.

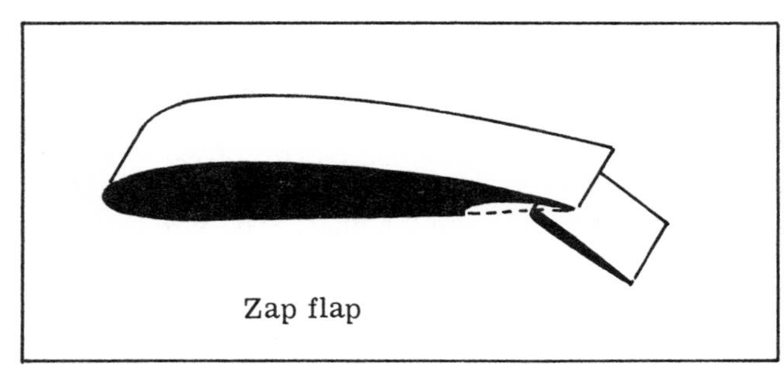

Fig. 1-15.

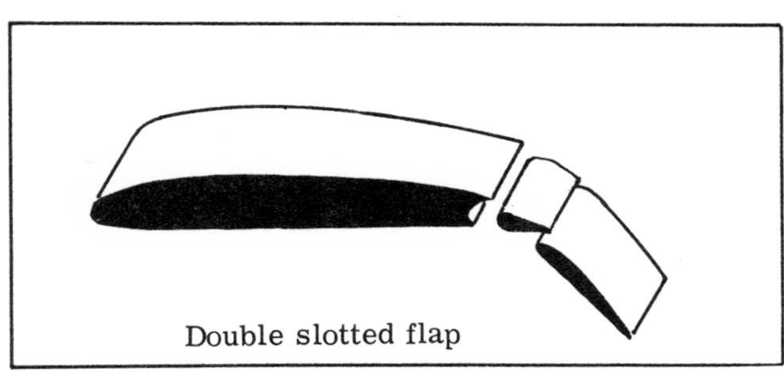

Fig. 1-16.

change with an increase in wing area — a good combination to lower the stall speed for landing but the system may be too complex and heavy for lighter planes. (Fig. 1-14)

4. Zap Flap — This is a split-type flap that increases wing area in the same way as the Fowler type. (Fig. 1-15)

5. Double Slotted Flap — By putting slots in the flaps, a combination of camber change and smoother flow is obtained. (Fig. 1-16)

SLOTS

The leading edge slot is a means of keeping a smooth flow at higher angles of attack than would be possible with an unslotted wing. Slots usually are placed near the wing tip to aid in lateral (aileron) control at the stall and to insure that the tips don't stall first, or they may be used along the entire span. You are familiar by now with the idea of a wing dropping during the stall and, like most pilots, probably prefer an airplane that has a good, straight-ahead stall break. (Fig. 1-17)

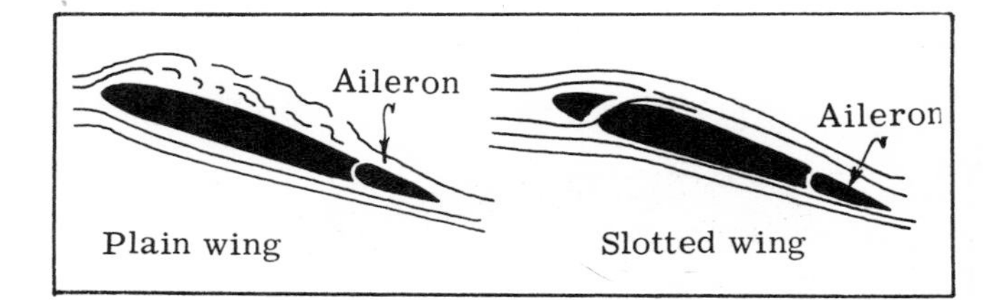

Fig. 1-17. Comparison of plain and slotted wing at higher angles of attack.

SLATS

Slats are movable leading edge vanes which form slots. The slot causes Drag at higher speeds and, as the slat can be retracted more or less flush with the leading edge, it is a boon to higher speed airplanes. Some jets use a "droop snoot" in combination with flaps to obtain a birdlike airfoil for lower landing speeds. (Fig. 1-18)

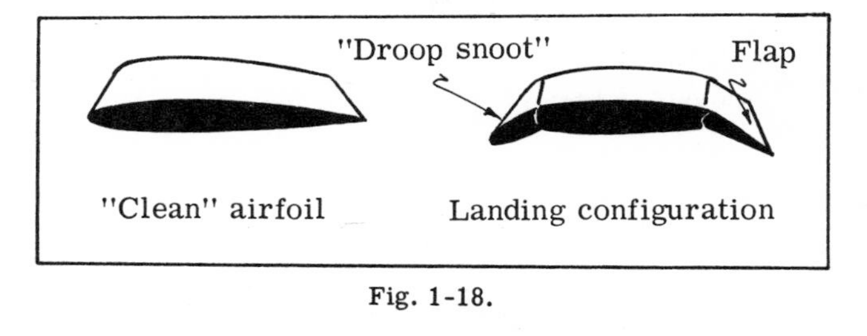

Fig. 1-18.

The high lift devices complement each other — that is, by adding flaps to a plain airfoil, the maximum coefficient of Lift is raised; by adding slots or slats to this flap-equipped wing, a further increase in $C_{L_{max}}$ is obtained.

BOUNDARY LAYER

You've probably heard this term many times. A good illustration of boundary layer can be seen on a dusty wing; you fly the airplane at speeds up to 200 knots but the dust isn't affected. The boundary layer effect is one factor that helps to keep the dust on. A definition of the boundary layer is that thin layer of air adjacent to the surface of a moving body and the velocity of the boundary layer air varies in speed from zero at the surface to the free stream velocity (true airspeed) at a certain distance from the surface. The thickness of a boundary layer varies for different conditions of velocity, surface roughness, etc., but normally may be considered in terms of very small fractions of an inch.

There are two types of boundary layers: (1) laminar or layered, smooth flow, and (2) turbulent. The laminar type creates much less skin friction Drag than the turbulent type and aeronautical engineers are particularly interested in maintaining a laminar boundary layer over as much of the wing and other components as possible at high speeds. Fig. 1-19 is a typical airfoil for a light trainer showing that both types are present.

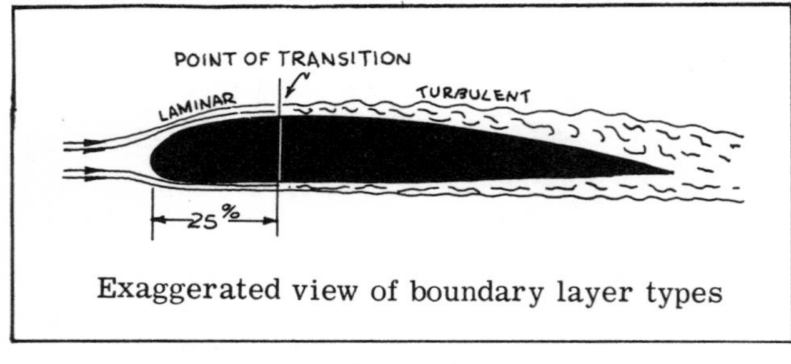

Exaggerated view of boundary layer types

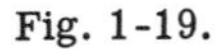
Fig. 1-19.

From a Drag standpoint then, it is advisable to have this transition from laminar to turbulent flow as far aft as possible, or have a large amount of the wing surface within the *laminar* portion of the boundary layer. As the transition usually occurs at approximately the thickest part of the airfoil (where the pressure is lowest) some airfoils are designed with the thickest parts at a position of 40 to 50 per cent of the chord instead of the usual 25 to 30 per cent. (Fig. 1-20)

These laminar types of airfoils are now being used on various high performance general aviation planes.

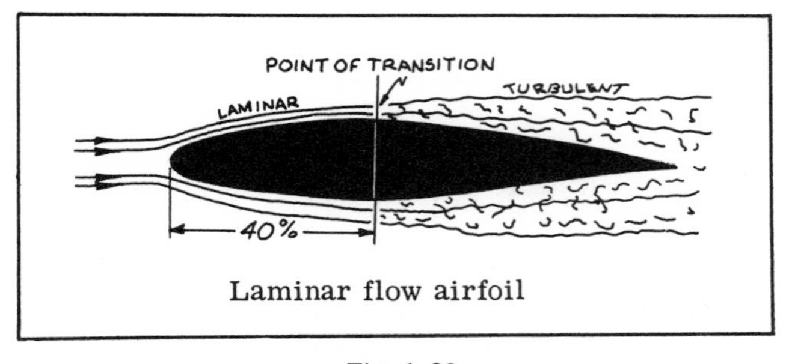

Laminar flow airfoil

Fig. 1-20.

While maintaining a laminar flow as long as possible is a decided advantage from a Drag standpoint, it doesn't always work quite so well for stall characteristics. One of the first steps in a stall is the separation of the boundary layer and the longer it remains intact the more delayed the stall (higher angle of attack and a greater $C_{L_{max}}$ at stall). The laminar layer tends to break down more suddenly than the turbulent layer so that the laminar flow airfoil usually does not have quite as good stall characteristics as might be found in older airfoil types (all other things being equal). The designer compromises between low Drag and good stall characteristics on the airplane using the laminar flow airfoil.

BOUNDARY LAYER CONTROL

For many years engineers have worked to find means of delaying boundary layer separation at higher angles of attack. Two methods are generally used: (1) Suction, which removes the boundary layer at various points of the airfoil, drawing it inward, and (2) Blowing, which adds energy to the boundary layer and in essence works as if the entire airfoil has a turbulent layer (with its resulting better stall characteristics because of later separation). Boundary layer control requires a great deal of energy which only can be furnished by adding Weight in the forms of pumps, piping and so forth. Jets use bleed-air from the compressor sections.

DESIGN OF THE WING

Every airplane is made up of compromises, and this is particularly noticeable in wing design. For speed, a tapered wing is better than a rectangular wing. But the tapered wing with no twist has poor stall characteristics as the tips tend to stall first. Common sense would tell you that the tapered wing has less Drag because of the lesser area near the tip – which results in less induced (vortex) Drag than on a rectangular wing of equal area (assuming the two planes have the same span loading). The elliptical wing (like that of the WW II *Spitfire*) is more efficient but does not have as good stall characteristics as the rectangular wing (if other factors are equal).

WING DESIGN AND THE STALL

There are several solutions to this problem of the stall. In every case the tips should stall *last*. You want lateral (aileron) control throughout the stall. An airplane that has bad rolling tendencies at the stall break is viewed with a jaundiced eye by pilots. The best stall pattern is to have the wing stall at the root area first, with the stall progressing outward toward the tips. This may be accomplished by several means:

WASHOUT OR TWIST

The wing may have a built-in twist so that the tip

has a *lower* angle of incidence (resulting in a lower angle of attack during the approach to a stall) and will be flying after the root section has stalled. Generally this difference in incidence is no more than 2 to 3 degrees from root to tip. The tips are said to have "washout" in this case. "Washed-in" tips would have a *higher* angle of incidence — which would hardly be conducive to pilot ease during the stall as the tips would naturally tend to stall first.

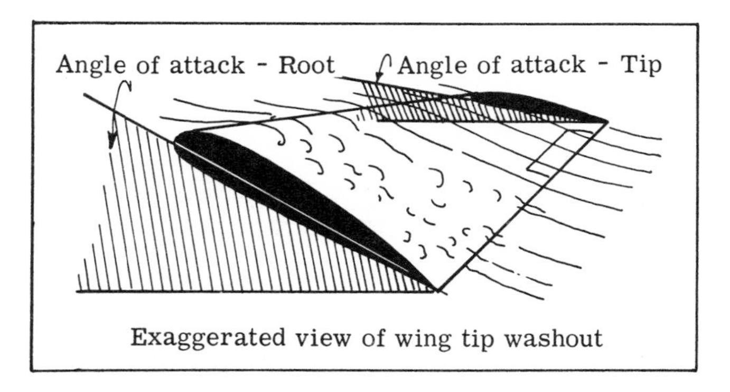

Fig. 1-21. Wing tip washout as a means of maintaining lateral control during the stall

SLOTS

Slots are not only considered to be a high lift device but are also used as a lateral control aid. Planes using slots usually have them only in that section near the wing tips, so that lateral control can be maintained throughout the stall, or at least so as not to have a wing drop suddenly with little warning, as may sometimes happen in a landing or practice stall. Unmodified tapered wings have this tendency. The slot gives the tapered wing airplane the benefit of added lateral control in the stall and tends to dampen any rolling tendency. If the tips stalled first (and it may be remembered that some airplanes have a different twist in each wing to counteract torque), a sizable rolling moment could be produced.

Some STOL (Short Take-Off and Landing) airplanes have full length slots or slats which make for good slow speed characteristics.

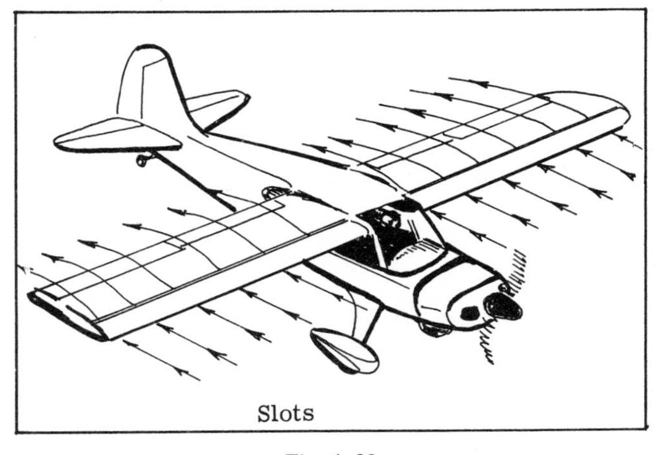

Fig. 1-22.

STALL STRIPS

Stall strips, or spin strips as they are sometimes called, are strips attached to the leading edge of the wing near the root. As the angle of attack increases, these strips break up the flow, which gives the desired effect of the root area stalling first.

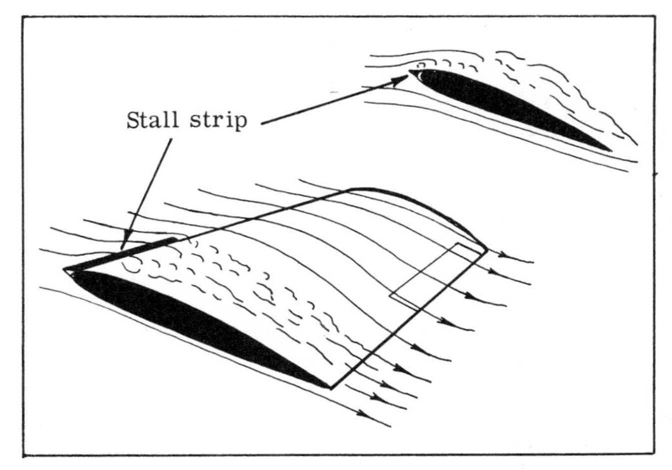

Fig. 1-23. The stall or spin strip.

SPANWISE AIRFOIL VARIATION

This high sounding title simply means that some wings may have a high speed type airfoil at the root and a low speed type at the tip. An extreme example would be a laminar flow airfoil at the root and a "birdlike" airfoil at the tip. The "birdlike" tip will be flying after the high speed section at the wing root has stalled. (Fig. 1-24) In some cases several of these wing design techniques may be combined. The fact that the root section stalls first tends to cause a flow disturbance that tends to result in tail buffeting and warning of the impending stall.

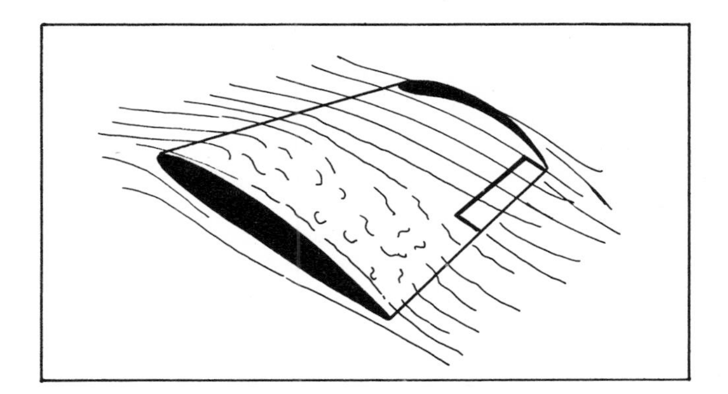

Fig. 1-24. Different airfoils at root and tip.

WINGTIP TANKS OR END PLATES

Many jets and several of the new high performance light planes use tip tanks. The aerodynamic effect of tip tanks or end plates is an increase in effective aspect ratio (span over average chord), and you'll see in the next section that a larger aspect ratio results in lower induced Drag. In most cases the tip tanks more than offset any additional penalty such as increased frontal area or skin friction area.

DRAG

Any time that a body is moved through a fluid such as air, Drag is produced. Airplane aerodynamic Drag is composed of two parts: induced Drag, or the Drag caused by Lift being created and parasite Drag (form Drag, skin friction, and interference Drag).

PARASITE DRAG

The factors affecting parasite Drag are similar to those affecting Lift. (Assume as each factor is discussed that the others remain constant.)

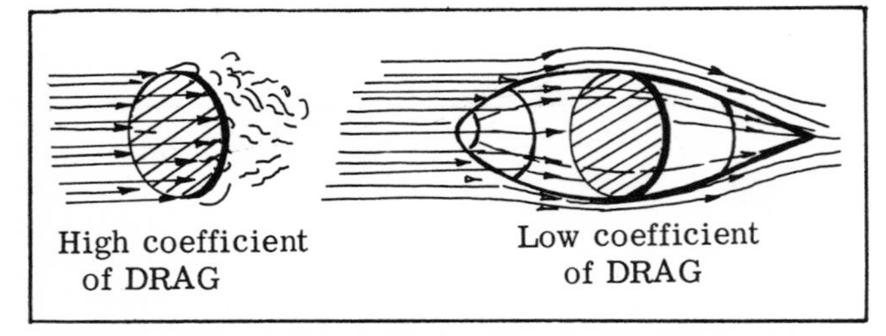

Fig. 1-25. A comparison of the Drag of a flat plate and a streamlined shape with the same cross sectional area at the same air density and velocity.

Coefficient of Parasite Drag — A relative measure of the parasite Drag of an object. The more *streamlined* an object the *lower* its coefficient of parasite Drag.

Air Density — The greater the density of the fluid moving past an object, the greater the parasite Drag, assuming the velocities are the same. You can prove this by noting the difference in effort required to move your hand through water and air at the same speeds.

Velocity — Double the airspeed and parasite Drag is quadrupled.

Area — Parasite Drag increases directly with the size of the object in the air stream. The engineers normally base the total Drag of an airplane on its wing area, so as to establish some basis for comparison between airplanes.

The Total Coefficient of Drag is the sum of C_{Di} (coefficient of induced Drag) and the C_{Dp} (coefficient of parasite Drag).

Or: $C_{D_{total}} = C_{Di} + C_{Dp}$ (More about C_{D_i} later).

Total Drag = $(C_{Di}+C_{Dp})S\frac{\rho}{2}V^2$.

FORM DRAG

This is the Drag caused by the frontal area of the airplane components. When you were a kid you no doubt stuck your hand out the car window when you went riding (until your parents noticed). When your hand was held palm forward, the Drag you felt was nearly all form Drag. When your hand was held palm down the Drag was caused mostly by skin friction. This can probably be best described by looking at a very thin flat plate. (Fig. 1-26)

You will note that in Fig. 1-26A the Drag existing is principally caused by the form of the plate whereas in Fig. 1-26B the largest part of the Drag is skin friction. This form Drag is the reason why streamlining is necessary in order to reach higher cruise speeds. (Fig. 1-27)

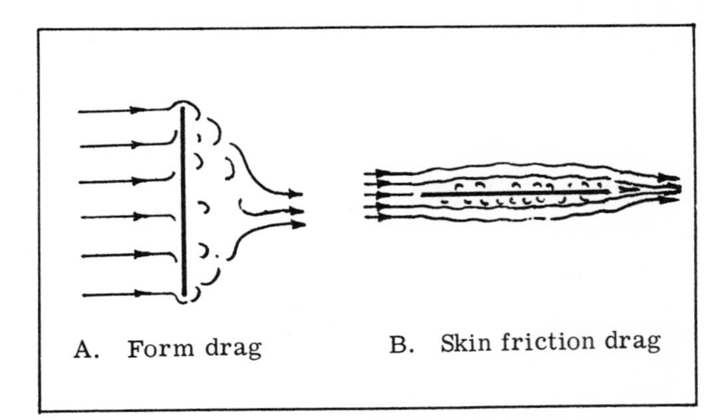

Fig. 1-26. Form Drag and skin friction Drag.

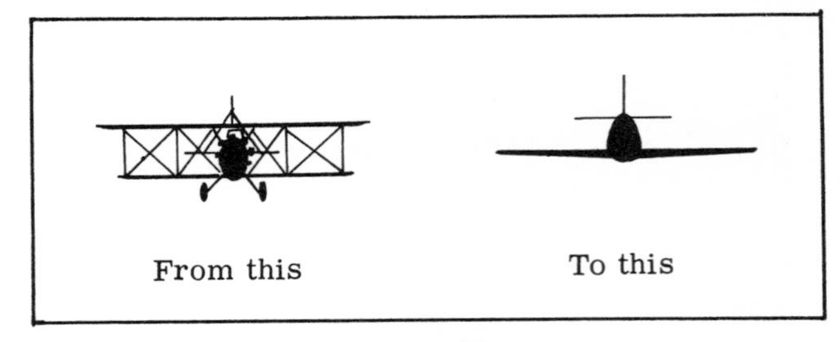

Fig. 1-27.

SKIN FRICTION DRAG

This is the Drag caused by the air passing over the airplane's surfaces. Flush riveting and smooth paint are good ways of decreasing skin friction Drag. A clean, polished airplane may be several miles an hour faster than another of the same model which is dirty and unpolished. Waxing and buffing will help an aerodynamically clean airplane because a large proportion of its parasite Drag is due to skin friction. Waxing the Wright Brothers' Flyer would have been a waste of elbow grease as far as getting a noticeable added amount of airspeed is concerned because the largest percentage of its Drag was form Drag.

INTERFERENCE DRAG

Interference Drag is caused by the interference of the airflow between parts of the airplane such as would be found at the intersection of the wings or empennage with the fuselage. This Drag is lessened by filleting these areas so that the mixing of the airflow is more gradual. (Fig. 1-28)

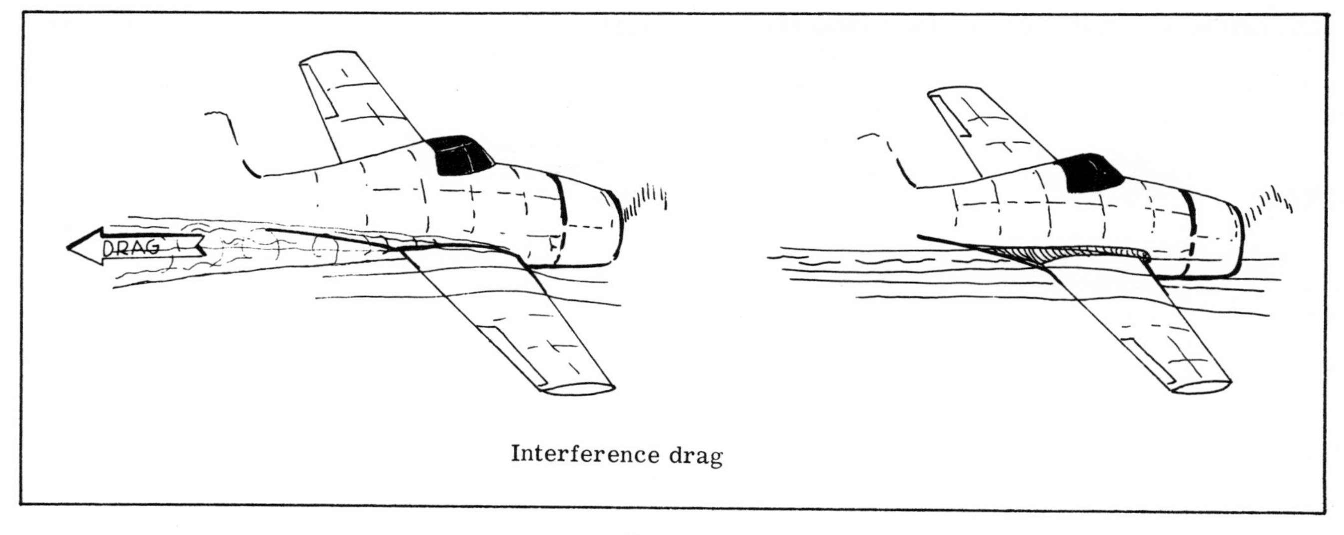

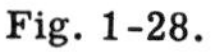

Fig. 1-28.

INDUCED DRAG

Induced Drag is a byproduct of Lift. As was covered in the section on Lift, there is a difference in the pressure on the top and bottom of the wing and, as nature abhors a vacuum (or at least tries to equalize pressures in a system such as unconfined air), the higher pressure air moves over the wing tip toward the lower pressure on top. Wing tip vortices result because as a particular mass of air gets over the tip the wing has moved on out from under it. (Fig. 1-29)

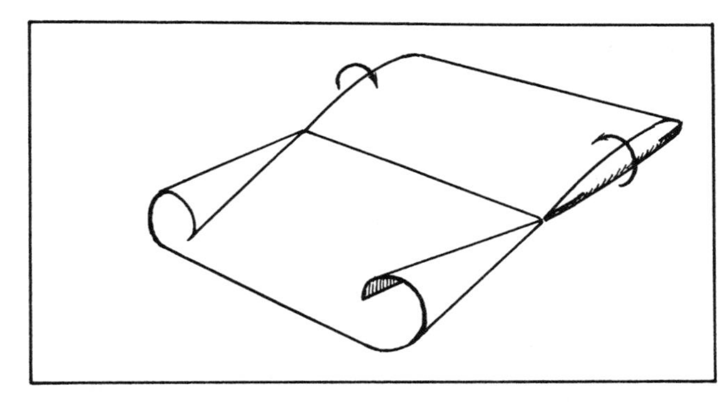

Fig. 1-29. Wing tip vortices.

Because of action of these vortices the relative wind passing the wing is deflected downward in producing Lift.

The downward deflection of the air means that the wing is actually operating at a lower angle of attack than would be seen by checking the airplane's flight path, because it's flying in an "average" relative wind partly of its own making. (See Fig. 1-30.)

Fig. 1-30 shows that the "true" Lift of the wing is operating perpendicular to its *own* relative wind rather than perpendicular to the air moving relative to the whole airplane (or the relative wind you, as a pilot, think about). Of course you know that the angle of attack is the angle between the chord line of the wing and its relative wind.

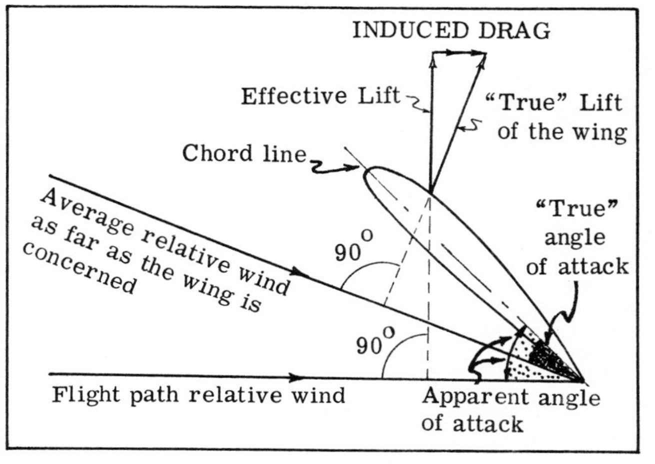

Fig. 1-30. Induced Drag.

As the coefficient of Lift (angle of attack) increases, the strengths of the wing tip vortices get greater with a resulting increase in downwash (and differences in angles of attack). This makes the difference between effective Lift and the wing's "true" Lift even greater which, as can be seen in Fig. 1-30, would make the retarding force, induced Drag, increase.

Keep in mind that wing tip vortices can be very powerful for large airplanes (which are producing many pounds of Lift) and are particularly vicious when they are flying at low speeds (high coefficients of Lift). As flaps increase C_L for a particular airspeed, following closely behind an airliner on approach can be an extremely exciting activity for a light plane.

Fig. 1-31 plots Parasite and Induced Drag versus Airspeed for a fictitious airplane at a specific Weight in the clean condition. As you can see, the total Drag at various speeds is made up of the sum of the two types. Induced Drag rises sharply as the airplane slows and approaches the stall speed (or more correctly as it approaches the maximum coefficient of Lift) and, at a point just above the stall speed, may be 80-85 per cent of the total drag.

Basically, induced Drag goes up at the rate of the

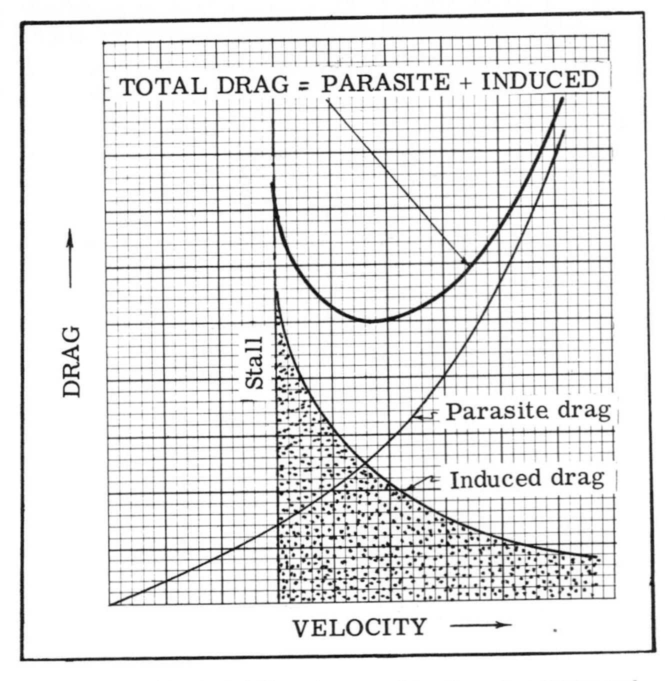

Fig. 1-31. Total Drag is a combination of parasite and induced Drags.

square of the coefficient of Lift. As can be seen in Fig. 1-31 it gets lower as the speed increases (as the angle of attack, or C_L, decreases). It will never disappear because as long as any Lift is being produced it's a resulting evil.

Induced Drag is inversely proportional to the aspect ratio of the wing; that is, longer thinner wings mean less induced Drag if indicated airspeed, airplane Weight and wing area are equal. (Aspect ratio is the ratio of span to average chord or, more correctly, the ratio of the span squared, divided by the wing area.) If the wing had an infinite span the wing tip vortices would naturally not exist and induced Drag would be zero. The power of the wing tip vortices is tied in directly with a term called "span loading," or the amount of Lift being produced per foot of wingspan, which is another way of talking about aspect ratio. Aircraft that require good aerodynamic efficiency rather than high speed, and operate the majority of the time at comparatively high coefficients of Lift, need high aspect ratio wings (sailplanes and the U-2). For airplanes such as jet fighters, operating at high speeds where induced Drag is small compared to parasite Drag, the aspect ratio must be low, both from an aerodynamic and a structural standpoint.

GROUND EFFECT

When the airplane wing operates within a certain distance of the ground the downwash characteristics are altered with a resulting decrease in induced Drag.

What happens is that when the airplane is close to the ground the strength of the vortices is decreased and the downwash angle is also decreased for a particular amount of Lift being produced. This means that the wing's true Lift and the effective Lift are working closer to each other (the angle between the two forces is less) and the retarding force (induced Drag) is less. (Check Fig. 1-30 again.)

When the airplane approaches the ground as for a landing, ground effect begins to really enter the picture at about a wingspan distance above the surface. Its effects are then increased radically as the plane nears the ground until at about touchdown induced Drag *can* possibly drop by about 48 per cent. All other things equal, the low wing airplane is more affected than the mid- or high-wing because the wing is closer to the ground. Fig. 1-32 shows the decrease in per cent of induced Drag in terms of span-height for airplanes of general configuration. (Remember the effects vary with aspect ratio.)

Basically, ground effect means that you are getting the same Lift for less induced Drag. As induced Drag can make up 80-85 per cent of the Total Drag at lift-off or touchdown, a 48 per cent decrease in induced Drag could mean a *decrease in total Drag of around 40 per cent* for a specific angle of attack. This amount of Drag decrease could fool a pilot into thinking on take-off that the airplane is ready to fly and climb out like a tiger; however he might discover after getting a few feet above the ground that he's in control of an anemic house cat. The power required to fly the airplane rises sharply as the induced Drag increases and a deficit in power would result in a sink rate. The pilot meanwhile is holding the same nose-attitude, trying to get some climb out of the lead sled he's suddenly inherited. As the airplane starts settling the angle of attack is increased because of the downward movement and, since the angle was at the raw edge to begin with, the airplane stalls and abruptly contacts the ground again (sounds of bending metal in the background).

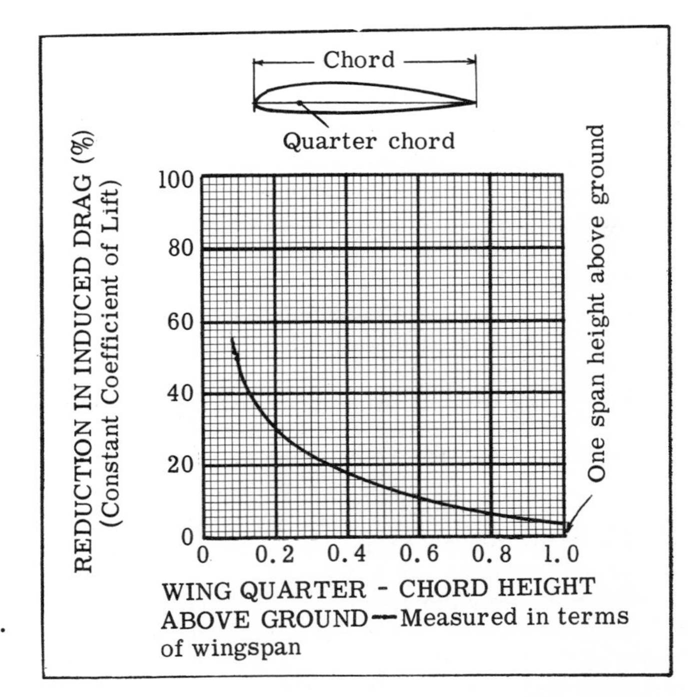

Fig. 1-32. Reduction of induced Drag with decrease in height above the ground.

Ground effect also has a bearing on the longitudinal (pitch) stability of the airplane. An airplane is more stable in ground effect, that is, the nose is "heavier" for any trim setting. In fact, ground effect is a major factor in deciding the forward center of gravity limits of the airplane. More up-elevator is needed near the ground because the wing downwash angle is decreased. The down-force exerted on the stabilizer-elevator is a mixture of free stream velocity, slipstream (which is weak in the power-off condition) and downwash from the wing. It may take anywhere from 4 to 15 degrees more up-elevator to obtain the max angle of attack at landing as compared to that in free flight at altitude, depending on the make and model of the airplane.

As the airplane is more nose-heavy in ground effect, an airplane loaded at (or past) the rearward C.G. limits might appear to be acceptably stable immediately following the take-off, but as it gained a few feet could tend to nose up, catching the pilot sleeping.

But the effects of ground effect on stability will be covered more thoroughly in Chapter 10.

You can see by looking at Fig. 1-32 that the wing quarter chord would have to be about $3\frac{1}{2}$ to 4 feet above the surface (this is about one-tenth of the span of most current light planes) in order to get a 48 per cent decrease in induced Drag.

The pilot's misunderstanding of ground effect has caused more than one landing accident when the airplane has "floated" the length of the runway while the pilot sat there with paralysis of the throttle hand thinking "it would settle on any day now."

You can use ground effect to aid you in acceleration to climb speed after take-off. Induced Drag is greater than parasite Drag at this point, so leave the gear down until you're sure the airplane is going to stay airborne (landing gear represents parasite Drag). Don't let yourself be fooled into lifting off before the plane is ready. Another reason for not getting the gear up too soon is that an engine failure could result in a gear-up landing with plenty of landing area ahead. Under normal conditions leave that gear down until you can no longer land on the runway ahead.

Ground effect has also enabled airplanes to get out of ticklish situations on take-off and landing and has been used by multiengine airplanes on overwater flights when an engine (or engines) failed. By taking advantage of this phenomenon they were able to keep flying under conditions that would have otherwise resulted in a ditching. Glassy water, which is best for ground effect can be hazardous as far as judging heights are concerned. You can see in Fig. 1-32 that a few feet can make a lot of difference as far as ground effect is concerned.

LIFT-TO-DRAG RATIO

One measure of an airplane's aerodynamic efficiency is its maximum Lift-to-Drag ratio. The usual engineering procedure is to use the term C_L/C_D since the other factors of Lift and Drag (density, wing area and velocity) are equally affected. (The C_D is the Coefficient of *Total* Drag.) This means that at a certain angle of attack the airplane is "giving more for your money." The angle of attack and ratio of this special point varies between airplanes as well as varying with a particular airplane's configuration (clean or dirty). Pilots hear the term "C_L/C_D maximum" and automatically assume that this point is found at $C_{L_{max}}$, or close to the stall. This is not true because, while the coefficient of Lift is large at large angles of attack, the total coefficient of Drag is *much* larger in proportion because of induced Drag. Following are figures showing how the Lift-to-Drag (or C_L/C_D) ratio varies for a sample airplane in the clean condition.

C_L	C_D	C_L/C_D Ratio	
0.935	0.080	11.7	Low Speed
0.898	0.075	12.0	
0.860	0.070	12.1	
0.820	0.065	12.6	
0.772	0.060	12.9	
0.725	0.055	13.2	
0.671	0.050	13.4	
0.620	0.045	13.7	
0.562	0.040	14.1	
0.500	0.035	14.3	
0.418	0.030	13.9	
0.332	0.025	13.3	
0.200	0.020	10.0	High Speed

The underlined figures show the values for C_L/C_D maximum.

The maximum Lift-to-Drag ratio is the condition at which maximum range and maximum glide distance will be found as will be covered in more detail later. It is found at the point of minimum Drag for the airplane.

THRUST

Thrust is the force exerted by a propeller, jet or rocket and is used to overcome aerodynamic Drag and other forces acting to retard the airplane's motion in the air and on the ground. It can be explained by one of Newton's laws of motion, "For every action there is an equal and opposite reaction."

The propeller is a rotating airfoil that takes a comparatively large mass of air, accelerating it rearward, resulting in the equal and opposite reaction of the airplane tending to move forward. The Thrust being exerted is proportional to the mass and the velocity of the accelerated air.

The jet engine takes a smaller mass of air and fuel and accelerates it to a faster velocity than does the propeller. The rocket takes an even smaller mass (of its fuel and oxidizer) and accelerates it to a very high speed.

THRUST AVAILABLE AND DRAG

The greatest Thrust for the *propeller-driven* airplane is found in the static condition; that is, when you are sitting on the end of the runway with the engine running at full power the propeller is producing the greatest Thrust. As the plane starts moving the Thrust force available decreases with an increase in speed.

For straight and level flight the Thrust available (pounds) and Drag (pounds) are considered to be equal if a constant airspeed is being maintained. For speeds in the area of cruise where the airplane's Thrust line is acting along the flight path (the nose is not "cocked up" as is the case for slower speeds) the assumption that at a constant airspeed the Thrust exerted equals the Drag is a valid one. The variations of the Four Forces in different maneuvers will be shown in the next chapter and this particular point will be covered.

Fig. 1-33 is a Thrust Available and Drag (or Thrust Required) versus True Airspeed for a fictitious high performance four-place airplane at sea level and at 10,000 feet (gross Weight).

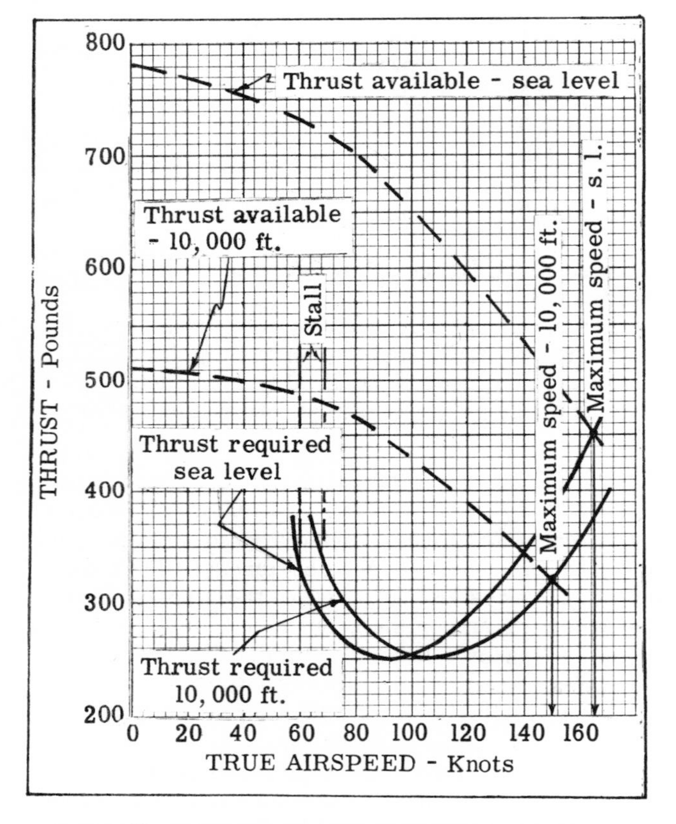

Fig. 1-33. Thrust Required and Maximum Thrust Available versus True Airspeed at sea level and 10,000 feet.

PROPELLER

The propeller is a rotating airfoil and, as such, is subject to stalls, induced Drag, and other troubles affecting airfoils. As you have noticed, the blade angle of the propeller changes from hub to tip with the greatest angle of incidence (highest pitch) at the hub and the smallest at the tip. (Fig. 1-34)

This twist is necessary because of the difference

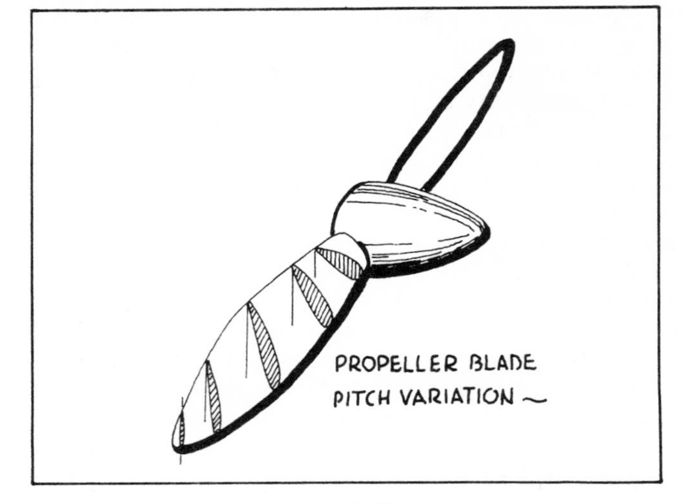

Fig. 1-34.

in the actual speed through the air of the various portions of the blade. If the blade had the same geometric pitch all along its length (say 20^0), at cruise the inner portion near the hub would have a negative angle of attack and the extreme outer portion would be stalled. This is hardly conducive to "get up and go" — so the twist is necessary.

Fixed pitch propellers are considered to come in two main categories: (1) Climbing or (2) Cruising. Regardless of which prop is on the plane, you always wish you had the other one. The climbing prop with its lower pitch, which results in a higher rpm and more horsepower being developed, gives efficient take-off and climb performance — but poor cruise characteristics. The cruise propeller, with its high pitch resulting in lower rpm (and less horsepower developed), has comparatively poor performance for take-off and climb but is efficient for cruise. There are two terms used in describing the angles of incidence and propeller effectiveness (1) Geometric pitch, or the built-in angle of incidence, the path a chosen portion of the blade would take in a nearly solid medium such as gelatin. (2) Effective pitch — the actual path the propeller is taking in air at any particular time.

The difference between the geometric pitch and the effective pitch is the angle of attack as seen in Fig. 1-35. The climbing propeller has a lower *average* geometric pitch than the cruising prop and revs up more, developing more horsepower. (Fig. 1-36)

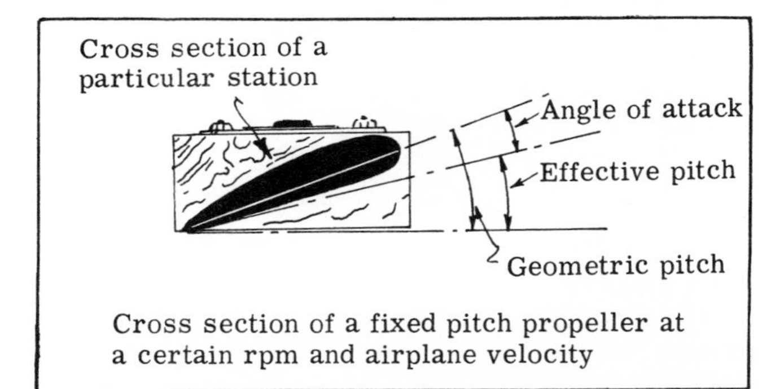

Fig. 1-35.

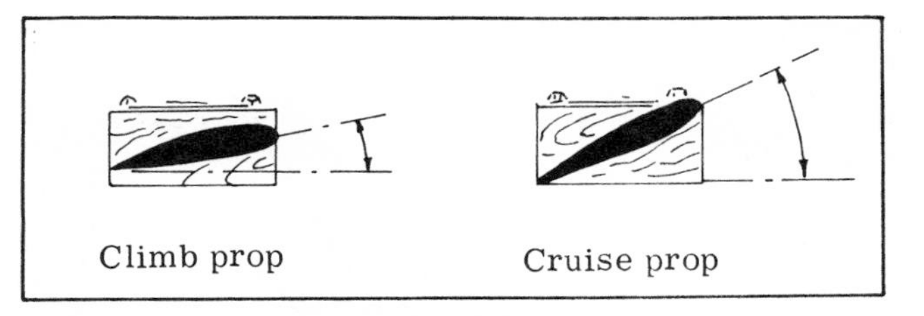

Fig. 1-36. Climbing and cruising propellers — A comparison of geometric pitch at the same station (same distance from hub).

Take an airplane sitting at the end of a runway with a cruise type propeller revolving at 2400 rpm. Fig. 1-37 shows that as the plane is not moving, the geometric pitch and angle of attack are the same and a large portion of the blade is stalled. A vector diagram would show the difference in the efficiency of a propeller at 0 mph and 100 mph (at 2400 rpm).

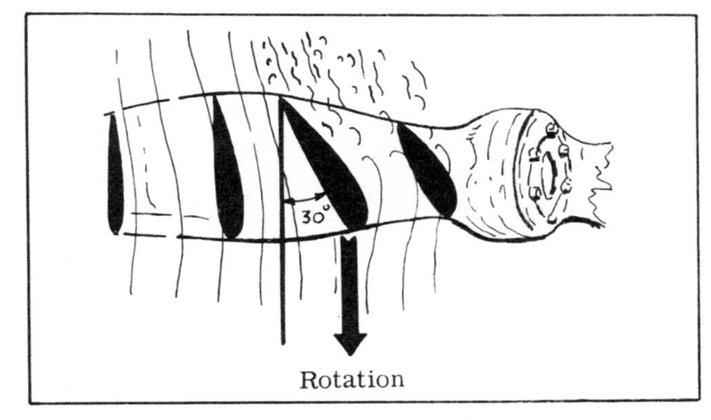

Fig. 1-37. Airplane stationary — portion of the blade stalled.

The velocity of a *particular station* of the blade can be readily found. Let's pick a station at a point slightly more than $1\frac{1}{2}$ feet from the hub center so that the point chosen travels 10 feet each revolution. At 2400 rpm the velocity at this point is 400 feet per second. Assuming a geometric pitch at this point of 30°, the prop stalls at an angle of attack of 20°, so this portion of the blade is stalled.

This would tell you then that the only part of the blade developing Thrust under this condition would be the portion having a geometric pitch of less than 20° — or the outboard portion of the blades. The Thrust pattern would look like that shown in Fig. 1-38.

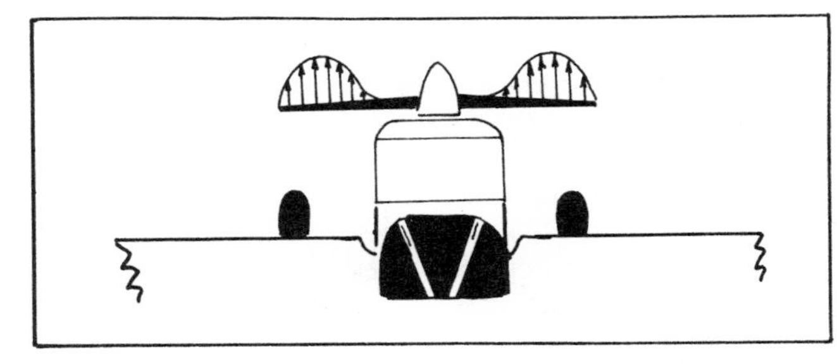

Fig. 1-38. Thrust pattern of the propeller of a stationary airplane.

For the same airplane moving at 89 knots, or 150 feet per second (again chosen for convenience), the result can be seen by vectors as is shown in Fig. 1-39. You will notice that the blade is operating at a more efficient angle of attack. As the propeller is an airfoil, it is most efficient at the angle of attack giving the greatest Lift/Drag ratio. A climb prop would be more efficient at low airspeeds because a greater portion of the blade is operating in this range. At high airspeeds it is operating at a lower angle of attack and is not producing efficient Thrust.

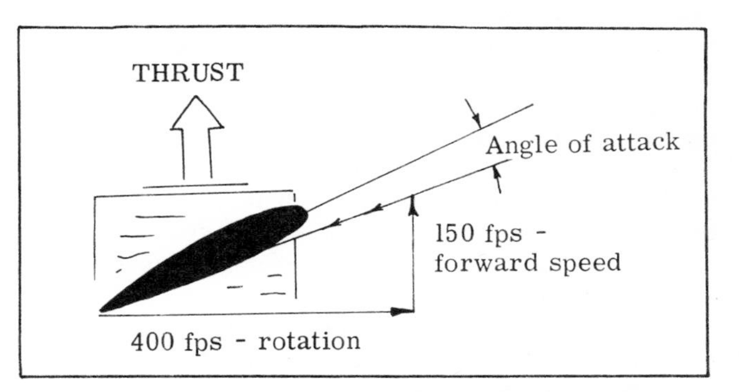

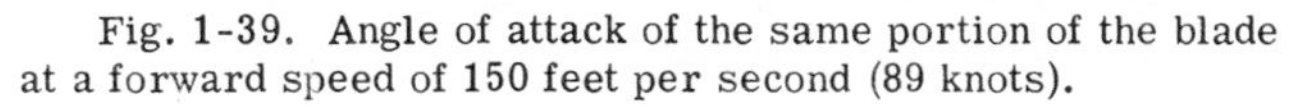

Fig. 1-39. Angle of attack of the same portion of the blade at a forward speed of 150 feet per second (89 knots).

At a higher pitch the Drag of the blades is higher and the engine is unable to get as high a number of rpm and develop as much horsepower as would be obtained with a lower pitch propeller. *The fixed pitch propeller is efficient only at one particular speed range.*

The answer to this is the variable pitch propeller which can be set to suit the pilot:

(a) For take-off and climb — LOW PITCH which results in HIGH RPM and high power. (The British call this "Fine Pitch.")

(b) For cruise — HIGH PITCH, which results in LOW RPM and efficient cruise. (The British call this "Coarse Pitch.")

In a variable pitch prop the rpm will still vary with airspeed as does the fixed pitch prop. In other words, if you set a cruise rpm of 2400, a climb or dive will cause it to vary. The constant speed prop, which is an "automatic" variable pitch propeller, will maintain a constant rpm after being set. Once the desired setting is made, changes in airspeed do not affect it.

THE CONSTANT SPEED PROPELLER (OIL-COUNTERWEIGHT TYPE)

The constant speed propeller setting is the result of a balance between oil pressure (using engine oil) and centrifugal force exerted by the propeller counterweights. This balance is maintained by the governor, which is driven by the crankshaft through a series of gears. The governor can be considered to be of two main parts: (1) The flyweight assembly and (2) the oil pump. The governor is set by the prop control(s) in the cockpit. Assume you have set the rpm at 2400 for cruise. The oil pressure and counterweight forces are equal because the flyweights in the governor are turning at constant speed and the oil valve to the prop pistons is closed with oil pressure locked in the propeller hub. Now assume that you pull the nose up. The airspeed drops, causing the prop and engine rpm to start to drop. The centrifugal force on the governor flyweights decreases because of the drop in rpm. The contraction of the flyweight causes a two-way valve to be opened so that increased pressure to the pistons moves the propeller to a lower pitch, allowing it to maintain 2400 rpm.

If the plane were dived, the prop would tend to overspeed and the resulting increase in centrifugal force of the governor flyweight would cause the oil valve to be opened so that the oil pressure in the propeller hub dome decreases. Centrifugal force on the propeller counterweights would then cause the prop to be pulled into a higher pitch; the increased blade angle of attack would result in more Drag and the propeller would not overspeed.

(a) LOWER PITCH (Higher rpm) — caused by added oil pressure.

(b) HIGHER PITCH (Lower rpm) — caused by centrifugal force on prop counterweights.

The operation of most noncounterweight propellers is reversed from that of the oil-counterweight types. The blade in creating Thrust creates a moment that tends to decrease its pitch (the same thing happens to a wing as will be covered in the next chapter). This is opposed by governor oil pressure which tends to increase its pitch.

Newer makes of propellers (see the listing under Recommended Texts at the end of the book) use compressed air or nitrogen in the dome to increase pitch and feather. This force opposes the pitch lowering tendency caused by the blade moment and the governor oil pressure. Because counterweights are not necessary a great deal of Weight is saved. In effect, the compressed air does the job of the counterweights.

Some of the electric propellers on light planes also have automatic controls which, when engaged, give the propeller constant speed properties. You set the optimum rpm and the constant speed propeller maintains it.

To get an idea of how a variable pitch or constant speed propeller can increase the efficiency of a propeller let's set up a hypothetical situation:

To simplify matters, assume that a certain engine can only be run at a certain rpm for cruise — no other can be used, and 2400 is a good round figure so that's the number for this problem. (The propeller is direct-drive and turns at this same rpm.)

Fig. 1-40 shows the efficiencies of a particular propeller at different pitch settings of the propeller blade (at a station three-quarters, or 75 per cent, of the radius from the hub) at various airspeeds.

The constant speed propeller will change its angle to maintain a constant rpm so as the airspeed increases the pitch setting will also increase automatically. The dashed line, or "envelope," shows that as the airspeed increases the efficiency remains fairly constant over the entire range. The propeller pitch changes as needed to keep a constant angle of attack. Notice that the solid line curves for different pitch settings have a comparatively narrow range of airspeed for peak efficiency. For this airplane a pitch (at the station at 75 per cent of the radius of the prop) of 15 degrees is only 80 per cent efficient (or higher)

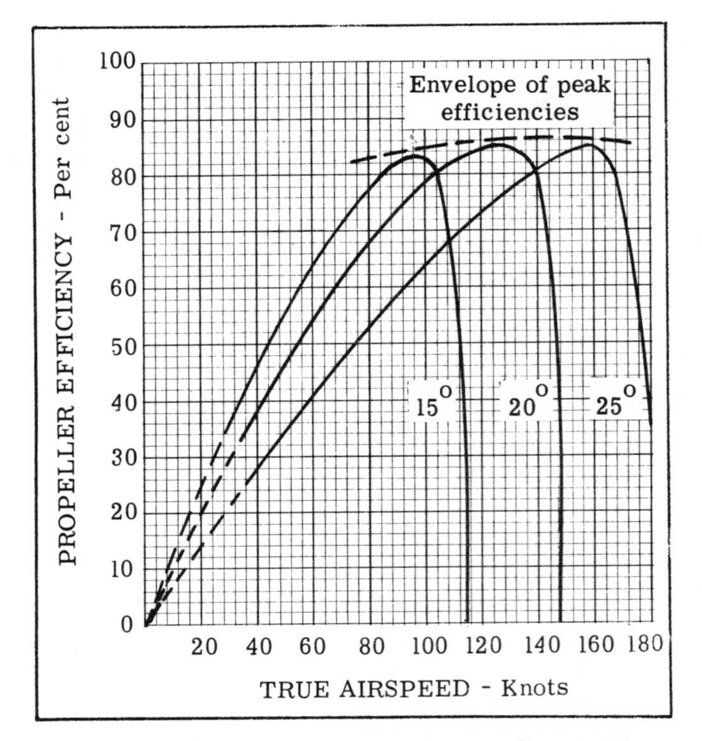

Fig. 1-40. Efficiency of a particular propeller at various blade settings and airspeeds (constant 2400 rpm).

between the speeds of 85-104 knots or has its peak efficiency at about 95 knots. At a pitch setting of 20 degrees the range of 80 per cent or higher efficiencies is from about 104 knots to 140 knots, with the peak at about 128 knots, etc. Shown are the ranges of efficiencies for fixed-pitch versions of that propeller at 15, 20 and 25 degrees pitch respectively. Notice in Fig. 1-40 that the efficiencies for these pitches drop rapidly at the upper ends of their speed ranges. The constant speed prop has practically an infinite number of pitch settings available within the airplane's operating speed ranges and this is shown by the envelope of peak efficiencies.

Use of the variable pitch and constant speed propeller also will be covered in later chapters.

TORQUE

You have long been familiar with "torque." It is the pilots' term for that force or moment that tends to yaw or turn the plane to the left when power is applied at low speeds. It is also the price paid for using a propeller. Knowing that the propeller is a rotating airfoil, you realize that it exerts some Drag as well as Lift. This Drag creates a moment that tends to rotate the airplane around its longitudinal (fuselage) axis opposite to prop rotation.

Although normally "torque" is thought of as being one force, it is, in fact, several forces combined.

1. Slipstream Effects — As the propeller turns clockwise, a rotating flow of air is moved rearward, striking the left side of the fin and rudder which results in a left yawing moment. The fin may be offset to counteract this reaction, with the fin setting built

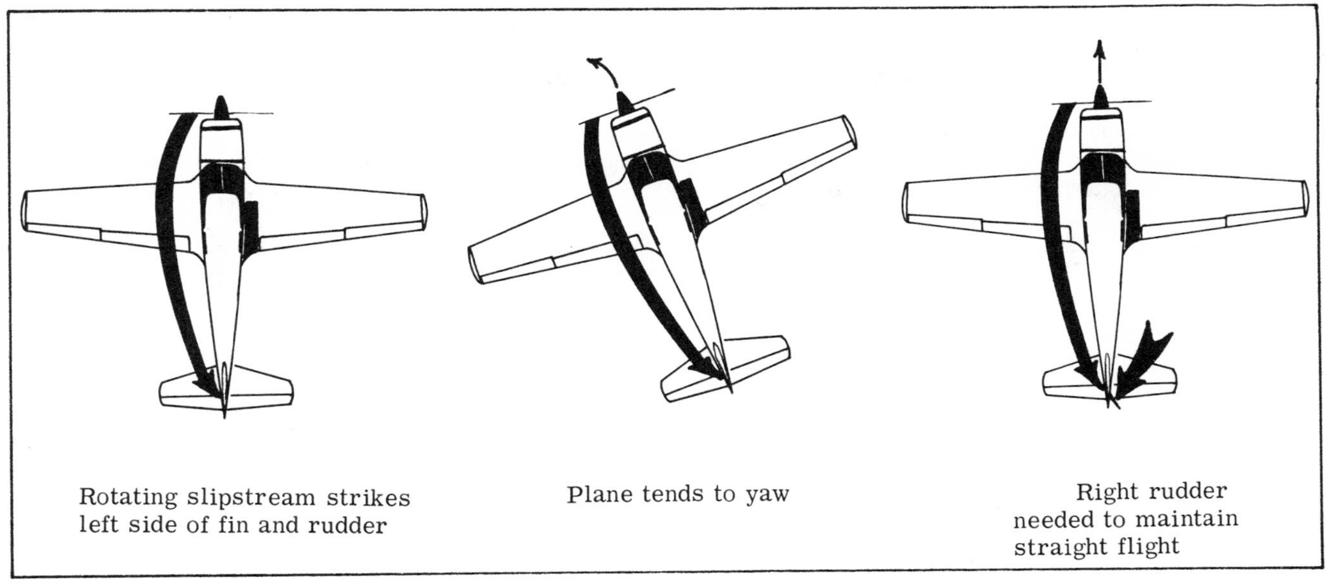

Fig. 1-41.

in for maximum effectiveness at the rated cruising speed of the airplane, since the plane will be flying at this speed most of the time (Figs. 1-41 and 42).

If it were not for the offset fin, right rudder would have to be held at all speeds. As it is, the balance of forces results in no yawing force at all and the plane flies straight at cruise with no right rudder being held. (Fig. 1-42)

Sometimes the fin may not be offset correctly due to tolerances of manufacturing, and a slight left yaw is present at cruise, making it necessary to use right rudder to keep the airplane straight. To take care of this, a small metal tab is attached to the trailing edge of the rudder and is bent to the left. The pressure of the relative air against the tab forces the rudder to the right. (Fig. 1-43)

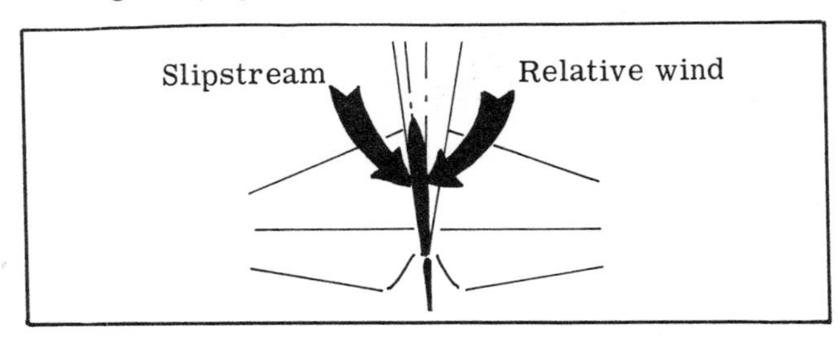

Fig. 1-42. The offset fin is designed so that the angle of attack of the fin is zero (the forces balance) at cruise.

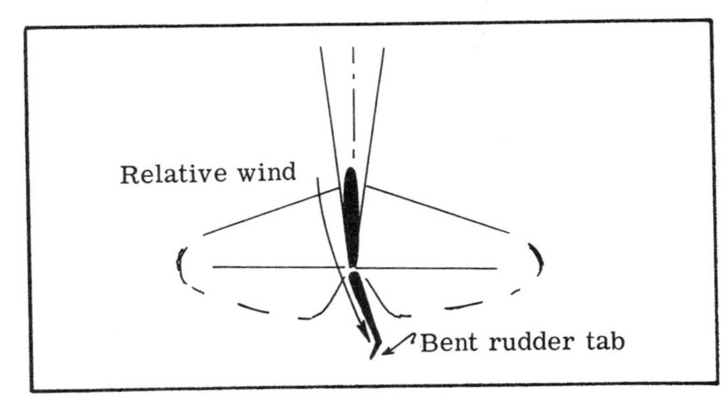

Fig. 1-43. The rudder trim tab.

On lighter planes this adjustment can be done only on the ground and may require several flights before a satisfactory setting is found. For heavier planes, a controllable rudder tab is used, allowing the pilot to correct for "torque" at all speed ranges. If the tab has been bent correctly (or has been set in the cockpit) for cruise, the torque and impact forces are balanced. Assuming no change in power, see what happens as the airplane's speed changes in Fig. 1-44.

Some manufacturers use an offset thrust line, or "cant" the engine to counteract torque at cruise and the airplane's reactions are the same as for the offset fin.

In a climb, right rudder must be held to keep the plane straight. In a dive, left rudder is necessary to keep it straight. In the larger planes the rudder tab may be set during flight for these variations from cruise.

In a glide there is considered to be no yawing effect. Although the engine is at idle and torque is less, the impact pressure on the fin is also less. *The slipstream effect is the most important of the "torque" forces working on the single engine airplane.*

2. Equal and Opposite Reactions — Newton's Law of "equal and opposite reactions" is only a minor factor in torque effects. The airplane tends to rotate in a direction opposite to that of the propeller's rotation. In some cases the left wing or wing tip area may have "washin" to compensate for this. Washin means that the angle of incidence is increased and the wing is bent up into the airstream for more Lift. "Washout" can be thought of as the wing turning down out of the relative wind for less Lift. These two terms are pretty hard to keep straight and pilots may mention one when actually they mean the other. Washin may also contribute very slightly to a left turning effect.

3. Asymmetric Loading or propeller disk asymmetric loading — a condition caused by the air not striking the propeller disk at exactly a ninety degree angle. This usually is encountered in a constant positive angle of attack such as in a climb or in slow

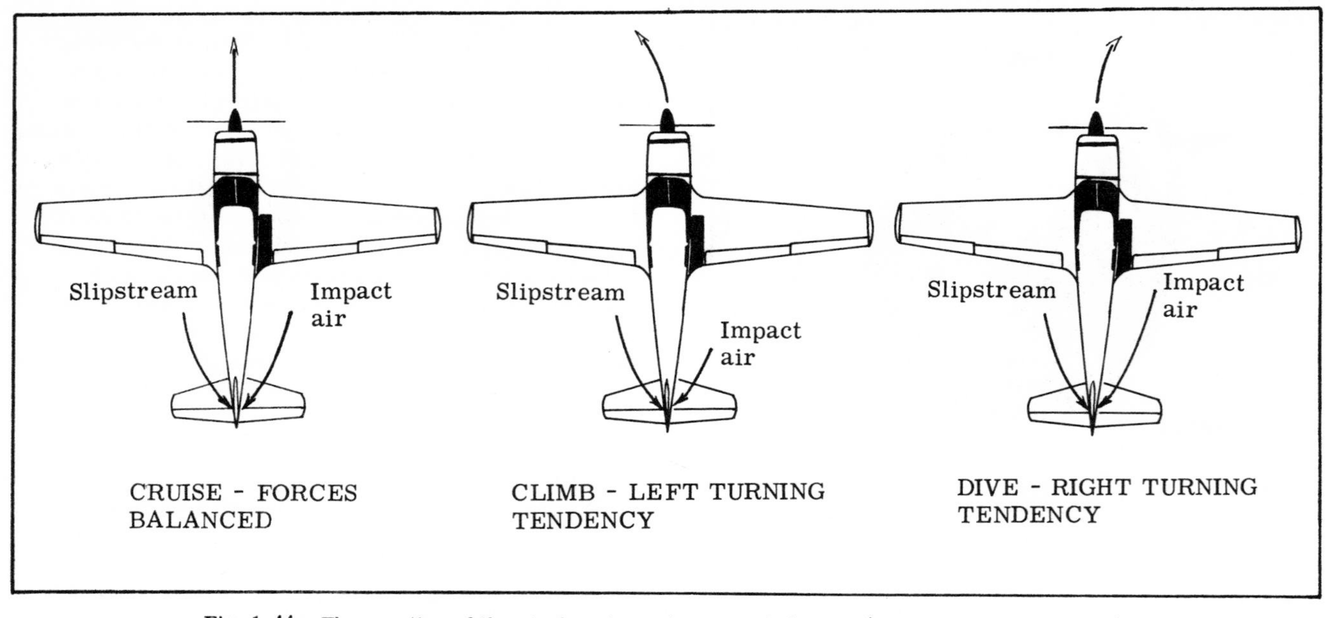

Fig. 1-44. The reaction of the airplane to various speed changes (constant power assumed).

flight. The down moving propeller blade, which is on the right side as seen from the cockpit, has a higher angle of attack and consequently a greater Thrust, which results in a left turning effect. This can be visualized by checking Fig. 1-45.

Precession is a gyroscopic property (the gyro will be covered more thoroughly in Chapter 3). If a force is exerted against the side of the gyro, it reacts as if the force had actually been exerted in the same direction at a point 90 degrees around the wheel. (Fig. 1-46)

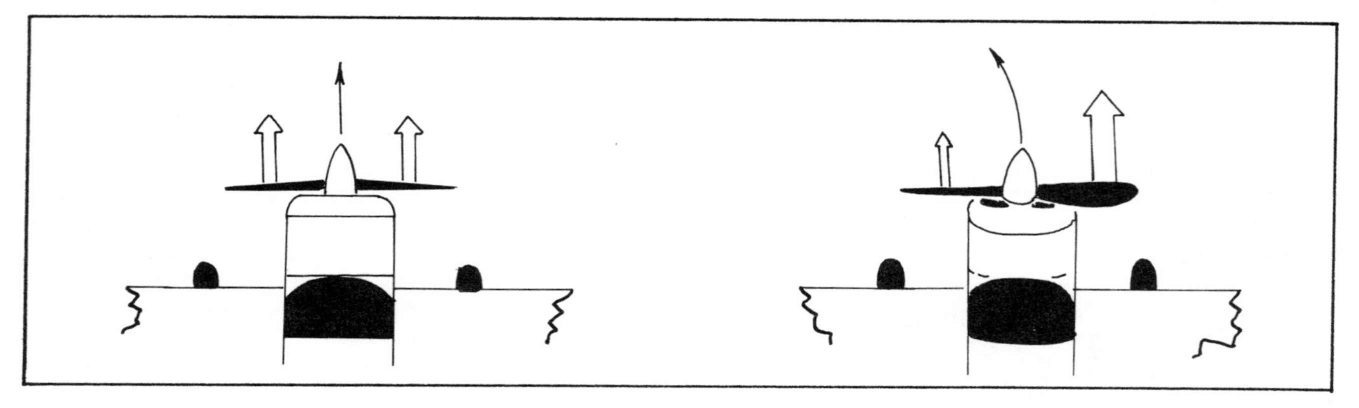

Fig. 1-45. Asymmetric disk loading effects.

To find the exact difference in the Thrust of the two sides of the propeller disk, a vector diagram must be drawn which includes the propeller blade angles, rotational velocity and the airplane's forward speed and angle of attack. If the airplane is yawed, the asymmetric loading effect is encountered. A left yaw would mean a slight nose-down tendency and a right yaw a slight nose-up tendency (you can reason this out). As climbs and slow flight are more usual maneuvers than are yaws, these will be the most likely spots for you to encounter asymmetric loading effects. (Commonly called "P-Factor," see page 24.)

4. Precession — you encounter precession when, in a tailwheel type airplane, you try to force the tail up quickly on a take-off run. The airplane wants to get away from you as it suddenly yaws to the left. Precession affects the airplane only *during* attitude changes.

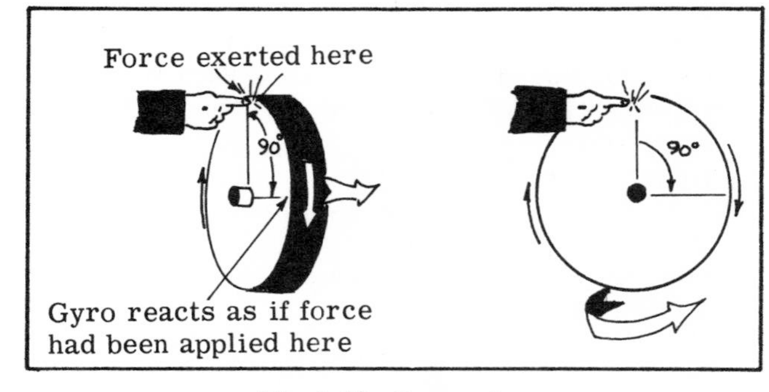

Fig. 1-46. Precession.

The propeller disk makes a good gyro wheel, as it has mass and a good rotational velocity. Another property of a gyro is "rigidity in space." This property is used in ships' gyros and aircraft directional gyros and artificial horizons. The rotating gyro

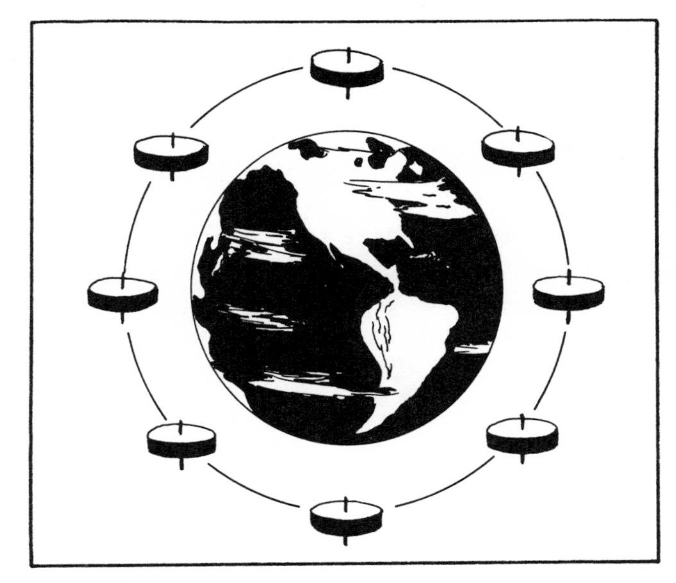

Fig. 1-47. The gyro wheel has the property of "rigidity in space."

tends to stay in the same plane of rotation in space and resists any change in that plane. (Fig. 1-47) If a force is insistent enough in trying to change this plane of rotation, precession results.

When you're rolling down the runway in the three-point position (in a tailwheel type plane), the propeller is in a certain rotational plane. You are using full power and are fighting the other torque effects (rotating slipstream, asymmetric loading, etc.). When you shove the wheel or stick forward and the tail rises, the plane reacts as if the force were exerted on the right side of the propeller disk from the rear. The result is a brisk swing to the left. You can further reason out the reactions of the propeller plane to yawing, pull-ups or push-overs. You're most likely, however, to notice precession effects on getting the tail up for take-off because the rudder is comparatively ineffective at this low speed and the sudden yaw is harder to control.

SUMMARY OF TORQUE

These forces or combinations of them make up what pilots call "torque," the force that tends to yaw the plane to the left at low speeds and high power settings. This discussion is presented so that you may get to understand your airplane better, but your real objective is to compensate for torque and to fly the airplane in the proper manner. There are a lot of darn good pilots flying around who wouldn't know what "asymmetric disk loading" was if they were hit in the face with it. The nose yaws to the left and they correct for it — and so should you.

THE POWER CURVE

FORCE, WORK AND POWER

In order to understand power and horsepower it would be well to discuss "force" and "work."

A *force* may be considered to be a "pressure, tension or weight." Thrust, Lift, Weight and Drag are forces and our present system uses the term *pounds* to express the value or strength of a force.

You can exert a *force* against a heavy object and nothing moves; that still doesn't alter the fact that force has been exerted. If the object doesn't move no *work* has been done as far as the engineering term is concerned (tell this to your aching back). So *work*, from an engineering standpoint is a measure of a *force* times *distance* (in the direction in which the force is being exerted). If a constant force of 100 pounds is exerted to move an object 10 feet, the deed has accomplished 1000 foot-pounds of *work*.

For instance, your *Zephyr Six* airplane weighs 2200 pounds and you must push (or pull) it by hand 100 feet to the gas pit. It does *not* require 220,000 foot-pounds of work to accomplish this as you could see by attaching a scale to the tow bar and checking the force required to move the airplane at a steady rate, and for example purposes we'll say that it requires a constant force of 55 pounds to keep the airplane rolling across the ramp. The job to be accomplished would require 5500 foot-pounds of work on your part (55 pounds times 100 feet).

If you lifted the airplane to a *height* of 100 feet at a constant rate the work done *would* be 2200 pounds x 100 feet or 220,000 foot-pounds. The force you would need to exert would be the Weight of the airplane, once you got it moving at a steady rate upward. And, obviously you would do less work pulling the airplane 100 feet than lifting it the same distance, because less force would be required to pull it than to lift it.

The amount of work done has nothing to do with time; you can take a second, or all week, to do the 5500 foot-pounds of work in pulling the airplane. This brings up another term — *power*. *Power* is defined as a time rate of work. If you pulled the airplane the 100 feet over to the gas pit in one second you would have been exerting a power of 5500 foot-pounds per second, or as it is often written, 5500 ft-lbs/sec. (And if you are strong enough to accomplish such a feat you don't need an airplane; just flap your arms when you want to fly.) Suppose that it takes 10 seconds to do the job. The power used would be $\frac{\text{work}}{\text{time}} = \frac{5500 \text{ ft-lbs}}{10 \text{ seconds}}$ = 550 foot-pounds per second or 375 mile-pounds per hour. You would have to exert one horsepower for 10 seconds to do the job. The most common measurement for power is the term "horsepower" which happens to be a power of 550 ft-lbs/sec.

<u>Thrust horsepower</u> is the horsepower developed by the fact that a force (Thrust) is being exerted to move an object (the airplane) at a certain rate. Remembering that a *force* times the distance it moves an object is *work* and when divided by time, *power* is found, the equation for Thrust horsepower is $\text{THP} = \frac{\text{TV}}{550}$. The T in the equation is the propeller Thrust, in pounds, and V is the airplane's velocity in feet per second. As velocity can be considered to be distance divided by time, TV (or T x V) is *power* in foot-pounds

per second. The power (TV) is divided by 550 to obtain the horsepower being developed. If you wanted to think in terms of *miles per hour* for V, the equation would be $THP = \frac{TV}{375}$, the 375 being the constant number used for miles an hour. For V in knots, $THP = \frac{TV}{325}$.

As far as the airplane is concerned there are several types of horsepower of interest.

Indicated Horsepower — This is the actual power developed in the cylinders and might be considered to be a calculated horsepower based on pressure, cubic inch displacement and revolutions per minute.

Brake Horsepower — is so named because in earlier times it was measured for smaller engines by the use of a braking system or absorption dynamometer such as the "prony brake." The horsepower thus exerted by the crankshaft was known as brake horsepower. (Another name for it is *shaft horsepower.*) The fact that there is internal friction existing in all engines means that all of the horsepower in the cylinders doesn't get to the crankshaft, so a loss of horsepower from indicated horsepower called "friction horsepower loss" results. The reciprocating engine is always rated in brake horsepower and, for instance, if your airplane has an unsupercharged (or normal aspirating) engine the specification will note that the engine is rated as a certain horsepower at full throttle at a certain rpm at sea level (meaning standard sea level conditions of air density). For supercharged engines the specification cites a specific manifold pressure and rpm at sea level and also at the critical altitude (above which even full throttle can't hold the required manifold pressure to get the rated horsepower). This idea will be covered in more detail in later chapters.

When the term "75 per cent power" is used, it means *75 per cent of the normal rated power, or max continuous available at sea level on a standard day* (59° F and a pressure of 29.92 inches of mercury). For instance, a particular engine has a take-off power rating of 340 hp and a *normal rating* of 320 horsepower. The take-off rating label means that the engine can be run at this power only for a *limited time* as given in the engine specifications. If you use the engine power chart and set up manifold pressure and rpm to get 75 per cent power for that engine, your horsepower will be 75 per cent of 320 (the normal rated power), or 240 horsepower.

For other engines, the take-off rating and normal rating are the same. That is, it develops a certain maximum amount of horsepower and it can be run continuously at this power if necessary. In effect, there is no take-off rating, or no special higher-than-normal, limited-time power setting.

Brake horsepower also increases slightly with intake ram effect but is normally considered to remain constant, hence the use of brake horsepower as a standard for setting power by the power chart.

Thrust horsepower — as discussed earlier, is considered to be a percentage of brake horsepower if the propeller efficiency is taken into account, or $THP = \eta\ BHP$. The term "η" (eta) is the propeller efficiency which runs *at best* to about 0.85 (85 per cent) for most engine-propeller combinations and varies with airspeed. (This was covered back in the section on the propeller.)

As an example of the various steps of getting from the ignition of the fuel-air mixture to the Thrust horsepower being developed, take a look at the following:

Work done in cylinders — *Indicated Horsepower*	325 hp
Friction and accessories-drive loss	- 25 hp
Resulting *Shaft* or *Brake Horsepower*	300 hp
Loss from propeller (80% efficient at a particular speed)	- 60 hp
Horsepower available as *Thrust Horsepower*	240 hp

Fig. 1-48 is a Power Available and Power Required curve for a fictitious four place airplane and shows the power in terms of both brake and Thrust horsepower. The engine is rated at 250 BHP.

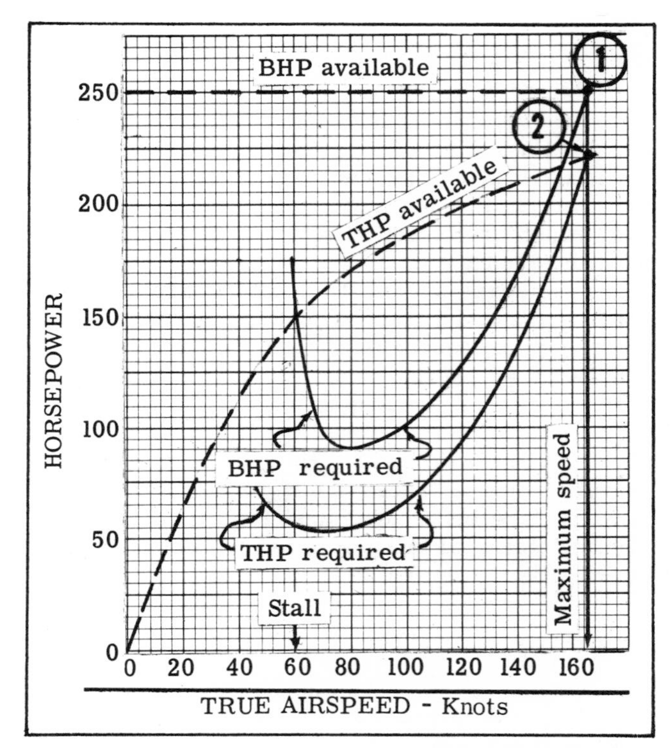

Fig. 1-48. Brake and Thrust Horsepower Available and Required versus True Airspeed for a particular airplane at sea level (gross Weight).

Fig. 1-48 shows the *maximum* horsepower available for the two types of power at various speeds. The point at which the maximum horsepower available equals the horsepower required establishes the maximum level flight speed of the airplane whether taken in terms of BHP or THP (points 1 and 2):

Notice how the available Thrust horsepower varies with airspeed. As the equation for THP is $\frac{TV(knots)}{325}$, you can see that although at a zero velocity the Thrust might be high, no horsepower is being developed because V equals zero, and a number times zero is still zero. As the airplane picks up speed

THP starts being developed and you will notice that it increases fairly rapidly at first. This is because Thrust is high at the lower speeds (check back to Fig. 1-33). THP increases at a lesser rate as the speed picks up (and Thrust decreases.)

You will also note that more brake horsepower than Thrust horsepower is required by the airplane at any speed, and this is because, as was mentioned, the propeller is not 100 per cent efficient and there's some loss. In other words, the engine has to produce more than enough effort to get the required amount of horsepower actually working to fly the airplane.

For the same airplane of Fig. 1-48 the amount of BHP necessary to get the THP required rises sharply at lower speeds because the propeller efficiency drops rapidly in that area. The Thrust horsepower required does not rise as sharply in this area as either the Drag or BHP required curves because, while the Drag is rising rapidly, the required THP is also a function of velocity ($THP = \frac{TV(knots)}{325}$ or $\frac{DV(knots)}{325}$) and the decrease in speed *tends* to offset the effects of the increase in Drag.

To give an idea of the comparison between brake and Thrust horsepower, suppose that instead of the prop an iron bar of equal Weight and Drag is attached on the hub. The engine would still be putting out a certain number of brake horsepower as measured by a dynamometer but the iron bar would be producing no Thrust and therefore no Thrust horsepower could ever be developed by the engine in that configuration — the efficiency (η) would be zero.

In thinking in terms of setting power, a curve for brake horsepower is more effective; for performance items such as climbs or glides, THP gives a clearer picture.

Fig. 1-49 shows some pertinent points on a Power Available and Power Required versus Velocity Curve as expressed in *brake horsepower for a fictitious airplane at the maximum certificated Weight of 3000 pounds at sea level in the clean* (gear and flaps retracted) *condition.*

The shaded area shows the areas of normal cruise power settings and airspeeds for this particular airplane. As you know, power settings most commonly used are from 60 to 75 per cent of the normal rated power. The majority of reciprocating engine airplanes avoid cruise (or continuous) power settings above 75 per cent because of increased fuel consumption and engine wear.

By now you've also noticed that the power required curve (for both THP and BHP) has a characteristic "U" shape similar to the Drag curve. This is because the power required to fly the airplane at a constant altitude varies with the Drag existing at different airspeeds — the values, of course, are different as one is expressed in pounds and the other in horsepower. If you had a Drag versus Airspeed curve for your airplane you could draw your own THP required curve by selecting particular airspeeds and using the Drags in the equation for THP.

Because of the varying power needed to maintain a constant altitude you'll notice in Fig. 1-49 that the airplane can fly at two speeds for the lower power settings. It is unlikely that in actuality it could fly at a slow enough speed to require close to 100 per cent power to maintain a constant altitude because the stall

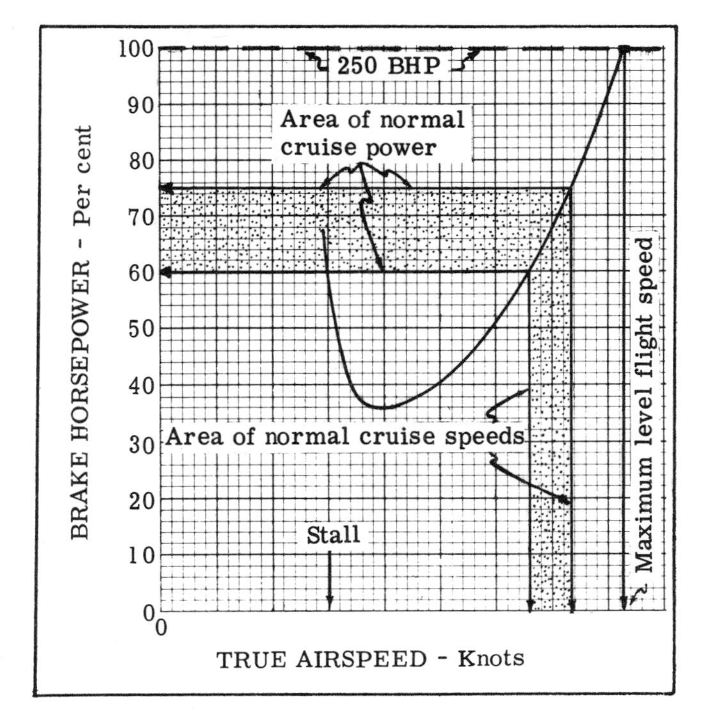

Fig. 1-49. Brake horsepower required and available versus airspeed for a particular airplane at sea level (gear and flaps retracted).

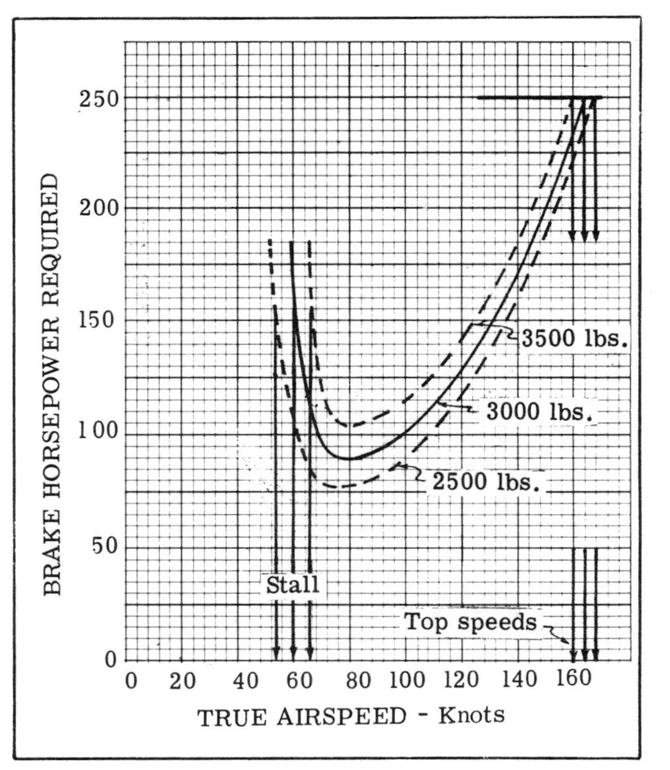

Fig. 1-50. The effects of Weight on the airplane of Figure 1-49.

characteristics of the airplane wouldn't allow it — the break would occur at some higher airspeed than that. If the stall could be delayed appreciably through use of, say, boundary layer control, it might well work out that it could fly at a slow speed where 100 per cent power is required to maintain a constant altitude.

The power required varies with Weight, altitude and airplane configuration. Fig. 1-50 shows the effects of various Weights on the power required for the airplane of Fig. 1-49. The solid line represents the curve of Fig. 1-49.

Added or subtracted Weight affects the induced Drag existing and power required at various airspeeds. (The engineers speak of this as induced power required.) You can also see that the stall speed is lower with less Weight and vice versa. *Notice that a variation in Weight has comparatively little effect on the max speed.* The effects of Weight are felt mostly where induced Drag is predominant.

Fig. 1-51 shows the effects of parasite Drag on the power required curve of Fig. 1-49. The new curve represents the power required (at the original Weight of 3000 pounds) with the gear extended. In this case, the maximum speed would be greatly affected because parasite Drag (or parasite power required) is the largest factor in that area. The cruise speed *would* be affected because it is also in an area of high parasite Drag. *The stall speed would not vary and comparatively small effects would be felt at the lower flight speeds where parasite Drag is low.*

The *total* power required equals induced power required *plus* parasite power required.

Fig. 1-52 shows *Thrust horsepower* required and available versus airspeed for a light twin at gross

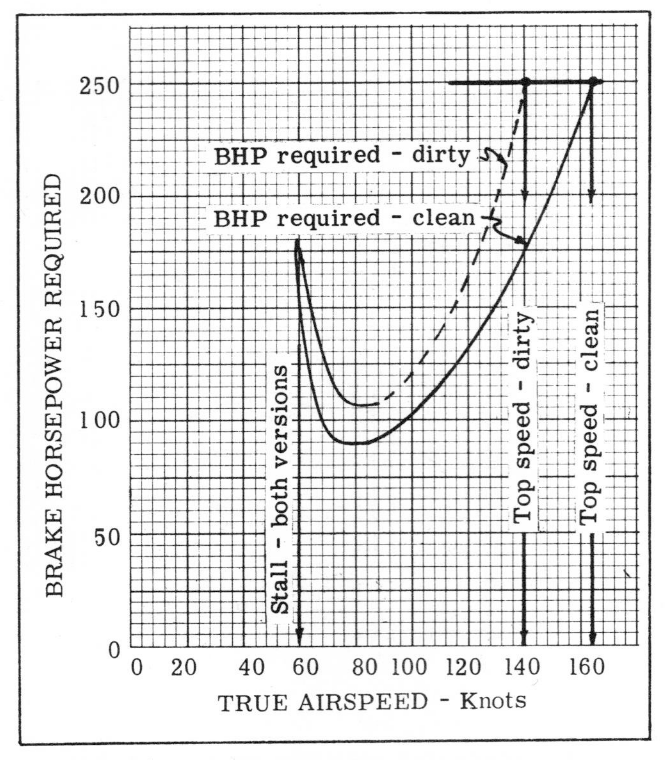

Fig. 1-51. The effects of parasite Drag (extended gear) on the airplane of Figure 1-49.

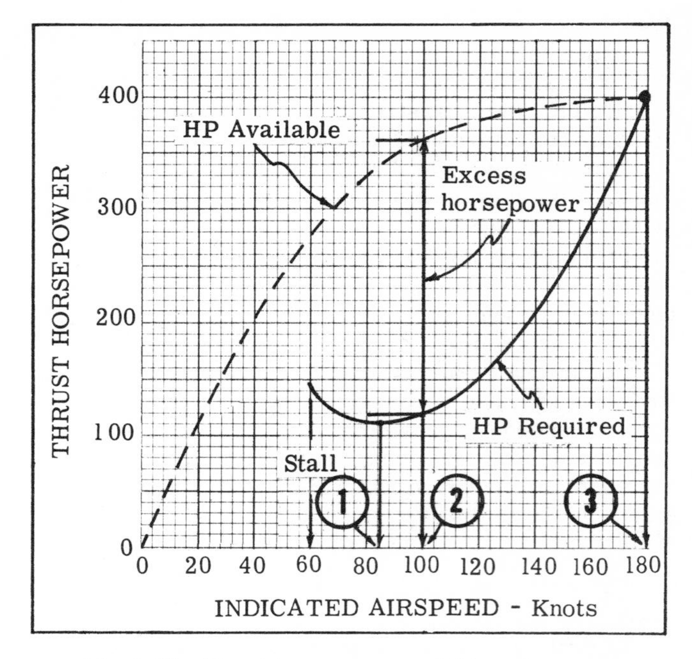

Fig. 1-52. Thrust horsepower available and required versus airspeed for a light twin at gross Weight at sea level. (Assume I.A.S. = T.A.S. at sea level.)

Weight at sea level. This curve is important in that the rate of climb of the airplane depends on the excess THP available. The maximum rate of climb, then, is found at the airspeed where the maximum excess THP is available because that is the horsepower working to raise the airplane. Notice that as you slow down past the point of minimum power required, Point (1), the THP required starts increasing again. The excess horsepower available depends on the characteristics of *both* the THP available and THP required curves. Point (2) shows the airspeed at which the maximum excess horsepower exists and is therefore the speed for a max rate of climb for this airplane at the particular Weight and altitude. Point (3) shows the maximum level flight speed at sea level.

Chapters 2 and 5 will go into more detail on climb requirements and how the excess THP is working in making the airplane climb.

JETS AND PROPS

The jet engine is considered to exert a constant Thrust at all airspeeds as compared to that shown for the propeller in Fig. 1-33. Therefore, the THP developed by the jet increases in a straight line with velocity (THP $= \frac{\text{TV mph}}{375}$ or $\frac{\text{TV knots}}{325}$).

Fig. 1-53 is a Power Required and Power Available versus velocity curve for the light twin of Fig. 1-52 when it is equipped with either jets or reciprocating engines.

Assume for simplification that the airplane could be equipped with either jet engines or reciprocating engines with no difference in parasite Drag or gross Weight. This means that the power *required* curve in Fig. 1-53 would be exactly the same for either version.

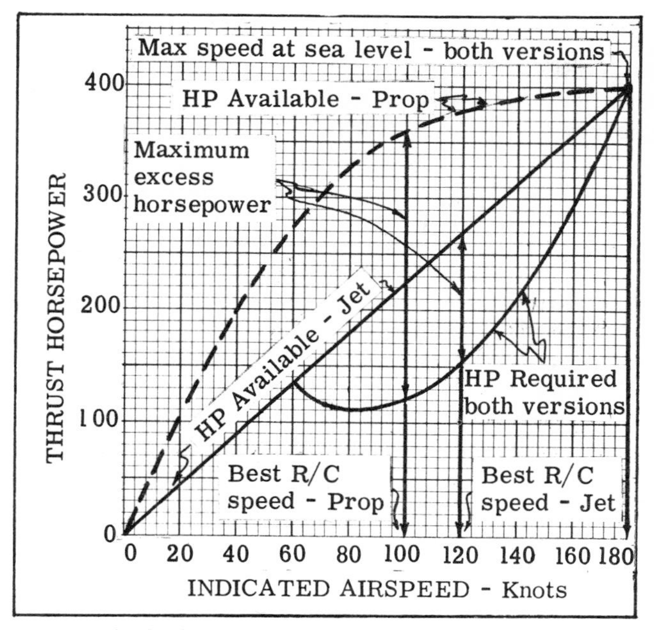

Fig. 1-53. A comparison of prop and jet versions of the light twin in Figure 1-52.

Suppose that the top speed of the reciprocating version of our fictitious airplane is 180 knots at sea level and it requires 400 (Thrust) horsepower to fly at this speed. (The top level flight speed is at that point at which the power required equals the total power available, you remember.) We'll also say that the jet version is equipped with two engines developing maximum Thrust at sea level of 361 pounds each, for a total of 722 pounds of Thrust. If the airplanes are at the same Weight and have the same parasite Drag, it can be shown (Fig. 1-53) that the top speed of the jet version is also 180 knots because we noted above that it required 400 Thrust horsepower to fly level at this speed and that just happens to be what our jet engines are producing at that speed:

Thrust Horsepower = $\frac{TV}{325}$; or THP =

$\frac{722 \text{ (lbs. Thrust)} \times 180 \text{ (knots)}}{325}$ = 400 (Thrust) horsepower. Or $\frac{722 \times 208 \text{(mph)}}{375}$ = 400 (Thrust) horsepower.

Thrust horsepower available curves are given both for props (dashed line) and jet engines (solid line). Note that the Thrust horsepower produced by the jet version is a straight line — it is directly proportional to velocity.

As was mentioned before, the rate of climb for any airplane is proportional to the excess horsepower available and several things can be noted in Fig. 1-53. The maximum rate of climb is found for the airplane at a speed of 100 knots when it uses reciprocating engines, and at the higher speed of 120 knots when the jet engines are installed.

The higher speed used for climb in a jet airplane is one of the hardest things for the ex-prop pilot to get used to. Notice in Fig. 1-53 that the performance of the jet-equipped version would be poor in the climb and low speed regime because the smaller amount of excess horsepower being available as compared to the props. In fact, at speeds close to the stall, a power deficit could exist in that airplane. If you tried to hurry the airplane off the ground at take-off you might get it too cocked up and find that you can't get, or stay, airborne. (This sometimes happens even in airplanes with a reasonable amount of power or Thrust available.)

Obviously this jet version is underpowered — even though it has a top speed the same as the propeller version of the airplane. Such low Thrust engines certainly would not be used for this particular airplane but are only cited here as an example. Of course, the jet engines would actually be more streamlined and the manufacturer would put higher-Thrust engines in, so that the jet version would be much faster. The big point here is that jet engines just aren't very practical in an airplane designed for a top speed of 180 knots.

Because of its relatively poor acceleration at low speeds, and because of the time required for the engine to develop full Thrust when the throttle is opened all the way from idle (it may take several seconds), a comparatively high amount of power is usually carried by the jet airplane on approach until the landing is assured.

The power curve will be covered in more detail as it applies to flight requirements throughout the book.

WEIGHT

The Weight of the airplane, as the Weight of any other object, always acts downward toward the center of the earth. Weight and Drag are the detrimental forces and are the main problems facing aeronautical engineers.

Weight acts toward the center of the earth so you can see by Fig. 1-54 that the Australians and other people in that area fly upside down as far as we're concerned. However, they seem quite happy about it, have been doing it for years and it's too late to mention it to them.

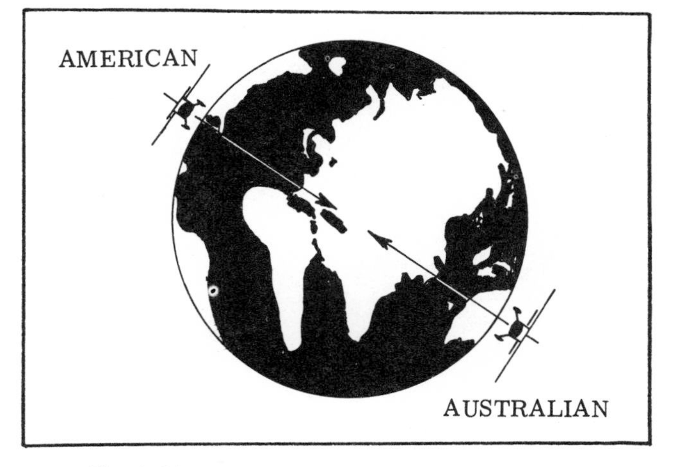

Fig. 1-54. As far as Americans are concerned, the Australians fly inverted all the time (and vice versa).

There are two airplane Weights that are used more than any others: (1) Empty Weight, or the Weight of the airplane without fuel, oil, occupants or baggage. The Empty Weight is the basic airframe and engine Weight and includes installed equipment such as radios. The regulations state that the Empty Weight shall also include unusable fuel and undrainable oil and hydraulic fluid.

(2) Gross Weight is the maximum allowable Weight for the airplane. The manufacturer's performance figures are usually given for the gross Weight of the airplane, although in some cases graphs or figures in Airplane Flight Manuals also show performance for several Weights below gross Weight. *The term as used in this book refers only to the maximum FAA certificated Weight.* This is the most commonly accepted use of the term. Airplane loading will be covered in Chapter 10.

LEFT-TURNING TENDENCIES

P-Factor has been given a great deal of credit for contributing to left-turning tendencies in situations where it has little, if any, effect on the airplane's yawing tendencies. For instance, the attitude of the tricycle gear on the take-off roll pretty well assures that the prop disk is perpendicular to the line of "flight"--yet the airplane still turns radically to the left. Let's see, there's no problem with "torque" (sure, the left wheel may be pressing on the runway a few pounds harder because of "equal and opposite reaction," but the turning effect is negligible). There is little or no difference in the angle at which the prop encounters the relative wind, so the rotating slipstream is the culprit here.

2. FLIGHT MECHANICS

The title of this chapter has nothing to do with the people who work on aircraft but covers an area of study of the forces and moments acting on the airplane in flight. While the Four Forces are fresh in your mind from the last chapter it would be well to see how they act on the airplane.

The term *force* was covered in the last chapter, but *moment*, as an engineering term, may be new to you. A moment normally results from a force (or weight) acting at the end of an arm (at a 90 degree angle to it) and is usually expressed as inch-pounds or foot-pounds. Fig. 2-1 shows a pair of equal moments acting in opposite directions.

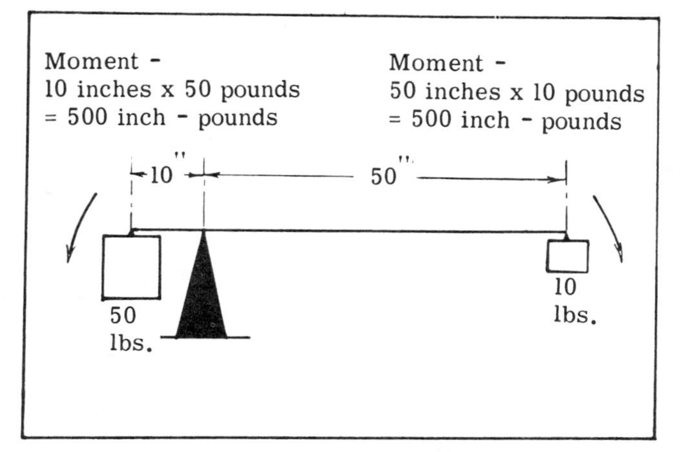

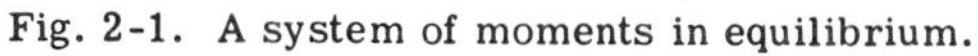
Fig. 2-1. A system of moments in equilibrium.

The airplane in steady state flight — that is, in a steady climb, glide, or in level unaccelerated flight (this includes steady level turns) — must be in "equilibrium"; the forces acting in opposite directions on the airplane must cancel out each other. (The same thing goes for the moments.)

A vector is an arrow used to represent the direction and strength of a force. You've had experience with vectors in working out wind triangles in navigation and also unconsciously discuss vector systems when you talk about "headwind" and "crosswind" components for take-off and landing. Fig. 2-2 shows a 30-knot wind acting at an angle of 30 degrees to a runway. As a pilot you use the runway centerline as a reference and consciously (or unconsciously) break the wind into components acting along and perpendicular to this reference axis.

You are interested in the component of wind acting across the runway (15 knots) and, if you were interested in computing the take-off run, you would use the headwind component, or the component of the wind acting down the runway (26 knots). You usually don't go so far as to figure out the exact crosswind component but note the wind velocity and its angle to the runway and make a subconscious estimate of how much trouble it might give you on take-off or landing. You set up your own axis and work with what would seem a most complicated system if people started talking about axes, vectors and components. What you do is break down the wind's vector into the two components of most interest to you. The same general idea will be used here for the forces acting on the airplane.

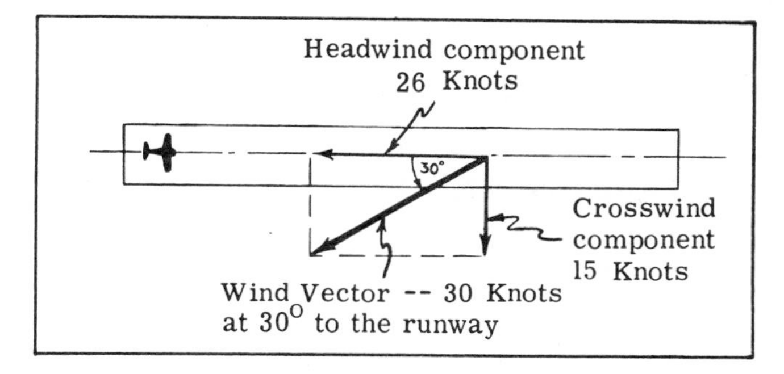

Fig. 2-2. A vector system as faced by the pilot during a take-off or landing in a crosswind. The wind is 30 degrees to the runway at 30 knots.

The reference axis for operating the airplane will be the flight path or line of flight, and the forces will be measured as operating parallel and perpendicular to it. (Fig. 2-3) For an airplane in a *steady state condition* of flight such as straight and level unaccelerated flight, a constant airspeed climb or glide, or a constant altitude balanced turn of a constant rate, the forces acting parallel to the flight path must be balanced. The same thing applies for those forces acting perpendicular, or at 90 degrees, to the flight path — they must cancel each other. Each of the vectors shown in Fig. 2-3 may represent the total of several forces acting in the direction shown.

The following must be realized in order to see the mechanics of flight:

(1) Lift always acts perpendicular to the relative wind (and, hence, perpendicular to the flight path). This is the "effective" Lift as discussed in the last chapter, or the Lift acting perpendicular to the actual path of the airplane.

(2) Drag always acts parallel to the relative wind (and flight path) and in a "rearward" direction.

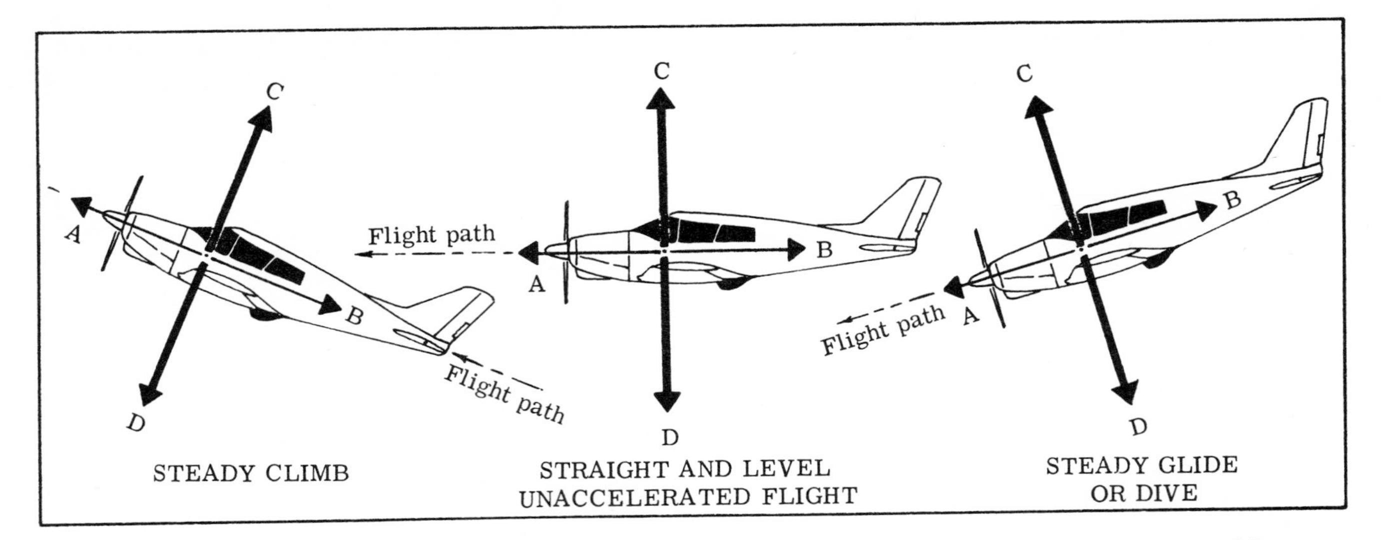

Fig. 2-3. In steady state flight the sum of the forces acting parallel to the flight path (A and B) must equal zero and the same must apply to those acting perpendicular. Minus signs may be given to the forces acting in a "downward" or "backward" direction.

(3) Weight always acts in a vertical (down) direction toward the center of the earth.

(4) Thrust, for these problems, will always act parallel to the center line of the fuselage. (In other words, at this point we'll assume no "offset" thrust line and that Thrust is acting parallel to the axis of the fuselage.)

THE FORCES AND MOMENTS IN STRAIGHT AND LEVEL FLIGHT

Take an airplane in straight and level *cruising* flight: The average airplane in this condition has a tail-down force in existence because it is designed that way (the need for this requirement will be covered in Chapter 10). It would be well to examine the forces and moments acting on a typical four-place airplane in straight and level flight at a constant speed at *cruise*.

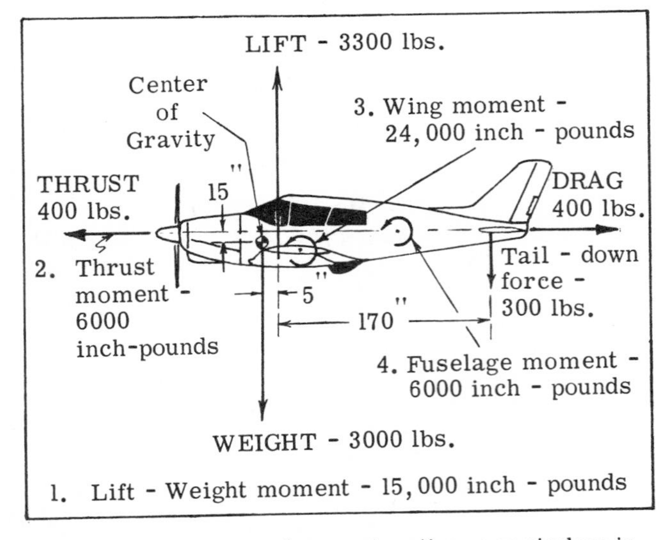

Fig. 2-4. Forces and moments acting on an airplane in steady straight and level flight.

For simplicity, rather than establish the vertical acting forces with respect to the center of gravity, which is the usual case, they will be measured fore and aft from the center of Lift. Assume at this point that Lift is a string holding the airplane "up" and its value will be found later (this is legal). The airplane in Fig. 2-4 weighs 3000 pounds, is flying at 154 knots I.A.S. and at this particular loading the center of gravity is 5 inches ahead of the "Lift line."

Summing up the major moments acting on the airplane (check Fig. 2-4 for each):

(1) The Lift-Weight moment — The Weight (3000 pounds) is acting 5 inches ahead of the center of Lift and this results in a 15,000 inch-pounds *nose-down* moment (5 inches x 3,000 pounds = 15,000 inch-pounds).

(2) The Thrust Moment — Thrust is acting 15 inches above the center of gravity and has a value of 400 pounds. The *nose-down* moment resulting is 15 x 400 = *6000 inch-pounds*. (The moment created by Thrust will be measured with respect to the C.G.) For simplicity it will be assumed that the Drag is operating back through the C.G. Although this is not usually the case, it saves working with another moment.

(3) The Wing Moment — The wing, in producing Lift, creates a nose-down moment which is the result of the forces working on the wing itself. Fig. 2-5 shows force patterns acting on a wing at two airspeeds (angles of attack). These moments are acting with respect to the aerodynamic center, a point considered to be located about 25 per cent of the chord for all airfoils.

Notice that as the speed increases (the angle of attack decreases) the moment becomes greater as the force pattern varies. The nose-down moment created by the wing increases as the *square* of the airspeed if the airfoil is not a symmetrical type. (There is no wing moment if the airfoil is symmetrical because all of the forces are acting through the aerodynamic center of the airfoil.)

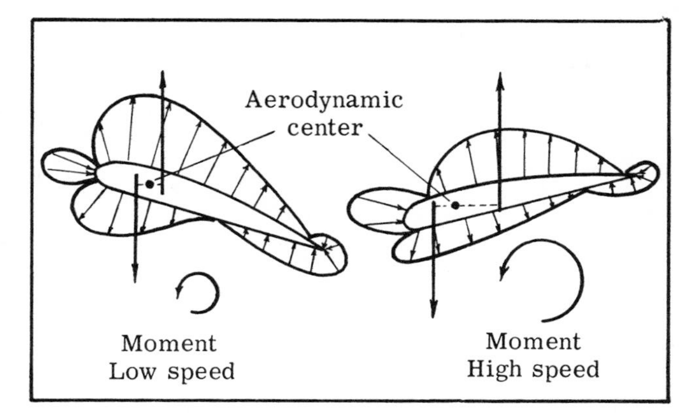

Fig. 2-5. The moments created by the unsymmetrical airfoil at two different airspeeds. The angles of attack and pressure patterns around the airfoil have been exaggerated.

For an airplane of the type, airspeed and Weight used here, a nose-down moment created by the wing of 24,000 inch-pounds would be a good round figure. Remember that this would vary with indicated airspeed. *Nose-down moment created by wing – 24,000 inch-pounds.*

(4) The fuselage may also be expected to have a moment about its C.G. because it, too, has a flow pattern and, for this example airplane type and airspeed, would be about *6000 inch-pounds nose-down.* (This is not always the case.)

Summing up the nose-down moments:

Lift-Weight Moment	=	15,000 inch-pounds
Thrust Moment	=	6,000 inch-pounds
Wing Moment (at 154 knots)	=	24,000 inch-pounds
Fuselage Moment (at 154 knots)	=	6,000 inch-pounds
Total Nose-Down Moment	=	51,000 inch-pounds

For equilibrium to exist there must be a *tail-down* moment of 51,000 inch-pounds and this is furnished by the tail-down force. Fig. 2-4 shows that the "arm" or the distance from the Lift line to the center of the tail-down force is 170 inches. So, the moment (51,000 in-lbs) and the arm (170 inches) are known and the force acting at the end of that arm (the tail-down force) can be found as $\frac{51,000 \text{ in-lbs}}{170 \text{ inches}} =$ 300 pounds. The airplane nose does not tend to pitch either way.

The *forces* must also be balanced for equilibrium to exist. Summing up the forces acting perpendicular to the flight path (in this case because the flight path is level, it can be said also that the *vertical* forces must be equal — in a climb or glide the forces acting perpendicular to the flight path will not be vertical as can be seen by checking Fig. 2-3). The "down" forces are the Weight — 3000 pounds, and the tail-down force — 300 pounds. The "up" force (Lift) must equal the down forces for equilibrium to exist so that its value must be 3300 pounds. Now the moments *and* forces acting perpendicular to the flight path are in equilibrium. As can be seen, Lift is not normally the same as Weight in straight and level unaccelerated flight. Of course, the center of gravity can be moved back to a point where no nose-down moment exists and no tail-down force is required. This, however, could cause stability problems which will be covered in Chapter 10.

In the situation just discussed it was stated that the airplane was at a *constant cruise* speed so that the force (in pounds) acting rearward (Drag) and the force (in pounds) acting forward (Thrust) are equal. (It is assumed that at higher speeds the Thrust line is acting parallel to the flight path so it can be considered to be equal to Drag.)

Then it can be said without too much loss of accuracy that in the cruise regime Thrust equals Drag and normally Lift is slightly greater than Weight when the forces are balanced.

But what about a situation where the airplane is flying straight and level at a constant airspeed in *slow flight*? Again the forces must be summed as shown in Fig. 2-6. Now the Thrust line is *not* acting parallel to the flight path (and opposite to Drag) and for purposes of this problem it will be assumed that it is inclined upward from the horizontal by 15 degrees.

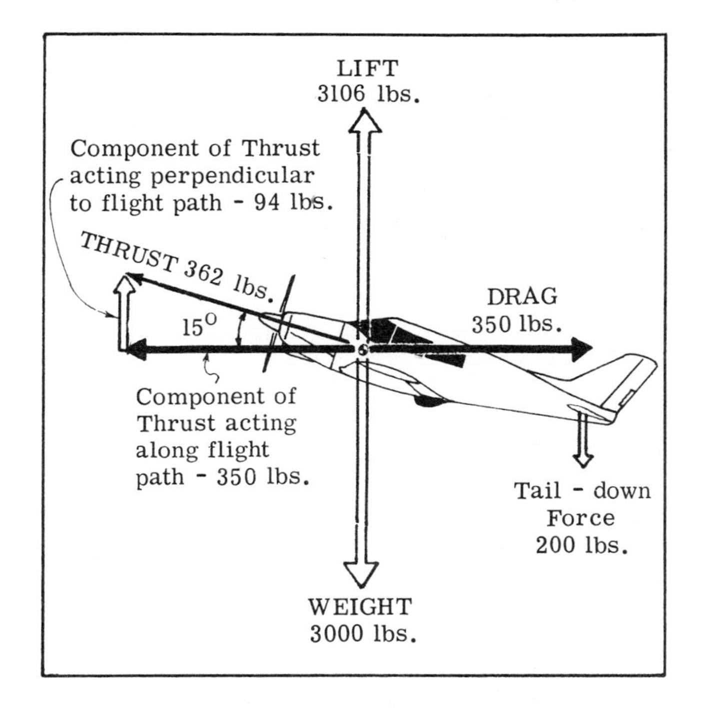

Fig. 2-6. The forces at work on the airplane in straight and level slow flight just above the stall. The vertical component of Thrust has been moved out ahead of the airplane for clarity. Because of the placement of the various forces it would appear that the moments are not in equilibrium. They will be assumed to be so for this problem.

As a pilot, for straight and level slow flight you set up the desired airspeed and use whatever power is necessary to maintain a constant altitude; you don't know the value of Drag, Thrust or Lift (and may have only a vague idea as to what the Weight is at that time, but for this problem it's 3000 pounds as before). The tail-down force will be assumed to be 200 pounds. (It will likely be less at this high angle of attack than for cruise.) In the problem of straight and level *cruising* flight it was just assumed that Thrust

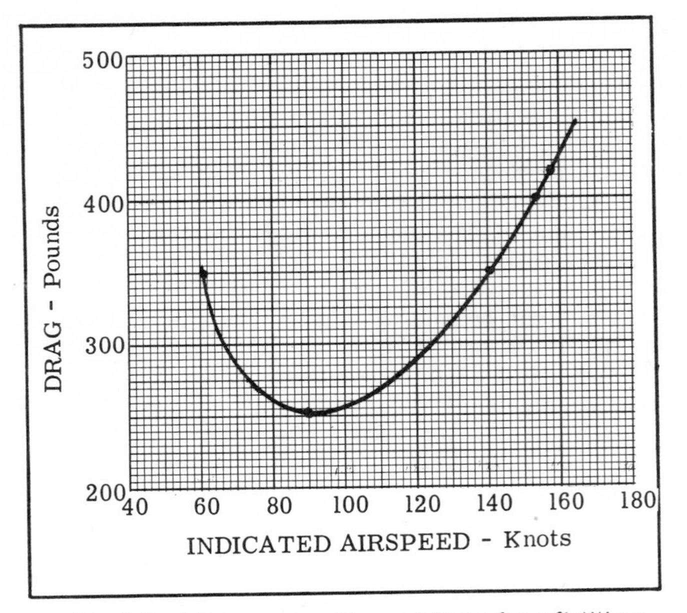

Fig. 2-7. A Drag versus Airspeed Curve for a fictitious four-place high performance, single engine airplane at gross Weight. The values are in the area expected of that current type airplane.

equaled Drag and we weren't particularly interested in the values. Look at Fig. 2-7 for a typical Drag versus Airspeed curve.

In summing the forces parallel to the flight path in slow flight with this airplane, there is Drag (350 pounds) and the component of Thrust acting opposite Drag, which must be 350 pounds also. No doubt you are already ahead of this in your thinking and realize that because it is inclined at an angle, the actual Thrust must be greater than Drag if its "forward" component along the flight path, is equal to Drag. You could look in a trigonometry table and find that at the 15 degree angle, the *actual* Thrust must be about $3\frac{1}{2}$ per cent higher or about 362 pounds as compared to 350 pounds of Drag. Thrust also has a component acting at a 90 degree angle to the flight path acting to help Lift. A check of a trigonometric table would show that this force is 26 per cent of the actual Thrust and has a value of about 94 pounds (which is a fair amount).

Now, summing the forces perpendicular to the flight path (the "up" forces must equal the "down" forces):

Forces "Down" = Weight + tail-down force = 3000 + 200 = *3200 lbs.*

Forces "Up" = Lift + Vertical component of Thrust = Lift + 94 lbs. = *3200 lbs.*

Lift, of course, is found as 3200 - 94 = 3106 lbs., using our arbitrary values. So Lift is less at low speed level flight (3106 pounds) than at cruise (3300 pounds), if you are talking strictly about each of the Four Forces. You don't worry about this in practical application but fly the airplane and set the power and airspeed to get the desired result.

As the vertical component of Thrust helps support the airplane the wings only have to support 3106 pounds rather than the full 3200 pounds (Weight and tail-down force) in slow flight and therefore the wing loading is less than would be expected. The airplane always stalls at a lower airspeed with power-on (for the same flap setting and Weight) than in the power-off condition. The effect of the slipstream across the wing helps lower the stall speed, too.

The greater that Thrust is in proportion to Weight the greater this effect. For instance, if the airplane had an engine-propeller combination capable of producing 3000 pounds of Thrust the airplane would be capable of "hanging on its prop" and in theory the power-on stall speed would be zero.

So, in summary, in straight and level flight in the *slow flight regime* it may be expected that: (1) the actual Thrust exerted by the propeller (in pounds) is greater than the Drag of the airplane and (2) Lift is less than at higher speeds. The location of the center of gravity and the angle the Thrust line makes with the flight path and other factors can have an effect on these figures, of course.

FORCES IN THE CLIMB

To keep from complicating matters, the tail-down force will be ignored for the first part of each section of flight mechanics. It exists, of course, and varies with C.G. and angle of attack (airspeed), but is comparatively small in most cases so Lift will be considered equal to Weight, at least at the beginning. We'll also assume that all moments are balanced and won't have to consider them further, and the Four Forces will be drawn as acting through a single point (the C.G.) of the airplane to avoid complicating the drawings.

One of the biggest fallacies of thinking among pilots is (as mentioned in the last chapter) that of thinking that the airplane climbs because of "excess Lift." For purposes of this problem the Drag (in pounds) of the example airplane will be 250 pounds at the recommended climb speed of 90 knots (Fig. 2-7). The values for Drag approximate a typical high performance, four-place, retractable gear airplane weighing 3000 pounds, but the figures have been rounded off.

Again, remembering that all forces (and moments) must be in balance for such equilibrium to exist, the following is noted:

Because the flight path is no longer level, Weight, for the first time, is no longer operating in a direction 90 degrees to the flight path. As the forces must be in equilibrium both parallel and perpendicular to the flight path, Weight must be broken down into the components acting in these directions (as you do with the wind when it is neither right down the runway nor straight across it). Check Fig. 2-8.

Fig. 2-9 shows the forces acting on the airplane in a steady state climb of 90 knots (I.A.S.). The airplane has an angle of climb of 8 degrees to the horizontal and requires an angle of attack of 6 degrees to fly at the climb airspeed of 90 knots. We are assuming that the angle of incidence is zero (the wing chord line is exactly parallel to the fuselage center line)

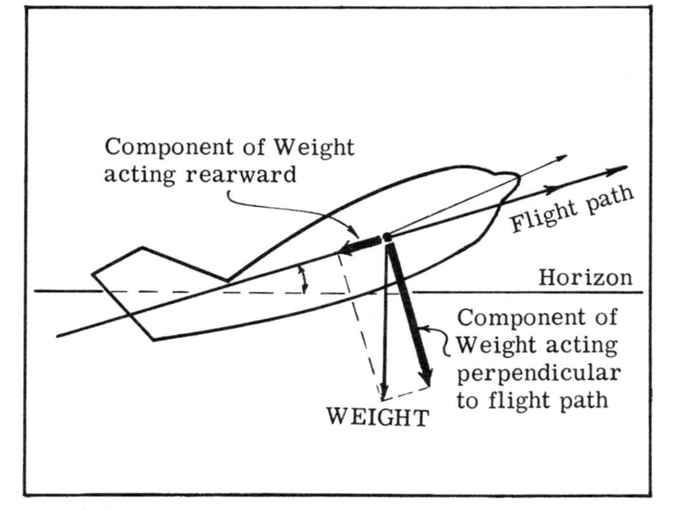

Fig. 2-8. As Weight is no longer acting perpendicular to the flight path it must be broken down into components as shown.

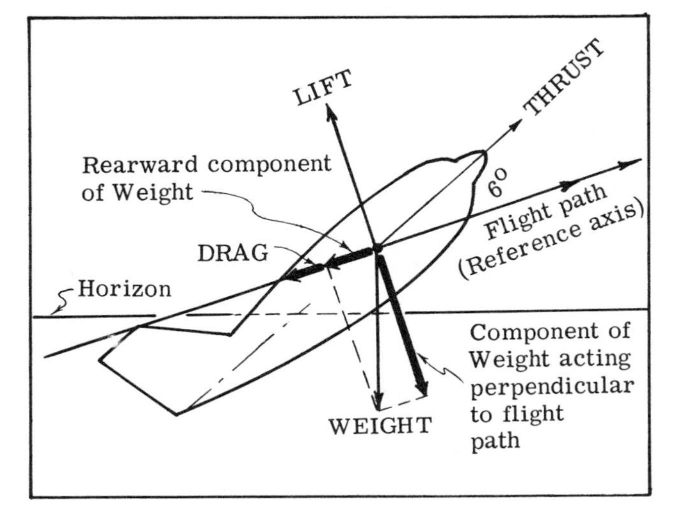

Fig. 2-9. The forces acting on an airplane in a steady climb.

and that the Thrust line is offset "upward" from the flight path by 6 degrees in this climb. In the following drawings the angles will be exaggerated and a simplified airplane silhouette used for clarity.

Summing the forces *parallel* to the flight path the following is noted:

The forces acting rearward along the flight path are (1) aerodynamic Drag (250 pounds) *plus* the rearward component of Weight which, by checking a trigonometric table, for the 8 degree angle of climb (in round numbers) is found to be 417 pounds. The *total* rearward acting force is aerodynamic Drag (250 pounds) + the rearward acting component of Weight (417 pounds), or *667 pounds.*

For the required equilibrium, or steady state climb condition, to exist there must be a balancing force acting forward along the flight path and this is furnished by Thrust. The fact that the Thrust line is offset upward from the flight path by 6 degrees further complicates the problem. Because of its inclination the actual Thrust produced by the propeller must be greater than 667 pounds in order to have that force acting along the flight path. The actual Thrust, you will note, is the hypotenuse of a right triangle, and you remember from your geometry that the hypotenuse of a right triangle is always longer than either one of its sides; one of the sides — the longer of the two — is the component of Thrust acting along the flight path, which must be equal to the rearward acting force(s).

Again, a check of a trigonometric table shows that to have 667 pounds along the flight path the *actual* Thrust must be about 0.55 per cent greater (a little more than one-half of one per cent) so that its value is three pounds greater or about 670 pounds. (A nit-picking addition, to be sure.) Fig. 2-10 shows that the forces acting *parallel* to the flight path at the climb speed of 90 knots and Weight of 3000 pounds are balanced.

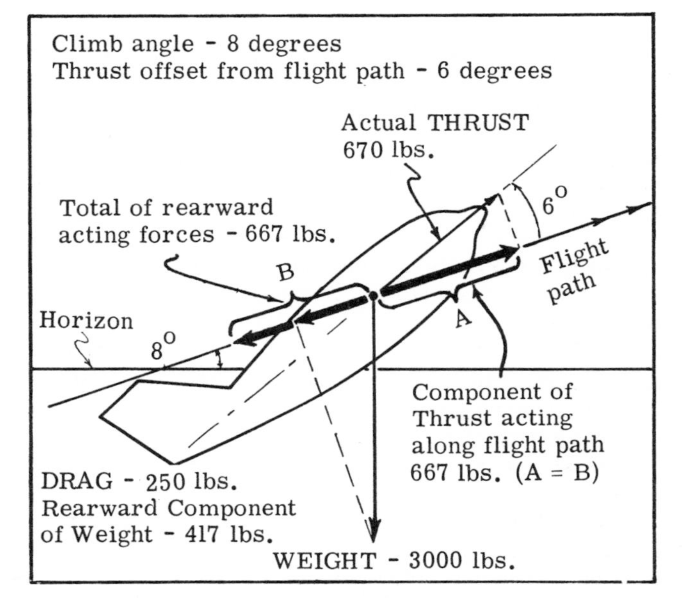

Fig. 2-10. A summary of the forces acting *parallel* to the flight path in the steady state climb.

Now to sum the forces acting *perpendicular* to the flight path:

The component of Weight acting perpendicular (or more or less "downward") to the flight path at the climb angle of 8 degrees turns out to be 2971 pounds according to the trigonometric table. As this is considered to be the only force acting in that direction (now that the tail-down force is being neglected) it must be balanced by an equal force (or forces) in the opposite direction. The two forces acting in that direction are (1) Lift and (2) the component of Thrust acting at 90 degrees, or perpendicular, to the flight path. (Fig. 2-11)

As Thrust is now a known quantity, we can solve for that component acting in the same direction as Lift. For a 6 degree angle of inclination the component for 670 pounds of Thrust is 70 pounds (rounded off). This means that Lift must have a value of 2901 pounds in this case (2971 - 70 = 2901 pounds) or Lift (2901 pounds) + Thrust Component (70 pounds) = Weight component (2971 pounds). The forces acting perpendicular to the flight path are balanced.

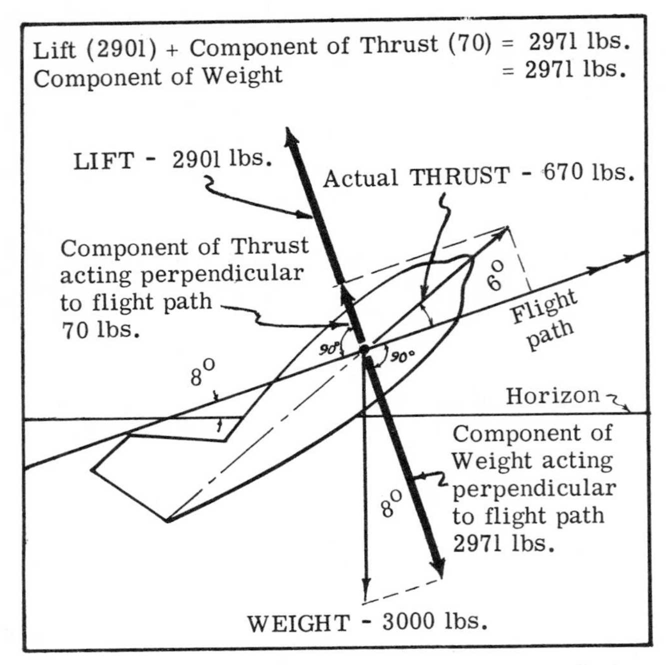

Fig. 2-11. A summary of the forces acting perpendicular to the flight path in the climb.

Lift (2901 pounds) is found here to be *less* than the airplane's Weight (3000 pounds) in the steady state climb. Thrust (670 pounds) is *greater* than aerodynamic Drag (250 pounds).

What happened to the idea that an airplane makes a steady climb because of "excess" Lift?

Even considering the tail-down force which for the airspeed, Weight and C.G. location of this airplane could be expected to be about 250 pounds, Lift is hardly greater than Weight. In any event, there is no "excess Lift" available — it's all being used to balance the tail-down force and the component of Weight acting perpendicular to the flight path. (Lift would have to be 2901 plus 250 pounds, or 3151 pounds.)

You remember from the last chapter that the Thrust horsepower equation is $THP = \frac{TV}{325}$ (the 325 is for the airspeed in knots) so that the Thrust horsepower being developed along the flight path is $\frac{667 \times 90}{325}$ = 184+ THP. The "V" in the equation is *true* airspeed and it will be assumed that the airplane is operating at sea level at this point so that the indicated climb airspeed of 90 knots equals a true airspeed of the same value (again assuming no instrument or position error so that indicated and calibrated airspeeds are the same).

The rate of climb of an airplane depends on the amount of *excess* Thrust horsepower available at a particular airspeed. This excess THP means the horsepower that is working to move the airplane vertically. The recommended best rate of climb speed is that one in which the greatest amount of excess thrust horsepower is available. The following equation may be used to determine the rate of climb.

Rate of climb (f.p.m.)

$$= \frac{\text{Excess Thrust horsepower} \times 33{,}000}{\text{Weight of airplane}}$$

Power is *force* times *distance per unit of time* and one horsepower is equal to 550 foot-pounds per second or 33,000 foot-pounds per minute. That's where the 33,000 in the equation comes in; it's set up for a rate of climb (vertical displacement) in feet per minute. Going back to the original idea for horsepower (in this case Thrust horsepower) the equation would be as follows: Thrust horsepower used to

$$\text{climb (excess THP)} = \frac{\text{Airplane Weight} \times \text{R.C. (fpm)}}{33{,}000}.$$

The Thrust horsepower required to climb is raising a certain Weight (the airplane) a certain vertical distance in a certain period of time.

But to find the rate of climb for the example airplane it would be well first to find out how much THP is required to fly the airplane *straight and level* at a constant altitude at sea level at 90 knots. As Weight in *level flight* will not have a component acting rearward to the flight path, the only retarding force is aerodynamic Drag, which was found to be 250 pounds. The Thrust component acting along the flight path must be equal to this, or 250 pounds. Assuming that the angle of attack and the angle that Thrust makes with the flight path is 6 degrees, to get this value the actual Thrust would be about 251 pounds but we'll round it off to the 250. (Fig. 2-12)

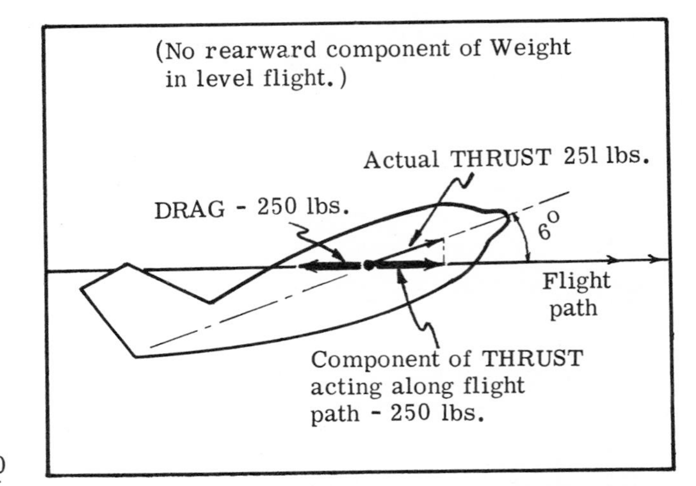

Fig. 2-12. The forces acting parallel to the flight path for the airplane flying *straight and level* at the recommended climb speed of 90 knots.

In the earlier look at the climb at 90 knots, 667 pounds of Thrust was being exerted along the flight path. This is 417 pounds more than required for level flight and is, in effect, the "excess Thrust" needed for a climb angle of 8 degrees at 90 knots. (The rearward component of Weight was 417 pounds.)

Solving for excess Thrust horsepower:

$$EHP = \frac{\text{Excess Thrust} \times \text{Velocity (knots)}}{325}$$

$$= \frac{417 \times 90}{325} = 115 \text{ THP}$$

Solving for rate of climb: $R.C. = \frac{EHP \times 33{,}000}{\text{Weight}}$

$$= \frac{115 \times 33{,}000}{3{,}000} = 1265 \text{ fpm.}$$

The brake horsepower required to get such performance for a 3000-pound airplane with the described characteristics could be estimated:

It can be assumed here that at the climb speed the propeller is 74 per cent efficient (efficiency varies with airspeed you remember from the last chapter) and that the Thrust horsepower being developed is 74 per cent of the horsepower (brake) being developed at the crankshaft. The *total* Thrust horsepower being used in the climb is $THP = \frac{T \times V}{325} = \frac{670 \times 90}{325} = 185$ THP (rounded off). The Thrust acting along the flight path was 667 pounds, but the *total* Thrust exerted was 670 pounds and this is what must be used to work back to the brake horsepower requirement.

This, then, is approximately 74 per cent of the horsepower developed at the crankshaft, so the brake horsepower required to get this performance for the fictitious airplane would be $\frac{185}{0.74}$ or approximately 250 brake horsepower. (Or 0.74 x 250 = 185.)

The rate of climb found is in the ball park for current four-place retractable gear airplanes. "Our" airplane may be cleaner or dirtier aerodynamically than others and all of this resulted from our arbitrarily selecting the aerodynamic Drag (250 pounds), angle of attack in the climb (6 degrees) and the climb angle of 8 degrees at the climb speed of 90 knots. The figures were picked to give a reasonable idea of how such airplane types get their climb performance. The 74 per cent used for propeller efficiency is also arbitrary, although the figure is close to that expected for the airplane type and speed discussed.

The more practical aspects of the climb will be covered in Chapter 5.

FORCES IN THE GLIDE

As you have probably already reasoned, anytime the flight path of the airplane is not horizontal, Weight has to be broken down into two components. The glide or descent at an angle of 8 degrees to the horizontal would have the same percentages of Weight acting perpendicular and parallel to the flight path as for the 8 degree angle of climb just mentioned — except that in the glide the component of Weight parallel to the glide path is not a retarding force but is acting in the direction of flight.

For this situation it is assumed that the power is at idle and *no Thrust exists*. The tail-down force will be neglected at first. The forces acting parallel to the flight path are (1) the component of Weight, which must be balanced by (2) aerodynamic Drag in order to keep the airspeed constant in the descent. For an 8-degree angle of descent the component of Weight acting along the flight path would be 417 pounds as for the climb — except that it's now working in the direction of motion. If an angle of descent of 8 degrees is arbitrarily chosen, then the aerodynamic Drag must equal the component of Weight acting along the flight path for a steady-state condition to exist. Looking back to Fig. 2-7 you see that this value of Drag exists at about 158 knots.

The more usual situation would be to use a power-off glide speed as recommended by the manufacturer. For this example 90 knots will be used as the recommended (clean) glide speed. We'll also ignore the effects of power decrease or windmilling prop on the Drag curve and say that aerodynamic Drag at 90 knots is 250 pounds as it was for the power-on climb. The speed of 90 knots may or may not be the best one for maximum glide efficiency (it depends on the airplane) but the niceties of that will be covered further on.

Using the reasoning used for the other steady-state flight conditions, Fig. 2-13 shows the forces acting parallel to the flight path in a power-off glide. (Again, the tail-down force is being neglected for simplicity.)

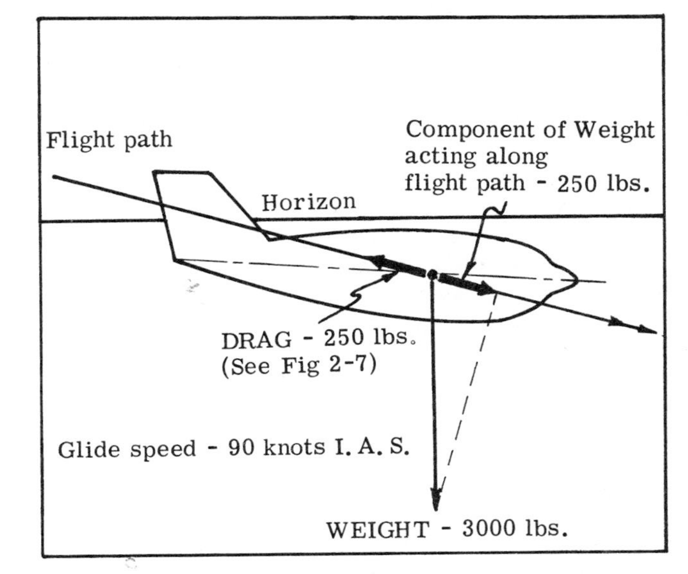

Fig. 2-13. The forces acting parallel to the flight path in the power-off glide at 90 knots.

Realizing that the component of Weight acting along the flight path must have a value equal to the 250 pounds of aerodynamic Drag, the glide path will be of a certain angle for this condition to occur and a check of a trigonometric table shows this to be a 4^{0} - 47' (4 degrees and 47 minutes), or nearly a 5-degree angle downward in relation to the horizon.

Knowing the glide angle, the forces acting 90 degrees to the flight path (Lift and the component of Weight acting perpendicular to the flight path) can be found. (Fig. 2-14)

That Weight component can be found by reference to a trig table as 2990 pounds and so Lift must also equal this value for a steady-state (or constant) glide under the conditions of ignoring the tail-down force.

For shallow angles of glide the variation of Lift from Weight is usually ignored. In this case, Lift is 2990 pounds to a Weight of 3000 pounds, a variation of about one-third of one per cent.

In the climb a final figure for Lift required at 90 knots (considering the tail-down force) was 3151

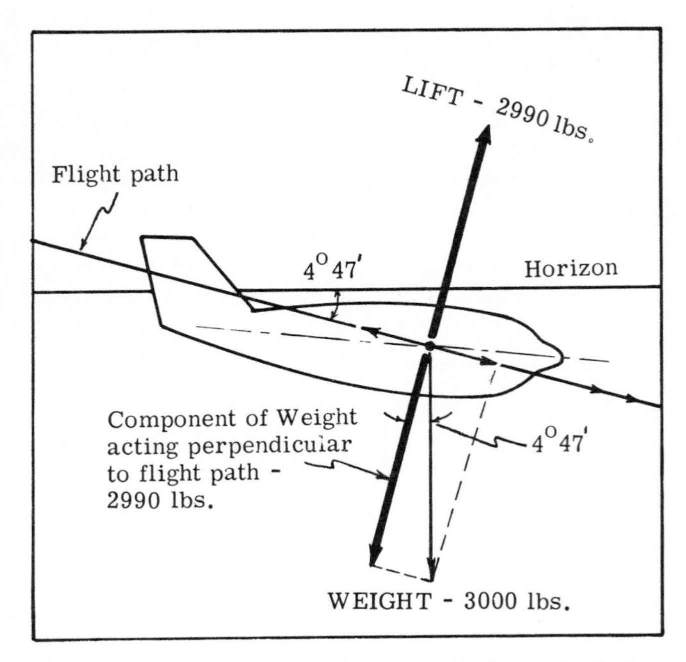

Fig. 2-14. The forces acting perpendicular to the flight path in the glide.

pounds. For the glide the tail-down force for this airplane would be in the vicinity of 225 pounds because of the lack of a moment created by Thrust. The component of Weight acting perpendicular to the flight path at the 90-knot glide was 2990 pounds and Lift required to take care of this would be 2990 + 225 or *3215 pounds. There are 64 more pounds of Lift in the glide than in the climb, or Lift would be greater in the glide than in the climb under the conditions established!*

The angle that Weight varies from being perpendicular to the flight path, is also the angle of glide or descent. If *aerodynamic Drag is cut to a minimum, the components of Weight acting parallel to the flight path can also be a minimum for a steady-state glide.* In other words, if the aerodynamic Drag could somehow be halved for this airplane the angle of glide would be halved, or it would descend at an angle of about $2\frac{1}{2}$ degrees to the horizontal *and would glide twice as far for the same altitude.*

As the airplane's Weight is considered to be constant for a particular instant of time the answer is that the farthest distance may be covered with the airplane flying at the angle of attack (or airspeed) at which the aerodynamic Drag is a minimum. For instance, assuming that at small angles of descent that Lift equals Weight (3000 pounds), the angle of glide of the example airplane is $\frac{\text{3000 pounds (Lift or Weight)}}{\text{250 pounds (aerodynamic Drag)}}$ = 12. The glide ratio for our example airplane at 90 knots is 12 to 1, or twelve feet forward for every foot down. And, as the point of minimum aerodynamic Drag (250 pounds at 90 knots I.A.S.) was used from Fig. 2-7 this would be the minimum glide angle (or *maximum distance* glide) for the example airplane. Anytime Drag is increased, the efficiency of the glide is *decreased.* If you used either a faster or slower glide speed than the 90 knots chosen, a check of Fig. 2-7 shows that Drag will increase — and the glide ratio will suffer.

One method of increasing Drag would be to glide with the landing gear extended (an increase in parasite Drag which would result in an increase in total Drag).

With the gear down a typical figure for Drag for an airplane of this type at 90 knots would be 300 pounds. The glide angle would be greater and the glide ratio would suffer.

Assume that the pilot starts gliding "clean" and the glide ratio is 12 to 1. The nose is at a certain attitude to get the 90 knots (and 4° 47' angle of descent) and for most airplanes of that type the nose will be approximately level.

He extends the gear and suddenly the forces acting parallel to the flight path are no longer in balance; Drag is greater than the component of Weight and the airplane would start slowing if the nose were kept at the same position. He, of course, has decided to glide at 90 knots as before so must drop the nose and change the flight path so that the component of Weight acting along the flight path would equal the 300 pounds of aerodynamic Drag. The new glide ratio at 90 knots with the gear down would be $\frac{\text{Lift (3000 lbs)}}{\text{Drag (300 lbs)}}$ = about 10 to 1, or the glide angle would be about 6 degrees relative to the horizon.

The method of finding the rate of sink of the airplane can be compared to that of solving for the rate of climb. The rate of sink, however, is a function of the *deficit Thrust horsepower* existing at the chosen airspeed.

$$\text{Rate of Sink (f.p.m.)} = \frac{\text{Deficit THP x 33,000}}{\text{Airplane's Weight}}$$

The aerodynamic Drag for the airplane is a force of 250 pounds acting rearward along the flight path at the airspeed of 90 knots (the airplane is clean and weighs 3000 pounds). The equivalent Thrust horsepower required to be acting in the direction of flight to equal the effects of Drag at 90 would be: THP = $\frac{\text{DV (knots)}}{325} = \frac{250 \times 90}{325}$ = 69. The combination of Thrust and velocity would have to equal 69 Thrust horsepower for level flight at 90 knots, or $\frac{TV}{325}$ = 69 THP. However, in this case Thrust is zero and, as you know, zero times any number (90 knots in this case) is still zero. So, there's no Thrust horsepower being developed by the engine and the airplane is 69 THP in the hole, or there is a deficit of 69 THP existing. The rate of sink can be calculated.

$$\text{Rate of Sink} = \frac{69 \times 33{,}000}{3000} = \text{760 feet per minute}$$

(rounded off).

This could be checked by looking at the situation in Fig. 2-13 again. The airplane is descending down a path inclined at an angle of 4° 47' at 90 knots forward speed. Converting the 90 knots to feet per minute it can be said that the airplane is moving down the path at a rate of 9130 feet per minute. It was already found that the glide ratio was 12 to 1 so that the feet down per minute would be one-twelfth of that traveled along the glide path, or about 760 feet per minute. (See the section below for an explanation.)

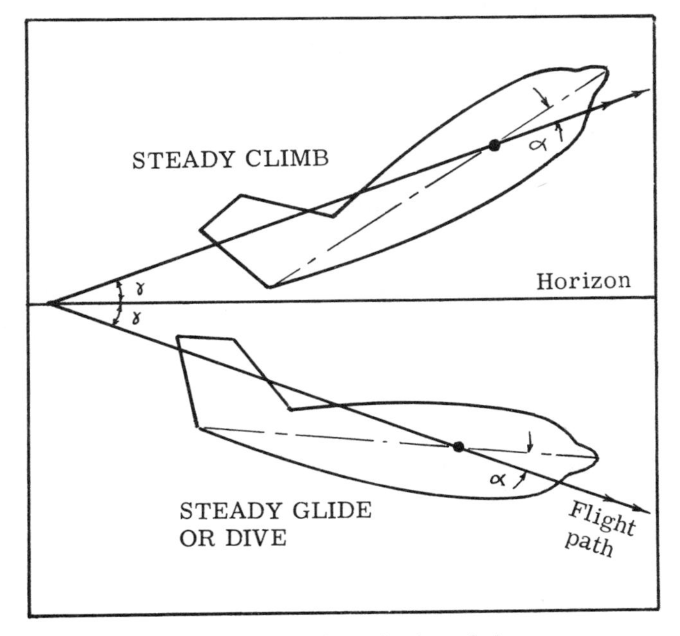

Fig. 2-15. Mathematical symbols.

NOTE TO THOSE INTERESTED IN THE MATHEMATICS

The assumption here for the shallow angle of 4^0 $47'$ *(*5^0*) is that the sine equals the tangent of the angle.* Fig. 2-15 gives the mathematical symbols used.

The flight path angle with the horizon is given the designation γ (gamma). This is used for both climbs and glides and all sine values are given as positive. Hence for the glide, for instance, the component of Weight acting along the flight path would be WSin γ. It was found that for the steady-state glide that this force had to equal the aerodynamic Drag at 90 knots (250 pounds). So that 250 = WSin γ (or Weight Times Sin γ); 250 = 3000 Sin γ, or Sin γ = 0.0833 (rounded off), $\gamma = 4^0 47'$.

The component of Weight acting perpendicular to the flight path at that angle of $4^0 47'$ would be W Cos γ, or 2990 pounds as found. (Lift must equal this figure for equilibrium to exist.)

The same principles were used for the climb angle (which you remember was arbitrarily established as 8 degrees) and for the components of Thrust acting parallel and perpendicular to the reference axis which is always the flight path line.

The symbol α (alpha) is used in this case as the angle of inclination of the fuselage axis to the flight path as well as the more common use of indicating the angle of attack. (The angle of incidence is zero here.)

FORCES IN THE TURN

There will be a chapter devoted to the turn later but a few comments might be in order now.

For normal flying (the Four Fundamentals) the turn is the only maneuver in which Lift is deliberately and maliciously made greater than Weight and is the only one of the Four Fundamentals in which "g" forces exist in a "steady state" condition.

For a balanced turn at a constant altitude the "up" forces must equal the "down" forces as in straight and level flight. For ease of discussion we'll ignore the tail-down force in this section on the turn.

Fig. 2-16 shows the forces acting in the balanced level turn.

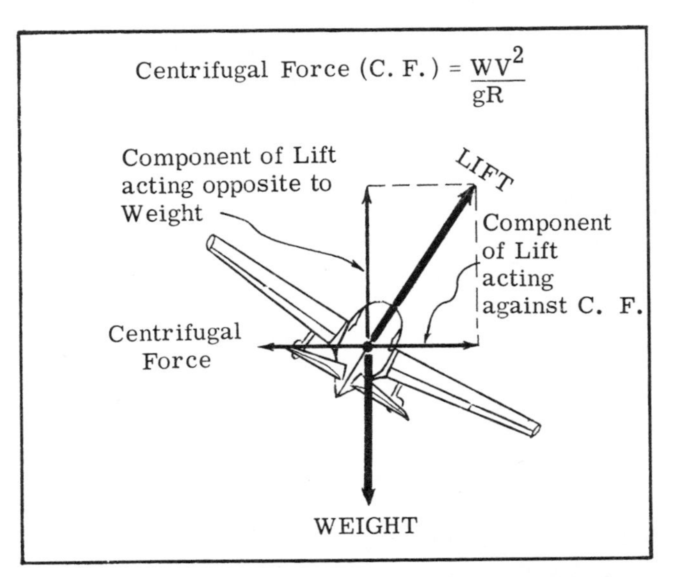

Fig. 2-16. Forces in a balanced turn of a constant bank.

As noted in Chapter 1, Lift acts perpendicular to the relative wind and *to the wingspan.* The latter consideration is of particular importance in discussing the turn.

As you know from turning a car, centrifugal force acts to move you toward the outside of the turn. In the turn should the car suddenly hit a spot of oil or ice it would move toward the "outside" of the circle (or more correctly, its path would be at a tangent to the circle). The centrifugal force created by the turn is offset by the *centripetal* force, or holding (friction) force of the tires against the road surface. When the oil spot or ice is encountered the friction (centripetal force) decreases suddenly and the car departs the beaten path for new adventures through somebody's hedge.

For the airplane in turning flight of a constant radius the centrifugal force caused by the turn is equaled by the component of Lift acting toward the center of the turn. (After the turn is established.)

As is shown in Fig. 2-16 the equation for Centrifugal Force is: $C.F. = \frac{WV^2}{gR}$.

W — Weight (pounds) of the airplane.
V^2 — The tangential velocity in feet per second, squared. (You can call it the true airspeed in feet per second, squared.)
g — The acceleration of gravity (32.2 feet per second, per second.)
R — Radius of turn, feet.

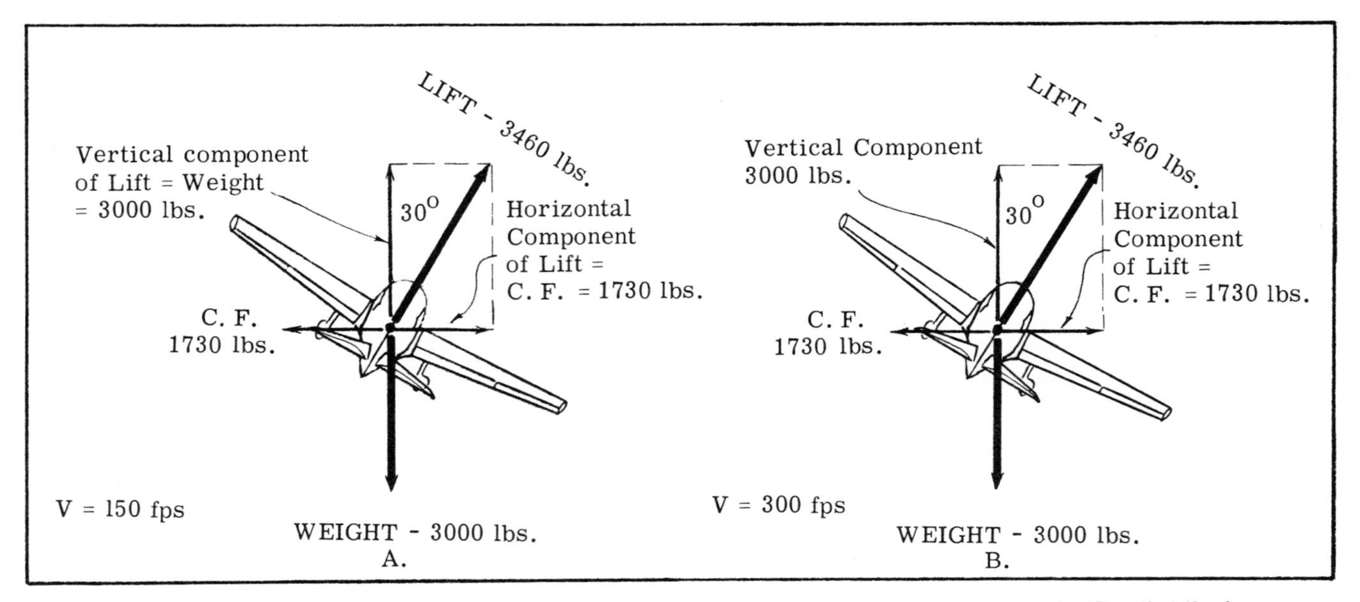

Fig. 2-17. Forces in the level turn for two airplanes flying at *different speeds* at a 30-degree bank. Note that the forces are the same.

You can see that the centrifugal force is doubled if the turning radius (R) is halved (everything else kept the same). If the velocity is *doubled* the radius of turn will be *quadrupled* to keep the same centrifugal force (everything else constant).

For instance, take two airplanes of the same Weight but one has a great deal more power and can maintain a level turn at twice the cruising speed of the other. Both will be banked at 30 degrees in a balanced level turn. (Fig. 2-17)

Both airplanes are pulling the same load factor or number of "g's" but there is quite a difference in their turn radii. (The load factor is function of the Lift-to-Weight ratio if the tail-down force is ignored; Lift in each case is 3460 pounds, or the airplanes are pulling $\frac{3460}{3000}$, or 1.15, g's.)

In the balanced turn, as was mentioned earlier, centrifugal force is equaled by the horizontal component of Lift, which for the bank of 30 degrees is found to be 1730 pounds (Fig. 2-17).

Everything is known except the radius of turn for each airplane under the conditions given. (W = 3000 pounds, g = 32.2 feet per second, per second, C.F. = 1730 pounds, V = 150 and 300 fps, R = ?.)

Solving for R of Airplane A, which requires a little algebraic shuffling;

$$R = \frac{WV^2}{gC.F.} = \frac{3000 \times (150)^2}{32.2 \times 1730} = 1210 \text{ ft.}$$

The turning radius of airplane A in a 30° banked turn at 150 fps, or about 90 knots is 1210 feet, or slightly less than a quarter of a mile. If it made a 360-degree turn the diameter of the turn would be a little less than one-half mile.

For airplane B, turning at a velocity of 300 fps, the radius of turn would be (rounded off):

$$\frac{3000 \times (300)^2}{32.2 \times 1730} = 4840 \text{ feet.}$$

The radius of its turn would be 4840 feet, or *four* times that of the airplane of the same Weight and angle of bank traveling at one-half the speed.

Suppose the pilot of airplane B wanted to make the *same radius of turn* as A, but still at the higher speed of 300 fps (about 180 knots). You've no doubt already figured the answer is that the airplane must be banked more steeply and this can be found by reshuffling of the centrifugal force equation and working back to the bank required to get a radius of turn of 1400 feet like Airplane A.

This time the centrifugal force acting at the speed of 300 fps at a radius of turn of 1210 feet is:

$$C.F. = \frac{WV^2}{gR} = \frac{3000 \times (300)^2}{32.2 \times 1210} = 6930 \text{ lbs.}$$

The horizontal component of Lift must equal this value of 6930 pounds. It's also a fact that the vertical component of Lift *must* be 3000 pounds (to equal Weight) so the Lift value and angle could be readily found by checking a trigonometric table. (Fig. 2-18)

It's found that the angle of bank must be about 66.6 degrees. The Lift value must be 7560 pounds, or the airplane's Lift-to-Weight ratio is $\frac{7560}{3000} = 2.52$ g's. The pilot's face is sagging downward trying to stay in the turning circle of the slower airplane (the pilot of which probably feels quite comfortable at his 1.15 g loading in the 30-degree bank). Fig. 2-19 shows the two airplanes in the turn.

In instrument flying the *time* to make a turn (rate of turn) is important and you'll find that the faster the airplane the steeper the bank must be in order to make a balanced standard rate turn of 3 degrees per second. Chapter 3 will go into this requirement in more detail.

The radius of turn does not depend on the Weight of the airplane. If airplane B turned at the same velocity as A (150 fps) but weighed 6000 instead of 3000 pounds its radius of turn would be the same because the centrif-

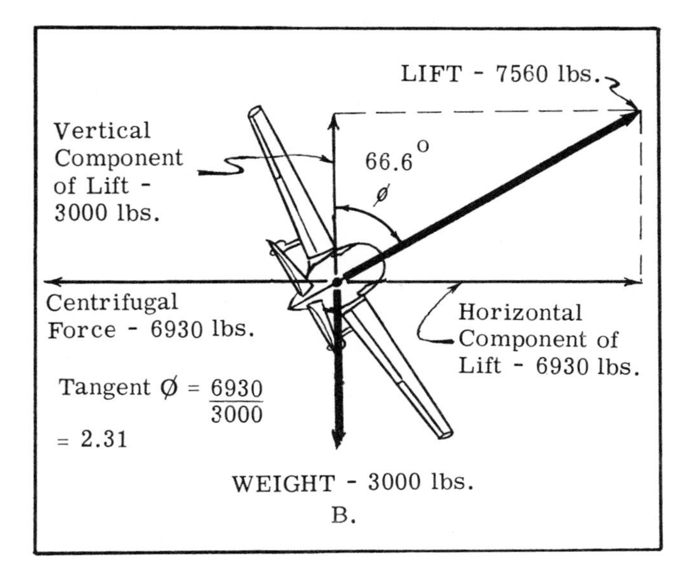

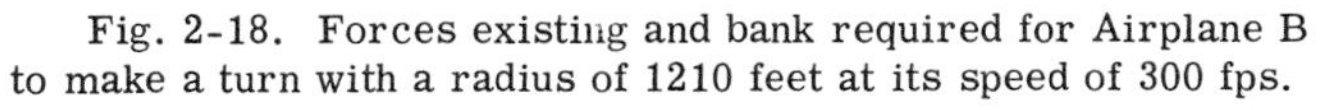
Fig. 2-18. Forces existing and bank required for Airplane B to make a turn with a radius of 1210 feet at its speed of 300 fps.

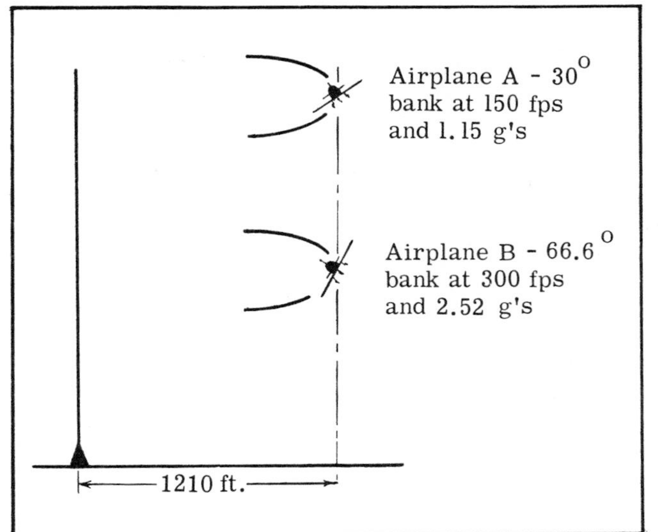

Fig. 2-19. A comparison of the two airplanes making the same radius of turn.

ugal force would also be doubled (in a balanced turn) and the ratio of Weight to C.F. would be the same. The Japanese *Zero* in World War II had a maximum Weight of 6500 pounds and the F4U-4 *Corsair* had nearly twice the maximum Weight (12,500 pounds). Because of the *Zero's* lower Weight (lower wing loading) as compared to the F4U-4 it could fly (and turn) at a *much slower speed.* U.S. pilots knew better than to try to outturn the *Zero.*

You could figure out the radius of turn for a jet weighing 10,000 pounds turning at 750 fps at 30 degrees of bank to be about 30,000 feet, or about *5.7 miles*.

A couple of things were neglected in this discussion of the Turn: The tail-down force was considered to be zero which it certainly would not be in a turn (you could expect it to appreciably increase in steeper turns) and the fact that because of the increased angle of attack in the turn, Thrust has components acting inward (helping centripetal force), and upward. However, putting all this into an illustration might result in a pretty confusing situation.

A climbing or gliding turn would be quite a picture in vectors, but you can see that the forces acting parallel and perpendicular to the flight path must be equal to zero for any steady state condition to occur.

SUMMARY OF THE CHAPTER

There is a lot of misunderstanding concerning the actions of the Four Forces in various maneuvers and the most common one is that "excess Lift" is what makes the airplane climb. This is not to deny that it will climb when an excess Lift exists, but in that case acceleration forces, or g's, will be exerted. Suppose you are flying along at cruise and suddenly exert back pressure on the wheel or stick. Assuming that you didn't overdo it and the wings are still with you, the airplane will *accelerate* upward (and then assume a normal steady-state climb if that's what you wanted). When you exerted back pressure the up-force (Lift) was greater than the down forces (Weight, etc.) at that time — you increased the angle of attack almost instantly and the airplane was still at the high cruise speed so the dynamic pressure "q" was still of the same high value.

The measurement of positive g's is the Lift-to-Weight ratio and at the instant of rotation Lift may be increased radically. Of course, as you know, an increase in angle of attack (coefficient of Lift) means an increase in Drag (induced) and Lift will tend to reassume its old value, depending on the new flight path. If you had wanted to climb at a certain airspeed (with a certain angle and rate of climb resulting) Lift would soon settle down to the required value. In flying the airplane you, as a pilot, decide what the airplane must do and keep this requirement by use of power, the airspeed and/or altimeter. When you have established a steady-state condition such as a steady climb or glide or straight and level flight, the forces settle down of their own accord. *You* balance the forces automatically by setting up a steady-state condition.

If the "up" forces (working toward the ceiling of the airplane cabin) are greater than the "down" forces (working toward the floor) you feel positive g's and feel "heavier" in the seat, this is the effect in a normal level turn, and the steeper the turn the greater the effects.

Probably a large number of stall-type accidents have occurred because the pilot unconsciously thought in terms of "increasing the Lift" to climb over an obstacle.

Another idea not often considered is that a tail-down force exists for the majority of airplanes throughout most of the range of flight speeds and loadings. Most laymen think that "little wing back there is always helping to hold the airplane up." In one sense perhaps it is, in that it is required for good stability, which is important to flight.

You can prove whether a tail-down force exists by a very simple experiment. In Chapter 1 the drawings showed that the wing-tip vortices curled over the tip toward the low pressure or "lifting" (top) side of the wing. Knowing this to be the case you can "see" this wing-tip vortex action by taping a ribbon or string with a light weight on the free end to each wing-tip and to the tip of each stabilizer. The rotation of the string will be as shown in Fig. 2-20.

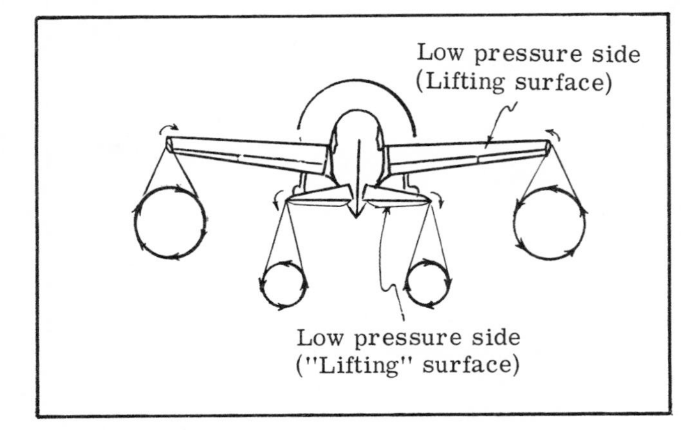

Fig. 2-20. An experiment for checking if a tail-down force exists.

Note in Fig. 2-20 that the string-weight combination is rotating in an opposite direction at the stabilizer tips as compared to the wings — the low pressure or "lifting" side is on the bottom — a tail-down force exists.

You need only to do one wing and stabilizer tip to see the action. Check the actions of the strings at cruise and watch the wing tip as the airplane is slowed (the angle of attack or coefficient of Lift is increased). You can readily see that vortex strength increases with a decrease in speed (increase in C_L).

By now you have likely figured that the airplane is actually flown by the Thrust-to-Drag relationship and that little is done in the way of controlling Lift in normal 1 g flight. By setting up a steady-state condition Lift takes care of itself and actually varies very little in wings level climbs and glides and straight and level flight. *Keep it in mind.*

Maybe it's been quite some time since you've had trigonometry, or maybe you and that subject have never met. It doesn't matter in following the basic idea of flight mechanics, and the references to a trigonometric table in this chapter were only done to give actual figures for the problems so that you can compare them. You could draw your own vectors (for instance, using one inch to equal 500 pounds) and measure off the resulting forces to get fairly accurate answers without the need for trigonometry. The main idea is to see the maneuvers covered.

3. THE AIRPLANE INSTRUMENTS

FLIGHT INSTRUMENTS AND AIRPLANE PERFORMANCE

The first two chapters were somewhat theoretical in their approach to the performance of the airplane and were presented to give background in preparation for this and the following chapters in Part I.

In order to get the most from your airplane you must have a good understanding of the airplane's instruments. You should know what affects the operation of the airspeed, altimeter, and other flight instruments. Whether you plan on going on to the commercial certificate or continuing to fly for personal business or pleasure, pride in your flying ability will cause you to want to learn more about your airplane.

There are certain relationships between the density, temperature, and pressure of the atmosphere. An airplane's performance depends on the density altitude, but density altitude has an interlocking relationship with temperature and pressure. For instance, cold air is more dense than warm air — so that a *decrease* in temperature (if the pressure is not changed) means an *increase* in density. If the pressure is *increased* (assuming no change in temperature) the density is *increased.* (More particles of air are being compressed into the same volume.)

EQUATION OF STATE

For those interested in the mathematics, the equation of state explains the exact relationships between the three variables: temperature, pressure, and density.

$$\rho = \frac{P}{1718T}$$

This form of the equation of state says that the air density (ρ), in slugs per cubic foot, equals the pressure, in pounds per square foot, divided by 1718 times the temperature in degrees Rankine.

A slug is a unit of mass as was mentioned in Chapter 1. The mass of an object in slugs may be found by dividing the Weight by the acceleration of gravity (32.2 feet per second, per second); hence a 161-pound man has a mass of 5 slugs $\left(\frac{161}{32.2} = 5\right)$.

The temperature in degrees Rankine may be found by adding 460 degrees to the Fahrenheit reading; for a standard sea-level day the temperature is 59° F. or 519° Rankine.

The density at sea level on a standard day may be found as follows:

$$\rho \text{ (air density)} = \frac{2116 \text{ (sea-level pressure - psf)}}{1718 \times 519}$$

This works out to be: $\rho = 0.002378$ slugs per cubic foot. Or, checking for pressure: pressure = density x temperature x 1718.

The equation of state and the symbols for density, pressure and temperature will be used further throughout the book.

The three factors directly affect each other. If you know your pressure altitude and the temperature, the density altitude can be found with a graph or computer.

PRESSURE FLIGHT INSTRUMENTS

As the name implies, these instruments operate because of air pressure or air pressure changes.

AIRSPEED INDICATOR

The airspeed indicator is nothing more than a specialized air pressure gage. The airspeed system is comprised of the pitot and static tubes and the airspeed indicator instrument.

An airplane moving through the air creates its own relative wind. This relative wind exerts a ram pressure in the pitot tube where its effects are passed on into a diaphragm which is linked to an indicating hand as shown in Fig. 3-1.

This relative wind force is calibrated in miles per hour, or knots, rather than pounds per square foot of pressure. The static tube acts as a neutralizer of the static pressure around the airplane and within the instrument, so that *only* the dynamic pressure is measured. For lighter planes the pitot and static tube inlets are together. But for greater accuracy, the static tube opening is placed at some point on the airplane where the most accurate measurement of the actual outside air static pressure is found. A usual spot is on the side of the fuselage somewhere between the wing and stabilizer. No doubt you've noticed these static pressure sources, accompanied by a sign, "Keep this hole free of dirt." These points are selected as being the places where the static pressure is least affected by the airflow about the airplane. It is difficult to find a spot on the airplane entirely free of static pressure error, so an expression *position error* is introduced. The proper

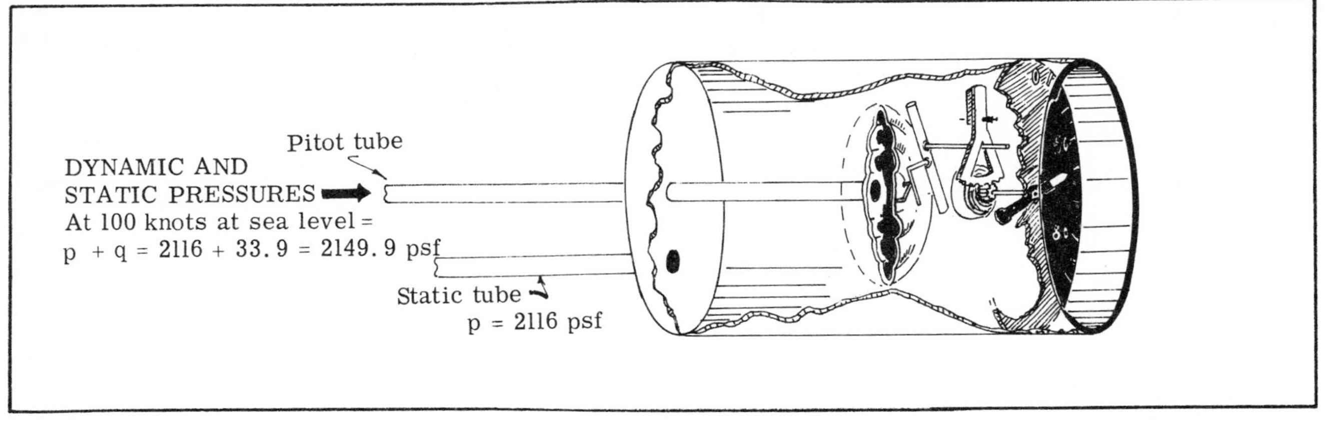

Fig. 3-1. Airspeed indicator.

placing of the static tube opening to minimize this error is responsible for no few ulcers in aircraft manufacturing. In addition to the position error in the system there is usually some error in the airspeed indicator instrument itself. This *instrument error* is another factor to contend with in airspeed calibration.

The pitot tube position is also important. It must be placed at a point where the actual relative wind is measured, free from any interfering aerodynamic effects. A particularly bad place would be just above the wing where the air velocity is greater than the free stream velocity.

Error is introduced into the airspeed indicator at high angles of attack or in a skid. You've seen this when practicing stalls. The airplane had a stall speed of 50 knots according to the Airplane Flight Manual, yet there you were, still flying (though nearly stalled) with the airspeed indicator showing 40 knots (or even zero). It wasn't because of your skill that you were able to fly the airplane at this lower speed, but the angle of the pitot tube to the airstream has introduced an error. While at first thought it would seem that the pitot tube, being at a fairly high angle of attack, would be the culprit this is not the case. For the airplane that stalls at the usual 15°-20° angle of attack the effect is small, but for STOL-type airplanes it could be a real factor in airspeed error.

Flight test airplanes use an elaborate extended boom with a swivel pitot head which results in much greater accuracy at high angles of attack. This is obviously not practical in cost or weight for normal light plane installation. The *major* source of error, however, is the static system; the pitot tube contributes a minor amount.

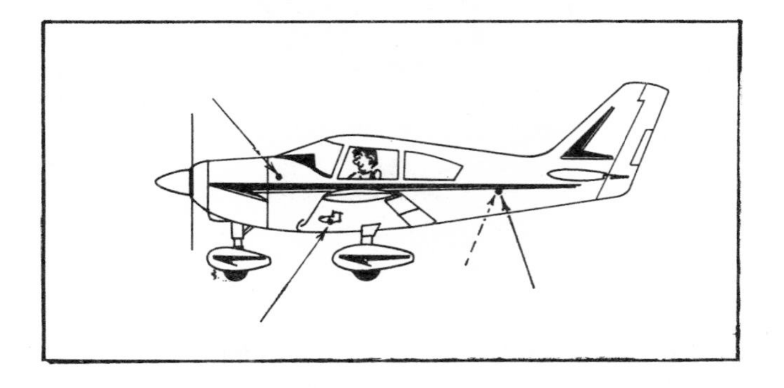

Fig. 3-2. Some typical static port locations (not all on the same airplane).

So, static pressure error can also be introduced at angles of attack or angles of yaw, the amount of error depending on the location of the static opening.

The dynamic pressure measured by the airspeed indicator is called "q" and has the designation $\frac{\rho}{2}V^2$. You will notice that dynamic pressure is a part of the Lift and Drag equations. This *dynamic pressure*, which is one half the *air density* (in slugs per cubic foot) *times* the *true air velocity* (in feet per second squared), is pressure in *pounds per square foot.*

At sea level on a standard day, at a speed of 100 knots (169 feet per second), "q" would be $\frac{(0.002378)}{2}(169)^2 = 33.9$ pounds per square foot. The light plane airspeed indicator is calibrated for standard sea level conditions with a temperature of 15° C. or 59° F. and a pressure of 29.92 inches of mercury, or 2116 pounds per square foot.

The perfect airspeed indicator would work as follows:

Pitot Tube measures dynamic and static pressure.

Static Tube equalizes static pressure or "subtracts" it from the pitot tube reading.

Airspeed Indicator indicates dynamic pressure only.

So the airspeed indicator only measures the dynamic pressure, which is a combination of density and airstream velocity (squared). As altitude increases, the air density decreases so that an airplane indicating 100 knots (or 33.9 pounds per square foot— dynamic pressure) at ten thousand feet actually has a higher true airspeed than the airplane at sea level *indicating* the same dynamic pressure (airspeed).

This airspeed correction for density change can be worked on your computer, but a good rule of thumb is to add 2 per cent per thousand feet to the indicated airspeed. A plane indicating 100 knots at ten thousand feet density altitude will have a true airspeed close to 120 knots. (By computer it's found to be 116 knots.)

There are airspeed indicators on the market today that correct for true airspeed; the true airspeed can be read directly off the dial. One type has a setup where the pilot adjusts the dial to compensate for altitude and the temperature effects (density altitude) and the needle indicates the true airspeed in the cruising range. Another more expensive instrument does this automatically for certain speeds well above stall or approach speeds. The reason for not having corrections at lower speeds is that the pilot might fly by his *true airspeed* and get into trouble landing at airports of high elevation and/or high temperatures. Perhaps an explanation is in order.

An airplane will always stall at the same indicated airspeed regardless of its altitude or the temperature (all other things such as Weight, angle of bank, etc., being equal). An airplane that stalls at an indicated 50 knots at sea level will stall at an indicated 50 knots at a density altitude of ten thousand feet because the airspeed indicator measures "q" and it still takes the same amount of "q" to support the airplane. However, his *true airspeed* at ten thousand feet would be approximately *60 knots.* If he were flying by a true airspeed instrument, he might get a shock when the plane dropped out from under him at an airspeed of 10 knots higher than he expected (it always stalled at 50 knots down at sea level — what gives?). Of course, the airspeed indicator isn't likely to be accurate at the speeds close to the stall anyway, but the pilot might not give himself enough leeway in the approach in this case. (These are unaccelerated stalls.)

There are several terms used in talking about airspeed:

(1) Indicated airspeed (I.A.S.) — the airspeed as read off the standard airspeed indicator.

(2) Calibrated airspeed (C.A.S.) — indicated airspeed corrected for instrument and position error.

(3) True airspeed (T.A.S.) — calibrated airspeed corrected for density effects.

Another term used in discussion of airspeeds is *equivalent airspeed,* which is calibrated airspeed corrected for compressibility effects. For the type of airplane you are likely to be flying and the altitudes at which you'll be operating, equivalent airspeed won't be a vital factor.

As the errors of the airspeed system are negligible in the major portion of the operating range and because many airplanes don't have airspeed correction cards, this book assumes I.A.S. = C.A.S.

AIRSPEED INDICATOR MARKINGS

The FAA requires that the airspeed indicator be marked for various important speeds and speed ranges. Fig. 3-3 shows required markings:

Red Line — Never Exceed Speed (V_{ne}). This speed should not be exceeded at any time.

Yellow Arc — Caution Range. Strong vertical gusts could damage the airplane in this speed range; therefore it is best to refrain from flying in this speed range when encountering turbulence of any intensity. The caution range starts at the maximum structural cruising speed and ends at the never exceed speed (V_{ne}).

Green Arc — Normal operating range. The airspeed at the lower end of this arc is the flaps up, gear up, power-off stall speed at gross weight, V_{s1}, (for most airplanes, the landing gear position has no effect on stall speed). The upper end of the green arc

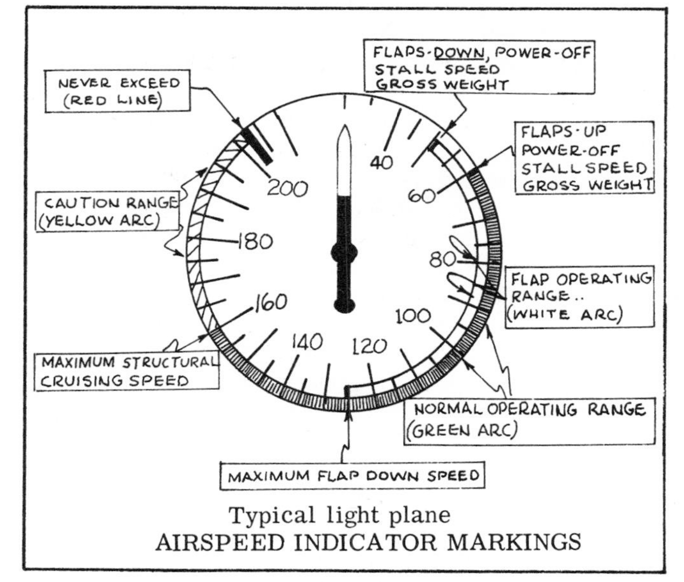

Typical light plane
AIRSPEED INDICATOR MARKINGS

Fig. 3-3.

is the maximum structural cruising speed, V_{no}, the maximum indicated airspeed where no structural damage would occur in moderate vertical gust conditions.

White Arc – The flap operating range. The lower limit is the stall speed at gross weight with the flaps in the *landing position* and the upper limit is the maximum flap operating speed.

These speeds assume no instrument error and are calibrated airspeeds. Later in this book these speeds will be referred to for various performance requirements.

ALTIMETER

The altimeter is an aneroid barometer calibrated in feet instead of inches of mercury. Its job is to measure the static pressure (or ambient pressure as it is sometimes called) and register this fact in terms of feet or thousands of feet.

The altimeter has an opening that allows static (outside) pressure to enter the otherwise sealed case. A series of sealed diaphragms or "aneroid wafers" within the case are mechanically linked to the three indicating hands. Since the wafers are sealed, they retain a constant internal "pressure" and expand or contract in response to the changing atmospheric pres-

Fig. 3-4. The sensitive altimeter.

sure surrounding them in the case. As the aircraft climbs, the atmospheric pressure decreases and the sealed wafers expand; this is duly noted by the indicating hands as an increase in altitude (or vice versa).

Standard sea level pressure is 29.92 inches of mercury and the operations of the altimeter are based on this fact. Any change in local pressure must be corrected by the pilot. This is done by using the setting knob to set the proper barometric pressure (corrected to sea level) in the setting window.

There are several altitudes that will be of interest to you: *True Altitude* is the height above sea level. *Absolute Altitude* is the height above the terrain. *Pressure Altitude* is the altitude read when the altimeter is set to 29.92. This indication shows what your altitude would be if the altimeter setting were 29.92— that is, if it were a standard pressure day. *Indicated Altitude* is the altitude read when the altimeter is set at the local barometric pressure corrected to sea level.

Density Altitude is the pressure altitude computed with temperature. The density altitude is used in performance. If you know your density altitude, air density can be found by tables, and airplane performance calculated. You go through this step every time you use a computer to find the true airspeed. You use the pressure altitude and the outside air temperature and get the true airspeed. Usually there's not enough difference in pressure altitude and indicated altitude to make it worthwhile to set up 29.92 in the altimeter setting window, so that the usual procedure is to use the *indicated* altitude.

The fact that the computer used pressure altitude and temperature to obtain density altitude in finding true airspeed didn't mean much as you were only interested in the final result. You may not even have been aware that you were working with density altitude during the process. Some computers also allow you to read the density altitude directly by setting up pressure altitude and temperature. This is handy in figuring the performance of your airplane for a high altitude and/or high temperature take-off or landing. The Airplane Flight Manual gives graphs or figures for take-off and landing performance at the various density altitudes. After finding your density altitude, you can find your predicted performance in the Airplane Flight Manual. Computers are not always available and the manufacturers sometimes furnish Conversion tables with their Airplane Flight Manuals. (Fig. 3-5)

Suppose you are at an airport at a pressure altitude of 6000 feet and the temperature is 80° F. Using the conversion chart you see that your density altitude is 8500 feet. (Fig. 3-5) Looking at the take-off curves

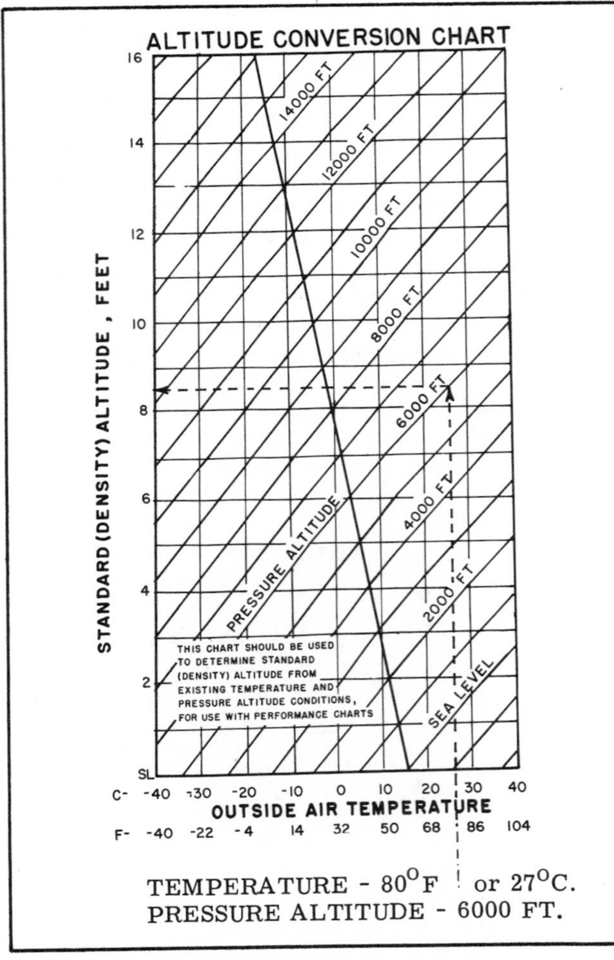

Fig. 3-5. Altitude conversion chart.

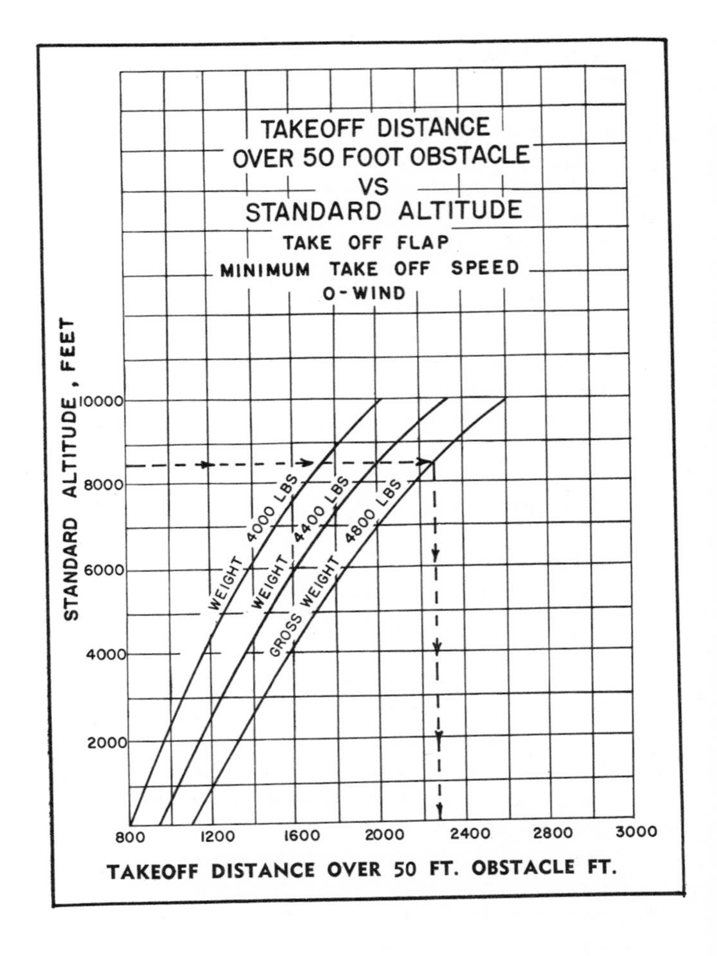

Fig. 3-6.

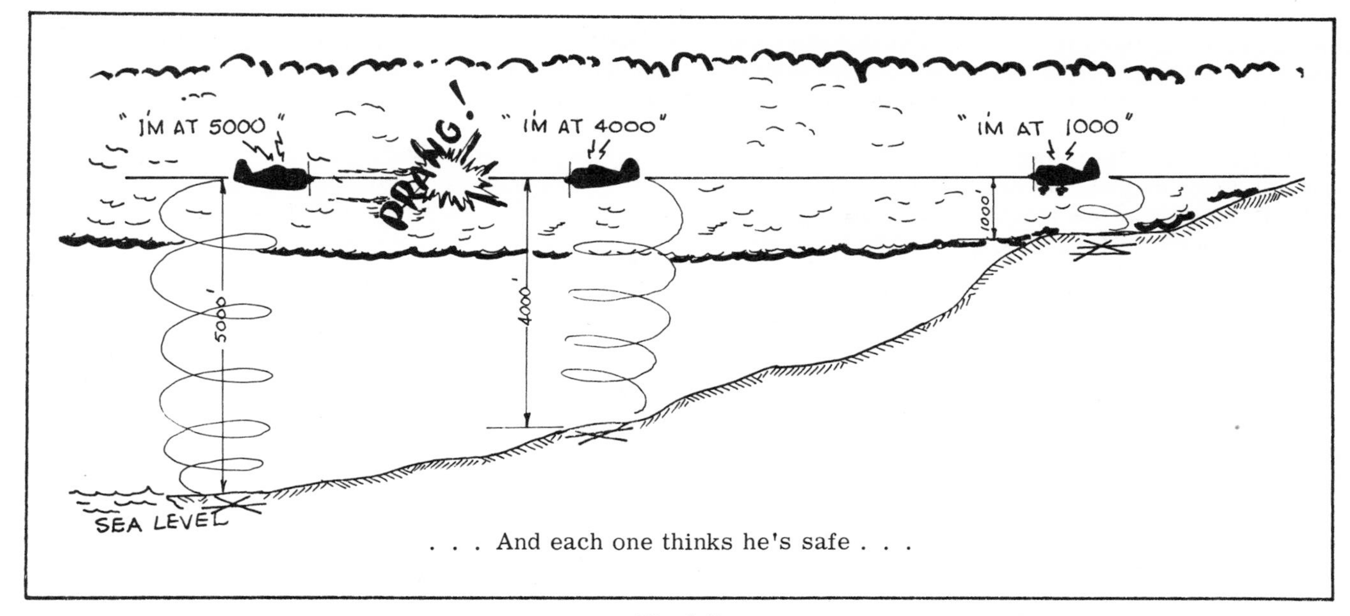

Fig. 3-7.

you can see that your expected distance to clear a 50-foot obstacle will be nearly 2300 feet at your gross Weight of 4800 pounds. (Fig. 3-6) This is more than double the distance at sea level and might be a handy fact to know.

You and other pilots fly indicated altitude. When you're flying cross country you will have no idea of your exact altitude above the terrain (although over level country you can check airport elevations on the way, subtract this from your indicated altitude, and have a barnyard figure). Over mountainous terrain this won't work, as the contours change too abruptly for you to hope to keep up with them. As you fly you'll get altimeter settings from various ground stations and keep up to date on pressure changes.

The use of indicated altitude for all planes makes good sense in that all pilots are using sea level as a base point. If each pilot set his altimeter at zero before taking off, you can imagine what pandemonium would reign. As you can see in Fig. 3-7 the controllers would have a hairy time trying to assign traffic altitudes for instrument flying.

ALTIMETER ERRORS

Instrument Error -- Being a mechanical contrivance the altimeter is subject to various quirks. If you set the current barometric pressure — corrected to sea level — for your airport (if you have a tower or Flight Service Station), the altimeter should indicate the field elevation when you're on the ground.

FAR 91.170 specifies that airplanes operating in controlled airspace (IFR) must have had each static pressure system and each altimeter instrument tested by the manufacturer or an FAA approved repair station within the past 24 calendar months.

Pressure Changes —When you fly from a high pressure area into a low pressure area, the altimeter "thinks" you have climbed and will register accordingly — even if you haven't changed altitude. You'll see this and will fly the plane down to the "correct altitude," and will actually be low. When you fly from a low to a high pressure area the altimeter thinks you've let down to a lower altitude and registers too low. A good way to remember (although you can certainly reason it out each time) is: HLH — High to Low, (altimeter reads) High. LHL — Low to High, (altimeter reads) Low.

You can see that it is worse to fly from a High to a Low pressure area as far as terrain clearance is concerned. You might find that the clouds have rocks in them. So get frequent altimeter settings as you fly cross country. (Don't try flying IFR unless rated.)

Temperature Errors — Going back to the equation of state for air where pressure = density x temperature x 1718, you see the relationship between temperature and pressure. For our purposes we can say that the pressure is *proportional* to the density times the temperature ($P \sim \rho T$), and get rid of the constant number (1718). Assuming the density to remain constant, for simplicity you can then say that pressure is proportional to temperature. Therefore, if you are flying at a certain altitude and the temperature is higher than normal, the pressure at your altitude is higher than normal. *The altimeter registers this as a lower altitude.* If the temperature is lower, the pressure is lower and the altimeter will register accordingly — *lower temperature, altimeter reads higher.*

You might remember it this way, using the letters H and L as in pressure changes. Temperature High, (altimeter reads) Low — HL. Temperature Low, (altimeter reads) High — LH. Or perhaps you'd prefer to remember HALT (High Altimeter because of Low Temperature).

The best thing, however, is to know that higher temperature means higher pressure (and vice versa) at altitude and reason it out from there.

The temperature error is zero at the surface point at which the setting was obtained and increases with altitude, so that the error could easily be 500 to

600 feet at the 10,000 foot level. In other words, you can have this error at altitude even if the altimeter reads correctly at the surface point at which the setting originated. Temperature error can be found with a computer as is shown in Fig. 3-8. For indicated altitude this error is neglected, but it makes a good question for an FAA written exam or flight test and it has been used!

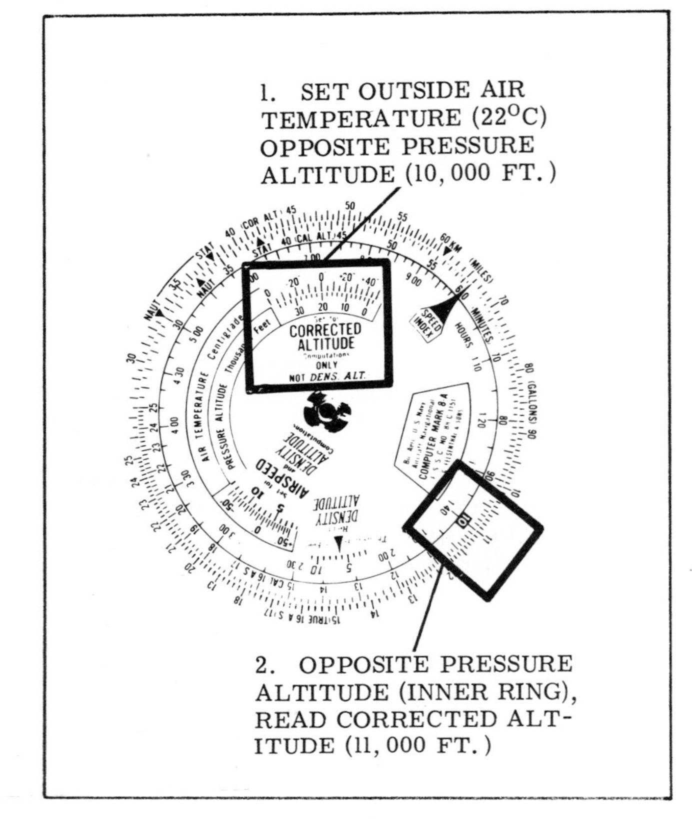

Fig. 3-8. Using the computer to find corrected pressure altitude.

These errors (particularly temperature errors which are normally ignored) affect everybody in that area (though slightly different for different altitudes) so that the altitude separation is still no problem. Temperature errors *could* cause problems as far as terrain clearance is concerned, however. An equation for finding true altitude is given in the Appendix.

ALTIMETER TIPS

You can convert your indicated altitude to pressure altitude without resetting the altimeter to 29.92 by looking at your altimeter setting. Suppose your altimeter registers 4000 feet and the current setting is 30.32 inches of mercury. Your pressure altitude is 3600 feet and is arrived at by the following: The pressure corrected to sea level is 30.32 inches, but to get pressure altitude the setting should be based on 29.92 inches. This shows that the actual pressure of 30.32 is *higher* than standard, therefore the pressure altitude is *less*. (Higher pressures at lower altitudes.) Using a figure of 1 inch per thousand feet you see that the pressure difference is 0.40 inches of mercury or 400 feet. The pressure altitude is 4000 minus 400 = 3600 feet. This will be as close as you can read an altitude conversion chart anyway. If the altimeter setting had been 29.52 your pressure altitude would be 400 feet higher (29.92 - 29.52 = 0.40 inch = 400 feet) or 4400 feet.

For Estimation of Pressure Altitude Without Resetting: If your altimeter setting is *lower* than 29.92, *add* 100 feet to your indicated altitude for each 0.10 inch difference. If your altimeter setting is *higher* than 29.92, *subtract* 100 feet from your indicated altitude for each 0.10 inch difference to get the pressure altitude.

The reason for this little mental exercise above is to get you familiar with working between pressure and indicated altitude. You may prefer to note your altimeter setting (so you can return the altimeter to the indicated altitude after getting the pressure altitude), and then set the altimeter to 29.92 to get the pressure altitude. After this is done you can return to the original indicated altitude setting.

Some sensitive altimeters have two small arrow heads on the outer edge of the face that move when the setting knob is turned. One thing you may not have noticed is that when the setting is 29.92 both these pointers are at zero. One arrow indicates in hundreds of feet and the other in thousands. These may be used to find the pressure altitude without resetting the altimeter to 29.92. Using the same altitude of 4000 feet as your indicated altitude, with an altimeter setting (corrected to sea level) of 30.32, you can see the indication of these pointers in Fig. 3-9.

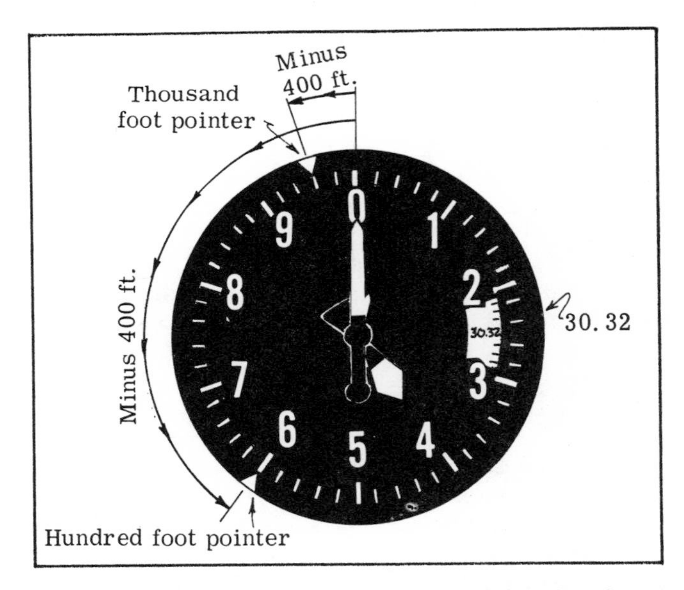

Fig. 3-9. If the thousand foot pointer is to the left, subtract the amount of altitude as shown by the movement of the hundred foot pointer (Indicated altitude = 4000 feet, pressure altitude 3600 feet).

Two pointers are necessary so that you'll know whether to subtract or add the indication. In Fig. 3-9, the hundred foot indicator wouldn't tell you whether to add 600 or subtract 400 feet to get the pressure altitude. The thousand foot marker has moved about 400 feet *down* (counter clockwise) and the *exact* figure down can be read at the one hundred foot pointer. In fact, you can read this difference to the nearest 20 feet if you are the precise type of pilot.

If the thousand foot pointer is to the left of zero on the altimeter, subtract the hundreds of feet the hundred foot pointer has moved in that direction. If the thousand foot pointer is to the right of the zero, add the hundreds of feet indicated (this one is easier because you can just read the hundred foot arrow head indication). Look at Fig. 3-9 again.

For computer work you are told to use the *pressure altitude* to find the true airspeed. For practical work use *indicated altitude* (current sea level setting) for true airspeed computations. Remember that the T.A.S. increases about 2 per cent per thousand feet so the most you will be off will be 2 per cent. That is, your sea level altimeter setting could possibly be 28.92 or 30.92 but this is extremely unlikely. So...

Assume that a total error of no more than one percent will be introduced by use of indicated altitude. For a 200 knot airplane this means you could be two knots off for true airspeed. But the instrument error or your error in reading the instrument could be this much.

One thing to remember concerning the altimeter which is useful for written tests and hangar flying sessions: *If you increase the number in the setting window* (by using the setting knob, naturally), *the altitude reading is also increased* — and vice versa. If you have the altimeter originally set at 29.82 as the sea level pressure while flying and get an altimeter setting of 30.02 from a station in your area, you'll find that in rolling in that *additional* 0.20 inch you've also given the altimeter an *additional* 200 feet of *indicated* altitude. This also follows from the earlier LHL idea; when flying from a Low (29.82) to a High (30.02) the altimeter reads Low (until you put it right by rolling in the added 0.20 inch in the setting window and adding another 200 feet to your indicated altitude).

RATE OF CLIMB OR VERTICAL SPEED INDICATOR

Like the altimeter, the vertical speed indicator has a diaphragm. But unlike the altimeter, it measures the *rate of change* of pressure rather than the pressure itself.

The diaphragm has a tube connecting it to the static tube of the airspeed indicator and altimeter (or the tube may just have access to the cabin air pressure in the case of cheaper or lighter installations). This means that the inside of the diaphragm has the same pressure as the air surrounding the airplane. Opening into the otherwise sealed instrument case is a capillary tube.

Fig. 3-10 is a schematic diagram of a typical rate of climb indicator. As an example, suppose the airplane is flying at a constant altitude. The pressure within the diaphragm is the same as that of the air surrounding it in the instrument case. The rate of climb is indicated as zero.

The plane is put into a glide or dive. Air pressure inside the diaphragm increases at the same rate as that of the surrounding air. However, because of

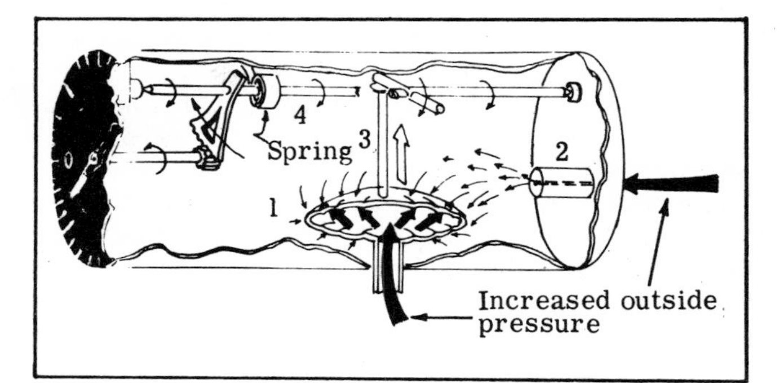

Fig. 3-10. Vertical speed indicator — as the airplane descends, the outside pressure increases. The diaphragm expands immediately (1). Because of the small size of the capillary tube, (2), the pressure within the case is not increased at the same rate. The link (3) pushes upward rotating the shaft (4) which causes the needle to indicate the proper rate of descent. The spring helps return the needle to zero when pressures are equal and also acts as a dampener.

the small size of the capillary tube, the pressure in the instrument case does not change at the same rate. In a glide or dive, the diaphragm expands, the amount of expansion depending on the difference of pressures. As the diaphragm is mechanically linked to a hand, the appropriate rate of descent in hundreds (or thousands) of feet per minute is read on the instrument face.

In a climb the pressure in the diaphragm decreases faster than that within the instrument case, and the needle will indicate an appropriate rate of climb.

Because in a climb or dive the pressure in the case is always "behind" the diaphragm pressure in the above described instrument, a lag of 6 to 9 seconds results. The instrument will still indicate a vertical speed for a short time after the plane is leveled off. For this reason the rate of climb indicator is not used to maintain altitude. On days when the air is bumpy, this lag is particularly noticeable. The rate of climb indicator is used as a check of the plane's climb, dive or glide rate. The sensitive altimeter is used to maintain a constant altitude.

There is a more expensive rate of climb indicator (Instantaneous Vertical Speed Indicator) on the market that does not have lag and is very accurate even in bumpy air. It contains a piston-cylinder arrangement whereby the airplane's vertical acceleration is immediately noted. The pistons are balanced by their own weights and springs. When a change in vertical speed is effected, the pistons are displaced and an immediate change of pressures in the cylinders is created. This pressure is transmitted to the diaphragm, producing an almost instantaneous change in indication. After the acceleration-induced pressures fade, the pistons are no longer displaced, and the diaphragm and capillary tube act as on the old type of indicator (as long as there is no acceleration). The actions of the acceleration elements and the diaphragm-capillary system overlap for smooth action.

It's possible to fly this type of instrument as accurately as an altimeter but the price may be out of the range of the owner of a lighter plane. In some cases it may cost well over twice as much as the old type of vertical speed indicator.

On lighter airplanes the pressure instruments obtain static pressure measurements from the air within the cabin. This introduces some error, as the cabin static pressure is usually slightly less than the actual outside air pressure because of the venturi effect of the airstream moving past the fuselage. A lower static pressure would mean that the altimeter would read a few feet higher and the airspeed indicator would read slightly fast. The rate of climb would be unaffected once the airplane had established some airspeed as it only measures *change of pressure.*

So that there'll be no confusion about the airspeed error: the above statement assumes that the airplane has a pitot tube on the wing or other point and has *no outside static tube,* but merely an opening in the back of the airspeed indicator instrument case. Your plane may have a pitot-static tube combination so that no cabin pressure errors are introduced.

For the type of flying done in lighter planes the above mentioned errors are usually ignored.

MAGNETIC COMPASS

The magnetic compass is a magnet that aligns itself with the Magnetic North Pole while the airplane turns around it. You are familiar enough with the compass so that an involved description isn't needed, but some hints on using the compass may be helpful.

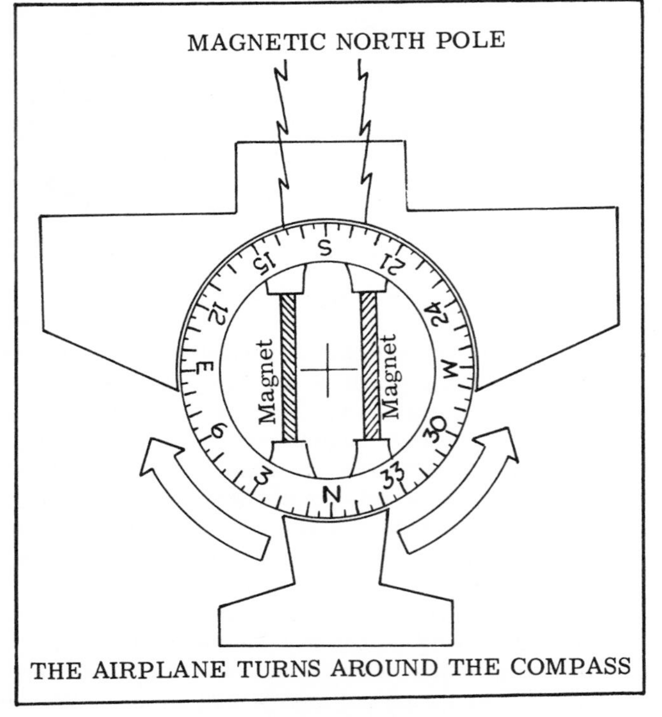

Fig. 3-11. The magnetic compass.

The magnets in the compass tend to align themselves parallel to the earth's lines of magnetic force. This tendency is more noticeable as the Magnetic North Pole is approached. The compass would theoretically point straight down when directly over the pole. (Fig. 3-12) The compass card is mounted so that a low center of gravity location fights this dipping tendency. Dip causes certain errors to be introduced into the compass readings and should be noted as follows:

Northerly Turning Error — In a shallow turn, the compass leads by about 30 degrees when passing through South and lags by about 30 degrees when passing through North. On passing East and West headings in the turn, the compass is approximately correct. (30 degrees is a rule of thumb for U. S. use.)

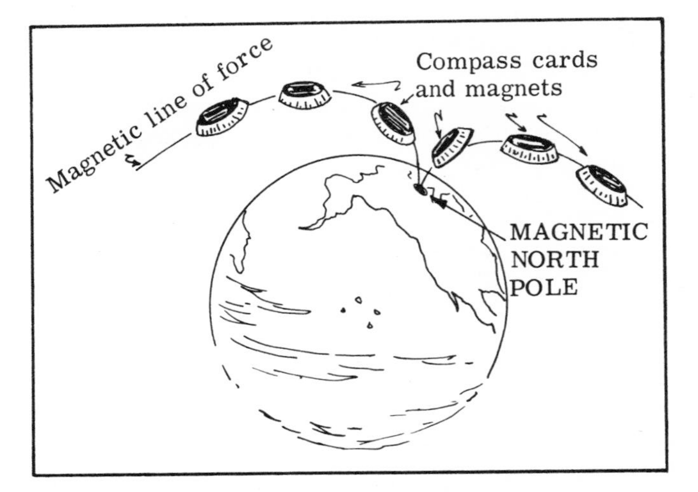

Fig. 3-12. The compass magnets tend to lie parallel with the earth's lines of magnetic force.

For instance, you are headed South and decide to make a *left* turn and fly due North. As soon as the left bank is entered, the compass will indicate about 30 degrees of left turn, when actually the nose has hardly started to move. *So, when a turn is started from a heading of South, the compass will indicate an extra fast turn in the direction of bank.* It will then hesitate and move slowly again so that as the heading of East is passed, it will be approximately correct. The compass will lag as North is approached so that you will roll out when the magnetic compass indicates 030 degrees (or "3").

Fig. 3-13 shows the reactions of the compass to the 180° left turn from a heading of South.

If you had made a right turn from a South heading, the same effects would have been noticed: an immediate indication of turn in the direction of bank, a correct reading at a heading of West and a compass lag of 30 degrees when headed North.

If you start a turn from a heading of North, the compass will initially register a turn in the opposite direction but will soon race back and be approximately correct as an East or West heading is passed. It will then lead by about 30 degrees as the airplane's nose points to Magnetic South. The initial errors in the turn are not too important. Set up your turn and know what to expect after the turn is started.

Here is a simple rule to cover the effects of bank (assuming a shallow bank of 20 degrees or less — if the bank is too steep the rule won't work).

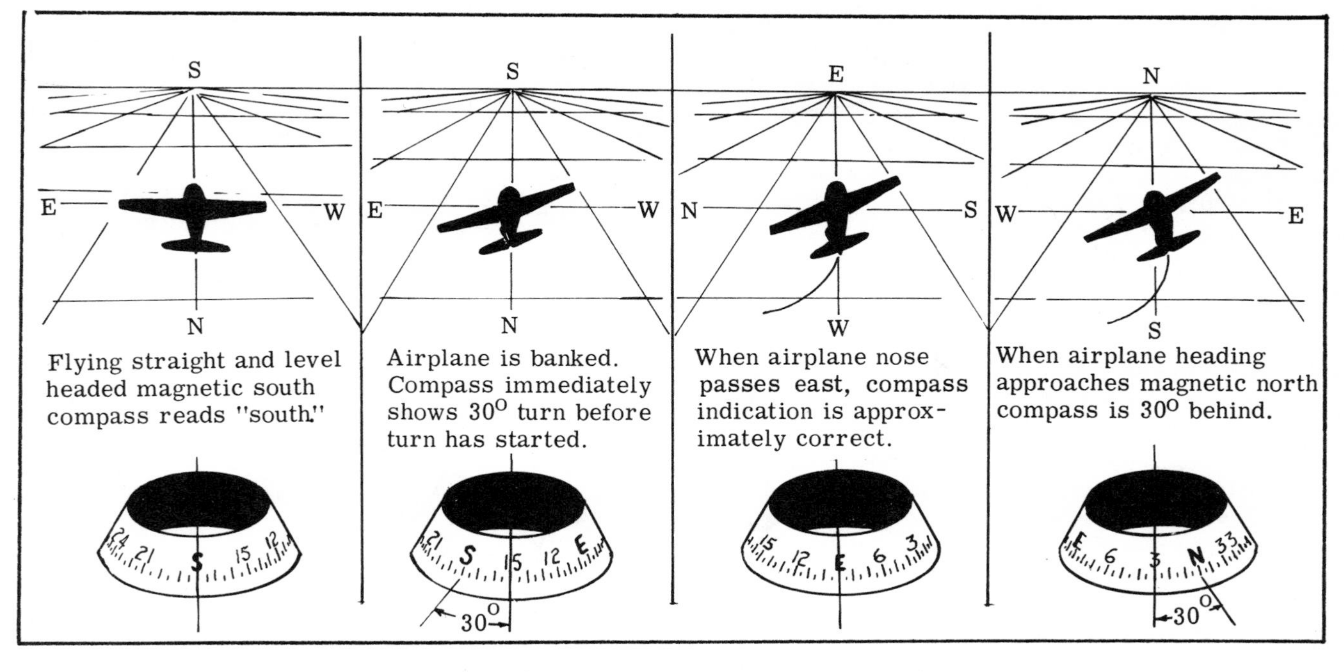

Fig. 3-13. Compass reactions to a turn.

NORTHERLY TURNING ERRORS—(NORTHERN HEMISPHERE)

North heading— compass *lags* 30° at start of turn, or in the turn.

South heading— compass *leads* 30° at start of turn, or in the turn.

East or West heading— compass correct at start of turn, or in the turn.

Just remember that North *lags* 30 degrees and South *leads* 30 degrees and this covers the problem. *Actually 30 degrees is a round figure; the lead or lag depends on the latitude, but 30 degrees is close enough for the work you'll be doing with the magnetic compass and is easy to remember.*

ACCELERATION ERRORS

Because of its correction for dip, the compass will react to acceleration and deceleration of the airplane. This is most apparent on East or West headings, where *acceleration results in a more northerly indication. Deceleration gives a more southerly indication.* You might check this the next time you're out just boring holes in the sky.

The magnetic compass reads correctly *only* when the airplane is in straight and level unaccelerated flight (and sometimes not even then). In bumpy air the compass oscillates so that readings are difficult to take and more difficult to hold. The fluid in the case (acid free white kerosene) is designed to keep the oscillations at a minimum, but the problem is still there.

VARIATION

The magnetic compass naturally points to the Magnetic North Pole, and this leads to the necessity of correcting for the "angle" between the Magnetic and Geographic North Poles. (Fig. 3-14)

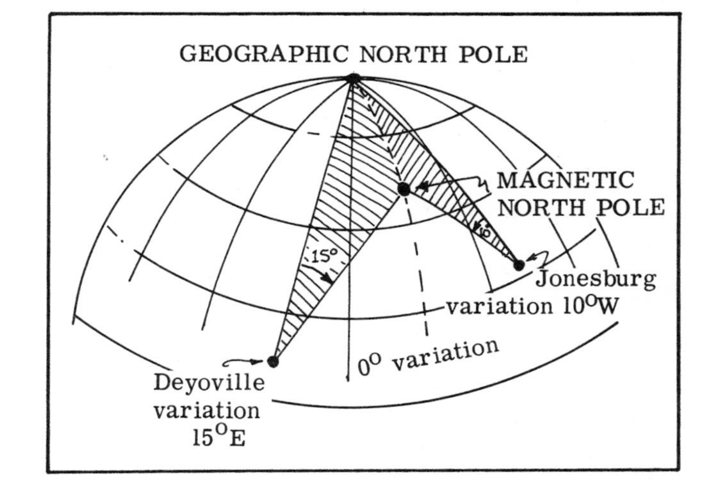

Fig. 3-14. Magnetic variation.

Normally a course will be measured from a midpoint meridian and will be the "True Course" or the course as referred to the True or Geographic North Pole. To get the magnetic course the following will apply. *Going from true to magnetic:*

East is least – subtract East variation as shown on the sectional or WAC Chart.

West is best – add West variation as shown on the sectional or WAC Chart.

The variation (15 degrees E or 10 degrees W) given by the isogonic lines means that the Magnetic North Pole is 15 degrees East or 10 degrees West of the True North Pole— from your position as far as the compass is concerned. Naturally, if you happen to be at a point where the two poles are in line, the variation will be zero. (Figs. 3-14 and 3-15)

Most pilots are getting cagey these days and measure their courses by laying a straight edge parallel to the course line through a nearby omni rose on the

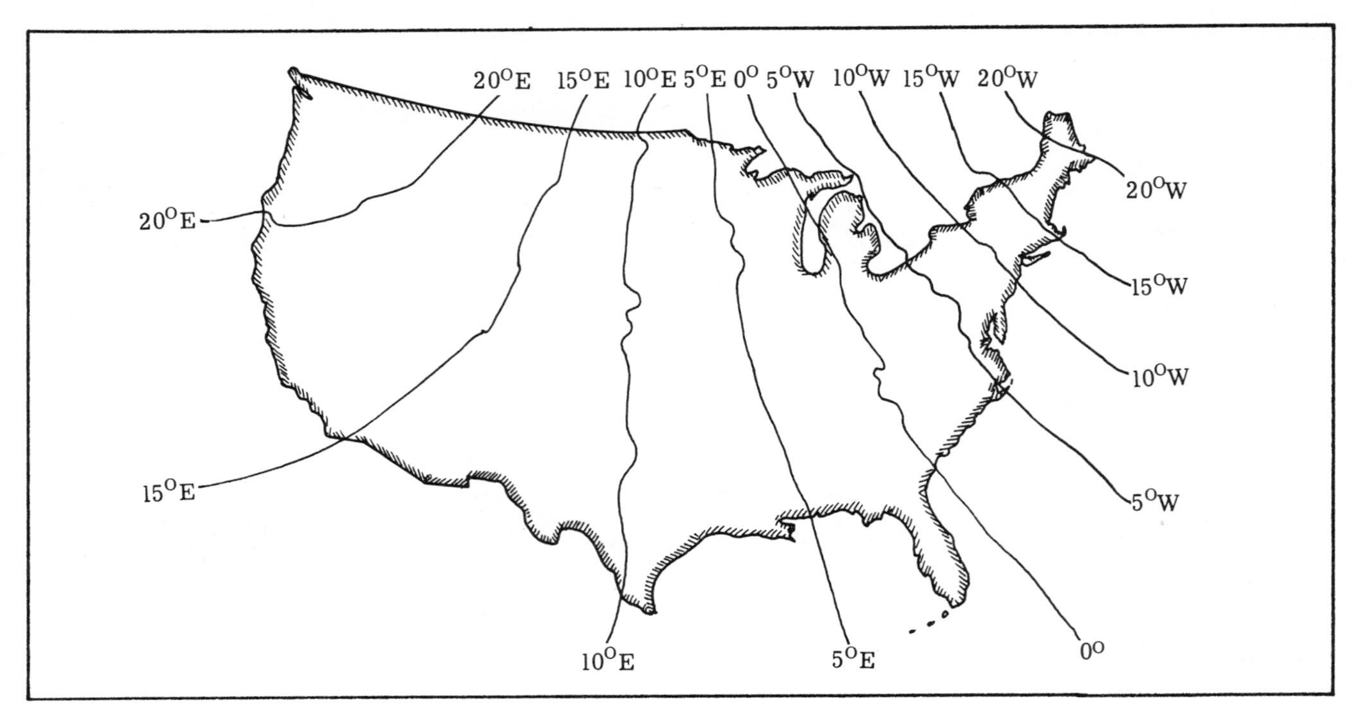

Fig. 3-15. Approximate location of isogonic lines in the U.S.

sectional chart. As the omni rose (or azimuth circle) is oriented on *magnetic north,* they've saved the step of measuring the true course on a meridian and adding or subtracting variation to get the magnetic course.

DEVIATION

The compass has an instrument error due to electrical equipment and metal parts of the plane. This error varies between headings and a correction card is placed near the compass, showing these errors for each 30 degrees. (Figure 3-16)

The compass is "swung," or corrected, on a compass rose — a large calibrated circle painted on the concrete ramp or taxiway away from metal interference such as hangars. The airplane is taxied onto the rose and corrections are made in the compass with a non-magnetic screwdriver. The engine should be running and normal radio and electrical equipment on. Attempts are made to balance out the errors — better to have all headings off a small amount than some correct and others badly in error.

In order to use the compass you must allow for corrections, and for navigation purposes the following steps would apply:

(1) *T*rue course (*or* heading) plus or minus *V*ariation gives *M*agnetic course (*or* heading).

FOR (MAGNETIC)	N	30	60	E	120	150
STEER(COMPASS)	358	27	58	89	119	150
FOR (MAGNETIC)	S	210	240	W	300	330
STEER(COMPASS)	182	213	244	275	302	329

Fig. 3-16. A typical compass correction card.

(2) *M*agnetic course (*or* heading) plus or minus *D*eviation gives *C*ompass course (*or* heading).

TVMDC or *T*rue *V*irgins *M*ake *D*ull *C*ompany; or *T*he *V*ery *M*ean *D*epartment of *C*ommerce (left over from the days when aviation was under the jurisdiction of the Department of Commerce).

The chances are that in your normal flying you've paid little attention to deviation and have been doing fine. But remember that if you plan on getting that commercial certificate there'll be some pretty good questions on the subject so it might be a good idea to start thinking about it again.

The magnetic compass has many quirks but once you understand them, it can be a valuable aid. One thing to remember — the mag compass "runs" on its own power and doesn't need electricity or suction to operate. This feature may be important to you some day when your other more expensive direction indicators have failed.

GYRO FLIGHT INSTRUMENTS

The gyro instruments depend on two main properties of the gyroscope; "rigidity in space" and "precession." Once spinning, the gyroscope resists any effort to tilt its axis (or plane of rotation). A spinning gyro, if in orbit around the earth, would present different sides to various countries. Take another look at Fig. 1-47. The gyro horizon and directional gyro operate on this principle. If a force is exerted to try to change the plane of rotation of a rotating gyro wheel, the gyro resists. If the force is insistent, the gyro reacts as if the force had been exerted at a point 90 degrees around the wheel (in the direction of rotation). See Fig. 1-46. Precession is the property used in the operation of the needle of the turn and slip indicator (or needle and ball as you may call it).

VACUUM DRIVEN INSTRUMENTS

For the less expensive airplanes the gyro instruments are usually vacuum driven, either by an engine driven pump or venturi system. A disadvantage of the venturi system is that its efficiency depends on airspeed, and the venturi tube itself causes slight aerodynamic Drag. Although a venturi system can be installed on nearly any airplane in a short while, the engine-driven vacuum pump is best for actual instrument operations, as it starts operating as soon as the engine(s) start. Multi-engine airplanes usually have a vacuum pump on each engine so that the vacuum driven instruments will still operate in the event of an engine failure. Each pump has the capacity to carry the system. The gyro instruments usually operate at a *suction* of 4.0 inches of mercury (29.92 inches of mercury is standard sea level pressure). The 4.0 inches of mercury shows a *relative* difference between the outside air pressure and the air in the vacuum system. The operating limits for the gyro horizon and directional gyro are normally from 3.75 to 4.25 inches of Hg of suction, whereas the turn and slip uses a lower suction of 1.8 to 2.1 inches Hg.

The combination of the heavy gyro wheel and high rpm results in efficient gyro operation.

Errors in the instruments may arise as they get older and bearings get worn, or the air filters get clogged with dirt. Low suction means low rpm and a loss in efficiency of operation.

ELECTRIC DRIVEN INSTRUMENTS

The electric-driven gyro instruments got their start when high performance aircraft such as jets began to operate at very high altitudes. The suction driven instruments lost much of their efficiency in the thin air and a different source of power was needed.

Below 30,000 feet either type of gyro performs equally well. It is common practice to use a combination of electric and vacuum driven instruments for safety's sake, should one type of power source fail. A typical gyro flight instrument group for a single piloted plane would probably include a vacuum driven gyro horizon and directional gyro, and an electric turn and slip. Large airplanes have two complete sets of flight instruments, one set of gyros vacuum driven and the other, electric.

An advantage of the electric instruments is that the gyro horizon is usually smaller than the vacuum driven type (which is larger than standard instrument size), leaving more room on the instrument panel for other instruments. Many of the newer electric horizons will not tumble, and aerobatics such as loops, rolls, etc., may be done by reference to the instrument.

GYRO HORIZON

The gyro horizon, or artificial horizon, operates on the "rigidity in space" principle and is an attitude instrument. The plane of rotation of the gyro wheel is horizontal and maintains this position, with the airplane (and instrument case) being moved about it. (Fig. 3-17)

Attached to the gyro is a black face with a white horizon line in front of it. When the instrument is operating correctly, this line will represent the actual horizon. A miniature airplane attached to the case moves with respect to this artificial horizon precisely as the real airplane moves with respect to the real horizon. This instrument is particularly easy to use as the pilot is able to "fly" the small airplane as he would the large airplane itself. A knob allows you to move the miniature airplane up or down to compensate for small deviations in the horizontal-line position.

There are limits of operation on the less expensive vacuum driven gyro horizons and these are, in most cases, 70 degrees of pitch (nose up or down) and 100 degrees of bank. The gyro will "tumble" above these limit stops and will give false information when the gyro is forced from its rotational plane. The instrument also will give false information during the several minutes required for it to return to the normal position after resuming straight and level flight.

"Caging" is done by a knob located on the instrument front. Because it is possible to damage the instrument through repeated tumbling, this caging is a must before you do deliberate aerobatics. The caging knob is useful also for quickly resetting the gyro horizon if it has tumbled. Some gyro horizons have caging knobs, some don't.

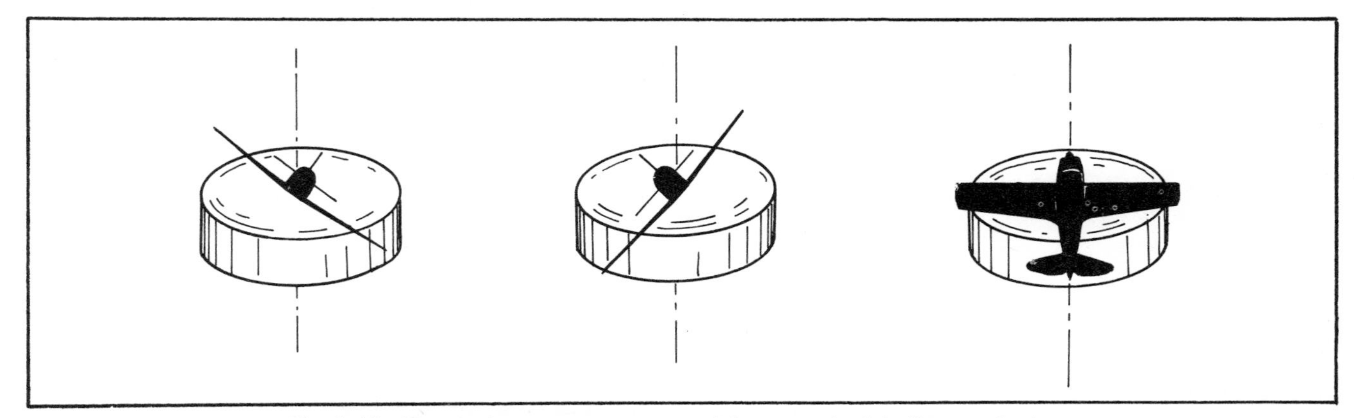

Fig. 3-17. The airplane maneuvers around the gyro wheel in the gyro horizon.

This instrument allows the pilot to get an immediate picture of the plane's attitude. It can be used to establish a standard-rate turn if necessary without reference to the turn and slip indicator, as will be shown. (The turn must be balanced for this to work.)

An airplane's rate of turn depends on its velocity and amount of bank. For any airplane, the slower the velocity and the greater the angle of bank, the greater the rate of turn. If you are interested in finding the angle of bank for standard-rate turn, the equation is:

$$\text{Tan}\phi = \frac{2\pi V}{gt}$$

Where ϕ = Angle of bank required.

$\pi = 3.1416$

- V = Velocity of your airplane in feet per second (multiply the velocity in miles per hour by 1.467 to get the velocity in feet per second).
- g = Acceleration of gravity (32.2 feet per second/ per second).
- t = Time in seconds required for a 360 degree turn (in this case a standard-rate turn, so it will be 120 seconds).

It is possible to combine $\frac{2\pi}{gt}$ into a single term as it is constant for our purpose here.

$$\frac{2\pi V}{gt} = \frac{6.2832V}{(32.2)(120)} = \frac{6.2832V}{3860} = 0.001625V \text{ (fps)} =$$

$$(0.001625)(1.467) = 0.00238V \text{ (mph)}$$

So that Tan ϕ = 0.00238V (mph). If V = 80 mph, Tan ϕ = 0.1910 (then, by reference to a trigonometric table),

ϕ = 10.8 degrees or 11 degrees.

Of course, a simple solution in any case is to set up a standard-rate turn with the turn and slip indicator and check the angle of bank. *A good rule of thumb to find the amount of bank needed for a standard-rate turn at various airspeeds is to divide your airspeed (mph) by 10 and add 5 to the answer.* For instance, airspeed = 150 mph; $\frac{150}{10}$ = 15; 15 + 5 = 20 degree bank required. This thumb rule is particularly accurate in the 100-200 mph range, as you can see by checking Fig. 3-18, and is as close as you'll be able to hold the bank at any speed range.

V mph	Degrees Angle of Bank ϕ
60	8
70	9.5
80	11
90	12
100	13.5
110	14.6
120	16
130	17
140	18.5
150	20
160	21
170	22
180	23
190	24.5
200	25.5
250	31
300	35.5
600	55
1000	67

Fig. 3-18. Bank required for various airspeeds for the standard-rate balanced turn.

For the airspeed in knots, divide it by 10 as for mph, but add one-half of the answer. For 130 knots, divide that number by 10 (13) and add one-half of the 13 to it (13 + one-half of 13 = 13 + 6.5 = 19.5 degrees.) Note that 130 knots equals 150 mph and a glance at Fig. 3-18 shows that 20 degrees is required for 150 mph. This thumb rule, like the one for mph, is most accurate in the 100-200 knot speed range.

The "V" in the equation for the bank required is the true airspeed and some small errors might be introduced at high altitudes where there are appreciable differences in indicated (calibrated) and true airspeeds. If you are indicating 150 mph at 10,000 feet standard altitude, a quick glance at the airspeed indicator would give an answer of 20 degrees required for a standard-rate turn. The true airspeed at 10,000 is 174 mph and actually a bank of 22.4 degrees would be required. For most cases, however, using the indicated airspeed for the thumb rule is a quick, if not completely accurate, method of getting the required angle of bank.

DIRECTIONAL GYRO

The directional gyro functions because of the principle of "rigidity in space" as did the gyro horizon. In this case, however, the plane of rotation is vertical. The directional gyro has a compass card or azimuth scale which is attached to the gyro gimbal and wheel. The wheel and card are fixed and, as in the case of the magnetic compass, the plane turns around them. (Fig. 3-19)

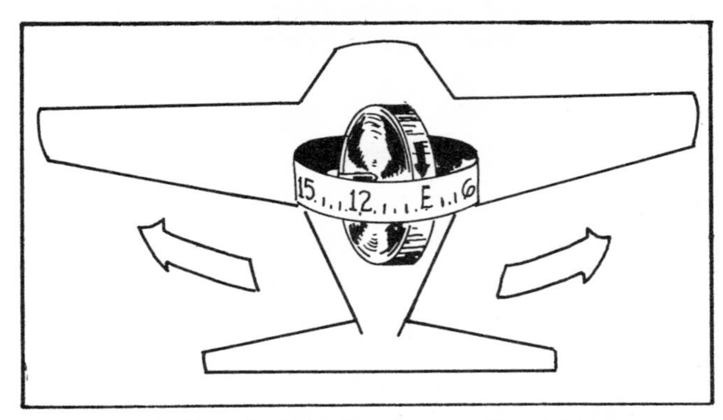

Fig. 3-19. The directional gyro. The airplane turns around the vertical gyro wheel.

The directional gyro has no magnet that causes it to point to the Magnetic North Pole, and must be set to the heading indicated by the magnetic compass. The directional gyro should be set when the magnetic compass is reading correctly. This is generally done in straight and level flight when the magnetic compass has "settled down."

The advantage of the directional gyro is that it does not oscillate in rough weather and gives a true reading during turns when the magnetic compass is erratic. A setting knob is used to cage the instrument for aerobatics or to set the proper heading.

A disadvantage of the older types of directional gyros is that they tumble when the limits of 55 degrees nose up or down or 55 degrees bank are reached. Although if you happen to be maneuvering (pitching or rolling) parallel to the plane of rotation of the gyro wheel (or around the gyro axis) this limitation does not apply.

The directional gyro creeps and must be reset with the magnetic compass about every 15 minutes. (Normal allowable creep is 3 degrees in fifteen minutes.)

More expensive gyros, such as are used by the military and airlines, are connected with a magnetic compass in such a way that this creep is automatically compensated for.

The greatest advantage of the directional gyro is that it allows you to turn directly to a heading without the allowance for lead or lag necessary with a magnetic compass.

TURN AND SLIP INDICATOR

The turn and slip indicator is actually two instruments. The slip indicator is merely a liquid-filled, curved glass tube containing an agate or steel ball. The liquid acts as a shock dampener. In a balanced turn the ball will remain in the center as centrifugal force offsets the pull of gravity.

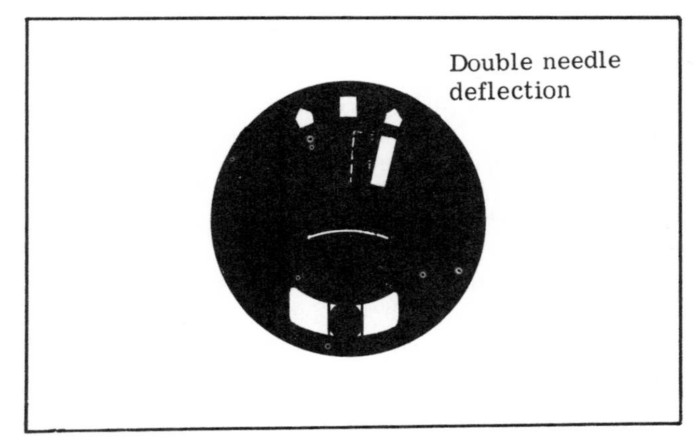

Fig. 3-20. A balanced right turn.

In a slip, there is not enough rate of turn for the amount of bank. The centrifugal force will be weak and this imbalance will be shown by the ball's falling down toward the inside of the turn. (You experience centrifugal force any time you turn a car — particularly in an abrupt turn at high speed as the force pushes you to the outside of the turn.)

The skid is a condition in which there is too high a rate of turn for the amount of bank. The centrifugal force is too strong, and this is indicated by the ball's sliding toward the outside of the turn. Usually a turn in an airplane is considered to be balanced if more than one-half of the ball is within the indicator marks. This will be covered more fully in Chapter 7.

The turn part of the turn and slip indicator, or "needle" as it is sometimes called, uses precession to indicate the direction and approximate rate of turn of the airplane.

Fig. 3-21. A slipping right turn.

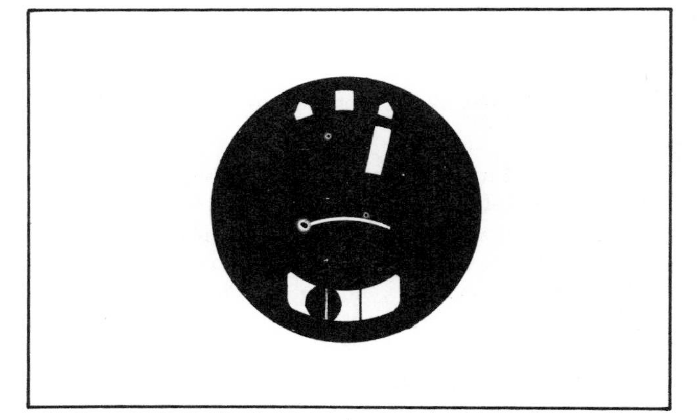

Fig. 3-22. A skidding right turn.

Shown in Fig. 3-23 is the reaction of the turn and slip indicator to a right turn. The case is rigidly attached to the instrument panel and turns as the airplane turns (1). The gyro wheel (2) reacts by trying to move in the direction shown by (3), moving the needle in proportion to the rate of turn (which controls the amount of precession). As soon as the turn stops, the precession is no longer in effect and the spring (4) returns the needle to the center. The spring resists the precession and acts as a damper, so the nose must actually be moving for the needle to move.

Most turn and slip indicators are calibrated so that a "standard-rate turn" of 3 degrees per second will be indicated by the needle's being off center by one needle width. (Fig. 3-20) This means that by setting up a standard-rate turn it is possible to roll out on a predetermined heading by the use of a watch or clock. It requires 120 seconds or 2 minutes to complete a 360-degree turn. There are types of turn and slip indicators calibrated so that a double-needle-width indication indicates a standard-rate turn.

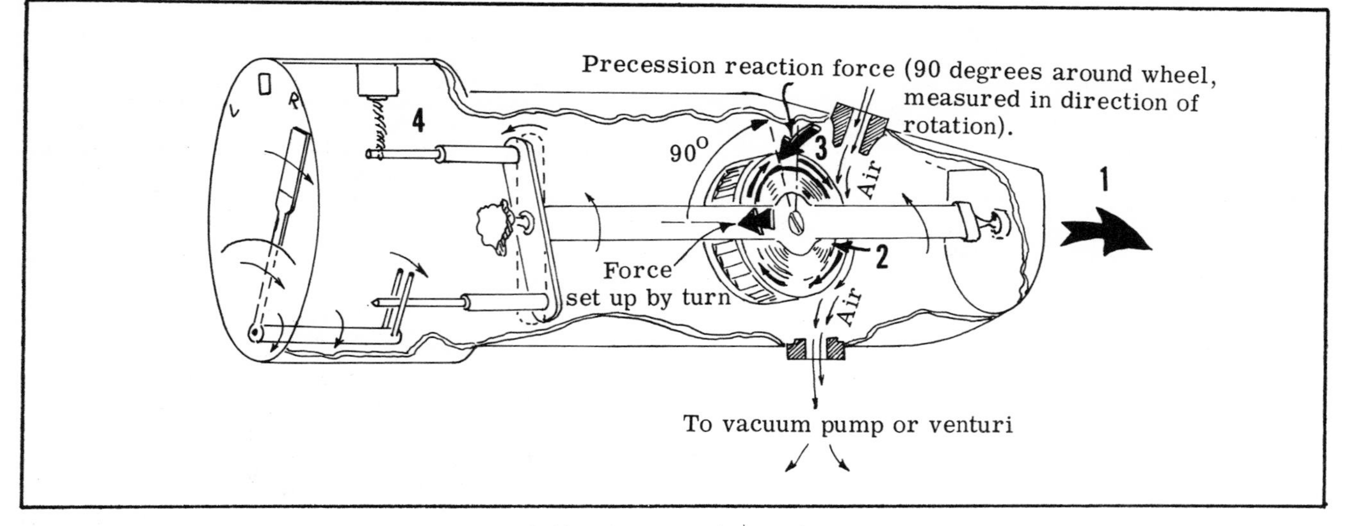

Fig. 3-23. The turn and slip indicator.

These are usually noted as such on the instrument (*four-minute turn*) or have a "doghouse" on each side of the center. If your heading is 070 degrees and you want to roll out on a heading of 240 degrees, first decide which way you should turn (to the right in this case). The amount to be turned is 240 minus 70 = 170 degrees. The number of seconds required at standard rate is $\frac{170}{3} = 57$. If you set up a standard-rate turn and hold it for 57 seconds and roll out until the needle and ball are centered, the heading should be very close to 240 degrees.

One of the most valuable maneuvers in coping with bad weather if you don't have an instrument rating, and sometimes if you do, is the 180-degree turn, or "getting the hell out of there." It is always best to do the turn *before* you lose visual contact, but if visual references are lost inadvertently, the 180-degree turn may be done by reference to the instruments.

The advantage of the turn and slip over other gyro instruments is that it does not "tumble" or become erratic as certain bank and pitch limits are exceeded.

A disadvantage of the turn and slip is that it is a rate instrument, and a certain amount of training is required before the pilot is able to quickly transfer the indications of the instrument into a visual picture of the airplane's attitudes and actions.

The gyro of the turn and slip, like the other gyro instruments, may be driven electrically, or by air, using an engine-driven vacuum pump or an outside-mounted venturi.

ENGINE INSTRUMENTS

TACHOMETER

For airplanes with fixed pitch propellers the tachometer is the engine instrument to be checked for an indication of power being used. The centrifugal tachometer operates on the same principle as a car speedometer. One end of a flexible shaft is connected to the engine crankshaft and the other connected to a shaft with counterweights within the instrument. The rate of turning of the crankshaft (and cable) causes expansion of the counterweight system. The instrument hand is mechanically linked to the counterweight assembly so that the engine speed is indicated in revolutions per minute.

For direct drive engines, the engine and propeller rpm are the same (Lycoming O-320, O-540, O-360). The geared engine (Lycoming GO-480, etc.) has different engine and propeller speeds and this is noted in the Airplane Flight Manual (the propeller rpm is *less* than the engine rpm). The tachometer always measures engine rpm and this is the basis for your power setting.

Another type of tachometer is the magnetic, which utilizes a flexible shaft which turns a magnet within a special collar in the instrument. The balance between the magnetic force and a hairspring is indicated as rpm by hand on the instrument face. This type of tachometer does not oscillate as sometimes happens with the less expensive centrifugal type.

A third type is the electric tachometer which depends on a generator unit driven by a tachometer drive shaft. The generator is wired to an electric motor unit of the indicator, which rotates at the same rpm and transmits this through a magnetic tachometer unit which registers the speed in rpm. This type of tachometer is also smoother than the centrifugal type.

MANIFOLD PRESSURE GAGE

For airplanes with controllable (which includes constant speed) propellers this instrument is used in combination with the tachometer to set up desired power from the engine. The manifold pressure gage measures the fuel-air mixture going to the cylinders and indicates the mixture pressure in inches of mercury.

The manifold pressure gage is an aneroid barometer like the altimeter but instead of measuring the outside air pressure, measures the *actual* pressure

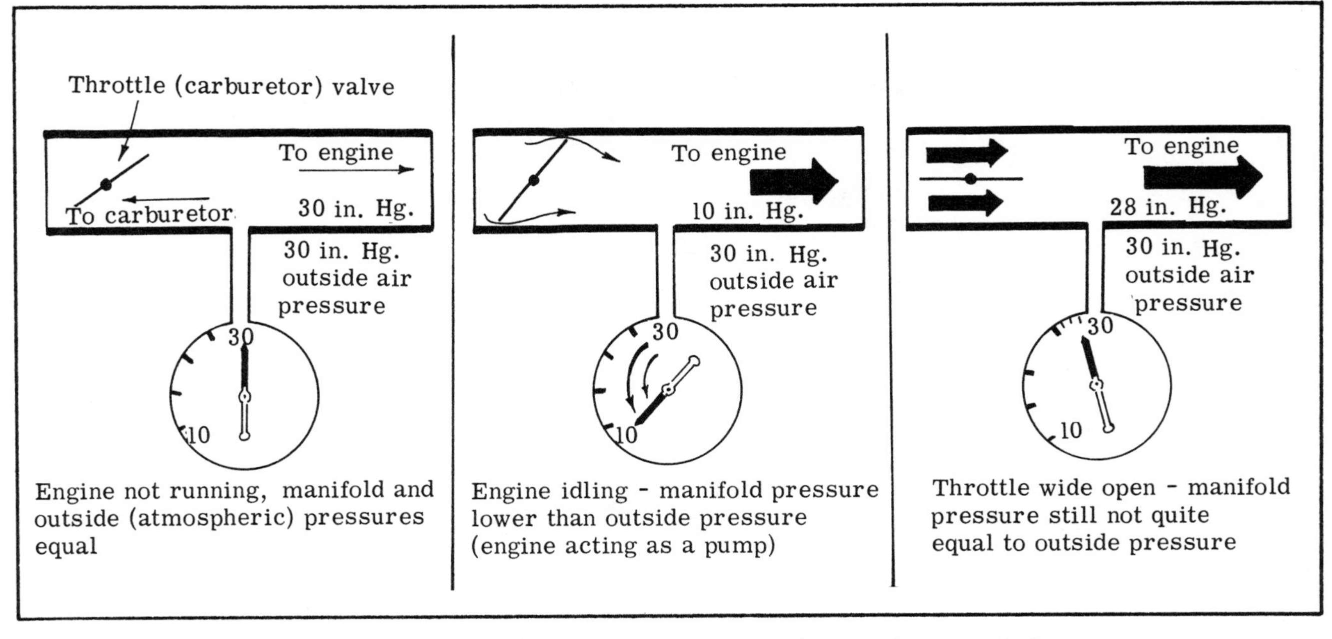

Fig. 3-24. The manifold pressure gage principle (unsupercharged engine).

of the mixture in the intake manifold. When the engine is not running the outside air pressure and the pressure in the intake manifold are the same, so that the manifold pressure gage will indicate the outside air pressure as would a barometer. At sea level on a standard day this would be 29.92″ Hg, but you can't read the manifold gage this closely and it would appear as approximately 30″.

You start the engine with the throttle cracked or closed. This means that the throttle valve or butterfly valve is nearly shut. The engine is a strong air pump in that it takes in fuel and air and discharges residual gases and air. At closed or cracked-throttle setting the engine is pulling air (and fuel) at such a rate past the nearly closed throttle valve that a decided drop in pressure is found in the intake manifold and is duly registered by the manifold pressure gage. As the engine starts, the indication of 30″ drops rapidly to 10″ or less at idle. It will never reach *zero* as this would mean a complete vacuum in the manifold (most manifold pressure gages don't even have indications of less than 10 inches Hg). Besides, if you tried to completely shut off all air (and fuel) the engine would quit running.

As you open the throttle you are allowing more and more fuel and air to enter the engine and the manifold pressure increases accordingly. (Fig. 3-24)

As you can see in Fig. 3-24, the unsupercharged engine will never indicate the full outside pressure on the manifold gage. The usual difference is 1 to 2 inches of mercury. The maximum indication on the manifold pressure gage you could expect to get would be 28 to 29 inches on take-off.

The supercharged engine has compressors that bring the air-fuel mixture to a higher pressure than the outside air before it goes into the manifold. This makes it possible to register more than the outside pressure, and results in more horsepower being developed for a given rpm, as horsepower is dependent on rpm and the amount of fuel-air (manifold pressure) going into the engine.

When the engine is shut down, the manifold pressure gage indication moves to the outside air pressure. The techniques in using a manifold pressure gage will be discussed later in Part II — "Checking Out in Advanced Models and Types."

OIL PRESSURE GAGE

The oil pressure gage consists of a curved Bourdon tube with a mechanical linkage to the indicating hand which registers the pressure in pounds per square inch. (Fig. 3-25) As is shown, oil pressure tends to straighten the tube and the appropriate oil pressure indication is registered. This is the direct pressure type gage.

Another type of oil pressure gage uses a unit containing a flexible diaphragm which separates the engine oil from a nonflammable fluid which fills the line from the unit into the Bourdon tube. The oil pressure is transmitted through the diaphragm and to the Bourdon tube by this liquid because liquids are incompressible. (Fig. 3-26)

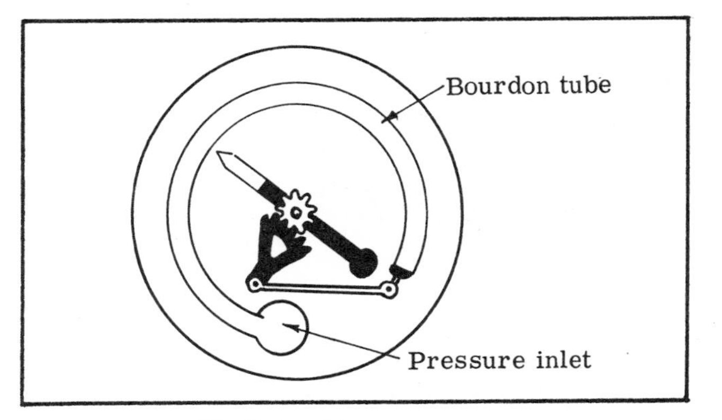

Fig. 3-25. Oil pressure gage.

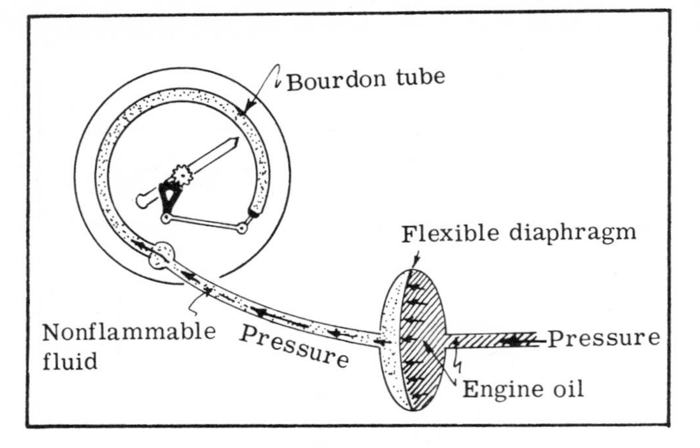

Fig. 3-26. Oil pressure gage utilizing a flexible diaphragm and nonflammable fluid.

OIL TEMPERATURE GAGE

The vapor-type is the most common type of oil temperature gage in use. This instrument, like the oil pressure gage, contains a Bourdon tube which is connected by a fine tube to a metal bulb containing a volatile liquid. Vapor expansion due to increased temperature exerts pressure — which is indicated as temperature on the instrument face.

Other types of oil temperature gages may use a thermocouple rather than a Bourdon tube.

CYLINDER HEAD TEMPERATURE GAGE

The cylinder head temperature gage is an important instrument for engines of higher compression and/or higher power. Engine cooling is a major problem in the design of a new airplane. Much flight testing and cowl modification may be required before satisfactory cooling is found for all airspeeds and power settings. The engineers are faced with the problem of keeping the engine within efficient operating limits for all air temperatures. An engine that has good cooling for summer flying may run too cool in the winter. Cowl flaps, which are controlled by the pilot, aid in compensating for variations in airspeed and power setting. Many of the newer high performance airplanes use "augmenter cooling" instead of cowl flaps. Air is drawn over the cylinders by venturi action of a tube around the exhaust stacks. (Fig. 3-27)

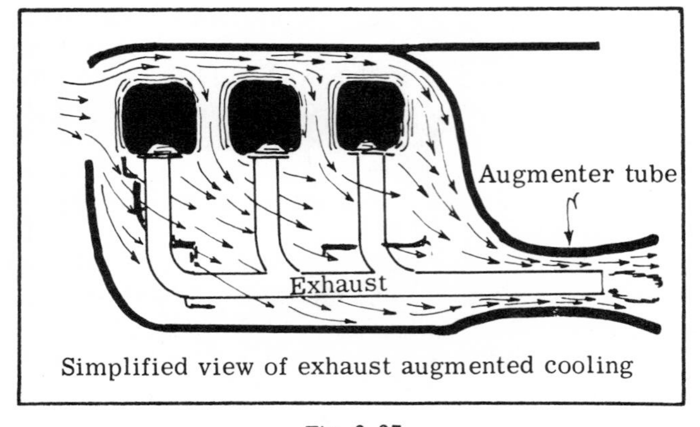

Fig. 3-27.

The cylinder head temperature gage usually warns of any possible damage to the engine before the oil temperature gage gives any such indication.

The "hottest" cylinder, which is usually one of the rear ones in the horizontally opposed engine, is chosen during the flight testing of the airplane. A thermocouple lead replaces one of the spark plug washers on this cylinder.

The cylinder head temperature gage uses the principle of the galvanometer. Two metals of different electrical potentials are in contact at the lead. As the electric currents of these two metals vary with temperature, a means is established of indicating the temperature at the cylinder through electric cables to a galvanometer (cylinder head temperature gage) which indicates temperature rather than electrical units.

Some pilots use cylinder head temperature as an aid in proper leaning of the mixture. Generally, richer mixtures mean lower head temperatures; leaner mixtures mean higher head temperatures, all other things (airspeed, power settings, etc.) being equal. But the engine may not be developing best power at the extremes. Too rich a mixture means power loss plus excessive fuel consumption, too lean a mixture means power loss plus the possibility of engine damage. Leaning procedures will be discussed in more detail in Part II.

FUEL GAGE

The cork float and wire fuel gages of earlier days have gone by the board. The corks sometimes got "fuel logged" and registered empty all the time. Worse, the wire sometimes got bent and the pilot had an unrealistic picture of fuel available. These indicators were followed by metal floats and indicators, and finally by the electric transmitter type now in popular use. (Check your fuel visually before the flight.)

The electric transmitter type may be considered to be broken down into the following components: (1) float and arm (2) rheostat-type control and (3) the indicator, a voltmeter indicating either fuel in fractions or in gallons. The float and arm are attached to the rheostat, which is connected by wires to the fuel gage. As the float level in the tank (or tanks) varies, the rheostat is rotated, changing the electrical resistance in the circuit — which changes the fuel gage indication accordingly. This is the most popular type of fuel measuring system for airplanes with electrical systems.

Frequent checks of the fuel gage is a good idea; sudden dropping of the fuel level indication may be caused by a serious fuel leak and you should know about this.

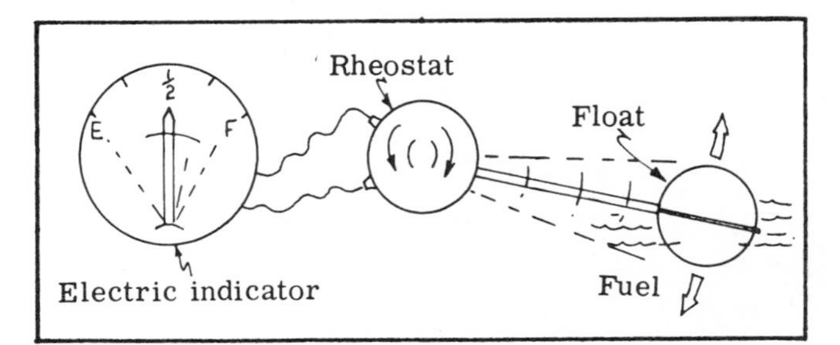

Fig. 3-28. Electric transmitter type fuel gage.

4. TAKE-OFF PERFORMANCE

The first measurement of a good pilot is headwork — does he think well in an airplane? Running a close second for the experienced pilot is his ability to get the most out of his plane when it is needed. A pilot who doesn't know what his airplane can do may either set such a high safety margin that performance suffers, or such a low margin as to damage an airplane and himself. Sometimes it takes a great deal of intestinal fortitude to do what's right — for instance, during a short field take-off with high trees at the far end. Logic and knowledge tell you that forcing the plane off too soon will cost take-off performance, while instinct pushes for you to get it off, *now*. It requires an understanding of airplane performance and a lot of argument with yourself sometimes to do what's right in a particular situation.

The following chapters on performance will be based on your understanding of the following:

(1) The air density in slugs per cubic foot is 0.002378 at sea level (standard day) and decreases with altitude.

(2) The sea level standard pressure is 29.92 inches of mercury, or 2116 pounds per square foot, and decreases at the rate of approximately one inch of mercury, or 75 pounds per square foot, per thousand feet. This is only an approximation, as can be seen from the Atmosphere Table in the back of the book. We will be considering only the lower part of the atmosphere (up to 10,000 feet) for this chapter.

(3) The standard sea level temperature is 59°F, or 15°C, and decreases $3\frac{1}{2}^{\circ}$F, or 2°C, per thousand feet (the temperature normal lapse rate is $3\frac{1}{2}^{\circ}$F or 2°C per thousand feet). Performance thumb rules and data are based on normally aspirated engines (no super-chargers) unless specifically stated.

TAKE-OFFS IN GENERAL

The take-off is usually the most critical part of the flight because of factors such as: (1) The plane is most heavily loaded at this point. (2) If the field is somewhat soft or has high grass or snow, the take-off suffers but the landing roll is helped (if it's not so soft as to cause a nose-up). Because of these two factors in particular, it's possible for you to get into a field from which you can't fly out.

TAKE-OFF VARIABLES

ALTITUDE AND TEMPERATURE EFFECTS

The air density decreases with altitude and, as you remember, air density is a factor of Lift.

Let's say that at the point of take-off, Lift just equals Weight (this is for a take-off at any altitude). It would also simplify our discussion to say that the plane lifts off at the maximum angle of attack. In most cases this doesn't happen — that is you don't "stall it off" — but it makes for easier figuring here, so we'll do it. So, the maximum angle of attack (without stalling) and wing area are the same for the take-off at any altitude, and the density is less at higher altitudes. Assuming the Weight of the airplane is the same that you had at a sea level airport, at higher altitudes you'll have to make up for the decrease in density by an increased true airspeed before the airplane can lift-off. The *indicated* airspeed will be the same for a high altitude take-off as it is at sea level, but it will take longer to get this indicated airspeed; the big reason being that the engine can't develop sea level horsepower. The result is that more runway is required with an increase in density altitude. This increase in take-off run can be predicted. As one pilot said after trying to take off from a short field at a high altitude and going off the end of the runway, through two fences, a hedge, across a busy highway and through a yard, "About this time I began to wonder if I was going to get off."

The atmospheric density does not decrease in a straight line like temperature. At 20,000 feet the density is about half that of sea level. At 40,000 feet the density is approximately half that of 20,000 feet, and so on, with the density halving about every 18-20,000 feet. Density is a function of pressure altitude and temperature.

The FAA has produced two density altitude computers, one for fixed pitch propeller equipped airplanes and the other for variable pitch propeller airplanes. The computers are used to check the take-off and climb performance of these two airplane types at higher density altitudes, and replace the older Koch Chart which was too conservative for the newer, higher performance general aviation airplanes.

Figure 4-1A is a graph for checking the effects of high density altitude on the take-off roll. An airplane with a fixed pitch prop (F) that takes 1000 feet to lift off at sea level would require 1700 feet of roll at a density altitude of 6000 feet (multiplier of 1.7). An airplane with a constant speed propeller (C) having a ground roll of 1000 feet would require 1620 feet at 6000 feet (multiplier 1.62).

Figure 4-1B is a graph for estimating the effects of density altitude on rates of climb for eleven general aviation airplanes. Note that those airplanes with a constant speed prop also give the better climb performance.

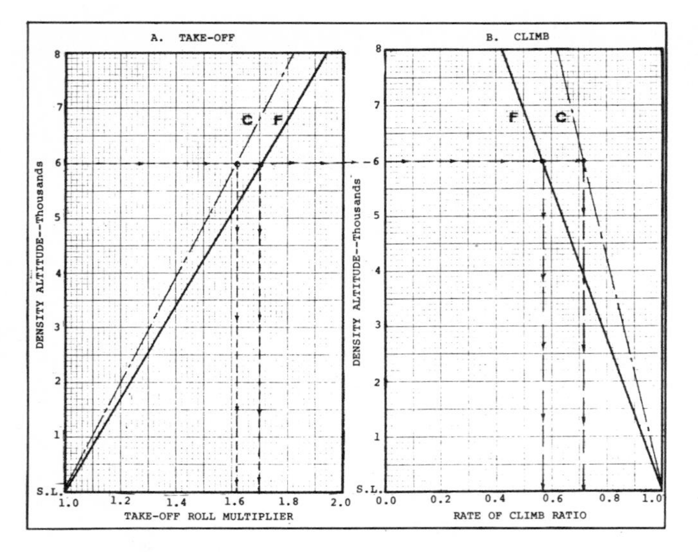

Fig. 4-1. A. A graph showing the take-off roll increase with density altitude as compared to sea level for airplanes with (C) constant speed and (F) fixed pitch propellers. The graph uses data from 11 current airplanes with power loadings of 10.5 to 16.0 pounds per horsepower (1967-1969 models at maximum certificated weight). As an example, the take-off roll (ground run) multiplier at 6000 feet will be 1.62 for a constant speed propeller equipped airplane, and 1.7 for the fixed pitch type. As might be expected, the constant speed propeller is more efficient. (Multiply the ground run at sea level by the factor at the chosen density altitude.)

B. The effects of density altitude on the rate of climb. The rates of climb for airplanes with fixed pitch (F) and constant speed prop (C) at 6000 feet density altitude are, respectively 0.57 (57 per cent) and 0.72 (72 per cent) of that at sea level.

How do you find the density altitude without a computer? Here are the temperatures for density (standard) altitudes from sea level to eight thousand feet (rounded off to the nearest degree):

Sea level--	59°F, 15°C	5000--	41°F, 5°C
1000--	55°F, 13°C	6000--	38°F, 3°C
2000--	52°F, 11°C	7000--	34°F, 1°C
3000--	48°F, 9°C	8000--	31°F, -1°C
4000--	45°F, 7°C		

Keep this in mind: For every ±15°F or ±8½°C variation from standard temperature at your pressure altitude, the density altitude is increased or decreased one thousand feet.

For instance, you are ready to take off and set the altimeter to 29.92. The pressure altitude given is 3000 feet and the outside air temperature is +22°C. The *standard* temperature at 3000 feet is +9°C, or the temperature is 13°C *higher than normal.* This higher temperature means that the air is *less* dense and the airplane is operating at a *higher* density altitude. This 13°C higher-than-standard temperature means adding another 1500 feet, for a *density* altitude of 4500 feet, which would be the altitude used in Figures 4-1A and 4-1B.

For a fixed pitch propeller-equipped airplane you expect to have a take-off roll of 1.53 times that at sea level and the rate of climb will be about 0.68 (68 per cent) of the sea level value as checked by Figure 4-1.

You should use the Airplane Flight Manual figures if they are available rather than using the FAA computers, Figure 4-1 or the following thumb rules.

A high temperature, even at sea level or a low altitude airport, can hurt the airplane's take-off performance. You can check by looking at the equation of state: $\rho = \frac{P}{1718T}$. If the temperature increases, the density decreases and vice versa (constant pressure). The relative humidity also affects performance. *Moist air, for the same temperature, is less dense than dry air.* "Common sense" would seem to tell you that water is heavier than air and the more water vapor present, the more dense the air should be. *This is not the case.* Remember — *the take-off figures in the Airplane Flight Manual are based on standard sea level air density, unless otherwise noted.*

Back in Chapter 3, in discussing altimeter errors, it was mentioned that wide variations from standard temperature can at your altitude cause errors in the altimeter indications. In the case of a take-off or landing at the airport giving the altimeter setting any errors due to variations from standard temperature will have been compensated for in the setting. (On take-off you will set your altimeter to field elevation, anyway.) Remember that the altimeter-temperature error is zero at the surface of the airport at which the setting is obtained (sea level is often used as an example but it's true for any airport elevation).

AIRCRAFT LOADING EFFECTS

Although aircraft loading will be covered in detail later, its effects on take-off performance should be noted. The worst possible situation for take-off is a short, soft field at high altitude on a hot day with an overloaded airplane under crosswind or no-wind conditions. Generally, the effect of Weight on a take-off run (all other things being equal) is that of $\left(\frac{\text{Present Weight}}{\text{Gross Weight}}\right)^2$. *The take-off distance figures given in the Airplane Flight Manual are based on the maximum certificated Weight of the airplane, unless otherwise noted.*

THE RUNWAY SURFACE

A soft or rough field, high grass, or deep snow can affect your take-off distance — but common sense has been telling you this for some time. The ground drag of your airplane caused by the runway surface is called "rolling resistance." The equation for this resistance is: R = μ(Weight minus Lift) where μ is the coefficient of friction for the particular runway surface being used. Some values of μ and their approximate effects on take-off distance are given in the following table.

The Airplane Flight Manual always bases take-off run figures on a hard surface, unless otherwise noted.

Fig. 4-2. Overloading means extra long take-off runs.

Runway Surface	μ	Required Take-off Roll
Concrete or Asphalt	0.02	Airplane Flight Manual Figure
Firm Turf	0.04	AFM Figure + 7%
Short Grass	0.05	AFM Figure + 10%
Long Grass	0.10	AFM Figure + 25%
Soft Field	0.10 - 0.30	AFM Figure + 25% to infinity

These figures are approximate and based on airplanes with a power loading of 10 to 15 pounds per horsepower — or $\frac{\text{Gross Weight}}{\text{Take-off Horsepower}} = 10$ to 16 pounds per horsepower. The figures will vary slightly with power loading. Airplanes that have a lower power loading — that is, more power per pound of Weight — are affected proportionately less by increased rolling resistance than a heavy airplane with little power.

The percentage of additional runway needed rises sharply for airplanes in this group after a μ of 0.15 is reached. In fact, none of the airplanes used as samples could *even move* at a μ of 0.30, although airplanes with a lighter power loading certainly could move — but not very fast.

As an example, a current airplane weighs 2900 pounds at gross, and the engine-propeller combination develops a static thrust of 780 pounds at sea level.

At a μ of 0.30 the rolling resistance of the airplane at the beginning of the roll is: μ (Weight-Lift) = 0.30 (2900-0) = 870 pounds, or *90 pounds more than the Thrust available at full power*. It won't move. That's why the word "infinity" was used in the table. Another point: definitions of short or long grass may vary between pilots. One of your flying buddies may consider grass as being short if it's less than waist high while you may think that the grass on a putting green is tall.

Because of this rolling resistance, for a soft field take-off you will want to get the Weight off the wheels as soon as possible.

You will note that at lift-off, the rolling resistance will be zero, since Lift will equal Weight at that point (the wheels will have no weight on them), as illustrated by Figure 4-3.

A problem that has risen to prominence since jets have come into wide use is the effects of slush on the take-off run. In many cases a comparatively thin layer of slush can hurt the take-off performance more than a thicker layer of snow. Slush is heavier than snow and tends to build up in front of the wheels.

Another factor on the take-off is aerodynamic Drag. Rolling resistance is the biggest problem at the early part of the run, with aerodynamic Drag becoming more of a factor as the speed increases.

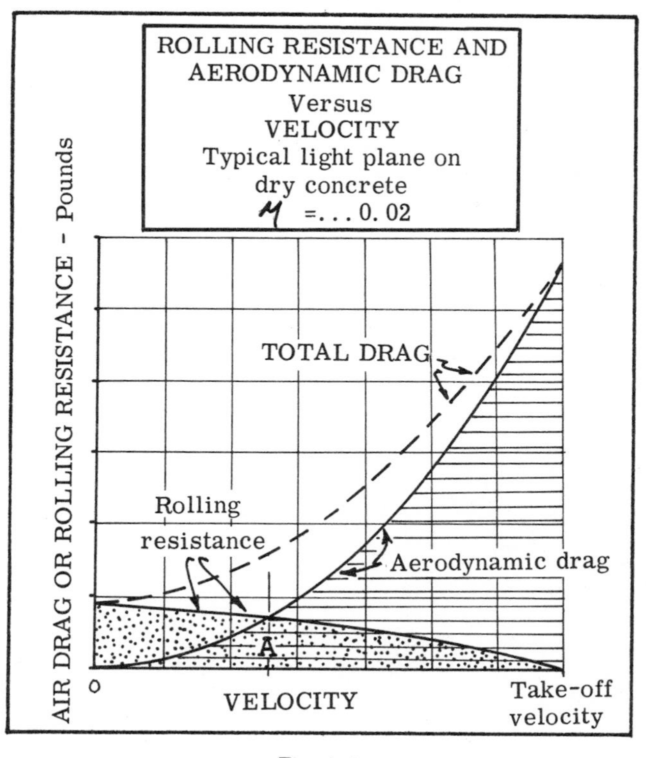

Fig. 4-3.

Shown in Fig. 4-3 is the rolling resistance and aerodynamic Drag during the take-off run. (Dry concrete.)

As you can see in Fig. 4-3, after point (A) the greater part of the retarding force is aerodynamic Drag.

For a soft field the sooner you can get the Weight off those wheels the better off you are as will be shown later. *The take-off figures in the Airplane Flight Manual are based on a hard surfaced runway unless otherwise stated.*

WIND CONDITIONS

The wind affects the take-off both in time and distance and, of these two, distance is the more important. Assuming that the wind is directly down the runway, an estimation of its effects may be given by Fig. 4-4.

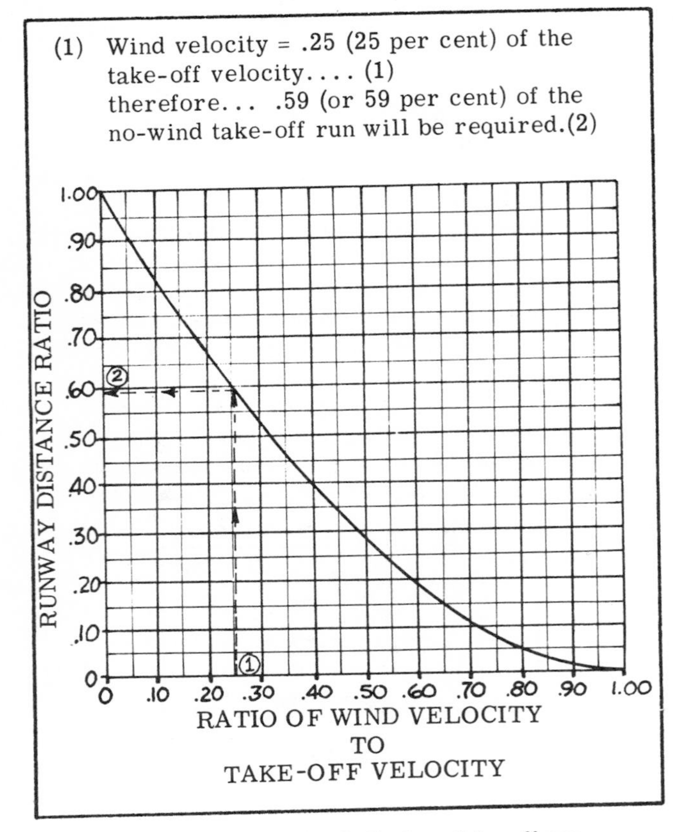

Fig. 4-4. Graph of wind effects on take-off run.

The graph shows that wind effects are not as straightforward as might at first be considered. For instance, if the wind were 25 per cent of the take-off speed, you can see by Fig. 4-4 that the take-off length required will be only 59 per cent of that required under no-wind conditions (common sense would seem to say that 75 per cent would be the answer). You gain more by having wind than you might at first consider. The wind used in the calculations is the headwind component, meaning that component of the wind that is directly down the runway. Fig. 4-5 gives the components for various angles that the wind makes with the runway. The figures are rounded off.

To use Figure 4-5 for take-off run calculations, you would multiply the headwind component (after finding the angle the wind makes with the runway) times the wind velocity. For instance, if a wind is 30 knots at an angle of 30° to the runway, you can find by referring to Fig. 4-5 that the headwind multiplier is 0.85. This figure multiplied times 30 knots is 25.5 (26) knots. You could also find the crosswind component for this particular situation by multiplying 0.50 times 30 which would give an answer of 15 knots.

The take-off figures in the Airplane Flight Manual are based on no-wind conditions unless specifically stated otherwise.

Ø The Angle That the Wind Makes With the Runway	Cos Ø Multiplier for Headwind Component	Sin Ø Multiplier for Crosswind Component
0° (straight down the runway)	1.0	0
10°	1.0	0.15
15°	0.95	0.25
20°	0.90	0.35
30°	0.85	0.50
40°	0.75	0.65
45°	0.70	0.70
60°	0.50	0.85
75°	0.25	0.95
90° (Direct Crosswind)	0	1.0

Fig. 4-5. Headwind and crosswind components for the wind at various angles to the runway (rounded off).

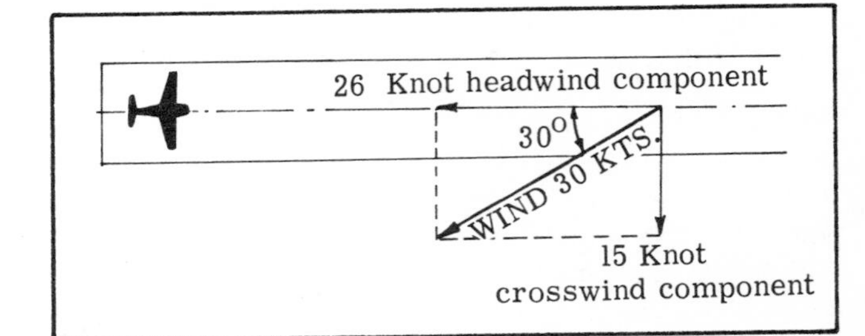

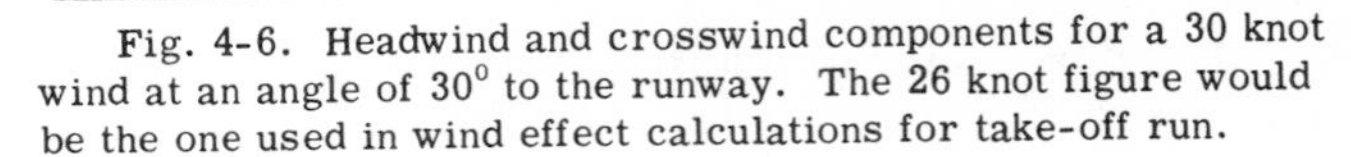

Fig. 4-6. Headwind and crosswind components for a 30 knot wind at an angle of 30° to the runway. The 26 knot figure would be the one used in wind effect calculations for take-off run.

RUNWAY SLOPE

This one is hard to handle. If you have figures on the slope of the particular runway you're using and are handy with mathematics (as well as having plenty of time), you can figure it out. Obviously if you are taking off uphill, more runway will be required; downhill, less. It's factors like runway slope that shoot some beautiful take-off calculations right out the window. For safety, add 10% for an uphill run.

There is always the question of whether to take-off uphill and upwind or downhill and downwind. This depends on the wind and runway slope, of course. If you are operating from the average hard surfaced airport, where slopes are held to within certain maximum allowable values it is better to take-off into the wind and uphill *if the headwind component is 10 per cent or more of your take-off speed.* On off-airport landings or at small airports where slopes could be comparatively steep you'll have to make your own decision according to the conditions.

PILOT TECHNIQUE

The figures for take-off distance in the Airplane Flight Manual are for "average techniques" (whatever this means). The pilots running take-off tests are experienced test pilots and get better performance than a 75-hour private pilot, but these published figures at least give some basis for comparison. The disheartening thing is that take-off techniques vary between pilots. In fact, it's very doubtful whether you could make two consecutive take-offs just alike even though runway and wind conditions were exactly the same.

Assume that you have carefully calculated pressure altitude and temperature effects, Weight, and wind effects before take-off — and then use sloppy techniques. The effect is as if you measured something carefully with a micrometer, marked it with a blunt crayon, and then cut it off with an axe.

Nevertheless, you *can* have some idea of the effects of variables on the take-off run and will have a rough estimate of how much distance you'll need.

Assume an airplane with a constant speed propeller grosses at 3000 pounds and according to the Flight Manual requires 1000 feet to break ground at sea level on a standard day (no wind). Suppose the pilot finds that the pressure altitude for the airport on that day is 3400 feet. (The airport may actually be only 3000 feet above sea level but due to local low pressure conditions the pressure altitude is higher than the actual elevation.) The temperature is 76° F and the wind is 30° to the runway at 10 knots. The airplane's power-off stall speed for the take-off configuration is 60 knots. The take-off speed is generally considered to be 1.2 times the power-off stall speed, so 72 knots will be the take-off speed to be used for these calculations. Assume that the runway is hard surfaced and level. The plane weighs 3300 pounds because, although he knows that it's illegal to be over gross, the pilot wants to take an extra passenger so he's wedged in the back seat (plus some extra baggage).

The following steps will apply:

1. *Pressure Altitude and Temperature Effects*— Note that the pressure altitude is 3400 feet (standard temperature for this altitude is 47° F or 8° C) and the temperature is 76° F (24° C) or 29° F higher than normal. Remember that each added 15° F or $8\frac{1}{2}$° C above standard equals another 1000 feet of density altitude. The density altitude is 2000 feet higher (rounded off) than the pressure altitude. *The airplane will perform at a density altitude of 5400 feet, no matter what the altimeter says.* Using Figure 4-1A you will find that the take-off roll multiplier is 1.56. *The take-off roll at this point will be 1560 feet* (call it 1600 feet).

2. *Weight Effects* —The airplane's Weight is $\left(\frac{\text{Present Weight}}{\text{Gross Weight}}\right)^2 = \left(\frac{3300}{3000}\right)^2 = (1.1)^2 = 1.1 \times 1.1 = 1.21$, $1.21 \times 1600 =$ *1936 feet now needed to break ground.*

3. *Wind Effects* — Assuming that he is going to take-off into the wind (although anyone who overloads an airplane so much might decide to take-off downwind just for the heck of it), the multiplier for the headwind component for a wind at an angle of 30° to the runway is 0.85. (Fig. 4-5) 0.85 x 10 = 8.5 knots. If your airspeed indicator is in mph, you can convert this wind speed to mph by multiplying by 1.15. 1.15 x 8.5 = 9.8 mph (call it 10 mph).

In order to use Fig. 4-4 the ratio of wind velocity and take-off velocity is needed. $\frac{\text{Vwind}}{\text{Vtake-off}} = \frac{8.5 \text{ knots}}{72 \text{ knots}}$ = 0.118 or 0.12. Referring to Fig. 4-4 you find that the ratio $\frac{\text{Take-off Distance (wind)}}{\text{Take-off Distance (no wind)}} = 0.80$. So finally: 0.80 x 1936 = *about 1450 feet runway required at a pressure altitude of 3400 feet, temperature of 76 °F, weight of 3300 pounds and a wind of 10 knots at an angle of 30 degrees to the runway. This doesn't take into account runway slope or bad pilot technique.* Another thing to remember is that tall grass, snow, or mud could possibly double this distance.

What about the climb? It's affected also. At sea level you can expect the distance (after lift-off) to clear a 50-foot obstacle to be about 80 per cent of the take-off roll. The sample airplane with a take-off roll of 1000 would likely require another 800 feet to clear a 50-foot obstacle. The ratio increases with increase in density altitude, but the airplane with the fixed pitch propeller suffers more.

Look at Figure 4-1B: At 5400 feet a constant speed propeller airplane will have a rate of climb of 0.74 (74%) of that at sea level or the distance required to clear a 50 foot obstacle is well increased. (More about this in the next chapter.)

Obviously you're not going to do a mathematical exercise every time you want to take-off, but you do want to have some idea of the factors involved. Carry each computation down to the next one.

RULE OF THUMB TAKE-OFF VARIABLES FOR UNSUPERCHARGED ENGINES

1. *Add 12 per cent to the take-off run, as given in the Airplane Flight Manual for sea level, for every thousand feet of pressure altitude at the take-off point.*

2. *Add 12 per cent to the above figure for every 15 ° F or $8\frac{1}{2}$ ° C above standard for the field pressure altitude.*

3. *Weight Effects*— The Airplane Flight Manual take-off figures are for the gross Weight of the airplane unless otherwise stated. The take-off run is affected approximately by the square of the Weight change. In other words — any given Weight change causes a two-fold change in take-off run. For example, a 10 per cent Weight change would cause a 20 per cent change in length of take-off run.

4. *Wind Effects*— The ratio of wind velocity (down the runway) to your take-off speed (in percentage)

subtracted from 90 per cent will give the expected ratio of *runway length needed to break ground.* If wind = 20 per cent of take-off velocity: 90-20 = 70 per cent. You'll use 70 per cent of the runway distance as given by the Airplane Flight Manual for no-wind conditions. This rule of thumb can be used if you don't have Fig. 4-4 handy. Consider any headwind component less than 5 knots to be calm and to have no effect on take-off run if you use the just mentioned rule of thumb.

THE NORMAL TAKE-OFF

TRICYCLE GEAR

For computing take-off distance, engineers use a speed at lift-off of 1.2 times the power-off stall speed. You know, as a pilot, that you can lift a plane off at stall speed, but this is usually reserved for special occasions such as soft field take-offs. The 1.2 figure is recommended to preclude your getting too deeply in the backside of the Power Required Curve.

If you pull the plane off in a "cocked up" attitude and try to climb too steeply, you will use a great deal of your power just keeping the plane flying, much less getting on with your trip.

This too nose-high attitude is particularly critical for jet take-offs. Jet pilots have pulled the nose too high during the take-off run and found themselves running out of runway with the plane having no inclination to make like the birds. In fact, a hundred-mile-long runway wouldn't be any better — you'd still be sitting there waiting for something to happen when there was no more runway left.

The prop plane has more horsepower to spare at low speeds, and will accelerate out of the bad situation more quickly. But accelerating takes time and distance, and if there are obstacles off the end of the runway, you may wish you had shown a little more discretion in your method of lift-off. This problem is more evident in propeller planes of high power loadings (power loading = $\frac{\text{Weight}}{\text{Horsepower}}$).

Again referring to "the back side of the power curve," it's important on approaches also. You can get your airplane so low and slow on final that by the time you can accelerate or climb out of this condition, you may have gone through a fence and killed somebody's prize Hereford bull a good quarter of a mile short of the runway. You can get out of this region, but it may take more distance than you can spare at the moment.

If you feel that the nose is too high on the take-off roll and the plane isn't accelerating as it should, then the nose must be lowered and the plane given a chance to accelerate — even though you don't particularly want to do this as there's not much runway left.

In an airplane that has an effective elevator and a not-so-effective rudder, it doesn't make for ease of mind if you yank the nosewheel off at the first opportunity, particularly in a strong left crosswind. On the other hand, nothing seems quite so amateurish as the pilot who runs down the runway hell-for-leather on all three wheels until the tires are screaming and then jerks it off. Just because it's a tricycle gear doesn't mean that it should be ridden like a tricycle. The best normal take-off in a tricycle gear plane is one in which after a proper interval of roll the nosewheel is gently raised and shortly afterward the airplane flies itself off. No book can tell you just when to raise the nosewheel for all the different airplanes. *In fact, no book can tell you how to fly as you are well aware of by now.*

Fig. 4-7. Sometimes you don't have time for computing.

Fig. 4-8. Haste can make waste.

Although retractable gear will be covered in detail later, there are a couple of points to be mentioned. Too many pilots have the idea that it makes them look "hot" to pull up the gear as soon as the plane breaks ground. Airplanes have been known to settle back on the runway after a take-off. The plane doesn't roll too well with the gear partially retracted. Don't retract the gear too soon, even if you are *definitely* airborne. That is, don't pull up the gear until you have reached the point where in case of engine failure the airplane can no longer be landed on the runway wheels *down*. It would be very embarrassing to use five hundred feet of a ten thousand foot runway and then have to belly it in when the engine quit because you didn't have time to get the gear back down. (Fig. 4-8)

COMMON ERRORS (TRICYCLE GEAR)

1. Holding the nosewheel on and jerking the airplane off the ground (a less violent technique similar to this is good for take-offs in strong crosswinds but some people do it under all wind and weather conditions).
2. The other extreme: pulling the nose up too high, too early, increasing chances of poor directional control in a crosswind, plus extending the take-off run. A typical case is one in which the pilot pulls the nosewheel off by brute force before it's ready, and has to apply lots of right rudder. The nose, not ready to stay up, falls back down with nosewheel cocked. This makes for funny feelings.

TAKE-OFFS FOR TAILWHEEL TYPES

Things being as they are today, it may be that you haven't checked out in a tailwheel airplane yet. For a long time there were few if any tricycle gear trainers and everybody learned to fly in airplanes with tailwheels. Now the trend is reversed and nearly all the new planes have tricycle gear. This section is presented to give you a little background should you get a chance to check out in the tailwheel type.

The take-off roll may be considered to be broken down into three phases.

PHASE 1

This phase usually gives the most trouble to the pilot checking out in this type. It's slightly harder to go from tricycle gear to tailwheel type than vice versa, but you'll have no trouble after the first few take-offs and landings.

The big problem in this phase seems to be the inability to see over the nose at the beginning of the take-off roll. You've been used to looking directly over the nose all through the tricycle gear take-off and this may be a habit hard to break. You'll have to get used to looking down the side of the nose in most cases, although some tailwheel airplanes have low nose positions comparable to tricycle types.

Ease the throttle open. This is important for the tailwheel airplane because (1) the rudder-tailwheel combination has less positive control at low speeds and the sudden application of power causes torque effects that could make directional control a problem at first and (2) the high nose makes it harder to detect this torque-induced movement and may delay your corrective action.

The tailwheel will be doing most of the steering with the rudder becoming more effective as the airspeed picks up. In a high-powered, propeller airplane it may take a great deal of rudder to do the job at the beginning of the take-off run.

The elevators should be left at neutral, or slightly ahead of neutral because you don't want to force the tail up too quickly and lose directional control.

You may for the first take-off or two have a tendency to "walk the rudder." You've had enough experience by now to recognize this mistake and correct it yourself.

PHASE 2

As the plane picks up speed, allow (or assist slightly) the tail to come up until the airplane is in the attitude of a shallow climb. When the tail comes up, tailwheel steering is lost and the rudder itself is responsible for keeping the airplane straight. This means that added rudder deflection must make up for the loss of steering of the tailwheel. Your biggest steering problems will be at the beginning of the take-off and the point at which the tail comes up.

If the tail is abruptly raised, our old friend "precession" has a chance to act. The rotating propeller makes an efficient gyro, and when the tail is raised, it is as if a force were exerted at the top of the propeller arc, from the rear. The airplane reacts as if the force had been exerted at a point 90° around the propeller arc, or at the right side. (Fig. 4-9) The precession force is added to the torque forces and could cause a left swerve if you are unprepared.

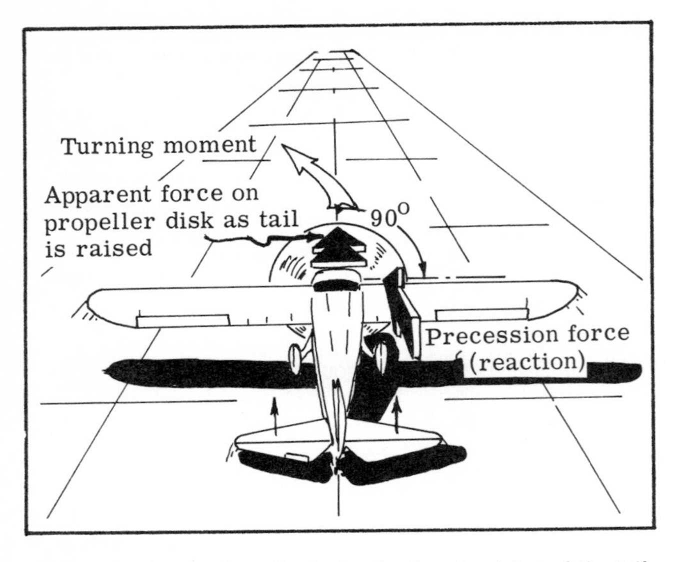

Fig. 4-9. Precession effects due to abrupt raising of the tail.

PHASE 3

The airplane is now in a shallow climb attitude and will fly itself off. However, you can help a little by slight back pressure. The rest of the take-off and climb are just like the procedures you've been using with the tricycle gear plane.

THE SHORT FIELD TAKE-OFF

The short field take-off is used in conjunction with the climb to clear an obstacle. Only the ground roll will be covered in this section; the maximum angle or obstacle climb will be covered in the next chapter "The Climb."

The airplane accelerates best when in the air if it is not "stalled off." The comparative amount of acceleration between the airplane at a given speed on the ground and airborne naturally depends on the surface. The airplane is ready to fly before the average pilot wants to, and he generally uses more runway than is necessary in becoming airborne. This is good under normal take-off conditions as well as in gusty air, as it gives a margin of safety at lift-off. *But* in the case of the short field take-off, you don't have the runway to spare.

Some pilots are firm believers in the idea of holding the plane on until the last instant and then hauling back abruptly and screaming up over the obstacle. This makes the take-off look a great deal more flashy and difficult than if the pilot had gotten the plane off sooner and set up a maximum angle climb. The rate of climb is a function of the excess horsepower available, as will be discussed in the next chapter. In the maximum angle climb, you are interested in getting *more altitude per foot of ground covered,* rather than the best rate of climb. This is a compromise between a lower airspeed and lower rate of climb in order to clear an obstacle at a specific distance. True, you're not climbing at quite as great a rate, but on the other hand, because of the lower speed, you're not approaching the obstacle as quickly either.

The recommended take-off procedure is close to the soft field technique. Get the plane airborne as soon as possible without stalling it off. In most cases the airplane will be accelerating as it climbs. Only in underpowered and/or overloaded airplanes will it be necessary to definitely level off to pick up the recommended maximum angle climb speed.

USE OF FLAPS

Manufacturers recommend a certain flap setting for the short field take-off because they have found that this flap setting results in a shorter take-off run and better angle of climb. For some airplanes, the manufacturers recommend that no flaps be used on the short field take-off and for best performance it would be wise to follow the recommendations as given in the Airplane Flight Manual.

There are two schools of thought on the technique of using flaps for a short field take-off. One idea is to use no flaps at all for the first part of the run and then apply flaps (generally full flaps) when the time seems ripe.

This technique generally disregards the fact that there is rolling resistance present. The first part of the roll usually is made with the airplane in a level flight attitude, the pilot counting on the sudden application of flaps to obtain lift for the take-off. There is no doubt that aerodynamic Drag is less in the level flight attitude than in a tail low attitude, but rolling resistance is the greatest factor in the earlier part of the run and this is often overlooked. If the field is soft, an inefficient and perhaps dangerous (particularly in the tailwheel type) condition may be set up because Weight is not being taken from the main wheels and a nosing-up tendency is present.

The ideal point at which to lower flaps may differ widely between pilots.

The effect of full flaps on the obstacle climb (unless the plane design calls specifically for the use of full flaps) results in a low Lift to Drag ratio — that is, an increase of proportionally more Drag for the amount of Lift required and the climb angle suffers.

There is usually a loss in pilot technique (particularly in the tailwheel type) during the flap lowering. The pilot has to divide his attention between the take-off and flap manipulation. This can be overcome as the pilot becomes more familiar with his airplane, but in most cases the flap handle is located in an awkward position or requires attention to operate. The flaps generally are designed to be operated at a point where the pilot can direct his attention to them if necessary — that is, *before* take-off, on the base leg or on final. The take-off itself requires more attention than these other procedures. The pilot attempting to deflect the flaps the correct amount during the take-off run, may get two notches instead of three, or over- or undershoot the desired setting if the flaps are hydraulically or electrically actuated.

The best method is to set the flaps at the recommended angle *before* starting the take-off run and then forget about them until the obstacle is well cleared. Fly the airplane off and attain, *and maintain,* the recommended climb speed. *After* the obstacle is cleared, ease the nose over slightly and pick up airspeed until you have a safe margin to ease the flaps up. Some pilots clear the obstacle, breathe a sigh of relief, jerk the flaps up and grandly sink back into the trees. On some airplanes the Airplane Flight Manual recommended maximum angle of climb speed is fairly close to the power-on stall speed. Although you had a good safety margin with the flaps down, when you jerk them up at this speed at a low altitude, you could have problems in gusty air. (A common error however, is to be overly cautious, with a loss in performance.)

SHORT FIELD TAKE-OFF PROCEDURE
(TAILWHEEL OR TRICYCLE TYPE)

1. Before take-off, use a careful pre take-off check and a full power run up to make sure the engine is developing full power. Don't waste runway, start at the extreme end. Set flaps as recommended by the Airplane Flight Manual.
2. Open the throttle wide (smoothly) as you release the brakes. (Don't wait until the throttle is completely open before releasing the brakes.)
3. Keep the airplane straight and avoid "rudder walking" as this slows the take-off.
4. Get the airplane airborne as soon as is safely possible without "stalling it off." This means that the nosewheel of the tricycle gear type is raised as soon as practicable (not *too* high) and on the tailwheel type the attitude is tail low, similar to that of the soft field take-off. In both cases, the airplane is flown off at a slightly lower than usual airspeed.
5. As soon as the plane is definitely airborne, retract the landing gear (if so equipped).
6. Attain and *maintain* the recommended maximum angle climb speed as given in the Airplane Flight Manual. Continue to use full power until the obstacle is cleared.
7. Assume a normal climb. (The climb will be discussed later.)

Some pilots argue for a 90° rolling take-off, but this is hard on tires and landing gear assemblies for very little, if any, gain. It's agreed that the fixed pitch prop is inefficient at low speeds (Chapter 1) but you might ground loop using the 90° run technique.

COMMON ERRORS

1. Poor directional control when the brakes are released.
2. Trying to hurry the plane off the ground — resulting in high Drag and actually slowing the take-off.
3. Holding the plane down after breaking ground, letting the airspeed pick up past the maximum angle of climb speed and losing climb efficiency.

THE SOFT FIELD TAKE-OFF

(Use flaps as recommended by the Airplane Flight Manual)

Maybe the only soft field take-off procedures you've used so far were the simulated ones practiced for the private flight test. But sooner or later you'll find yourself in a spot where the field may be too soft for a normal take-off.

Generally speaking, mud, snow, and high grass can be considered to fall into the category where a special take-off technique is required. This same technique is useful on a rough field where it's better to get the plane off as soon as possible to minimize chances of damaging the landing gear.

TRICYCLE GEAR

Keep the airplane rolling. If you stop to think things over in the middle of the take-off area, you may find yourself watching the wheels slowly sinking in the muck. It may take full power to even move once you've stopped. The propeller will pick up mud and gravel and throw it into the stabilizer. This doesn't help either the prop or the stabilizer.

Rolling resistance is high, so you will want a tail-low attitude on the take-off run to help overcome this resistance.

1. The airplane is kept rolling, full throttle is applied and the wheel (or stick) is held back for two reasons: (a) to get the Weight off the nosewheel which will decrease rolling resistance as well as lessen chances of the nosewheel hitting an extra-soft spot and being damaged and (b) to increase the angle of attack as soon as possible so that the Weight on the main wheels is minimized.
2. As soon as the plane is definitely airborne, lower the nose and establish a normal climb.

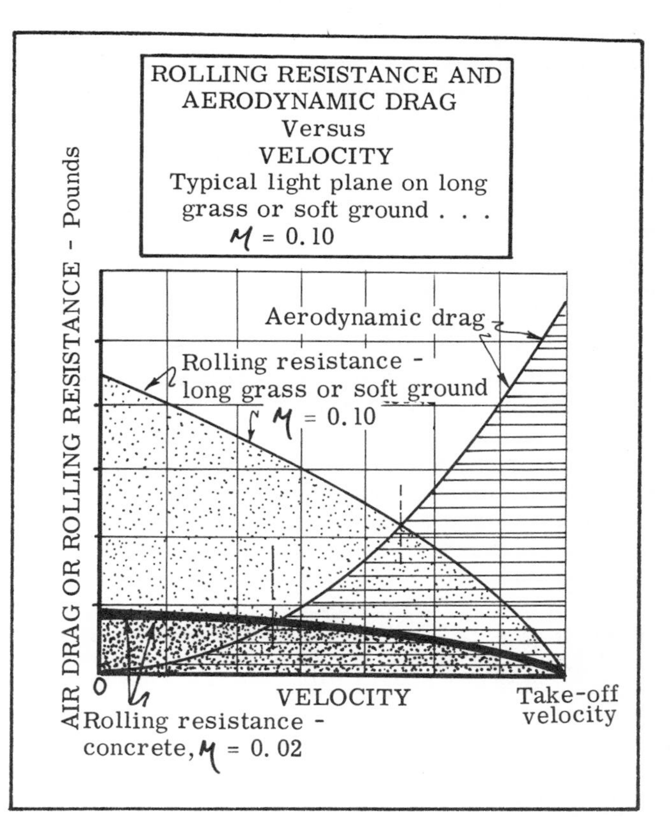

Fig. 4-10. A comparison of rolling resistance for an airplane on dry concrete and on long grass or soft ground. Aerodynamic drag is equal in each case.

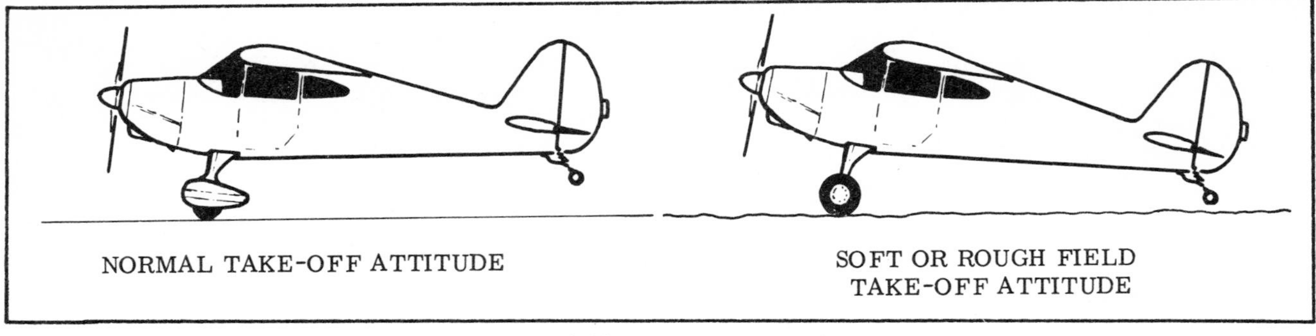

Fig. 4-11. Take-off attitudes.

COMMON ERRORS (TRICYCLE GEAR)

1. Not keeping the airplane moving as it is lined up with the take-off area, requiring a great deal of power to get rolling again and increasing the possibility of prop damage.

2. Not enough back pressure. Most pilots tend to underestimate the amount of back pressure required to break the nosewheel from the ground at lower airspeeds.

Of course, you may be able to get the nose too high and suffer the same results as found in the normal take-off under these conditions — that is, no take-off at all.

TAILWHEEL TYPE

The soft field take-off is a little more touchy with the tailwheel airplane because of the greater chances of nosing up if a particularly soft spot, deeper snow, or higher grass is suddenly encountered. The tailwheel airplane has some advantage in that there is no large third wheel to cause added rolling resistance. The tailwheel is small and has comparatively little Weight on it to cause rolling resistance. *But* this could cause a nose up in a situation where the tricycle gear plane would have no particular trouble. (It could break the nosewheel, however.)

You had soft field take-offs on the private flight test and, if you used a tailwheel type plane, have a pretty good idea of the technique.

As with the tricycle gear plane, the plane should be kept rolling onto the runway (invariably someone else is coming in for a landing and you'll have to wait anyway, but if you do, try to pick a firmer spot so you won't get mired down). This means that the pretake-off check should be run at a good spot so that after ascertaining that there is no traffic coming in, you can get on the take-off area and get about your business.

THE TECHNIQUE

1. Keep the airplane rolling onto the runway and apply full power in the same manner as for a normal take-off. Don't ram it open!

2. Keep the wheel (or stick) back to stop any early tendencies to nose over. Then:

3. Ease the tail up to a definitely tail *low* position so that (a) the plane doesn't have a tendency to nose up and (b) the angle of attack is such as to get the plane airborne (and the Weight off the main wheels) as soon as possible.

4. As soon as the airplane is definitely airborne, lower the nose and establish a normal climb.

COMMON ERRORS

1. Holding the stick back too firmly and too long, causing added tailwheel rolling resistance as well as aerodynamic Drag.

2. Getting the tail too high, with the danger of nosing up.

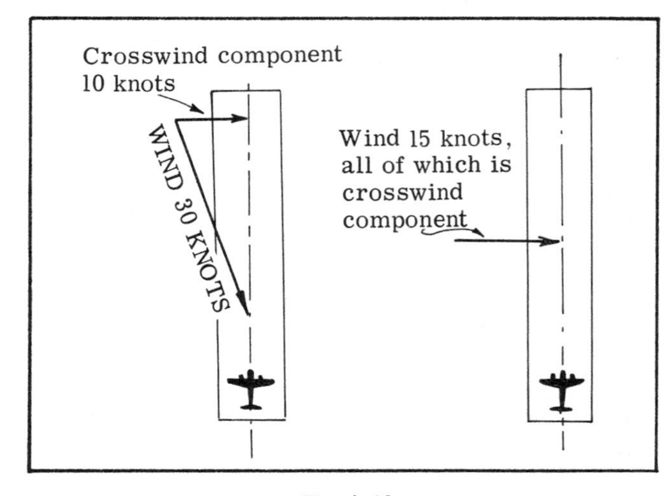

Fig. 4-12.

CROSSWIND TAKE-OFFS

While crosswind take-offs are primarily a function of pilot technique, the matter of airplane performance in the crosswind take-off needs some discussion. The crosswind component is the wind that is actually trying to push you sideways. As can be seen in Fig. 4-12, the crosswind component can be small in a strong wind, large in a weak wind, or very strong in a strong wind.

TRICYCLE GEAR

The wind's effect on both the tricycle gear and tailwheel airplane on the ground generally is the same. That is, the airplane tends to weathercock

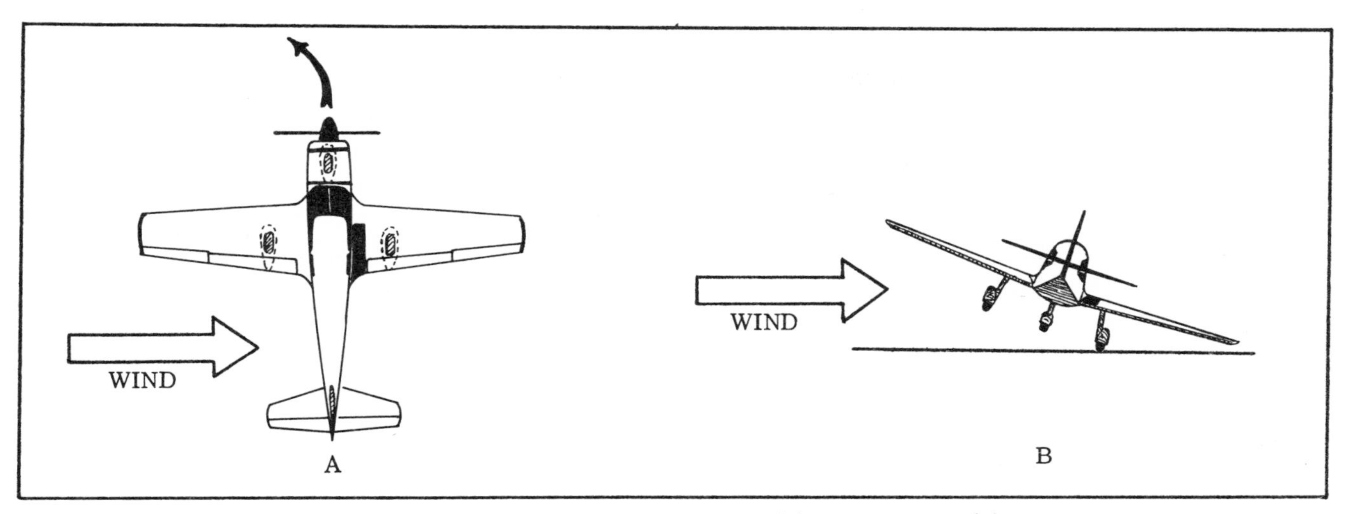

Fig. 4-13. Airplane tends to weathercock (A) as well as lean (B).

into the wind and lean over as well. (Fig. 4-13)

The tricycle gear plane does not have as strong a tendency to weathercock because the nosewheel is large and there is more Weight on it, so there is a greater ground resistance as compared to the tail-wheel type.

In order to make a smooth crosswind take-off you'll have to overcome these wind effects. Some light planes are hard to control when taken off in a strong left crosswind. The weathercocking tendency plus torque effects make it extremely difficult to have a straight take-off run.

PROCEDURE

Line up with the centerline of the runway, or on the downwind side if you think weathercocking will be so great as to make it impossible to keep the plane straight. If the wind is this strong, however, you might be better off to stay on the ground.

Assume, for instance, that there is a fairly strong left crosswind component (10 knots).

It will require conscious effort on your part to apply and hold aileron into the wind. This is probably the one most common fault of the relatively inexperienced pilot. It doesn't feel right to have the wheel or stick in such an awkward position. It is natural that the correction be eased off unconsciously shortly after the run begins. This may allow the wind to lift that upwind wing.

Another common error is the other extreme. The pilot uses full aileron at the beginning of the run and gets so engrossed in keeping the airplane straight that, when the plane does break ground, the full aileron may cause the airplane to start a very steep bank into the wind. This does little for the passengers' peace of mind and is actually useful only if the pilot is interested in picking up a handkerchief with his wing tip.

Notice in Fig. 4-14 that the aileron into the wind actually has two effects: (1) to offset the "leaning over" tendency (the ailerons have no effect in a 90° crosswind until the plane gets moving, though) and (2) the Drag of the down aileron helps fight the weathercocking tendency. Of course, most airplanes these days have differential aileron movement (the aileron goes further up than down) and the down aileron Drag has little effect on the tricycle gear plane because of its take-off attitude. Still, you'll have to fight the leaning tendency and the ailerons will be the answer.

This will be a take-off where you definitely should not rush the airplane into flying. The crosswind take-off feels uncomfortable. It's perfectly natural to want to get the plane airborne and stop all this monkey business. Keep the plane on the ground until you are certain it is ready to fly, then lift it off with definite, but not abrupt, back pressure. If the plane should skip into the air, try to keep it flying, if possible. It will start drifting as soon as you are off the ground, and it won't help the tires and landing gear assembly to hit again when you're moving sideways.

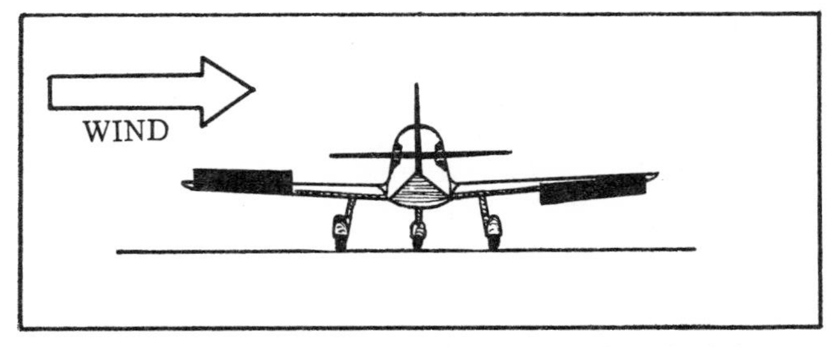

Fig. 4-14. Tricycle gear attitude and use of ailerons to compensate for a crosswind.

In other words, don't try to ease the nosewheel off as in a normal take-off. The weathercocking effect in a large crosswind component may be more than you can handle with rudder alone (no nosewheel help). Again, this is particularly true in a left crosswind. In a right crosswind, torque and weathercocking tend to work against each other. The airplane should be kept on all three wheels until at the lift-off point.

The idea of keeping the plane on the ground longer applies to the tailwheel type for the same reasons as for the tricycle gear airplane. The attitude of the plane during the run will be slightly more tail high than for the normal take-off (which for the tailwheel type, you remember, was the attitude of a shallow climb).

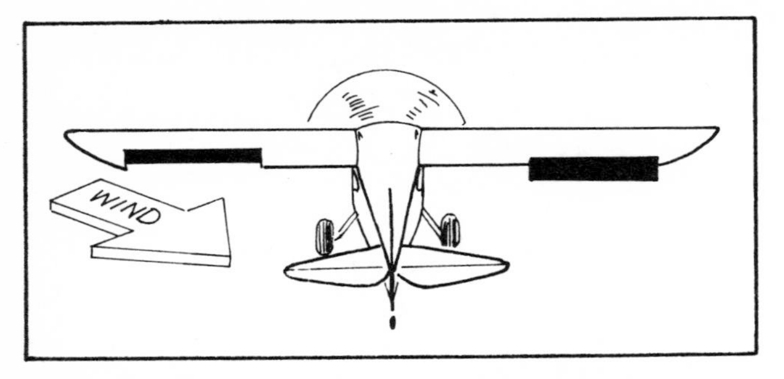

Fig. 4-15. Aileron deflection at the beginning of the crosswind take-off (tailwheel type).

A common failing of pilots in both types of airplanes is that a poor drift correction is set up on the climb-out. There are the hardy but misguided souls who still believe that holding rudder into the wind or holding the upwind wing down is the best way to correct for a crosswind *on the climb-out.*

You know by now that the average plane, if trimmed properly, will make its own take-off, but this may not be the most efficient procedure, particularly under abnormal conditions. Many pilots have never flown their airplanes at gross Weight until one day they attempt it under adverse conditions—and leave an indelible impression on some object off the end of the runway.

Study and know the Airplane Flight Manual. If in doubt, read it again and *above all, always give yourself a safety factor, particularly on Take-offs.*

5. THE CLIMB

An airplane's rate of climb is a function of excess horsepower available. This can be approximated by the equation: Rate of Climb (feet per minute) = $\frac{\text{Excess Horsepower x } 33{,}000}{\text{Airplane Weight}}$ or R/C fpm = $\frac{(\text{EHP})(33{,}000)}{\text{W}}$. (This is Thrust horsepower.)

Looking again at a horsepower required and horsepower available (Thrust horsepower) versus velocity curve for a particular altitude (we'll choose sea level for simplicity) you can note the following: (Fig. 5-1)

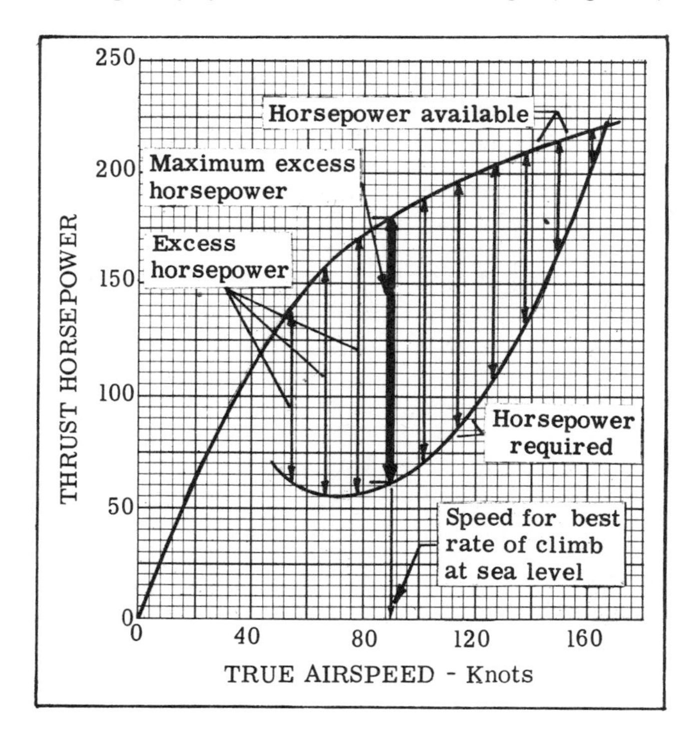

Fig. 5-1. THP Available and THP Required curve for a high performance general aviation airplane at sea level. (Indicated airspeed equals true airspeed at sea level.) The best rate of climb is found at the speed where there is the greatest amount of excess Thrust horsepower available. (Airplane Weight – 3,000 lbs.)

At some particular velocity, the difference between the Thrust horsepower available at recommended climb power and the horsepower required is the maximum. By looking at this graph for a particular airplane, you would immediately know the best rate of climb speed at sea level. By having a series of these curves up to the airplane's ceiling, you can find the speed for best rate of climb for any altitude.

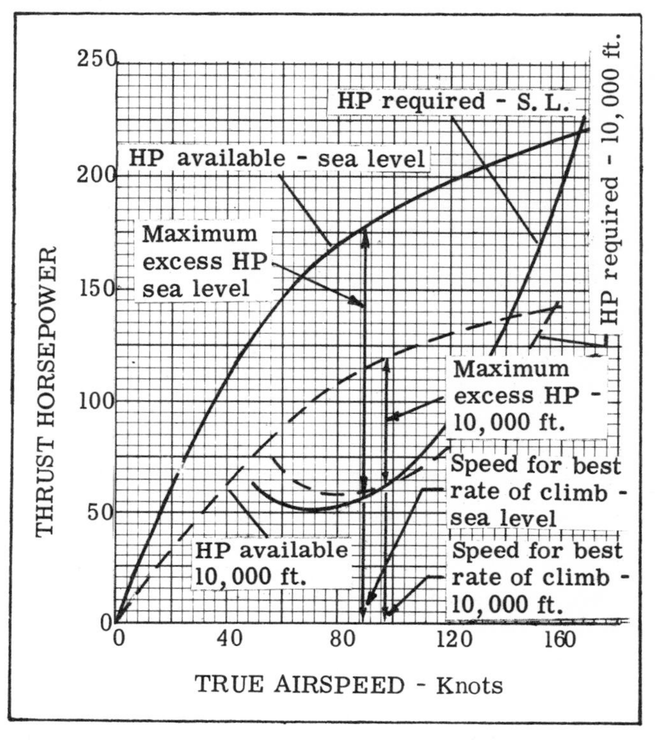

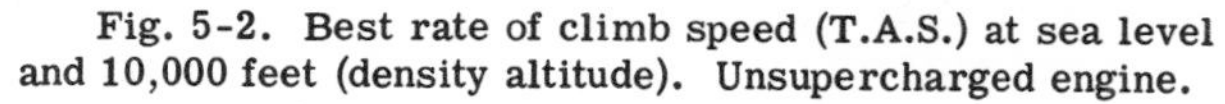

Fig. 5-2. Best rate of climb speed (T.A.S.) at sea level and 10,000 feet (density altitude). Unsupercharged engine.

(Fig. 5-2) Taking a look at Fig. 5-2 you can see that at 90 knots at sea level a power of about 57 THP is required to fly the airplane straight and level and 177 THP is the available power, or an excess of 120 THP exists. Using the equation for rate of climb: R/C = $\frac{\text{EHP x } 33{,}000}{\text{Weight}}$. (The Weight for this airplane is 3000 pounds.) R/C = $\frac{120 \text{ x } 33{,}000}{3000}$ = 1320 fpm at sea level. The climb angle is found to be $8^{0}15'$ relative to the horizon. If you remember back in Chapter 2 the rate of climb for this fictitious airplane was found to be 1265 fpm because a climb angle of an *even* 8 degrees was set up for the climb speed of 90 knots and the problem was worked "backwards." The extra 55 fpm here shows that the airplane was slightly underrated by that method (and numbers were rounded off).

Checking the rate of climb at 10,000 feet at the best climb speed (given as T.A.S. here but the pilot would use I.A.S., which computed for the T.A.S. of 97 knots at 10,000 feet would work out to be about 84 knots). The THP required is about 62 and the THP available can be seen as 120 from Fig. 5-2, an excess of 58 THP. $R/C = \frac{58 \times 33{,}000}{3000} = 638$ fpm, a reasonable figure to be expected at that altitude for the airplane of this Weight and THP required.

Some Airplane Flight Manuals include graphs of the best rate of climb speeds for various altitudes (indicated or calibrated airspeeds). (Fig. 5-3)

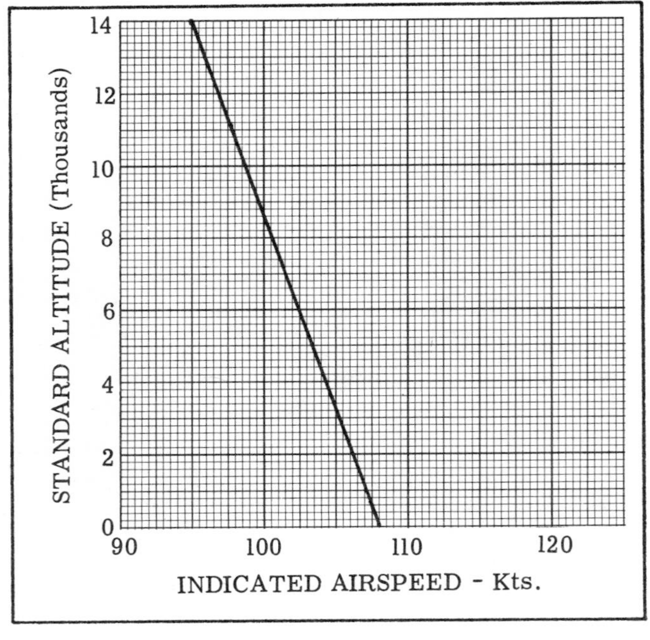

Fig. 5-3.

In cases where a graph like Fig. 5-3 is unavailable for lighter planes such as two-place trainers, which have a simplified type of Airplane Flight Manual, a rule of thumb may be applied. The standard sea level recommended best rate of climb speed is available in these abbreviated Flight Manuals and the idea of maintaining a nearly constant true airspeed may be used. That is, the true airspeed and the calibrated airspeed are the same under sea level standard conditions. You can find the indicated airspeed by *subtracting* 1% per thousand feet, from the published figure. Note that in order to maintain the same true airspeed at various altitudes the rule of thumb would be to subtract 2% per thousand feet from the sea level indicated climb speed but the best rate of climb *true* airspeed increases about 1% per thousand feet. (We're assuming no instrument error.)

The 1% per thousand feet decrease in *indicated* airspeed takes care of this. It must be repeated that the rule of thumb is for light trainers, but it also works pretty well for heavier airplanes. You couldn't go too far wrong by maintaining the same indicated airspeed throughout the climb in the trainer, although a slight loss in efficiency would result. This type of plane normally does not operate much over 5000 feet anyway.

There are two *ceilings* commonly mentioned: (1) *Service Ceiling*— that altitude at which the rate of climb is 100 feet per minute while (2) *Absolute Ceiling* is the absolute altitude the plane can reach, where the rate of climb is zero. These ceilings are normally based on gross Weight but can be computed for any Weight. If you want to come right down to it, the absolute ceiling as a part of a climb schedule at gross Weight could never really be reached. In the first place, the airplane would be burning fuel and getting lighter as it climbed so wouldn't reach the absolute ceiling at the correct Weight and secondly, even if the Weight could be kept constant the situation would be somewhat like that of the old problem of the frog two feet from a wall, who jumps one foot the first time, six inches the second, halving the length each hop. In theory, he would never get there. For the airplane, that last *inch* up to the absolute ceiling would take a very long time. The rate of climb for single engine airplanes and light twins with both engines operating is 100 feet per minute at the normal service ceiling but the light twin service ceiling with *one engine inoperative* is listed at the altitude at which the rate of climb is 50 fpm.

The service and absolute ceilings can be approximated by extrapolation. Measure the rate of climb at sea level and at several other altitudes and join these points as is shown in Fig. 5-4.

For instance, you could set your altimeter to 29.92, noting the outside air temperature and the rate of climb for several altitudes. You can then convert your pressure altitude to density altitude and, using a piece of graph paper, determine your absolute and

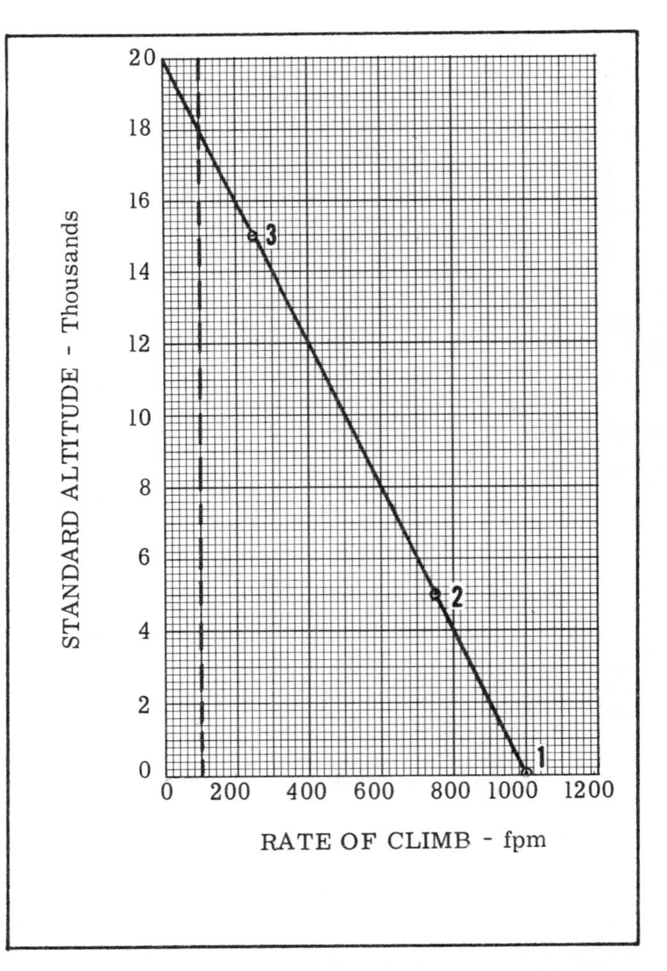

Fig. 5-4. The principle of establishing a rate of climb graph.

service ceilings, correcting for the difference in Weight from the gross Weight. You're not likely to do this, but it would give you some idea of the rate of climb of your plane for various standard altitudes if you are interested. The Airplane Flight Manual lists the rate of climb at sea level and also the service and absolute ceilings (at gross Weight). You can use this information to check your rates of climb at various altitudes by making a graph such as in Fig. 5-5. As an example, suppose your Airplane Flight Manual gives a rate of climb at sea level of 1000 feet per minute and a service ceiling of 18,000 feet. You know that the service ceiling rate of climb is always 100 feet per minute and can now set it up as in Fig. 5-5.

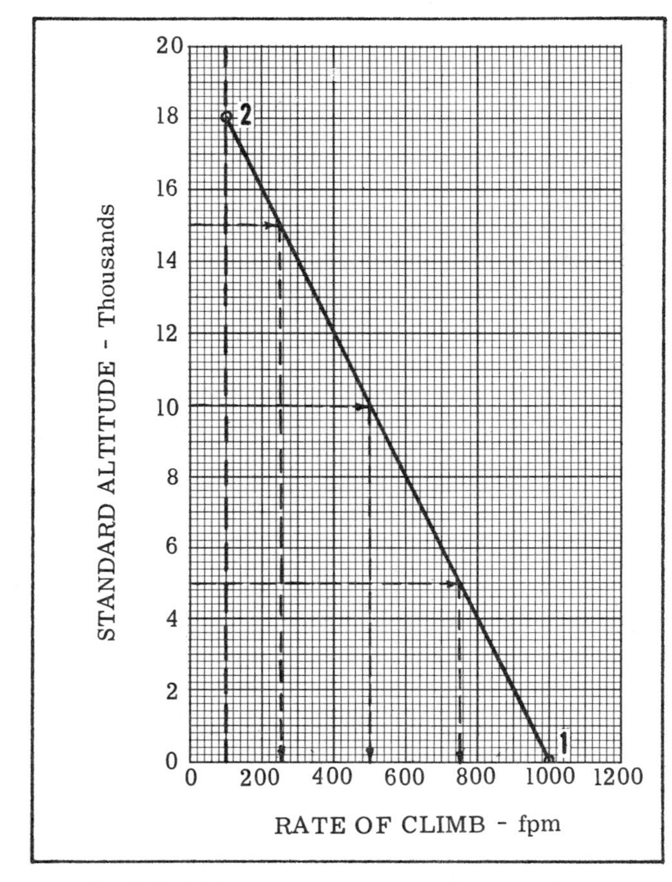

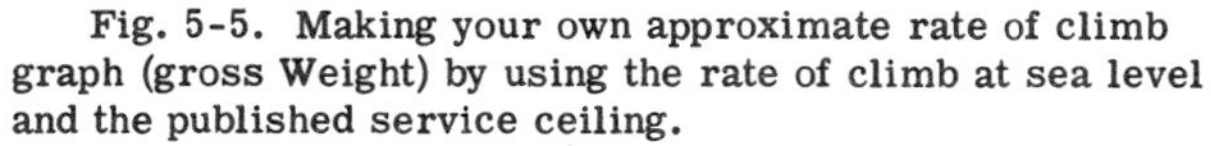

Fig. 5-5. Making your own approximate rate of climb graph (gross Weight) by using the rate of climb at sea level and the published service ceiling.

You can set up points 1 and 2, and by connecting them with a straight line can pick off the rate of climb for any standard altitude. As is shown in Fig. 5-5, your expected rate of climb is about 500 feet per minute at 10,000 feet, etc.

You can also work out a rule of thumb for your airplane. In the example, your rate of climb drops from 1000 fpm at zero altitude to 100 fpm at 18,000 feet. This means a drop of 900 fpm in 18,000 feet and a little division shows a rate of climb drop of 50 fpm for every thousand feet, so that a good approximation of your expected rate of climb at various altitudes is: 1000 fpm at sea level, 950 fpm at 1000 feet, 900 fpm at 2000 feet and so on. You can work out the

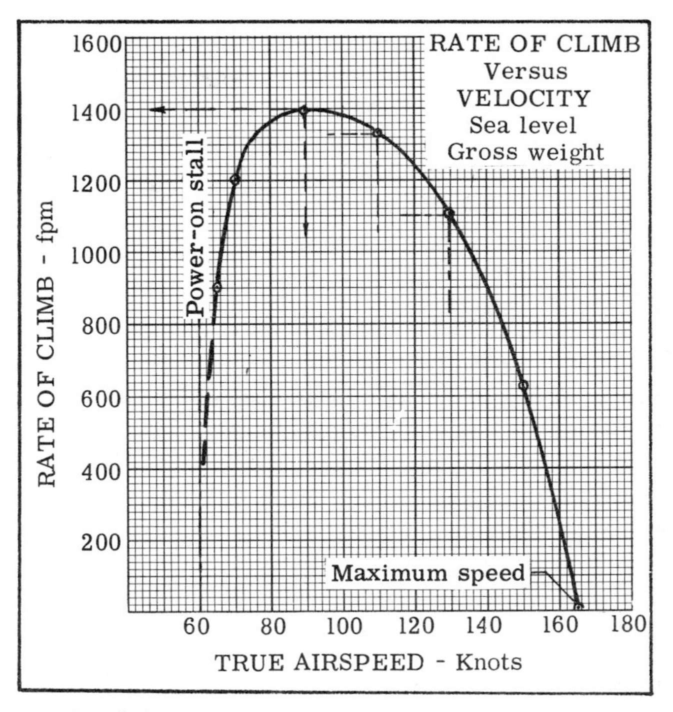

Fig. 5-6. The rate of climb curve is made by noting the rate of climb at various airspeeds and joining the points.

figures for the plane you are flying (but it's doubtful if they would work out as evenly as this "fixed" problem).

Your rate of climb varies inversely with Weight so that if you are flying at a lighter than gross Weight (and know the present airplane Weight) you can correct the Flight Manual figures approximately by the following: $R/C \text{ (corrected)} = R/C \text{ (published)} \times \frac{\text{Gross Weight}}{\text{Present Weight}}$. You can use common sense and reason that if your present Weight is 80% of the gross Weight, the rate of climb will be $\frac{100\%}{80\%} = 1.25$. Your rate of climb is about 1.25 times that of the published figure. (It's not *quite* as simple as that, however.)

The rate of climb depends on excess Thrust horsepower available and this depends on the *density altitude*. Your rate of climb is affected by pressure altitude *and* temperature (which combine to give you density altitude).

There are two important speeds that concern the climb: (1) The speed for the best rate of climb and (2) the speed for the maximum angle of climb.

These speeds are obtained by flight test. Generally the manufacturer will measure the rates of climb at various airspeeds, starting from just above the stall to the maximum level flight speed. This is done for several altitudes and the rate of climb is plotted against the true airspeed and a curve is drawn for each. (Fig.5-6) Corrections are made for Weight changes during the testing process.

You can see that the rate of climb would be zero at speeds near the stall and at the maximum level flight speed because of there being no excess horsepower available. Looking back at Fig. 5-1 you can see that the excess THP drops off to nothing at the

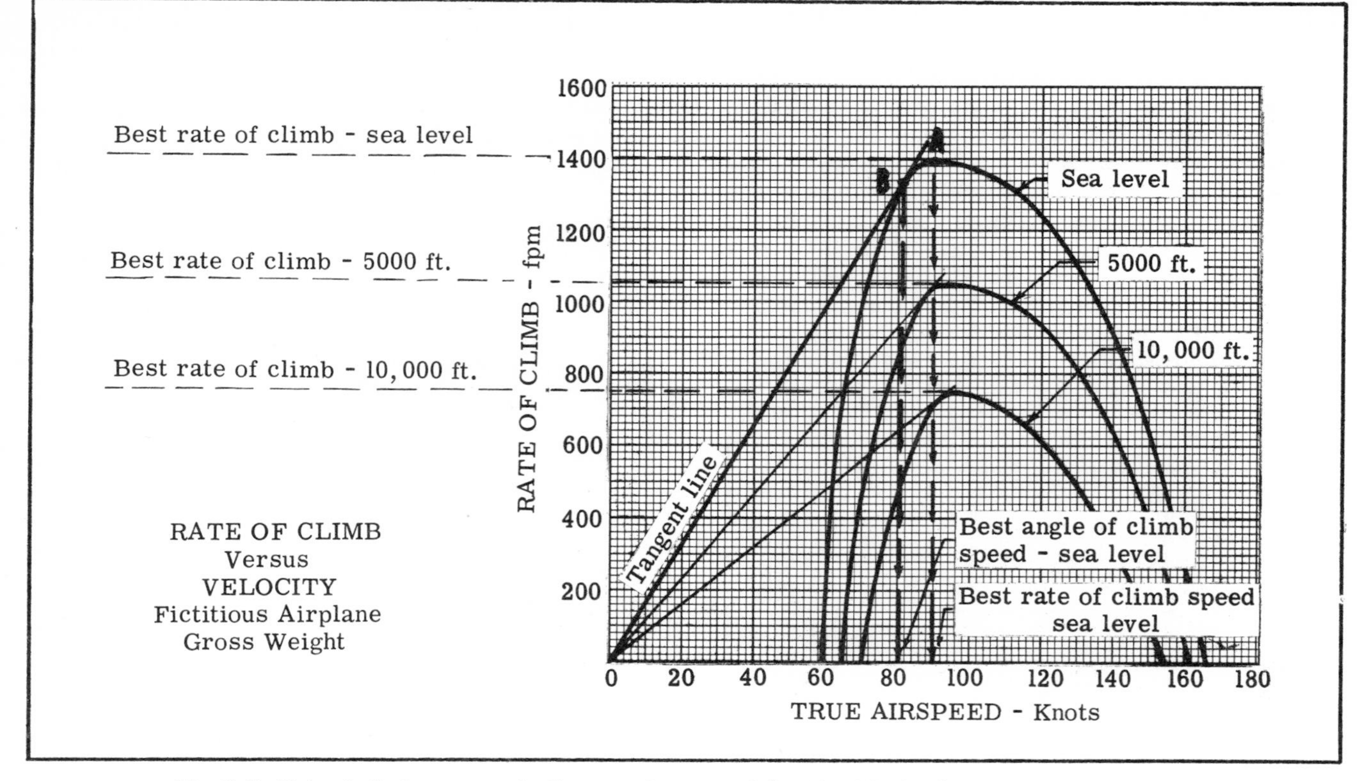

Fig. 5-7. Rate of climb versus velocity curve for several density altitudes for a particular airplane.

maximum level flight speed as all the horsepower is being used to maintain altitude at that speed. At the lower end of the speed range the same thing occurs but the stall characteristics of the airplane may not allow such clear-cut answers in that area.

By looking at the resulting graph after these figures have been reduced to standard conditions and plotted, you can learn the speed for best rate of climb and for the maximum angle of climb. (Fig. 5-7)

The best rate of climb is at the peak of the curve. Reading the velocity below point (A), you can find the speed for the best rate of climb. This is the published figure for sea level in the Airplane Flight Manual.

A line is drawn from the Origin (O) on the graph tangent to the curve. Mathematically speaking this will give the highest ratio of climb to velocity (which means the same thing as the maximum altitude gain per foot of forward flight). The velocity directly below point (B) is the published figure for maximum angle of climb at sea level. Each airplane make and model is tested to find these recommended speeds. The tangent lines for the other two altitudes are shown in Fig. 5-7 and you can see that the angle of climb decreases with altitude as might be expected. This is important to remember in that a 50-foot obstacle that can be cleared easily at the max angle climb at sea level could be a problem at airports of higher elevation (and/or higher density altitude).

Fig. 5-7 is shown in terms of true airspeed to give clearer picture of altitude effects. If the points of maximum angle were connected by a straight line and the same thing was done for the points for the best rate of climb the lines would in theory converge at the same airspeed at the *absolute* ceiling of the airplane (zero rate of climb). In other words, the curves would get smaller and smaller until the "curve" for the absolute ceiling would be a point at some airspeed (and the rate of climb would be zero— all of the power available would be needed to keep from losing altitude at that one and only airspeed). The airspeeds for max angle and best rate get closer to each other as the altitude increases. The required indicated airspeed for best rate *decreases* with altitude, the required I.A.S. for max angle *increases* with altitude and they will (in theory) be the same as the airplane reaches its absolute ceiling. The best rate of climb is always found at a higher airspeed than that for the max angle of climb. (Fig. 5-8)

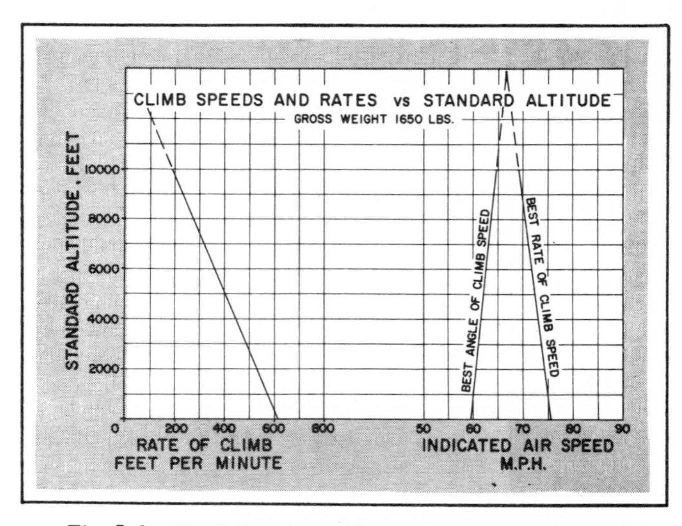

Fig. 5-8. Climb Speeds and Rates versus Standard Altitude for a two-place trainer. *(Piper Aircraft Corp.)*

Fig. 5-8 is a graph of Climb Speeds and Rates versus Standard Altitude for a two-place trainer. For this particular airplane the service and absolute ceilings are given as 12,000 and 14,000 feet respectively. Note how the indicated airspeeds for best (or max) angle and best rate of climb speeds converge at the absolute ceiling. (The same condition would apply if the climb speeds were drawn in terms of T.A.S.)

NORMAL CLIMB

The normal climb is the best rate of climb without overworking the engine. The manufacturer recommends a certain airspeed and power setting for the normal climb. Pilots sometimes get impatient during a prolonged climb and start to cheat a little by easing the nose up. This does nothing more than decrease the rate of climb and strain the engine by decreasing the relative wind's cooling effects. You can review Fig. 5-7 and see that when the speed is varied in either direction from the peak at (A), the rate of climb is not at a maximum. You could lower the nose an equivalent amount and the rate of climb would not suffer any worse than if you raised it — and the engine would be a lot better off!

After take-off, the landing gear is raised and, as the speed approaches the best rate of climb speed, the flaps are raised and the power retarded to the recommended climb setting. As a rule of thumb for airplanes with unsupercharged engines, knock off 1% per thousand feet from your climb *indicated* airspeed. For airplanes up through the light twins this can be considered to be one knot per thousand feet.

CRUISE CLIMB

This is a climb that results in a good rate of climb as well as a high forward speed, and is from 10 to 30 knots faster than the recommended best rate, or normal climb speed. The cruise climb is ideal for a long cross country where you want to fly at a certain altitude but don't want to lose much of your cruise speed getting there. Of course, if it's bumpy at lower altitudes you may want to use the best rate of climb up to smooth air and a cruise climb from that point up to your chosen altitude. An advantage to the cruise climb is that because of the higher airspeed (and better cooling) the engine can be leaned during the climb and more economical operation will result.

Looking back at Fig. 5-7 (the sea level curve) you can see that by climbing at 110 knots the rate of climb is *decreased* by about 60 fpm or about $4\frac{1}{2}$ per cent, while the forward speed (110 as compared to the best rate of climb speed of 90 knots) is *increased* by 22 per cent — a good advantage if you are interested in going places as you climb. Notice that as the climb speed increases past this value the rate of climb begins to drop off at a faster and faster rate. One method of setting up a cruise climb condition for your airplane (particularly for the cleaner, higher performance type) would be to add the difference between the recommended max angle and max rate speeds to the max rate speed. In other words, if the recommended max rate speed at sea level and gross Weight is 90 knots and the max angle speed (same conditions) is 75 knots you would add the difference — 15 knots — to the speed for max rate and come up with a cruise climb speed of 105 knots. Speaking simply, you're operating on the opposite side of the climb curve from the max angle speed but not at such a speed that the rate of climb has dropped radically. (Notice on the curves in Figs. 5-6 and 5-7 that as the airspeed increases above that for best rate the rate of climb decreases at a greater and greater rate per knot.) The speed for cruise climb found this way is at best an approximation and you should, as always, use the manufacturer's figure if available. Lower the cruise climb speed about one per cent per thousand feet as was done for the max rate climb.

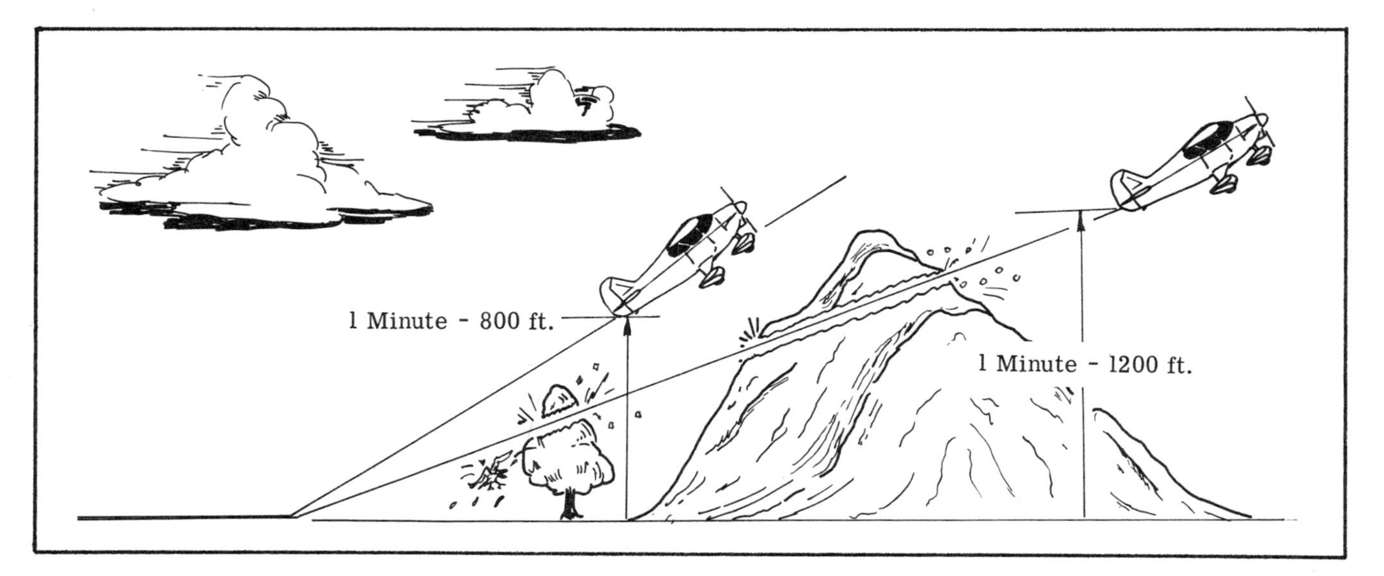

Fig. 5-9. Exaggerated comparison of max angle and best rate of climb.

MAXIMUM ANGLE CLIMB

This climb usually is not an extended one and seldom continues for more than a couple of hundred feet altitude. As soon as the plane is firmly airborne, retract the landing gear, attain and maintain the recommended max angle climb speed. Keep the engine at full power. You'll need all the horsepower you can get, and the short length of time that full power will be used won't hurt the engine. Leave the flaps alone until the obstacle has been well cleared. After you have sufficient altitude and the adrenalin has stopped racing around, raise the flaps (if used), throttle back and assume a normal climb. For the best *angle of climb add* one-half per cent per one thousand feet (about one-half knot) to the *indicated* airspeed.

The max angle of climb is found at the speed at which the maximum excess Thrust is found. It will be found at the speed at which the greatest amount of Thrust component is available to move the airplane upward as compared to its forward motion. The max *rate* of climb was a function of *time* (the max angle is not) and so was dependent on the excess Thrust *horsepower* working to move the airplane upward at a certain *rate*.

6. CRUISE CONTROL–RANGE AND ENDURANCE

CRUISE CONTROL IN GENERAL

Cruise control is an area too often ignored by pilots and unfortunately the performance charts that come with the airplane are often classed with the writing on the walls of a Pharaoh's tomb. This chapter is a general coverage of cruise control and specific *methods* of setting up power will be covered later.

Fig. 6-1 is a typical True Airspeed versus Standard (density) Altitude Curve for a high performance retractable gear, four-place airplane. Point (1) is the maximum level flight speed and, like all airplanes with unsupercharged engines, is found at sea level where the maximum amount of power is available to allow the airplane to maintain altitude at such a high speed.

The normal cruise settings usually vary between 55 and 75 per cent, with 65 and 75 per cent being the most popular. Most engine manufacturers recommend that no continuous power settings of over 75 per cent be used because of increased fuel consumption and added engine wear. You remember from Chapter 1 that these percentages of power are based on normal rated power, or the maximum *continuous* power allowed for the engine (brake horsepower).

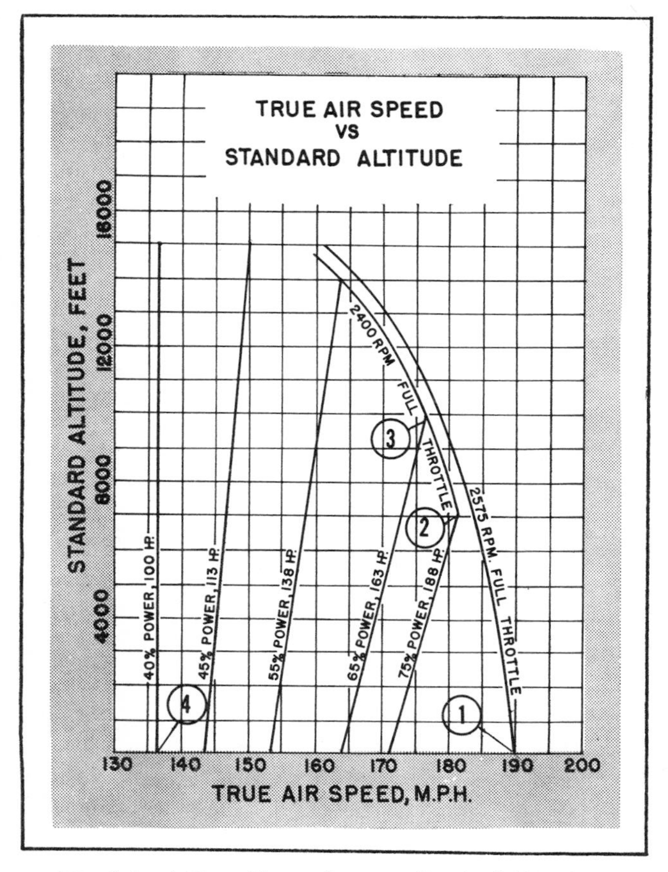

Fig. 6-1. A True Airspeed versus Standard Altitude Chart for a four-place high performance airplane. *(Piper Aircraft)*

Looking at Fig. 6-1 it is noted that for the higher cruise power settings true airspeed is gained with altitude so that if a power setting of 75 per cent is desired it would be best to fly at an altitude of 7000 feet to get the most miles per hour per horsepower – assuming that outside factors such as ceiling, IFR assigned altitude requirements and winds aloft are not considered – Point (2). Above 7000 feet for this airplane even full throttle will no longer furnish the required manifold pressure to maintain 75 per cent power. If 65 per cent power is used note that this setting can be maintained to about 10,000 feet – Point (3) – and this would be the best altitude for that setting. As the desired cruise setting decreases, the best altitude increases, but the loss of time in climbing to the optimum altitude could offset true airspeed gains, particularly on shorter trips. Note that as the cruise power setting decreases from 75 per cent the T.A.S. gain *per thousand feet* also decreases until at 40 per cent – Point (4) – there is no gain shown with altitude increase.

At the higher power settings it can also be expected that because the T.A.S. increases with altitude for a specific power (and fuel consumption) it follows that the range is also increased with altitude for a fixed percentage of power being used – except for one condition, that of maximum range, which will be discussed later.

ESTABLISHING THE CRUISE CONDITION

As the desired altitude is reached there are three main techniques used by pilots in establishing the cruise: (1) As soon as the altitude is reached the power is immediately retarded from climb to cruise setting, (2) Maintaining climb power after level-off until cruise speed is attained and (3) Climbing past the altitude a couple of hundred feet and diving to attain cruise speed and then setting power.

The first technique is the least effective method of attaining cruise (as far as time is concerned) and also means the resetting of power after the area of cruise speed is reached. Because the airplane is slow (at climb speed) when power is set, the increased airspeed as cruise is approached will mean that for the fixed-pitch propeller the rpm will have increased past that desired and, for the constant-speed type, ram effect will have increased the manifold pressure above the original setting. Acceleration to cruise speed is necessarily slow as compared to (2), assuming a constant altitude as was noted.

Climbing past the altitude and diving down to aid in establishing cruise (3) is sometimes used for cleaner airplanes but it is questionable as to whether this method has a perceptible advantage over leaving climb power on until cruise is reached since any comparison of the methods must include starting the timing as the airplane initially passes through the cruise altitude in the climb-past-and-dive-technique.

A term used by pilots is called getting the airplane "on the step," an idea likely taken from seaplane operations. Pilots sometimes say, "I wasn't doing too well at first but as the flight progressed I began to get on the step." Actually they had not encountered a mysterious phenomenon but were following predictable aerodynamic laws.

The "step," by the popular definition, is a condition in normal cruise in which the pilot, by lowering the nose slightly, is able to get several more knots than predicted by the manufacturer. Fig. 6-2 shows a Power Required versus Airspeed curve for an airplane at sea level. Note that there *are* two speeds available for most of lower power settings but for settings used for cruise (55 to 75 per cent) the two speeds are far enough apart so there is no question about the correct airspeed. For instance, if you're carrying 55 per cent power and are maintaining, say, 65 knots — when the airplane normally cruises at 140 at that setting -- you'll know that all is not well. However, if you are carrying 35-40 per cent and indicating 80 knots when you *could* be indicating 90 knots the problem is not so obvious.

But getting back to the idea of "getting on the step" as the flight progresses, Fig. 6-3 shows the power required for a particular airplane at gross

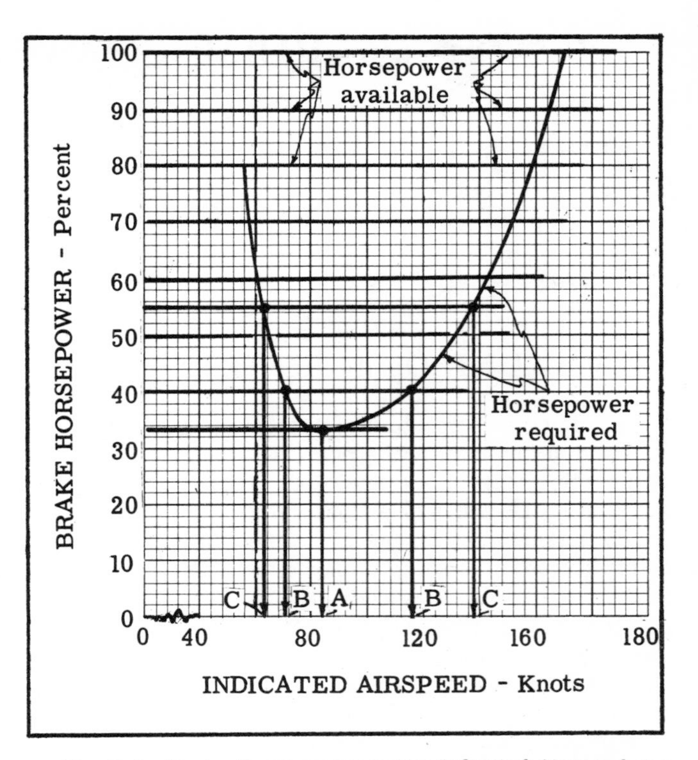

Fig. 6-2. Brake Horsepower versus Indicated Airspeed at a *constant* altitude for a fictitious high performance airplane. This would be what the pilot would obtain by maintaining a constant altitude at various airspeeds. He could make a note of what manifold pressure and rpm was necessary for each airspeed and check the percentage of power he had used after he gets on the ground and refers to a power-setting chart. For instance, "C" represents the two speeds available for this airplane using 55 per cent of normal rated power.

Weight and when it is nearly empty of fuel. The airplane *will* indicate a higher airspeed at the lighter Weight and this is expected and predictable. The difference in airspeeds depends on the ratio of fuel Weight to airplane Weight. Fig. 1-50 also shows this idea.

Down in the area of max range and max endurance, at 35 to 45 per cent power, the speeds are not so far

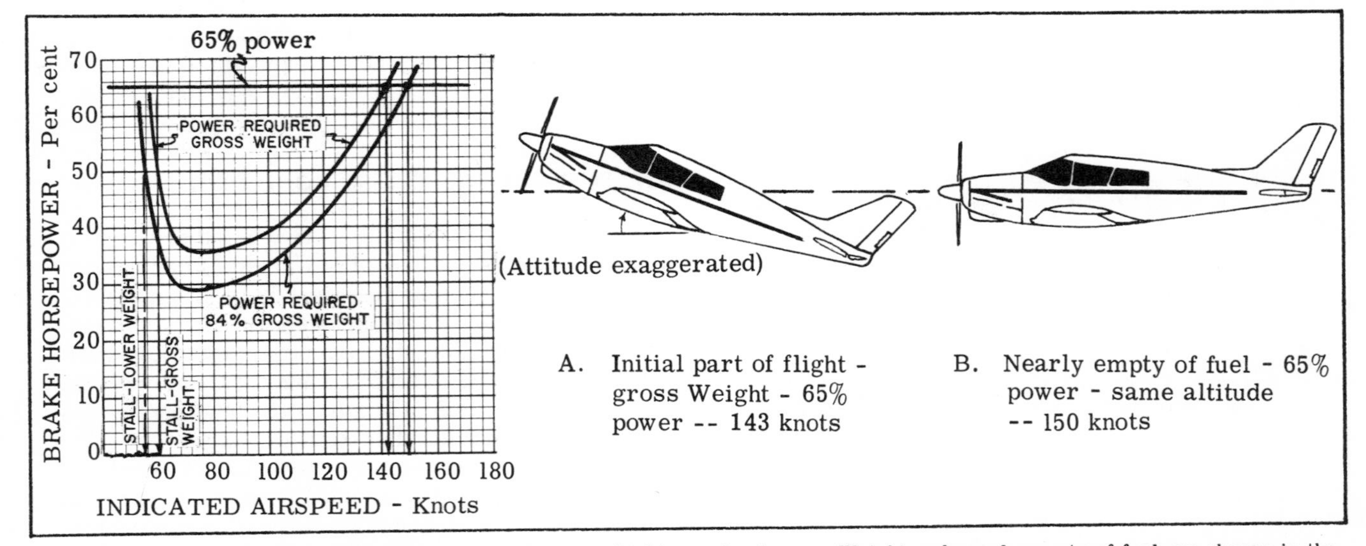

Fig. 6-3. Comparison of BHP required and airplane flight speeds at gross Weight and nearly empty of fuel, no change in the power setting of 65 per cent.

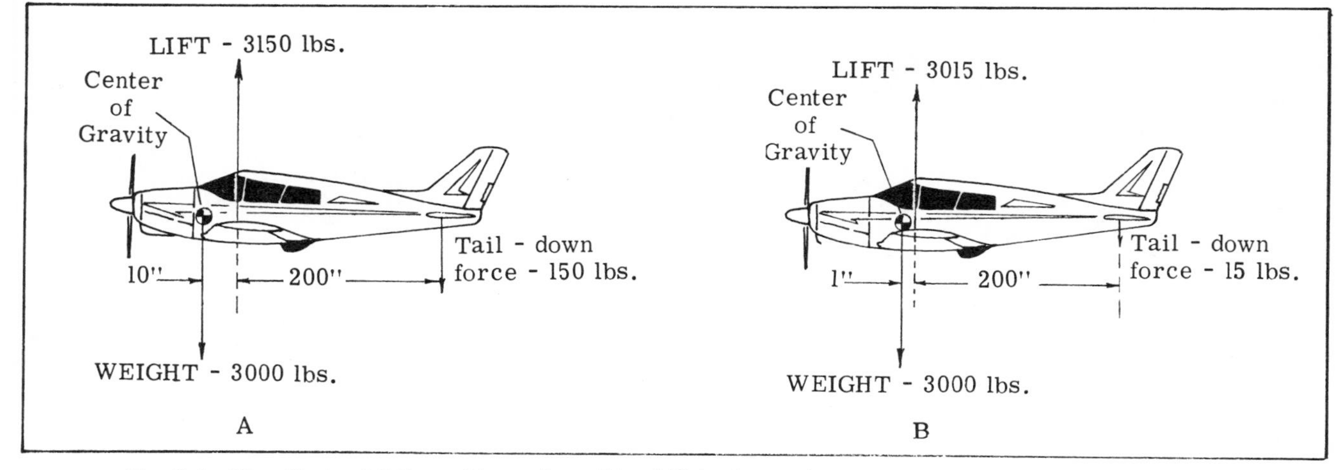

Fig. 6-4. The effects of C.G. position and resulting Lift (and power) required for an airplane in cruising flight.

apart and the pilot might be flying at the low speed when he could be cruising several knots faster (Fig. 6-2). The principle is still the same: the pilot can fly at the higher speed rather than the low, but he *cannot* fly at an even higher speed than the highest airspeed allowed at one of the intersections of the power required/power available curve. The pilot cannot, for example, go faster than the greater of the two speeds as shown by B using 40 per cent power and still maintain a constant altitude. The pilot's problem is that he may not be aware of the higher speed available to him, since both speeds are likely to be close together; the airplane would be quite happy to maintain that airspeed and altitude until Weight changes began to have their effect.

If the pilot is unaware of the two available speeds, he may experiment by lowering the nose, and lo and behold, he seems to have gotten on "the step." It might be hard to convince him he hasn't beaten the game, but still, he has obtained only what is obtainable from the laws of aircraft performance.

Another fallacy of cruising flight, and one often put forth by those of the step school of thought, is that the airplane should be loaded for a forward center of gravity. One reason for this theory of loading might well be that with the C.G. forward, the airplane wants to run "downhill." The pilot doesn't let it lose any altitude, goes the theory, but still benefits in added speed from this downhill running tendency. This is close to that old theory that you don't need an engine in an automobile if the front wheels are smaller than the rear ones.

Actually, the airplane with the C.G. as far back as safely possible is faster. Take two airplanes of the same model and weight, carrying the same power. The one with the more aft C.G. will cruise slightly faster. The airplane with the aft C.G. will not be as longitudinally stable as the other, and will be more easily disturbed from its trim. This doesn't mean that the airplane becomes unsafe; flight characteristics could be quite reasonable once the airplane is in smooth air and trimmed. If you went too far with this idea, however, serious problems of instability could arise.

Just how can aft-loading add to performance? Take a look at an average high performance airplane (Fig. 6-4). You'll find it has a tail-down force at cruise, the result of the center of gravity location for any particular airplane/Weight combination. The center of Lift is considered to be fixed as shown by both planes in Fig. 6-4. This produces a nose-down moment which must be balanced by a tail-down moment furnished by the horizontal tail surfaces.

A moment is usually measured in inch-pounds and is the result of distance times a force or weight. For this problem, we will measure the moments around the center of Lift rather than around the center of gravity as was done back in Chapter 2.

Airplane A is so loaded that the C.G. is ten inches ahead of the center of Lift. The airplane weighs 3000 pounds, so that the nose-down moment is ten inches x 3000 pounds or 30,000 inch-pounds. The tail-down moment must equal this (or the airplane will be wanting to do an outside loop) and the center of the tail-down force in this case is 200 inches behind the center of Lift. This calls for a force of 150 lbs. (the 30,000 inch-pounds must equal 200 inches times the force, which is 150 lbs.). When the moments are equal, the airplane's nose does not tend to pitch either way. (The other moments covered in Chapter 2 will be ignored for this problem.)

If equilibrium exists and Airplane A is to maintain level flight, the *vertical* forces must also be equal; *up* must equal *down* in other words. Down forces are Weight (3000 lbs.) and the tail-down force (150 lbs.) — a total of 3150 lbs. down. Lift must equal this same value, so that 3150 lbs. of Lift are required to fly a 3000 lb. airplane.

Airplane B has a more aft center of gravity. It is loaded so that the C.G. is one inch ahead of the center of Lift. The nose-down moment here is one inch times the 3000 pounds or 3000 inch-pounds. The center of the tail-down force is 200 inches behind the center of Lift, so that a down force of 15 lbs. is required (3000 inch-pounds in this case, is a product of 200 inches times the force, which must therefore be 15 lbs.) To sum up the vertical forces: 3000 plus 15 equals 3015 lbs. of Lift required. Airplane B has to fly at an angle of attack and airspeed to support only 3015 lbs.

Both airplanes would weigh 3000 pounds if placed

on a scale, of course, but airplane A weighs five per cent more as far as the combined angle of attack and airspeed are concerned. That airplane also requires more power to fly at a constant altitude, and as both airplanes are carrying the same power setting, the heavier airplane A would cruise more slowly. To believers riding on airplane B and noting that it flies faster than airplane A, it might well seem that their airplane is "on the step" — yet all that is happening is that both airplanes are merely following predictable aerodynamic laws.

Relying solely on airspeed indicators can lead to another oversight: temperature and its effects.

Say a pilot is flying a certain airplane on a Canadian winter morning at a pressure altitude of 4000 feet, carrying 65 per cent power: His true airspeed for that power will be 186 mph. The outside air temperature is -10° F (-23° C). This gives him a density altitude of sea level. As the density altitude is sea level, the indicated airspeed will be 186 mph (assuming no instrument or position error). He flies to Florida the next day and starts operating there at the same pressure altitude of 4000 feet and the same Weight at 65 per cent, but the OAT is 75° F (+24° C). His density altitude is 6000 feet, giving a true airspeed of 196 mph. In getting this *T.A.S. of 196,* however, he is indicating only *179 mph.* By just comparing indicated airspeeds, the pilot would come up with the fact that yesterday — at the same pressure, altitude, power setting and Weight — he was indicating *7 mph more.* Yesterday, it might seem, he was "on the step"; today, he just couldn't seem to make it. Even once, perhaps, he tried to ease the nose down today to get on the step, but just lost altitude, so he's more convinced than ever that the "step" is a mysterious thing only found once in a while. Quite possibly, he noted the variation from standard temperature in setting up his manifold pressure, but fell nevertheless into that oft-repeated trap of comparing indicated airspeeds only.

A LOOK AT MAXIMUM RANGE CONDITIONS

It has been established that maximum range, as far as the aerodynamics of the airplane is concerned, is a function of a particular coefficient of Lift, the one at which the Lift-to-Drag ratio is at a maximum. This C_L (or angle of attack) is constant for a particular airplane (and configuration) and does not vary with Weight. However, the average airplane does not have means of measuring C_L or angle of attack so the pilot must decrease his indicated airspeed as Weight decreases in order to maintain the required optimum angle of attack. Look at the problem in terms of the power curve as shown in Fig. 6-5.

A line drawn from the Origin (O), tangent to the Power Required curve gives the speed for best range at this altitude for each Weight. Assuming that the altitude remains constant, the T.A.S. (and I.A.S.) must be constantly decreased with decreasing Weight in order to get the absolute maximum range. This means that the power setting (BHP) is reduced from,

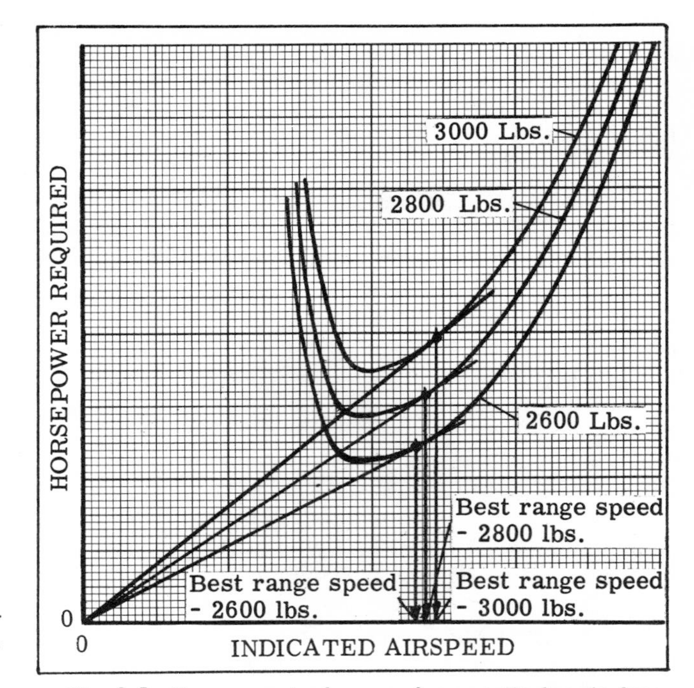

Fig. 6-5. Power required curves for a particular airplane at various Weights. Notice that for maximum range the airspeed must be decreased as Weight decreases.

say, 40 per cent to 35 per cent over the period of the flight.

From an aerodynamic standpoint the tangent line should be drawn on the *Thrust* Horsepower Required curve. The speed found would be point of maximum efficiency of the airplane, and in theory at least, would coincide with the speed for the minimum Drag on the Drag versus Airspeed curve for the airplane (Maximum Lift-to-Drag ratio). However, there are other factors involved such as propulsive efficiency (η) and brake specific fuel consumption (pounds of fuel used per BHP per hour). For instance, if your engine-propeller combination happened to have an extremely low efficiency at the speed found by using the THP required curve, a compromise must be found. As an exaggeration, suppose at the speed found as the aerodynamically ideal one for best range the efficiency of the engine propeller combination is only 33 per cent or *three* BHP is required to get *one* THP, but at a slightly higher speed the efficiency jumps up to 85 per cent. A compromise would be in order. The same thing applies to the brake specific fuel consumption. Later Fig. 6-8 will illustrate that the bfsc will change with power and this might also be a factor. Basically, max range would be found at a condition at which the combination $(\frac{L}{D} \times \frac{\text{Efficiency}}{\text{Bsfc}})$ is a maximum. You may have to fly at a *lower* Lift-to-Drag ratio in order to *increase* efficiency or *decrease* bsfc and get an overall greater range than would result from sticking to the max L/D speed and ignoring low efficiency or high bsfc at that speed. The manufacturers take this into consideration when they publish max range figures. In order to show the expected power settings for various airspeeds for max range and endurance Figs. 6-5, 6-6, 6-9 and 6-10 are actually based on BHP as you can see by the shape of the curves.

You can see that the normal cruise speed is well above that of the maximum range speed and is a compromise between speed and economy. For instance, if you desire a 10% increase in speed, it requires a 33% increase in horsepower (horsepower required is proportional to the velocity cubed) or hp $\backsim V^3$ (1.10 x 1.10 x 1.10). Naturally everybody would like to get from A to B as rapidly as possible, but this is neither aerodynamically nor economically feasible.

One thing you'll find is that after careful planning for winds and other factors you may save some time en route — only to lose it at the other end by a delayed landing or ground transportation troubles. On a two-hundred nautical mile trip, a plane that cruises at 150 knots takes 16 minutes less than one that cruises at 125 knots and these savings can easily be lost at the destination. Of course the longer the trip the greater the time savings, but more time is lost at airports than is generally considered.

A study of current general aviation airplanes shows that a rule of thumb may be used to approximate the maximum range airspeeds for gross Weight. The following ratios are based on three major airplane types and are the ratio of the maximum range speed to the flaps-up, power-off stall speeds (wings level) as shown by the bottom of the green arc on the airspeed indicator.

Single engine — fixed gear — ratio	1.5 x power-off stall speed
Single engine — retractable gear	2.0 x power-off stall speed
Twin engine — retractable gear	1.7 x power-off stall speed

Suppose you find in the Airplane Flight Manual or by rule of thumb that the indicated airspeed is 120 knots for max range of gross Weight, how do you take care of the Weight change as fuel is burned? Looking at the Lift equation and working around to solving for V: $V = \sqrt{\frac{2W}{S\rho C_L}}$. (Assume here for simplicity that Lift equals Weight.) C_L is to be held constant as it is the one found for max L/D, or max range, and the wing area (S) does not change nor does the density (ρ) for a constant altitude, so the only variable in the square root symbol is W, or Weight. (The "2" is also a constant, as always.) This means that as Weight changes, V must also change as the *square root* of the Weight change. Put in simple terms, V must change one-half the percentage of that of the Weight. If Weight decreases 20 per cent, the V (or airspeed) must be decreased by 10 per cent to keep the required constant C_L. If the gross Weight of the airplane is 3000 pounds and the required airspeed for max range is 120 knots, the airspeed at 2400 pounds (a decrease in Weight of 20 per cent) should be 108 knots (a decrease in airspeed of 10 per cent). This works for the maximum expected decrease from gross to minimum flyable Weight for most airplanes.

You will find that a higher percentage of normal rated power is required to maintain a required I.A.S. at higher altitudes, but as shown by Fig. 6-6 the true airspeed will be increased so that the ratio of miles per gallon remains essentially the same. Why do you try to use the same indicated airspeed for all altitudes (assuming equal Weights)? It goes back basically to the Lift equation of earlier chapters, $L = C_L S \frac{\rho}{2} V^2$. The C_L, or coefficient of Lift, is the fixed one found for max range (or max Lift-to-Drag ratio). The Weight, or Lift required, is the same for a particular time, the wing area (S) is the same, hence the final $\frac{\rho}{2}V^2$ must be the same under all conditions if the C_L is to remain constant. The dynamic pressure "q", or $\frac{\rho}{2}V^2$, is that as measured by the airspeed instrument as indicated airspeed.

The ratios given are the approximate I.A.S. for maximum range for a chosen altitude. For most airplanes with unsupercharged engines, the altitude for best *normal* cruise is roughly 8000 feet standard altitude (call it seven to ten thousand). This is the altitude range where wide open throttle is necessary to get the 65 to 75 per cent recommended power setting for normal flying. But getting back to the idea of obtaining max range: Once you have the ratio for your type of airplane and have found the maximum range speed for sea level, you can *use this figure as an approximate indicated airspeed* for your chosen flight level. This would give you the best range under no-wind conditions for that particular altitude at gross Weight. This holds true up to the altitude where you are no longer able to get the necessary power to maintain altitude at the recommended I.A.S. The drawback to this is that you may burn some fuel getting up to this altitude, but if the trip is a long one, it might be worthwhile time-wise because of the increased true airspeed.

For maximum range, you would climb to the chosen altitude and set up an indicated airspeed at the correct max range to stall speed ratio (1.5, 1.7 or 2.0, depending on your airplane type). You would then set up the power necessary to maintain altitude at this indicated airspeed and would lean the mixture (leaning will be discussed later).

Fig. 6-6 shows the relationship between the sea level maximum range speed and that at some altitude expressed in terms of true airspeed. Your project is

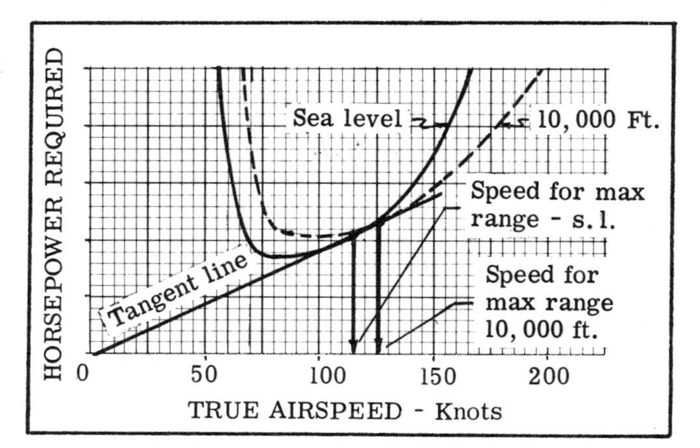

Fig. 6-6. Best range speeds (T.A.S.) for a particular airplane at sea level and at 10,000 feet.

to maintain the same indicated airspeed at higher altitudes. You are unable to maintain the *normal* cruise indicated airspeed to very high altitudes with an unsupercharged engine because you start with a high percentage of power to begin with (65-75 per cent) and cannot obtain this above a certain altitude. If you start at 40-45 per cent as would be required to maintain the lower maximum range speed, you will be able to use the required power at a higher altitude. (Fig. 6-7)

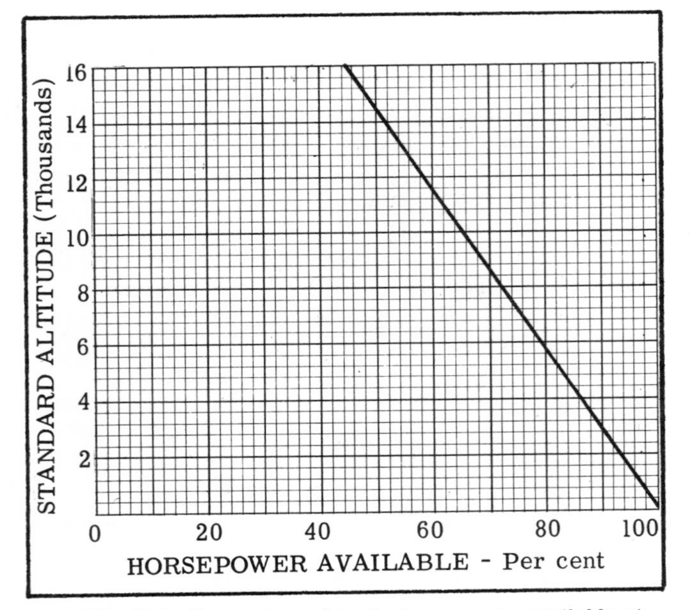

Fig. 6-7. Percentage of brake horsepower available at various altitudes — typical unsupercharged engine.

The average pilot is more interested in saving time than extending range. There is no question that the max range speed may be only 60-70 per cent of the normal cruise speed. If you are interested in getting somewhere in a reasonable length of time, you are willing to operate at a higher, though costlier, speed.

These ratios are presented for you to remember in case they are needed. Suppose you are at a point where you have overestimated your range under normal cruise conditions. It's night and there is some bad country to cross before getting to a lighted airport. You can extend the remaining range by pulling the power back and judiciously leaning the mixture until you are able to maintain altitude at an *indicated* airspeed of 1.5, 1.7 or 2.0 times your power-off, flaps-up stall speed *as can be seen on the lower end of the green arc on the airspeed indicator.*

These ratios are not completely accurate but they do give you an approximate speed to extend your range if necessary. Such variables as propeller and engine efficiency at lower speeds, specific fuel consumption at various power settings, and other factors may result in slightly different airspeeds than given here. Fig. 6-8 shows a specific fuel consumption graph for a typical light plane engine.

The maximum range speed at gross Weight will be found at power settings of approximately 40-45 per cent as given on the power chart in your airplane

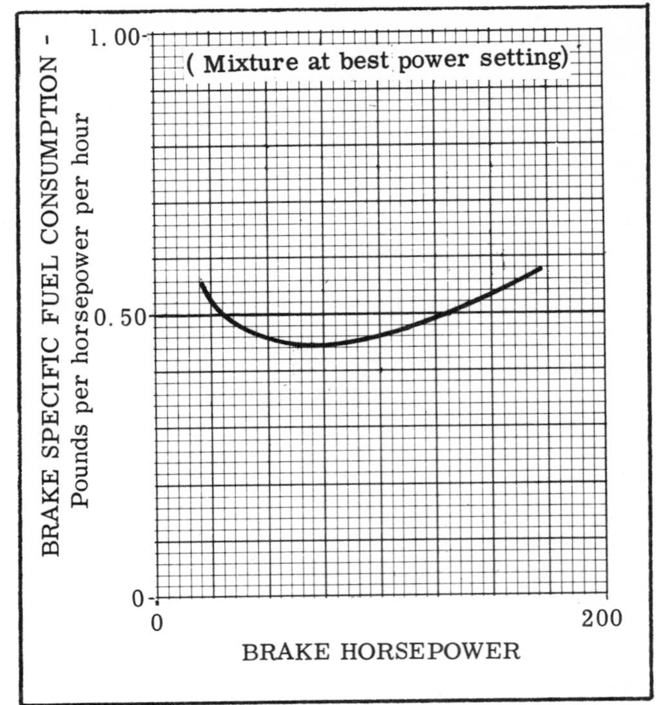

Fig. 6-8. Specific fuel consumption curve for a typical reciprocating engine as is used in light planes.

(unsupercharged engines). A variation of 5 knots from these figures will make very little difference in max range.

In review: Maximum range conditions depend on the Lift-to-Drag ratio, propulsive efficiency and brake specific fuel consumption (bsfc). The speed at which the max L/D is found as far as the aerodynamics of the airplane is concerned might be one at which the propeller efficiency is low and/or bsfc is proportionally higher. So, actually max range is found at the airspeed at which the term $\frac{C_L}{C_D} \times \frac{\text{prop efficiency}}{\text{Bsfc}}$ is greatest. The max range speed for various aircraft is found by flight testing and the rules of thumb were taken from data of current airplanes, which takes the above-mentioned variables into consideration.

WIND EFFECTS ON RANGE

There's no doubt that wind affects the range of your airplane as well as your groundspeed. Obviously, with a tailwind the range is greater than under no-wind conditions and much greater than under headwind conditions — the difference depending on the wind. So, what's new?

What's new is that you can compensate for wind effects by varying power. Suppose that you are in the situation as cited previously — the night is dark and it looks like it'll be touch and go making the next available airport. After pulling the power back you remember that there is a 10-15 knot headwind. This hurts, but there's something you can do about it. You can increase the indicated airspeed about 10 per cent for moderately heavy headwinds. *You will never get as much range with a headwind as under no-wind conditions but will be doing the best possible in this situ-*

ation. What you have done is decreased the time spent in this predicament.

Conversely, if you have a moderate tailwind, you can stretch your range even farther by subtracting about 5 per cent from the speed given for maximum range under no-wind conditions. Figure 6-9 shows the graphic presentation.

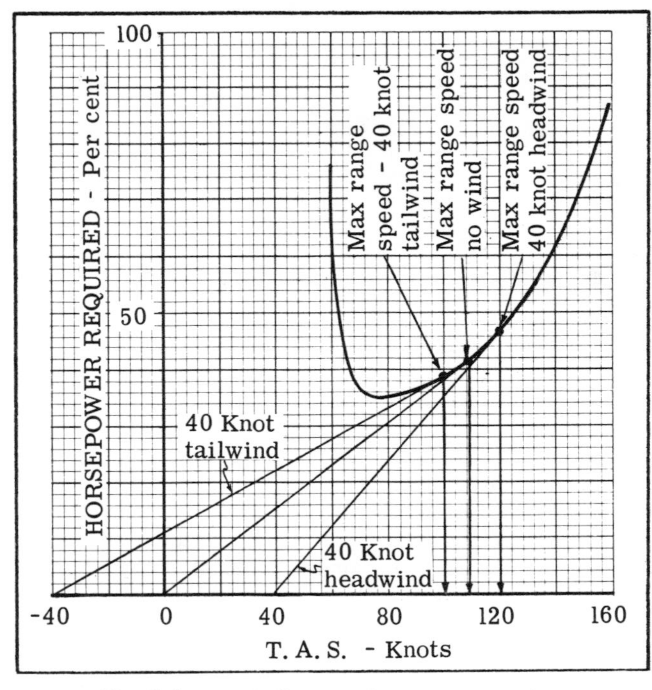

Fig. 6-9. Wind effects on best range speed.

SUMMARY OF RANGE

These ideas are presented to give you a greater safety factor, not to get you to push the range of the airplane to the limit. You may never use these points but they could come in handy if you get into a bind and *have* to stretch it. It will require a great deal of will-power to throttle back when you feel like increasing power so as to "get there before you run outa gas."

The added time required to get to the airport may make you a nervous wreck, but you'll make it under conditions that wouldn't allow you to otherwise.

ENDURANCE

The maximum endurance of an airplane is seldom needed but when it is, it's needed very badly! You should be familiar with the idea of maximum endurance, particularly if you plan on getting an instrument rating, because under extreme conditions you may have to hold over a certain fix for one or two hours and then discover the destination field has gone below Instrument Flight Rules minimums and you have to go to an alternate airport. Things could get binding, particularly if you haven't tried to conserve fuel but have been boring holes in the soup all this time on full rich mixture and cruise power settings.

Or take a VFR situation: You're coming into a large airport after a long trip. It's below VFR minimums but the weather is good enough for you to get a controlled VFR into the field, but there's a lot of instrument traffic (and maybe other controlled VFR pilots like yourself). The tower orders you to hold outside the control zone and control areas. The local weather is 1000 overcast and $1\frac{1}{2}$ miles — good enough for you to fly underneath "elsewhere" but not to enter a control zone without clearance, so you hold just outside of the control zone and "await further clearance." Here, too, would be a good place to use maximum endurance unless you have plenty of fuel.

For reciprocating engines, maximum endurance is found at sea level. Refer to the Power Required versus Velocity Curve of Fig. 6-10.

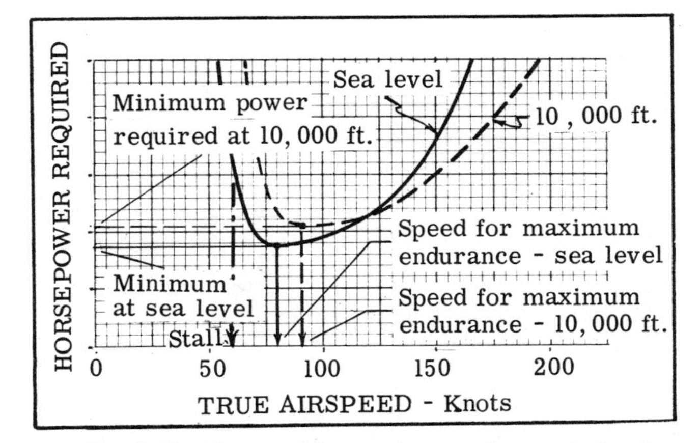

Fig. 6-10. The speed for maximum endurance is found at the point where power required is the least.

You are interested in minimum gas consumption while still maintaining altitude. The low point of the power curve is where the least power is required and the point where gas consumption is a minimum if you properly lean the mixture. For most airplanes (including unsupercharged light twins) this will require a power setting of 30 to 35 per cent of max continuous power. Few power charts go this low. The minimum brake horsepower required (in percentages) depends on the aerodynamic cleanness and power loading of the airplane. An aerodynamically dirty, heavy airplane with comparatively little power might need a *minimum* of 50 to 55 per cent or more of its available power to maintain altitude.

After examining Fig. 6-10 you can readily see that contrary to what may have been expected, the maximum endurance speed is *not* the lowest speed possible for the airplane to fly without stalling. Many pilots, when confronted with the need for maximum endurance for the first time, automatically assume that the lower the speed, the less power required.

Horace Endsdorfer, private pilot, flew around just above a stall for two hours one day when he needed maximum endurance and was worn out keeping the airplane under control. Needless to say, he was one disgusted pilot when he learned that he was burning only slightly less gas than cruise — and working like heck all the time. (Fig. 6-11) It should have struck him as being odd that he had to use so much power to maintain altitude in the holding pattern.

Manufacturers' recommended figures on speeds

Fig. 6-11.

for maximum endurance are not always available. In many cases this is true because of the airplane's low speed handling, and the constant speed propeller governing characteristics at low rpm. A compromise may be desired such as a slight loss of endurance by increasing airspeed to have smoother operation.

A few manufacturers publish the figure and the others leave it up to the pilot. The only trouble is, as the pilot does not have access to a power required chart for his airplane, he may not know what this speed is and may either fly around at a speed just above stall or figure to heck with it, and circle around at cruise power. In either case, he's like the fellow who had carved on his tombstone, "Not dead, only sleeping." He isn't fooling anybody but himself! If you have an approximate idea of, say, the ratio between the power-off, flaps-up stall speed and the maximum endurance speed, you may experiment until you find the exact speed, power setting and mixture that will result in the lowest gas consumption. At any rate, you'll be somewhat better off.

If the information is not available in your Airplane Flight Manual, then the following ratios may give an *approximate* figure for your type of airplane. For a single engine with fixed gear, use a 1.2 ratio of maximum endurance speed to power-off, flaps-up stall speed (bottom of green arc on airspeed indicator). For a single engine with retractable gear or a twin with retractable gear, use a ratio of 1.3.

The speeds found by this method are Indicated Airspeeds and will apply at all altitudes — although the lower the better for reciprocating engines. This doesn't mean get down to twenty feet above the ground when you could be at five hundred or a thousand feet. It does mean that if you have a choice, an altitude of one thousand feet is better for maximum endurance than ten thousand.

To set up maximum endurance, take the following steps:

1. Throttle back and slow the plane to the recommended indicated airspeed.

2. Retard the prop control (if so equipped) until you get the lowest rpm possible and still have a reasonably smooth prop operation. In the case of a constant speed prop this means no "hunting" by the propeller as it reaches the lower limits of governor control. Low rpm means low friction losses.

3. Set the throttle so that the recommended speed and a constant altitude are maintained.

4. Lean the mixture as much as possible without the possibility of damaging the engine. You may then have to experiment again with the throttle to obtain the optimum setting for your particular condition (hot day, cold day, etc.).

Some of the above steps may not apply, for example, if you are flying a light trainer that has neither a mixture control nor a variable pitch prop.

The point is that you are trying to maintain altitude with the minimum fuel consumption. This means minimum power (*not* minimum speed) and judicious leaning of the mixture. You'll be using about 30-35 per cent power.

If you have to hold or "endure," it's usually over a particular spot, and this means *turns*. Keep your turns as shallow as possible, without wandering over into the next county, because steeper turns mean increased back pressure to maintain altitude — and this is a speed killer. Making steep turns usually results in constant variation of throttle as altitude or speed is lost and regained at a cost of added power and fuel consumption. Of course, if you start out high enough, you may just be lucky enough to get cleared into the airport before you run out of altitude. However, if you are on Instrument Flight Rules, it's not considered cricket to go blindly barging down into the next fellow's holding pattern.

Wind has no effect on endurance, but *turbulence* decreases it.

You may find the approximate max endurance speed for your airplane by flying at various airspeeds, maintaining a constant altitude. Fig. 6-12 shows an actual test of an airplane with a fixed pitch propeller and fixed gear (flaps-up, power-off stall speed— 56 knots).

Indicated Airspeed (knots)	rpm
101	2700
96	2600
90	2500
86	2400
78	2300
74	2200
70	2150
65	2075
61	2250
57	2300

Fig. 6-12.

Notice that 65 knots is the indicated airspeed at which minimum power is required. The ratio of max endurance speed to flaps-up, power-off stall speed is $\frac{65}{56} = 1.16$, reasonably close to the predicted ratio figure of 1.20 for this type of airplane. (The 1.20 figure would assume a maximum endurance speed of 1.20 x 56 or 67 knots.)

You could also find the max endurance speed for an airplane with a constant speed propeller by leaving the rpm at the lowest smooth value and making a table like Fig. 6-12, noting manifold pressure required rather than rpm. Don't use too high a manifold pressure with low rpm, as the engine could be damaged. Actually, since the max endurance speed normally falls within 50 per cent of the stall speed, this is the only area that needs to be investigated. Make a note of the lowest manifold pressure, rpm, and airspeed at this value for future use. You may find that the airplane doesn't handle well at that speed and may want to add a few miles an hour to take care of this. At any rate you'll have an approximation if needed. If the test is run at less than gross Weight, remember that the required speed should be slightly higher at gross.

7. A REVIEW OF THE TURN

The airplane turns because it's banked—this you were taught when you first started to fly. You remember from Chapters 1 and 2 that the Lift vector is always perpendicular to the wingspan and to the relative wind. If you are in wings-level flight, Lift is opposite to Weight, and in level flight, Lift equals Weight. Actually the Lift exerted by the wings will be slightly greater than Weight at cruise, as was discussed in Chapter 2, but for the purposes of this chapter they can be considered to be equal and the tail force will be ignored.

FORCES IN THE TURN

Weight always acts downward toward the center of the earth, but as you bank the airplane, Lift is no longer opposite to Weight. (Fig. 7-1) Assume an airplane weighs 3000 lbs., and has a Lift of 3000 lbs. In

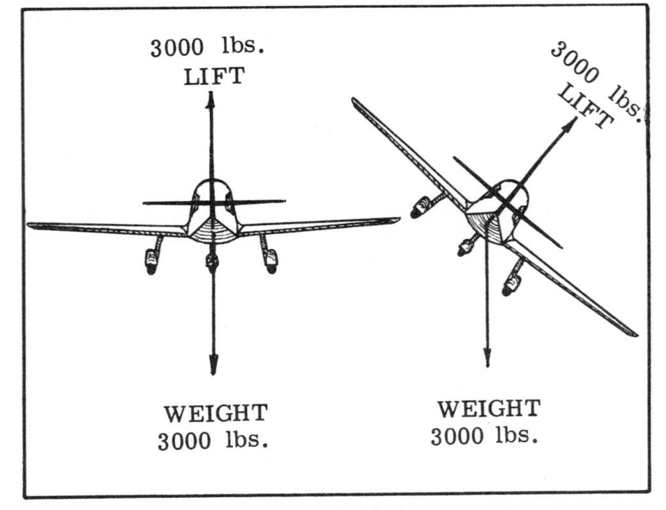

Fig. 7-1. The Lift vector is 90 degrees to the wing span and therefore does not always act opposite to Weight.

the turn, the Lift vector can again be considered to be broken down into horizontal and vertical components as was done in Chapter 2.

A 60° bank has been chosen for simplicity. As you can see in Fig. 7-2, the vertical component, acting opposite to Weight, is no longer equal to Weight, but in a 60° bank is now one half of the Lift value, or 1500 pounds. This means that the vertical forces are no longer in balance, and in this condition the plane must lose altitude.

Centrifugal force tends to pull the plane to the outside of a turn. (Centrifugal force is not a *force* at

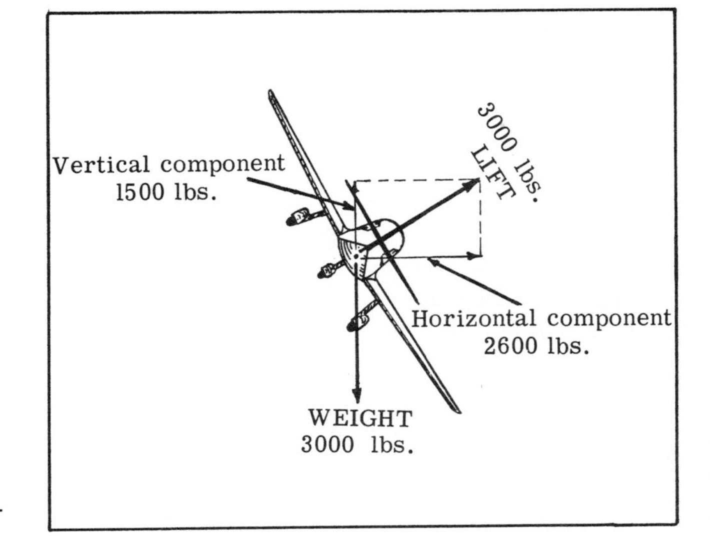

Fig. 7-2. In a 60 degree bank the vertical component of Lift has been halved (no back pressure exerted).

all, but is an *apparent force* resulting from the fact that inertia tends to make the airplane fly straight instead of following the circular path of the turn.) The horizontal component of Lift acts as a centripetal force, and exerts a holding action which counteracts the centrifugal force (in a balanced turn). A more familiar example of these forces in action is that of a small boy twirling a rock around on a string. Centrifugal force (and temptation) keeps trying to force the rock outwards and off in the general direction of the neighbor's picture window. Centripetal force (and discretion) is the holding force exerted on the zooming rock by the boy's hand through the string.

LOAD FACTORS

If you make a turn without applying back pressure, the airplane starts losing altitude. You learned this when you were first introduced to the turn. Holding top rudder doesn't help hold altitude. In fact, it causes you to start slipping because you decrease the force holding you in the turn.

You were taught to hold the nose up with back pressure. By increasing back pressure you increase the angle of attack which results in an increase in the Lift vector so that the vertical component and Weight are again equal. In a 60° bank this means that Lift has to be doubled. (Fig. 7-3) For a 3000-pound airplane Lift would have to equal 6000 pounds. The wings must support 6000 pounds and the wing loading (pounds per square foot) has been doubled.

In normal level flight the load factor is "one."

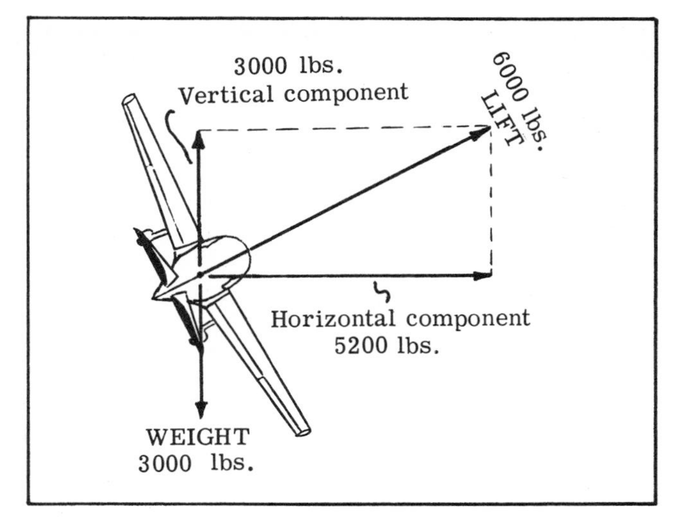

Fig. 7-3.

You have this force of one "g" acting on you under normal conditions on the ground. Now you have suddenly been required to double this force in the 60° bank. You *and* the airplane have a load factor of two, or two "g's," acting on both of you. If you are a 97 pound weakling, you will weigh 194 pounds in a properly executed 60° banked turn.

LOAD FACTORS AND THE STALL SPEED

Taking a look at the equation for Lift:

$\text{Lift} = C_L S \frac{\rho}{2} V^2$, and in level flight, assuming that Lift = Weight it becomes:

$\text{Weight} = C_L S \frac{\rho}{2} V^2$: ($C_L$ = Coefficient of Lift; S = Wing Area; ρ = Air Density; and V^2 = the Velocity, fps, squared).

It can be found that the stall speed of an airplane depends on the square root of wing loading (the ratio of Weight to wing area). Solving for Vstall:

$$V\text{stall(fps)} = \sqrt{\frac{W}{C_{L\max} S \frac{\rho}{2}}};$$

$$V\text{stall is proportional to } \sqrt{\frac{W}{S}}.$$

Assuming that C_L max and ρ are constant for any airfoil and altitude, the stall speed changes with the square root of the wing loading or $\sqrt{\frac{W}{S}}$. In the 60° banked turn (with back pressure applied to maintain altitude) the wing loading is doubled and the load factor is 2. Therefore, the stall speed will be increased by the *square root* of the load factor; $\sqrt{2} = 1.414$ times the wings-level stall speed. If the plane stalls at 60 knots with the wings level, the stall speed in a 60° bank will be 1.414 times 60, or about 85 knots. This ratio of the square root of the load factor applies to the stall speed in feet per second or miles per hour; the new stall speed would still be 1.414 times the old one.

As the turn gets steeper the load factor increases until for a 90° bank the load factor is infinite. The airplane cannot maintain altitude in a vertical bank—it will stall first, or worse, the load factors may cause structural failure *before* it stalls if you roar into the turn at a high speed.

However, if you don't hold any back pressure, the load factor is not increased and the stall speed is not increased, but the airplane will not maintain altitude.

You remember that in 720° power turns you were told to open the throttle to climbing power because the increase in power gave a greater margin of airspeed above the stall. In the level steep turn you held back pressure, increasing the angle of attack which slowed the airplane (you probably noticed that the airspeed dropped off 5 to 10 knots). You were being caught from both ends of the airspeed range. The stall speed was increased because of the increased load factor and your airspeed decreased because of the increased angle of attack necessary to maintain altitude.

Fig. 7-4 shows the power required curves for a fictitious airplane in wings-level flight and in a 30 degree bank at a constant altitude. Notice that in the 30 degree bank more power is required for any airspeed and the stall speed is computed to be 1.075 times that of wings-level flight. For *any given power setting* the I.A.S. and stall speed are much closer together in the banked condition. As the bank angle increases the power required to maintain a constant altitude increases at a much greater rate. For most of the larger airplanes it is not necessary to increase power to make a good 720 degree power turn in a 60 degree bank; for the lighter trainers it's a good idea to use extra power.

Fig. 7-5 shows the relationship between the angle of bank and the stall speed increase just mentioned. Instead of showing the load factor for various angles

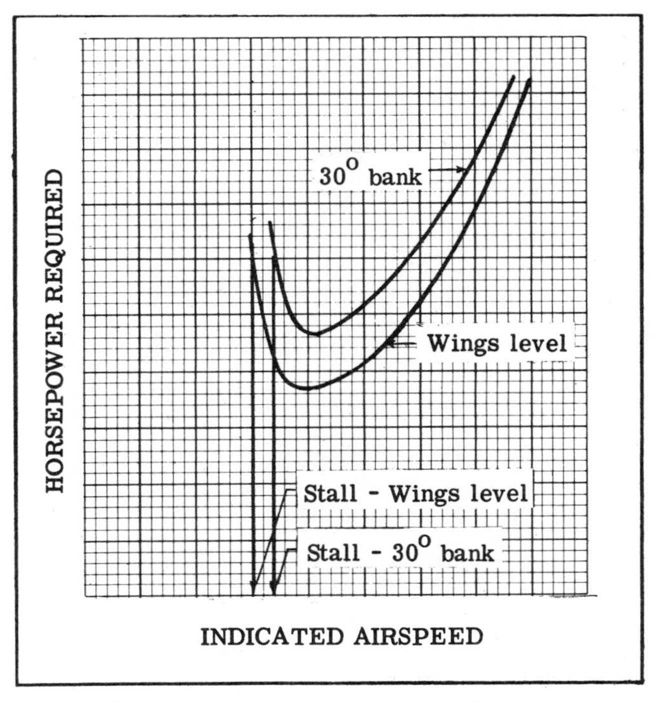

Fig. 7-4. Power Required curves for a fictitious airplane in wings level flight and a balanced turn of 30 degrees bank.

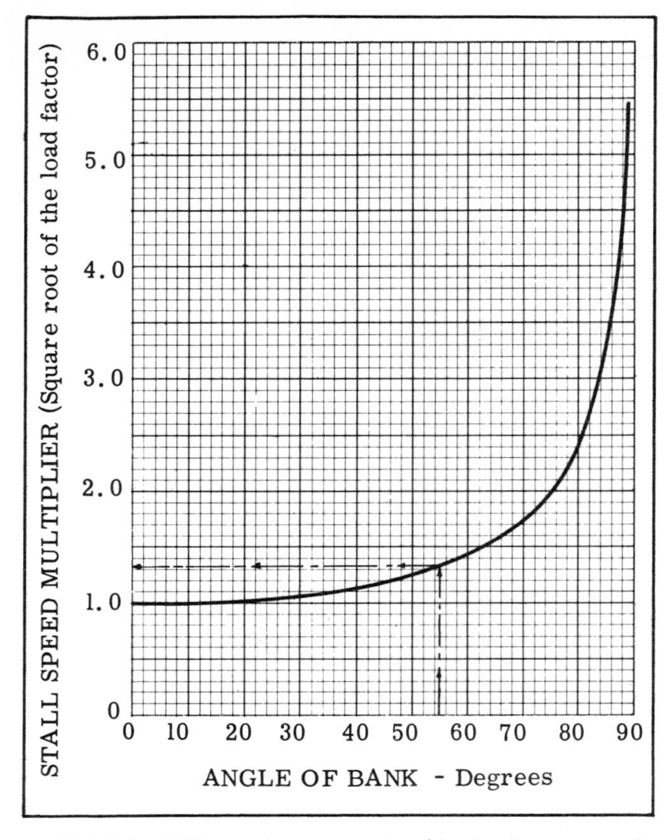

Fig. 7-5. Stall speed versus angle of bank. As an example, if the airplane is maintaining altitude in a 55 degree bank, the load factor produced will cause the stall speed to increase by a factor of 1.32. Assuming the airplane stalls at 60 knots in level flight, its stall speed in a 55 degree bank will be 1.32 x 60 = 78 knots.

of bank, the square root of the load factor is used so that you can directly arrive at the new stall speed without taking square roots.

Fig. 7-6 shows the forces acting on an airplane in a balanced turn at a constant altitude. All the forces are balanced so that the airplane has a constant rate of turn and altitude.

The load factors discussed here are positive load factors and are attained by pulling the stick or wheel back. A negative load factor is one that is caused by pushing forward on the stick or wheel. Positive or negative load factors can be imposed on an airplane by vertical gusts as well as by the pilot's handling of the elevators.

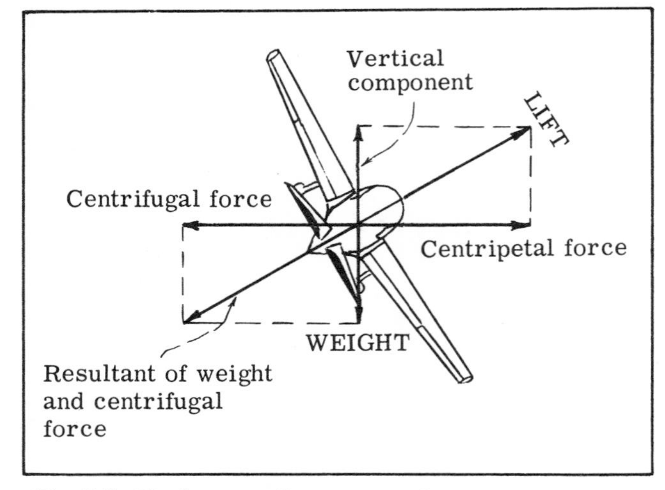

Fig. 7-6. The forces acting on an airplane in a balanced turn.

UNBALANCED TURNS

If you are skidding, the rate of turn is too great for the amount of bank and the forces are no longer in perfect balance. The centrifugal force is too great and tends to throw you and the ball in the turn and slip to the outside of the turn. (Fig. 7-7)

A skid can be induced by the use of bottom (inside rudder) or by aileron opposite to the turn (with no rudder). In the case of bottom rudder, the bank will increase because the outside wing is speeded up and has more Lift than the inside one. The use of opposite aileron, of course, tends to decrease the bank and this plus adverse aileron yaw effects send the ball (and you) to the outside of the turn. As you remember from Chapter 1, induced Drag is a function of Coefficient of Lift. There is a difference in C_L in the two wings caused by the aileron deflection and the inside wing has more Drag and the turning rate increases temporarily.

A slip is caused by too steep a bank for the rate of turn, and can be caused in two ways: (1) aileron into the turn (with no rudder) or (2) the application of top rudder. (Fig. 7-8) There is *not enough* centrifugal force in action for the angle of bank. Both skids and slips are the result of improper rudder use.

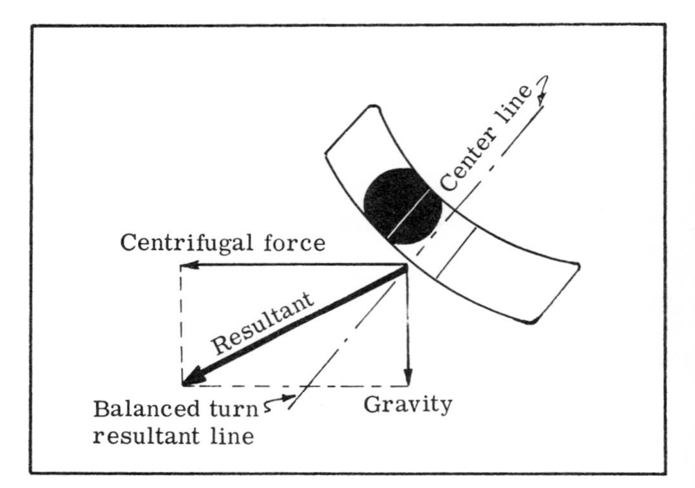

Fig. 7-7. A skidding right turn.

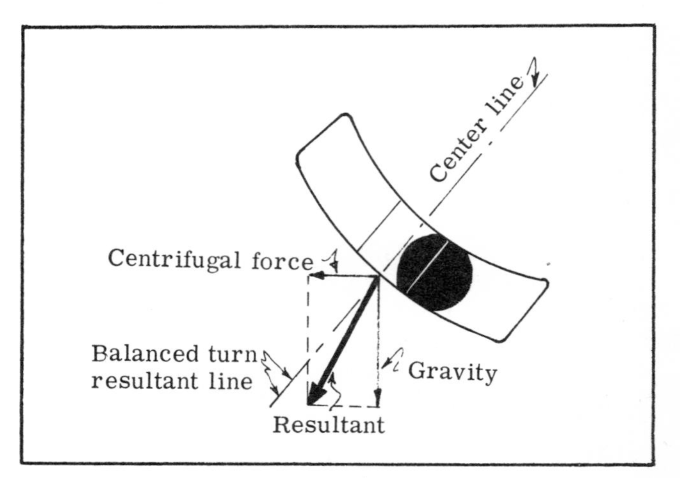

Fig. 7-8. A slipping right turn.

8. GLIDES

Comparatively little attention is paid to the airplane's glide characteristics by pilots these days. The trend has been toward power approaches for all airplanes no matter how light, and pilots sometimes have been caught short by an engine failure.

For airplanes with higher wing loadings, power approaches are usually necessary in order to avoid steep angles of approach to landing. If the approach angle is steep and the airspeed low, you may find that the airplane will "rotate" for landing but will continue downward at an undiminished (or even greater) rate of descent, making a large airplane-shaped hole in the runway. (Fig. 8-1) The stall characteristics of the swept wing make it particularly susceptible to this type of trouble.

Because of this and the fact that most jet engines give poor acceleration from idle settings, jets generally use comparatively high power during the landing approach.

This chapter discusses the airplane's clean glide characteristics.

Back in the old days when engines were not as reliable as they are now, the pilot's knowledge of his airplane's power-off glide characteristics was of supreme importance. Nearly all approaches were at engine idle and were made so that should the engine quit at some point during the process, the field could still be made. Even now, applicants for the commercial certificate are required to land beyond and within a certain distance from a point on the runway. They are allowed to use flaps, or slip to hit the spot — in earlier times even these aids were taboo.

It behooves every pilot to occasionally make power-off (idle) approaches to keep in practice should he have an engine failure at altitude.

Two glide speeds will be of interest to you: (1) airspeed for minimum rate of sink and (2) airspeed for farthest glide distance. The two conditions are not the same, although they might appear to be at first glance.

These figures are arrived at by flight test. Glide the airplane at various airspeeds and plot the rates of sink for each airspeed and altitude. The graph for one altitude and Weight looks like Fig. 8-2.

THE MINIMUM SINK GLIDE

Point A on Fig. 8-2 represents the velocity at which rate of sink is a minimum. Point B is that at which the max distance glide is found. You remember that the rate of climb is a function of excess Thrust horsepower and the rate of sink is a function of deficit Thrust horsepower. The less horsepower you "require," the less the rate of sink in the power-off condition. The best velocity for this is at Point A for a particular airplane. (Check Fig. 8-3 also).

You remember from Chapter 1 that the power required curve would be moved by the effects of Weight or altitude. It can also be affected by a change in parasite Drag. So, while in theory the speeds for the minimum sink and maximum distance glides should be the same as that for maximum endurance and maximum range respectively, for propeller air-

Fig. 8-1.

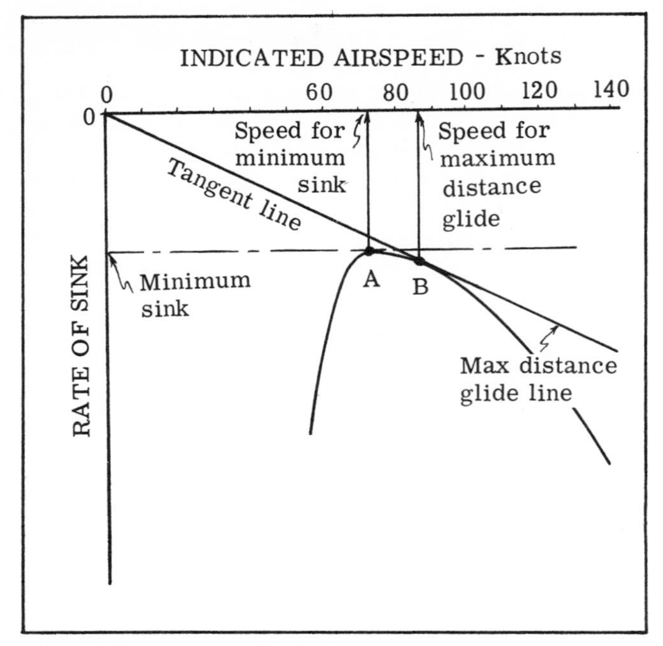

Fig. 8-2. Rate of sink versus velocity; a particular altitude and Weight for a fictitious airplane in the clean condition.

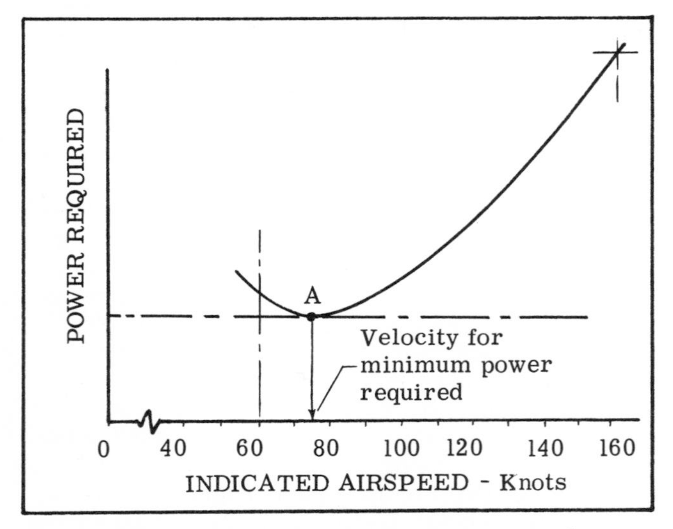

Fig. 8-3. Power required versus velocity curve (Thrust horsepower).

planes certain practical factors are involved. When you are flying at the maximum endurance (or max range) speed you naturally have power on. Power, even the comparatively small amount used for endurance, normally increases the efficiency of the airplane by furnishing a slipstream across the wing center section. The airspeed ratios arrived at in the chapter on range and endurance are based on power-on configurations. With the power off, Thrust and slipstream effects are missing. If the propeller is windmilling, parasite Drag increases sharply (a windmilling prop is like a barn door out front). To maintain the new, *lower* Lift-to-Drag ratio the airplane must fly at a slower airspeed in order to get the best performance in this less efficient condition. For the light twin the props should be feathered if the engines are out of action.

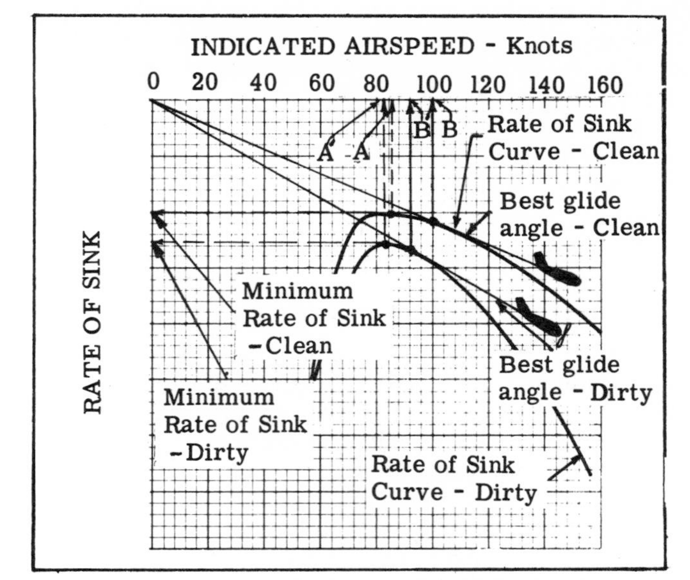

Fig. 8-4. Rate of Sink curves for an airplane in the clean and dirty condition (prop windmilling in low pitch). Both curves for same altitude and Weight.

Fig. 8-4 is a comparison of the Rate of Sink curves for an airplane in the "clean" condition (prop feathered or removed) and with the prop windmilling in low pitch. Both curves are based on the same Weight and altitude. By looking at the curves you can see the effects of increased parasite Drag of the windmilling prop on the rate of sink curves. A and B represent respectively the speeds of minimum sink and max distance glides for the clean airplane; A′ and B′ represent the same speeds for the dirty airplane. Notice that for the dirty condition the airspeed for max distance glide must be decreased and is closer to its minimum sink speed than for the clean airplane. Parasite Drag varies with airspeed so that a lower speed is necessary to help keep it to a minimum in the dirty configuration. *Incidentally, all of the rate of sink curves in this chapter are exaggerated, particularly at the lower end of the speed ranges, in order to more clearly show the theory.*

There's another tie-in between minimum sink and max endurance — you'll do better at low altitudes for both; the minimum sink rate will be less at lower altitudes.

The glide is one of the most difficult factors of airplane performance to pin down for rule of thumb purposes. Minimum sink speed will be in the vicinity of the speed for maximum endurance but somewhat lower because of the effects discussed earlier. The glide properties not only vary between airplanes of the same general classification (single engine fixed gear, etc.) but will also vary for the same airplane, depending on propeller blade setting if a variable pitch prop is used. As far as the propeller affects the glide, the following lists some propeller settings and their effects on the glide:

Very bad — Prop Windmilling, Low Pitch
Better — Prop Windmilling, High Pitch
Best — Prop Feathered (applicable to multiengine only)

Of course, you can usually stop the propeller by slowing up to about the stall speed, but you may not feel like doing stalls with a dead engine at low altitudes.

Here are some approximations of the airspeed for minimum sink as a ratio to the flaps-up, power-off stall speed (I.A.S.).

Single engine fixed gear (Flaps up, prop windmilling)	1.1
Single engine retractable gear (Gear and flaps up, prop windmilling in high pitch)	1.2
Twin engine retractable gear (Gear and flaps up, *props feathered*)	1.3

The minimum rate of sink condition will probably be used only in an emergency situation (you have engine failure at night and don't know the terrain below). The minimum rate of sink will be low enough so that you will have a good chance in flat territory where you don't have something solid like a stone wall to hit. This would be the best approach for an engine failure at night over water, marsh grass or snow where altitude would be hard to judge. Notice that you won't have much safety margin in gusty air.

MAXIMUM DISTANCE GLIDE

Point B on Fig. 8-2 gives the speed for the maximum glide distance (or maximum forward distance per foot down). This happens to be the airspeed for the max Lift-to-Drag ratio, which you remember was also the speed for maximum range. The speed for maximum range was with power; for the maximum glide distance, power effects are not present. Because of this, the recommended airspeed will be somewhat lower.

This type of glide will be of more use, particularly in a single engine airplane where engine failure can give you some "small concern." *The more Drag your airplane has, the less its glide ratio,* so naturally the gear should be up (if possible) and the flaps retracted.

This maximum distance glide speed, in the case of a forced landing, would be used to make sure that you get to the field and the "key point" or "key position box," after which you will set up the familiar approach speed for the final part of the problem. A possible forced landing situation might be like this: You are on a cross-country in a single engine plane at 3000 feet when the engine quits. Carburetor heat, switching tanks, turning on the electric boost pump or other corrective measures cannot remedy the situation and you are faced with landing whether you want to or not. There's a decent-looking field over to one side that will allow you to make an into-the-wind landing.

The first time this happens (or the tenth, or the hundredth) you are "shook."

The best procedure would be to set up the maximum distance glide speed as you turn toward the field. This will mean that you will be slowing the airplane to this speed from cruise. This will be almost automatic as you unconsciously try to maintain altitude — the trouble is that many pilots tend to overdo it.

Some Airplane Flight Manuals may give you the airspeed for maximum distance glide, but many do not. Use the Airplane Flight Manual recommended speed if it is available. Here are some approximations for max distance glide speeds (I.A.S.) at gross Weight for various types of airplanes taken from Airplane Flight Manual recommended max distance glide speeds.

Single engine fixed gear (Flaps up, prop windmilling)	1.3
Single engine retractable gear (Gear and flaps up, prop windmilling in high pitch)	1.4
Twin engine retractable gear (Gear and flaps up, *props feathered*)	1.5

Notice that the single engine retractable gear airplane is most affected by the windmilling propeller. The max range speed is about 2.0 times the flaps-up, power-off stall speed but the max glide distance speed is only 1.4 times the reference stall speed. The difference is greatest for this type as it is normally cleaner than the other two groups and the windmilling propeller (parasite Drag) affects it more as was shown in Fig. 8-4.

The feathered propellers on the light twin cause its max glide distance speed to be comparatively closer to the max range speed than the other two groups because the parasite Drag is not increased so radically due to windmilling.

So, while the maximum Lift-to-Drag ratio for the airplane in its cleanest condition may be 14 or 15 to 1; with a windmilling propeller the maximum ratio may be cut down to 9 to 1 or less. You are trying to maintain the best airspeed for this new ratio. You have to do the best with what you have. Remember that cowl flaps cause Drag, also.

Fig. 8-5 shows an exaggerated comparison between a minimum sink and a maximum distance glide.

About "stretching the glide," there is *one indicated airspeed* for maximum distance for your airplane at a given Weight and configuration. Any deviation from this means a lesser distance per foot of altitude. The airspeed for maximum glide distance decreases with Weight decrease. *The maximum glide ratio is the maximum Lift-to-Drag ratio for the airplane in the glide condition and is independent of Weight.* This means you can glide the same distance at gross Weight as at a near empty Weight — but in order to do this you'll use different airspeeds for different Weights. (Fig. 8-6)

As the glide ratio is that of the Lift-to-Drag ratio, cleaner airplanes will get more feet forward per foot of altitude. Compare a jet fighter weighing 20,000 pounds and a J-3 Cub weighing 1200 pounds, and have each pass through 10,000 feet at its *particular max distance glide speed* (throttle at idle). Which will glide farther from 10,000 feet? The jet would likely glide about 50 per cent farther than the Cub under the conditions cited — the Weights were just put in to cloud the issue. The jet would likely have a max

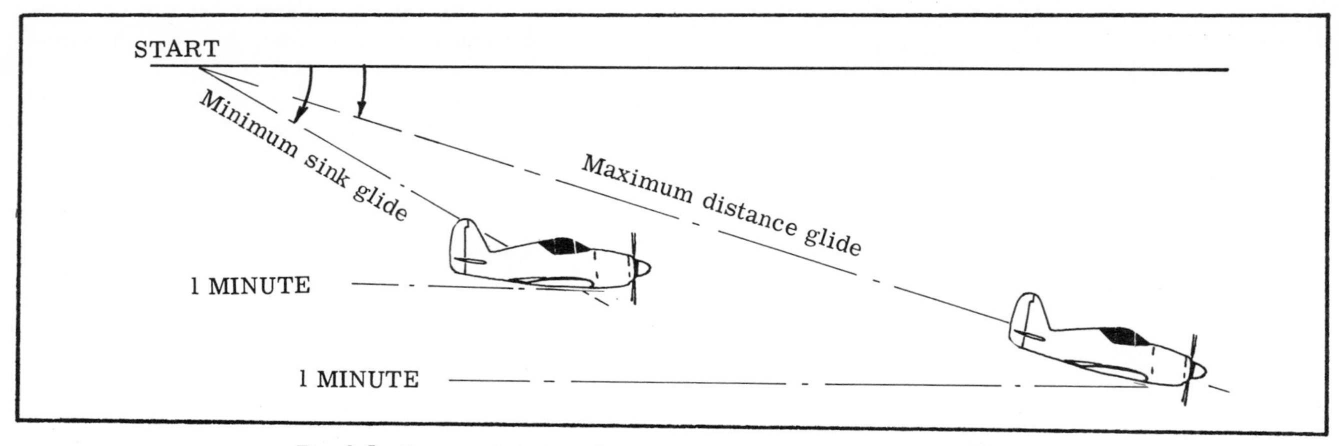

Fig. 8-5. Exaggerated view of minimum sink and maximum distance glides.

Lift-to-Drag ratio of 15:1 whereas the Cub would likely fall in the area of a max L/D ratio of about 10:1. A max distance glide speed of 200 knots would be reasonable for some of the earlier jet fighters while the Cub's best glide speed would likely be in the neighborhood of 45 knots at gross Weight.

The jet's glide angle would be only about two-thirds as steep as that of the Cub but it would be moving down the shallower slope four and a half times as fast. The sum total would be that the fighter would have reached the ground long before the Cub— but it would end up much farther away. (Fig. 8-5 could also be seen as an example of the two airplanes just mentioned, the Cub naturally being the airplane on the left.)

Take a look at Fig. 8-7 which is the Rate of Sink vs. Velocity (indicated airspeed) for a particular airplane. While it might appear that gliding too slowly

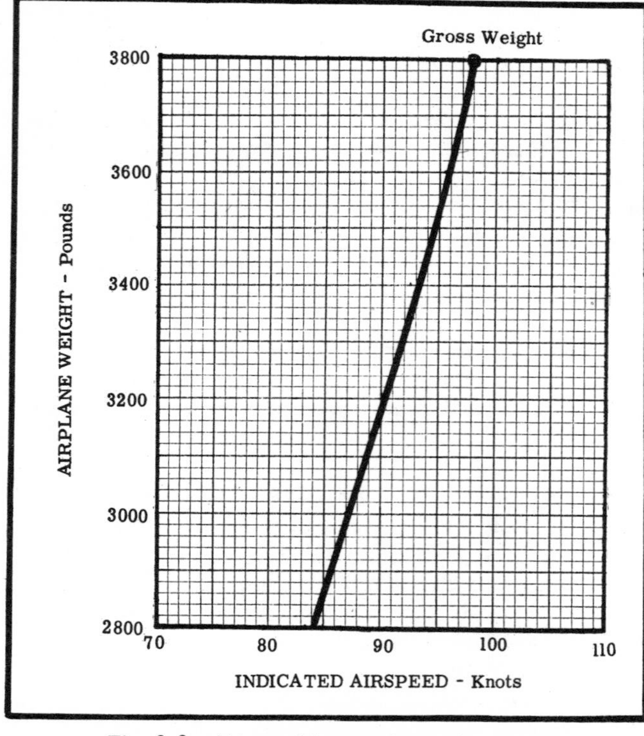

Fig. 8-6. Airspeed for maximum distance glide versus Weight for a particular airplane.

is better than gliding too fast, the glide angle for 70 knots is the same as that for 160 knots. Fig. 8-7 shows that for the two airspeeds, although the glide angle is the same at 70 and 160 knots, there is a great deal of difference in the rates of sink. The lower airspeed would give a lesser rate of sink, but *as far as the distance covered is concerned both speeds would be bad.*

Back to stretching the glide: suppose a pilot has an engine failure and is trying to make a field. He doesn't know the maximum distance glide speed for his airplane (which is 100 knots— the one used for the graphs) so he uses a speed of 70 knots. Looking at Fig. 8-7 you can see that he is definitely *not* getting the maximum distance and he soon sees that it will be very close— if he's going to make the field at all.

So, like a lot of pilots, he tries to stretch the glide by pulling the nose up until he's indicating 60 knots. Fig. 8-8 shows that the 10 knots slowdown has resulted in a much steeper glide angle and higher rate of sink— now he surely won't make it! He'd have been much better off to have *added* 10 knots and held 80 knots.

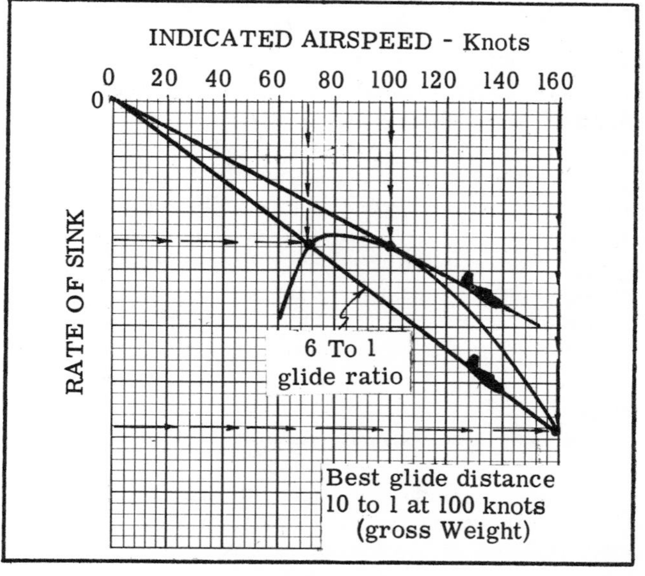

Fig. 8-7.

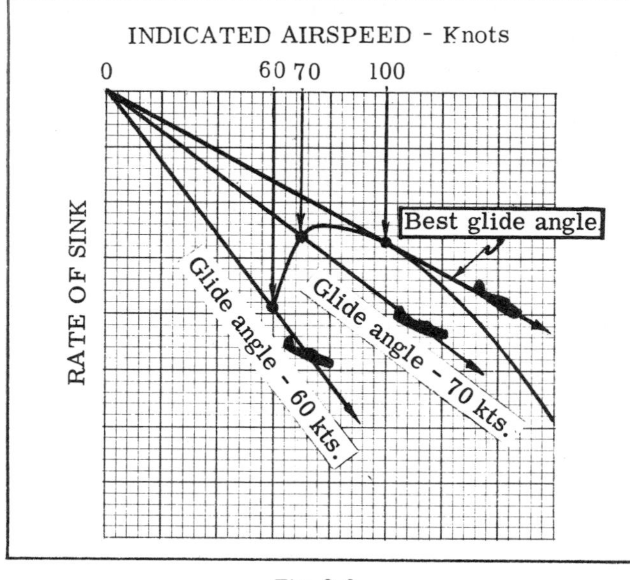

Fig. 8-8.

As the graphs show, the closer to the stall you get, the more the glide ratio and the sink rate are affected by a change in airspeed.

Know the max distance glide speed for your airplane and stick with it — don't try to stretch the glide.

ALTITUDE EFFECTS ON THE GLIDE

It was mentioned earlier that Weight had no effect on the maximum distance glide if the condition of maintaining the max L/D angle of attack is followed. (And with no angle of attack indicator your only course is to vary the airspeed with Weight change to maintain the constant angle of attack.)

Altitude does not change the maximum distance glide *angle* if the proper indicated airspeed is maintained for the current Weight. Fig. 8-9 shows a Rate of Sink versus *True* Airspeed curve for an airplane at sea level and some altitude at the same Weight. The curves have been "stretched" apart for clarity. Notice that the line which represents the maximum distance glide is tangent to both curves. The difference is that the true airspeed is greater at altitude — the airplane, although indicating the same, is moving down the slope at a greater rate and hence has a greater rate of descent.

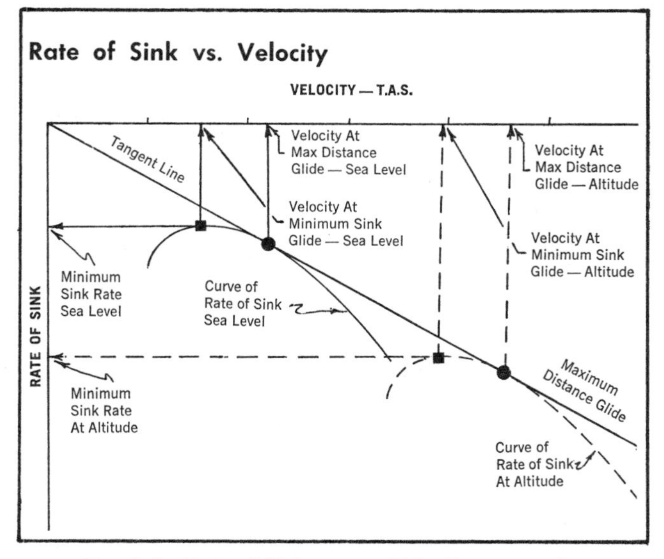

Fig. 8-9. Rate of Sink versus Velocity curves for a particular airplane at two altitudes (same Weight).

There have been arguments about whether a higher approach speed should be used for landings at airports of higher elevations. Assuming that the airplane weighs the same in both instances, it will stall at the same *indicated* airspeed at altitude as at sea level, but its *true* airspeed will be much higher at touchdown, hence it will use more runway at the airport of higher elevation (but the landing roll is a subject for another chapter).

Now, it's agreed that the airplane will stall at the same indicated airspeed at higher altitude so there should be no problem — except that the rate of descent is greater for that same approach speed (I.A.S.). (Fig. 8-9)

If you are in the habit of crossing the fence at an I.A.S. just above a stall you might find that at higher elevations you'd require just a touch more power than usual to sweeten the landing because of this greater rate of descent.

The chances are good that you wouldn't even notice the difference on landing except at very high altitudes and, as it's a matter of judgment or "eyeballing," would handle it with no problem. Adding airspeed would increase the landing roll as will be covered in the next chapter.

A problem could be encountered on a short-field approach at a higher elevation (and higher density altitude). If the pilot chops power and starts sinking, the increased sink rate as compared to sea level (for the same I.A.S.) might fool him. Added to this would be the fact that there is less horsepower available (for unsupercharged engines) to stop the sink rate at the higher altitude. The best thing would be to exercise more care to avoid getting into such a condition at higher altitudes.

WIND EFFECTS ON THE GLIDE

Wind affects the glide distance (and angle) for a particular airspeed as you've noticed, particularly on power-off approaches. From a practical standpoint it is unlikely that in an engine-out emergency you would want to take the time to worry about working out a new max distance glide speed for a head- or tailwind. The theoretical side of the problem is that you would add a few knots of indicated airspeed to the best glide speed for a moderate headwind and subtract airspeed for a moderate tailwind, basically the same idea as was discussed in the chapter on maximum range. It has been shown that, for instance, increasing the airspeed to take care of a headwind (or decreasing for a tailwind) makes only a slight difference in glide distance in these conditions for normally expected winds. So you would most likely be better off in an actual emergency using the no-wind glide speed for headwind or tailwind conditions, rather than further complicating an already complicated situation. You have other things to do such as picking a landing spot, trying to locate the trouble in the cockpit, and deciding whether to land gear-up or down.

Fig. 8-10 shows the effects of a headwind or tailwind on the glide angle for a particular airspeed.

You may note the similarity to the wind triangles you did earlier in your training; here the only difference is that the vectors are in a vertical plane. The length of the no-wind glide vector represents the T.A.S. and the others represent the glide angles and "ground speed" for the winds given. The glide angles have been exaggerated for clarity.

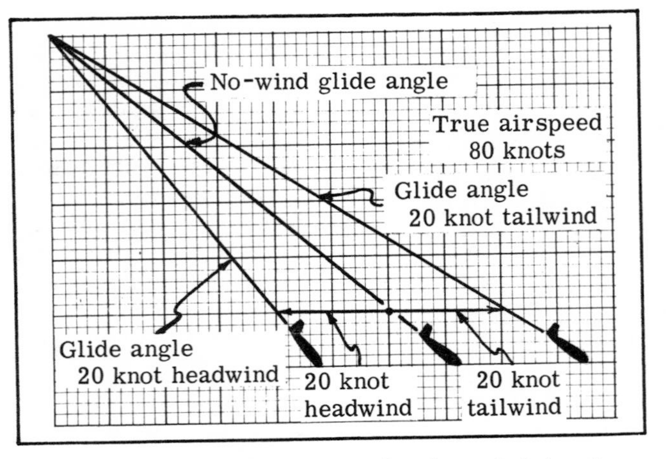

Fig. 8-10. The effects of a headwind or tailwind on the glide angle of an airplane at a particular true airspeed.

9. LANDINGS

APPROACH

THE NORMAL APPROACH

The rule that a good landing is generally preceded by a good approach is a true one. If you approach at too high an airspeed (adding five knots for the wife and two knots for each of the kids and maybe a little because it's Sunday), you'll use more runway than is necessary and won't be flying the airplane efficiently. Murgatroyd Sump, private pilot, has a beautiful wife and eleven fine children at home. It is usually necessary to shoot him down to keep him from using the whole five thousand feet of runway (Murgatroyd just doesn't fly into a field any shorter). He hasn't read his Airplane Flight Manual for a recommended approach speed nor does he know that a rule of thumb for normal approaches is an indicated airspeed of approximately 1.3 times stall at the flap setting (no-flaps) he uses. If he used full flaps, then it would be 1.3 times the figure given at the lower end of the white arc. No flaps — he would use the bottom of the green arc for setting up his approach speed. Also, although his plane has flaps, Murgatroyd originally trained in an airplane that didn't have flaps and old habits die hard. Besides, he tried using flaps once and it felt "funny," so he didn't like it.

THE USE OF FLAPS DURING THE APPROACH

For a normal approach and landing, use as much flaps as is consistent with the wind conditions. For strong and/or gusty winds use less or, perhaps, no flaps. Some airplanes have steep glide angles with no power and full flaps, and it may be preferable to use some setting less than full flaps when planning a power-off approach.

Using flaps will help you maintain the recommended approach speed, whereas with no flaps you will nearly always be too fast on final and at the start of the landing. With flaps, you will land at a lower airspeed, using less runway and making it easier on the tires. This makes a big difference if you're operating out of an airport with a hard-surfaced runway.

A major area of disagreement between pilots is *when* to put the flaps down on approach. Some pilots advocate putting the flaps down in increments. If they plan on using full flaps, they put down one-fourth flaps turning base, increase the setting to one-half just before turning on final, and then make the setting to full flaps on final. This makes for a great deal of activity in the cockpit — when your attention should be directed outside. It also means varying power, airspeed and approach angle. If you are flying a jet airliner with a copilot, this is fine, but we're talking in terms of single-pilot airplanes.

Have the final flap setting completed on base and do it in one move so that you can put your attention to using wheel, rudder and throttle — not the flap handle. Set your flaps, set up power (if necessary) and fly the airplane. This means that no matter what flap setting you plan on using on the landing, one-fourth or full flaps, this should be completed before you reach a mid-base leg position.

Another reason for having the final flap setting completed on base is that airspeed is easier to control. If you are too fast on final the sudden application of flaps may result in altitude gain and the possibility of being too high. Unless you have a very long final (and this is bad) you won't have time to get set up in attitude and airspeed. Generally the base leg is slightly faster (about 5-10 knots) than the final. (This is no absolute law -- you can use final approach speed all the way around if you wish.)

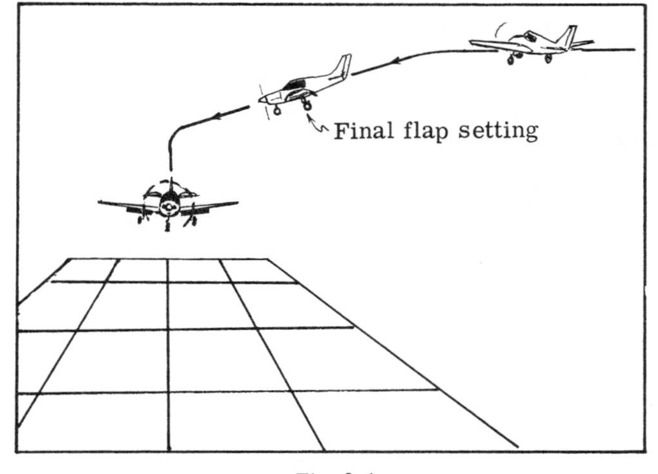

Fig. 9-1.

If you have to make a go-around you'll want to clean the airplane up as expeditiously as possible. Gear and flaps create Drag (and require horsepower). There is an argument against raising the flaps too fast in a critical situation right at the stall but the average pilot new to their use tends to be too timid in bringing them up in such a case. For many airplanes the addition of full power will just about offset the difference in stall speeds between flaps-up and full flaps down. Some Airplane Flight Manuals recommend raising of flaps before raising the gear in a go-around. At any rate, don't be so particular about seeing how slowly and smoothly you can raise the flaps that you fly into some object off the far end of the runway. When you check out in a new airplane do some simulated go-arounds at altitude in the full-dirty condition; pick a "base" altitude for the ground and try different techniques (flaps up slowly, then gear; gear up and then flaps slowly; flaps and gear up immediately, etc.) The check pilot will also have recommendations for the best technique for that particular airplane. Add power first in every case to stop or decrease the sink rate and then use the recommended clean-up procedure.

There are two terms in flap usage that are often confusing to the initiate (and to the pros also). These are "milking" and "dumping" the flaps. "Milking" usually is associated with easy and incremental operation, whereas "dumping" is a more hasty movement of the flaps. There is little agreement as to which direction these movements must take. If you were a copilot and were ordered to "milk the flaps" from a half setting, which way would you go? It would depend on your situation, of course. If on final it would *probably* mean "milk 'em down." On a take-off or go-around it would *probably* mean "milk 'em up." "Milking" usually is associated with incremental *retraction* but "dumping" can mean anything (and usually does). Get in the habit of thinking of what you want. Asking for a particular setting like, "Give me half flaps" is better than saying "Dump 'em." Your future copilots will appreciate it more and so will you.

TRAFFIC PATTERN

Nothing looks worse or delays traffic more than a drawn-out final approach. You'll be operating into some pretty busy airports and traffic controllers don't appreciate some guy in a light plane with an approach speed of 70 knots who happens to be making a three mile final. The results are cumulative. The pilot in the faster airplane will have to make an even longer final to keep from running over you and the cycle begins; each plane following must go farther before turning final and YOU are the instigator of all this.

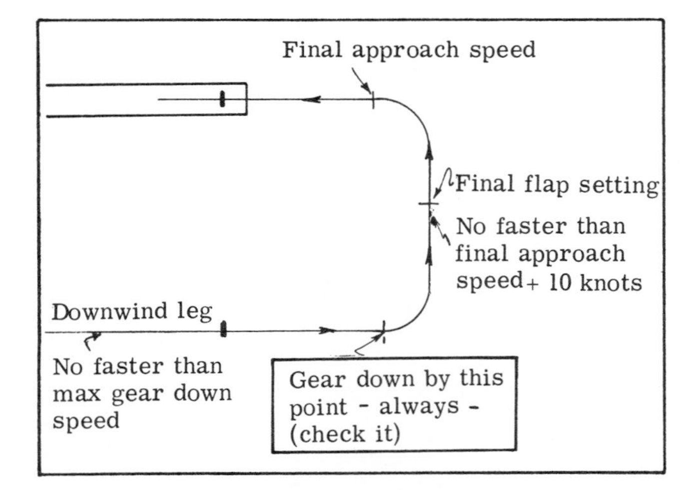

Fig. 9-2. Normal landing pattern.

Fig. 9-2 shows a typical landing procedure for a retractable gear plane using flaps.

There'll be times at a busy airport when you'll be rushed (there's a Connie back there that looks as if he'll chew up your flippers any minute). *But always have the gear down by the time you turn on the base leg.* This must become an ingrained habit. In some cases, in order to expedite the approach, you may not use flaps and will come in "hot." The runways will be long enough at these big airports to handle you without flaps but not without landing gear. If you persist in cluttering up busy runways with airplanes resting on their bellies, you won't be welcome after a while.

THE LANDING

LANDINGS IN GENERAL

The idea with any normal landing is to have the airplane touch down as slowly as possible, consistent with wind conditions. This is sometimes forgotten by pilots who fly by themselves for many hours. When they start practicing for the commercial flight test they find that old habits are hard to break.

The most prevalent misconception about landings among private pilots, even among those who learned to fly on tailwheel airplanes, is that the tricycle gear requires special technique in landing. Maybe it should be put the other way -- they think that *no* technique is required. They may listen to the instructor during the check-out, but sooner or later they get the habit of landing the tricycle gear in a too-nose-low attitude. Some even go so far as to land on all three wheels at the same time on all occasions and wind conditions. This means that the airplane is not stalled and lands at a much higher speed than is necessary.

You can think of the landing distance in terms of kinetic energy that must be dissipated before the airplane is stopped. Kinetic Energy = $\frac{1}{2}MV^2$, where M = mass of the particular airplane (Weight in pounds divided by 32.2); and V^2 is the touchdown velocity (in feet per second) squared.

While your landing distance is directly affected by Weight (double the Weight and you *double* the energy to be dissipated), the effect of velocity is even more pronounced (double the velocity at landing and the energy to be dissipated is *quadrupled*).

An approximation for landing roll distance can be obtained from: $\text{Landing Roll} = \frac{(\text{Velocity of Landing})^2}{-2a}$ where -a is equal to a deceleration of 7 feet per second, per second. This is for airplanes on a concrete runway with normal braking. Converting this to miles per hour, or knots:

Ground Run (no wind) = $0.225 V_L^2$ (mph) or $0.3 V_L^2$ (knots). Fig. 9-3 shows some comparative ground roll figures for various airplanes.

You will notice that the Airplane Flight Manual figures are usually lower than those arrived at using the equation. The Airplane Flight Manual figures

Airplane	Flaps-down stall speed		Landing Roll — Airplane Flight Manual Figures	Landing Roll — Equation Figures
	mph	knots	feet	feet
1	50	43	360	565
2	56	49	350	705
3	61	53	600	835
4	55	48	560	680
5	49	42	500	540
6	54	47	500	655
7	43	37	350	415
8	62	54	900	870
9	56	49	750	710
10	74	64	620	1230
11	59	51	535	780
12	71	61	700	1130
13	82.5	72	1000	1530

Fig. 9-3. Landing rolls, sea level, no wind (gross Weight).

come from flight tests by test pilots who are old pros. Pilot technique can make a lot of difference on take-off and landing. Maximum range and endurance airspeeds are precomputed figures, but even there a pilot who is more skillful in leaning his engine will get more out of his airplane. Pilot technique shows up the most on take-offs or landings. The test pilots can get these published figures for landing, but can you? Their braking may be greater than you normally use. Airplane manufacturers are in a highly competitive business and they will get the best performance possible. The given equation is an approximation — but don't cut your planning too closely.

VARIABLES AFFECTING THE LANDING ROLL

ALTITUDE EFFECTS

The landing is not affected as greatly by altitude as the take-off. Engine performance is not a critical factor on the landing as it is usually at idle so the altitude effect generally can be more easily predicted.

In the equation, Ground Run = $0.225VL^2$ (mph), or $0.3VL^2$ (knots), the VL (Landing Velocity) is the true airspeed. At sea level *true airspeed* and *indicated airspeed* are the same (assuming your airspeed indicator is exactly right — which doesn't usually happen). Remember that the airplane will stall at the same wings level indicated airspeed at *all altitudes* (assuming same Weights) but the true airspeed will increase *2 per cent* per thousand feet. This means that if you stall the airplane at sea level and then at ten thousand feet the indicated airspeed at the "break" will be the same at both altitudes, but your actual speed with reference to the air at ten thousand is 20 per cent (10 x 2 per cent) faster than at sea level. In calm air this means that you'll also contact the ground at landing 20 per cent faster — which results in a longer ground roll. VL goes up 2 per cent per thousand feet but this figure is squared in the landing equation so that *the effect on landing run is to add 4 per cent per thousand feet for density altitude effects.* For every thousand feet of density altitude above sea level add 4 per cent to the landing run as given for sea level standard conditions in the Airplane Flight Manual.

TEMPERATURE

While it would be best to compute density altitude and use this directly for landing computations, it involves the use of a conversion table. It may be easier to compute for altitude and temperature effects separately. A rule of thumb may be of some help.

Landing rolls are not as greatly affected by temperature changes as take-off runs because of the very small amount of power used, whereas take-offs depend on full power available from the engine. If you know the pressure altitude, you know the standard temperature for this altitude (from 59° F subtract $3\frac{1}{2}$° F for every thousand feet of pressure altitude). For a pressure altitude of 6000 feet the standard temperature should be 59 - 21 = *38°F. For every 15°F above the standard temperature for this altitude add 4 per cent to the landing run as computed for pressure altitude effects* (subtract 4 per cent for every 15° F below this figure).

Remember that a non-standard temperature affects the pressure altitude indication of the altimeter. However, for the thumb rule used here it is normally ignored.

As an example, suppose your airplane uses 800 feet for a landing run at sea level, in no-wind conditions. You are landing at an airport at a 6000 foot-pressure altitude and the last sequence report gives the surface temperature as 48° (Fahrenheit).

(a) The temperature is 10° above normal for the field altitude standard (38°). The pressure altitude is 6000 feet so this means an increased landing run of 6 x 4 = 24 per cent for altitude effects: 1.24 x 800 = 992 feet. Added to this figure is 2 2/3 per cent (call it 3) for the extra 10° of temperature; 0.03 x 800 = *24 feet.* Your landing roll will be 216 feet longer landing at this airport. The 2 2/3 per cent figure was arrived at by the ratio $\frac{10^\circ\ F}{15^\circ\ F} = \frac{X\%}{4\%}$; X = 2 2/3 per cent. (Total roll = 1016 feet.)

(b) If you had an altitude conversion chart you would find that at a pressure altitude of 6000 feet and a temperature of 48° F, the density altitude would be 6800 feet. You would then use straight altitude effects: 6.8 x 4% = 27.2 per cent; 1.272 x 800 = *1018 feet.* In either case you are close enough to be in the ball park (or the airport).

In step (a) above you converted to the correct density altitude the hard way by working with pressure altitude and temperature separately. You assumed the density altitude to be 6000 feet and then corrected this assumption for temperature effects.

If you know the field elevation of the destination airport and are able to get the altimeter setting and temperature from a sequence report, you could work out your probable landing run on the way in if you think it's going to be a close squeeze on landing.

Take the same airport discussed earlier. We stated that the pressure altitude was 6000 feet. Here's one way we could have arrived at that figure: The field elevation is 5700 feet and the latest altimeter setting for the area is 29.62. This setting is corrected to sea level. If the field elevation and pressure altitude had been the same, the altimeter setting would have been 29.92 (remember that setting an altimeter to 29.92 gives the pressure altitude). But this altimeter setting is .30 of an inch *low* for the pressure altitude setting of 29.92. This means that the pressure altitude at the destination is approximately 300 feet higher than the field elevation. The pressure altitude is 5700 plus 300, or 6000 feet (another canned problem here). If you had ignored the three tenths of an inch or 300 feet of pressure altitude, what would have happened? Suppose you call the field elevation the pressure altitude. The altitude effects would have been 5.7 x 4 = 22.8 per cent; 1.228 x 800 = 982 feet. Correcting for temperature as before (plus 3 per cent); 24 + 982 = 1006 feet, a difference of 12 feet from the first calculation. If the destination altimeter setting is within a half an inch of 29.92 inches (29.42 to 30.42), use the field elevation for your pressure altitude correction and then correct for temperature. The altimeter setting is nearly always within the above stated limits. Even if the corrected altimeter setting was *one inch* off it would only mean a difference of 4 per cent (about 40 feet) in the landing run in the above problem. You can use up 40 feet or considerably more by poor pilot technique; *so for an approximation of pressure altitude the field elevation works fine for landing.* For simplicity, the effect of temperature on the landing roll can be ignored unless the temperature is extremely high or low for the landing altitude. An approximation you can use to correct for altitude and higher temperature effects calls for you to *add 5 per cent per thousand feet of field elevation. This saves a lot of computing.* Actually it's pretty ridiculous to work it out to the nearest 6 feet (1006 feet) and it was done only to show the arithmetic involved.

AIRPLANE WEIGHT

The effect of Weight on landing roll is generally considered to be straightforward — increase the airplane Weight 20 per cent and the landing roll increase is close to this figure. Airplane Flight Manual figures show that there is a difference of about 5 per cent between the two figures. That is, if the Weight is decreased 20 per cent, the landing roll is decreased only about 15 per cent. Or, if the Weight is increased 20 per cent, the landing roll is increased about 15 per cent. *But for estimation purposes the percentage of Weight change equals the percentage of approach and/or rolling distance change.* Landing roll is directly proportional to Weight at landing.

RUNWAY CONDITION

Added rolling resistance in the form of high grass, soft ground or snow naturally shortens the landing roll. The effect of increased rolling resistance on the landing is to help in all cases — unless it becomes so great as to cause a nose-up. As you will probably be using brakes in the latter part of the roll, no set figures can be given here.

WIND

The wind affects the landing roll exactly as it does the take-off run, and the rule of thumb, 90% - $\frac{\text{Wind Velocity}}{\text{Landing Speed}}$ = percentage of no-wind runway, can be used. If the wind velocity is 20 per cent of your landing speed, you'll use 70 per cent of the published figure for no-wind conditions (90% - 20% = 70%). Fig. 4-4 also can be applied.

BRAKING

For normal landings, use aerodynamic braking by holding the nose wheel off (in the tricycle gear type) and leaving the flaps down. Aerodynamic Drag, $D = C_D S \frac{\rho V^2}{2}$, is a function of the square of the velocity. As you slow down on the roll to one-half your landing speed, the aerodynamic Drag is approximately one-fourth of that at touchdown. Aerodynamic Drag is not as important as wheel braking. Use aerodynamic Drag for what you think is about one-fourth of the expected landing roll, then lower the nose and use brakes as needed. In lowering the nose you increase the rolling resistance by decreasing Lift, remembering that rolling resistance = μ (Weight - Lift). The less the Lift, the greater the rolling resistance.

Some pilots start applying brakes as soon as they touch down. The brake effectiveness is not at its best because the airplane still has some Lift (though not enough to support the airplane) and this usually results in skidding and a lesser braking effect.

Once you've lowered the nose, flaps can still give aerodynamic Drag, so for normal landings leave them down, particularly if you're using full flaps. For cleaner airplanes, *full* flaps help more in aerodynamic Drag than they hinder the braking action by any Lift furnished by their being down. Leave 'em down throughout the roll and save your brakes. For short fields, get the flaps up shortly after landing, as soon as you feel you've gotten the most Drag out of them. In this case you are not interested in taking care of the brakes but want to stop in as short a distance as possible. *Hold the wheel full back as you brake.*

To get the most out of your brakes, you will want to apply them as much as possible *without skidding.* Not only will skidding result in the possibility of blown tires, but it will give much less braking action than found under proper brake usage.

Some of the larger airplanes have anti-skid devices so that if the wheels start to skid the brakes are automatically relaxed, even though the pilot continues to hold full force against the pedals. If the device is working properly, this may mean up to twice the braking effectiveness as compared to a pilot's braking by "feel."

In the winter it's common to see at the end of weather sequence reports — "Snow (or ice) on the runway, braking action fair (poor, good)." In the situation where braking action is poor, such as on frost-covered, wet or icy runways, it's best not to count on the brakes. Although skidding sidewise is not as critical for an airplane as a car (it says here),

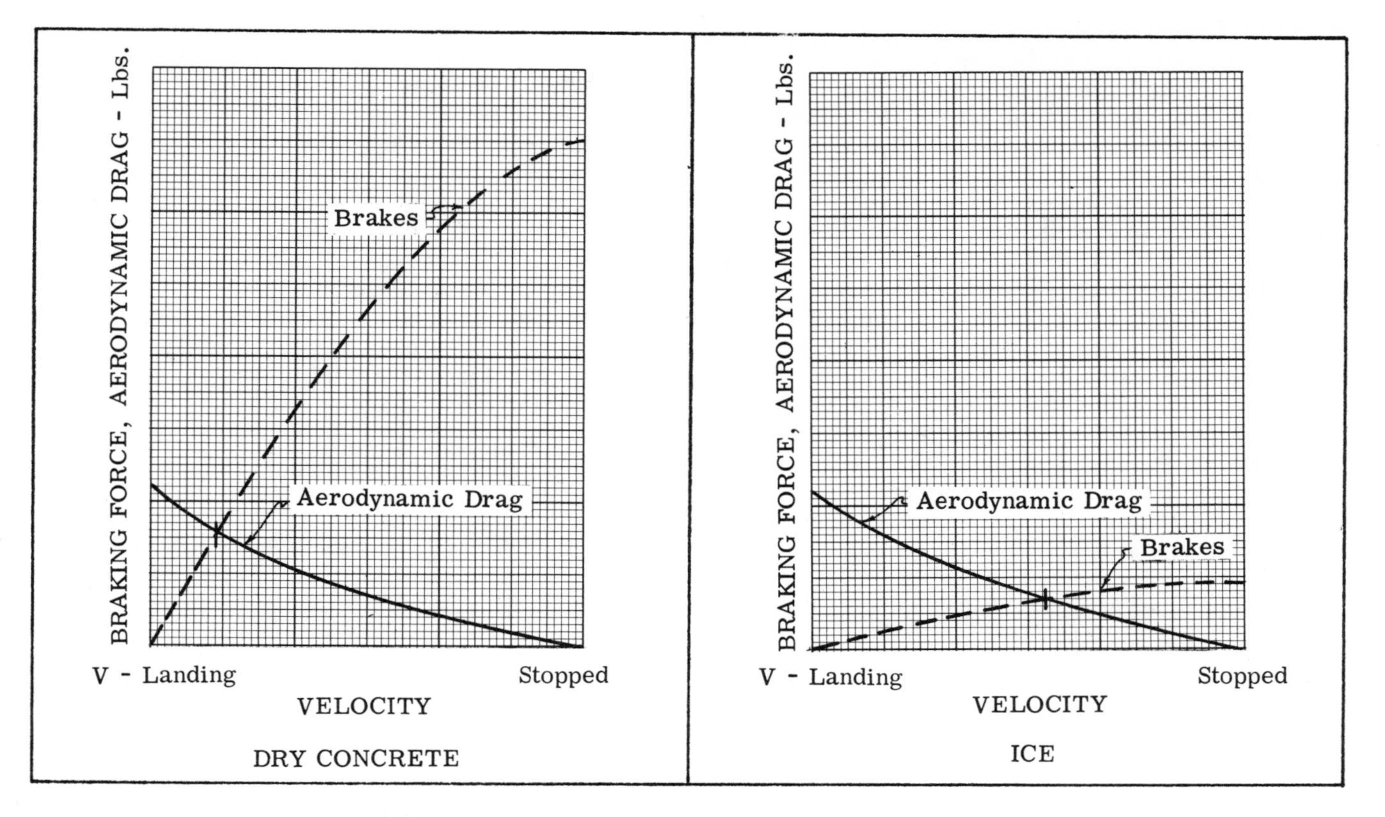

Fig. 9-4. A comparison of brake effectiveness on dry concrete and ice.

improper brake action could result in the airplane's skidding sidewise on an icy runway and if a clear spot of runway is hit, landing gear failure could result. Also, if you apply full brakes on ice, even headed straight, and suddenly hit a bare spot of runway, a tire may blow or, in the tailwheel type airplane, a nose-up could occur.

The brake effectiveness on ice is a great deal less than that for dry concrete, so aerodynamic Drag will be the big factor for an icy runway. Plan on it!

Fig. 9-4 shows some comparison of rolling resistance caused by braking and aerodynamic Drag for concrete and ice-covered runways for a particular airplane. Note that the aerodynamic Drag is the same, as it is assumed that the airplane touches down at the minimum speed and at the same Weight both times.

RUNWAY SLOPE EFFECTS

Little can be said here except that if the slope is great, it is better to land uphill and downwind (unless the wind is very strong).

PILOT TECHNIQUE

As in take-offs, here is the item that can shoot all your careful computations. Landing too fast, poor brake usage and other goof-ups can cause you to lose all you would have gained by helpful factors. The only answer is to get an occasional check ride in your airplane with an instructor, and know the variables that can affect your landings.

SHORT FIELD LANDING

When landing area is critical you want to land as short as possible without damaging the airplane. You want a safety margin of speed, but not enough to cause floating, because every foot of runway counts.

Power is used to control your approach path at the recommended speed. Fly a wider pattern so you won't be rushed or have to make steep turns at this near critical airspeed. The power approach angle will be one or two degrees shallower than the normal power-off approach for your airplane and you'll need more room.

A rule of thumb for short fields uses an approach speed of no more than 1.3 times the *power-off, full flaps down stall speed* (the bottom of the white arc on the airspeed indicator); you'll be using full flaps for a short field landing, and this ratio will give a safety factor if you should think you'll be too high and suddenly chop the throttle.

For gusty air this speed should be increased by 5 to 10 knots, as sudden changes in wind velocity can affect your airplane and you are interested in *not* suddenly finding yourself with a critically low airspeed at a bad time.

Obstacle approaches require a steeper angle of descent in that you must clear the obstacle and still land as short as possible. The danger here is that you approach at a low airspeed, and after the obstacle is passed a steep angle of descent is continued toward the ground. You may possibly find that there is not enough airspeed to allow you to flare. The airplane is rotated quickly but stalls. A sudden short burst of power just before touching can be used to cushion the landing if you get too slow. Don't leave the power on too long or you'll use too much runway.

PROCEDURE

1. Start the approach from a slightly wider downwind leg.
2. Have full flaps set and attain recommended short field approach speed on base. If you plan a long final, wait until after the final turn to set up your recommended airspeed.
3. Control the approach angle with throttle after turning on final. Don't be a throttle jockey, use minor adjustments.
4. Use power as necessary to make the spot. You'll have to use power all the way to the ground if you get low and slow.

Fig. 9-5. The obstacle approach.

SOFT FIELD LANDING

APPROACH

The approach for the soft field landing is usually a normal one — only the actual touchdown is different from other landings. Of course, in an emergency situation you may be running low on fuel and have to land in a pasture — which may be both short and soft. This would require a short field approach and a soft field landing. Never make an approach to a soft field at a higher than normal approach speed because the airplane will float and usually is "put on" at a higher than minimum speed (pilots get impatient, it seems).

If you make a short field type of approach to a soft field, you can pick the firmest possible landing spot. There may be parts of a muddy field where the grass cover is better or the snow is not drifted as deeply on a snow-covered field.

LANDING

The same principle applies on a soft field landing for both the tricycle gear airplane and the tailwheel type. Touch down as slowly as possible and with a higher nose attitude than for the normal landing. This means the use of power during touchdown. (Fig. 9-6)

If you know beforehand that you'll have to land in snow or on a soft field, it would be a good idea to have the speed fairings removed before the trip. Speed fairings look good and in most cases help out on speed a little, but they will get clogged up in short order on a muddy field or in snow. (If conditions are such that they are liable to get clogged up on the landing run itself, things can get pretty hairy.) The reason speed fairings weren't mentioned in more detail on the soft field take-off is that if the field is really soft you won't get to the take-off area anyway.

Fig. 9-6. The soft field landing, tailwheel and tricycle gear airplanes.

After touching down, keep the wheel or stick *full back* and use power as necessary to stop any nosing up tendency.

The point is to keep the tail on the ground on the tailwheel type, and keep as much Weight as possible *off* the nosewheel on the tricycle gear airplane as long as possible. The slipstream, in a single engine airplane particularly, can aid in this.

GUSTY AND CROSSWIND LANDINGS

APPROACHES IN GUSTY WIND CONDITIONS

In gusty wind conditions the approach must be flown 5-10 knots faster than for a normal approach. Little or no flaps should be used so that when the airplane is landed it won't be so apt to be lifted off again by a sudden sharp gust. If the wind is strong your landing run will be short anyway, so the higher approach speed and lack of flaps won't particularly hurt the landing roll. The approach is naturally the same for tricycle gear or tailwheel type.

WHEEL LANDINGS

The wheel landing is the best means of landing the tailwheel airplane in strong and/or gusty wind conditions in that the plane contacts the ground at a low angle of attack. You are literally flying the plane onto the ground.

Two-place trainers and other light planes of low wing loading generally do not require power to make the landing; in fact the use of power makes the problem more knotty. A "power juggler" may find himself using up more runway than is necessary.

Fig. 9-7. The wheel landing.

PROCEDURE

Make an approach about 5 knots faster than for a normal glide. The landing transition is made at a lower height for two reasons: (1) The airplane must touch down at a higher speed and (2) the attitude will be only slightly tail-low — not three-point. For the lighter planes, power is only used to control the approach — not to make the landing. Make your path a curved one that is tangent to the runway at the touch-down point. (Fig. 9-7)

The correct procedure is to land in a slightly tail-low attitude by "rounding off" the glide properly. After the plane has touched, apply slight forward pressure to keep the tail up — and maintain a low angle of attack. If you are too hasty in bringing the tail down, the chances are good of your becoming air-borne again — and you will probably have to take it around again.

Impatience will be your biggest problem on the wheel landing, particularly if you're gliding too fast. The airplane is skimming a few inches above the runway and you may try to "put it on." This results in some fancy "crow hopping" and usually means you'll have to open it up and take it around. Then there are pilots who hold the plane off too long and wind up making a half three-point, half wheels and all-bouncing type of landing when the plane settles fast on the front two wheels.

As the plane slows, continue to hold more forward pressure until you run out of elevator and the tail moves down, then hold the wheel (or stick) back to keep it on the ground.

Some airplanes have comparatively poor directional control at lower speeds with the tailwheel off the ground. There just may not be enough rudder effectiveness to do the job. In these airplanes it's best to maintain the forward pressure held at touchdown — don't push forward. As the speed decreases the tail will come down before you lose rudder control — then you may move the wheel smoothly back to the full aft position. This is a good technique in a strong crosswind for all types of tailwheel airplanes.

The crosswind correction for a wheel landing is the same as for the three-point landing. Lower the wing and hold opposite rudder as needed. Land on one wheel, the other will come down immediately. Hold aileron into the wind and apply rudder as needed to keep it straight. Remember that your airplane may be placarded against landing in crosswind components above a certain value.

COMMON ERRORS

1. Too fast an approach — the plane floats.
2. Too slow an approach or leveling off too high — the plane settles fast on the main gear and bounces.
3. Getting impatient — trying to put it on.

A bounce usually means taking it around, but you can lower the nose, apply power and re-land if there is enough runway left.

Keep in mind that wheel landings and soft field conditions don't mix very well.

GUSTY WIND LANDING FOR TRICYCLE GEAR

The approach in gusty air is the same for both types of airplanes and the landing technique is very similar. (Fig. 9-8)

You remember that the recommended procedure for normal landings for tricycle gear is to land on the main wheels and hold the nosewheel off during the initial portion of the landing roll. For strong, gusty wind conditions, however, holding the nose up may result in a sudden gust lifting the airplane off again. The best technique for this wind condition is to touch down in a nearly level flight attitude. If you used flaps get them up immediately after touchdown to lessen any chance of a gust picking you up. The objection to picking up the flaps during the landing roll

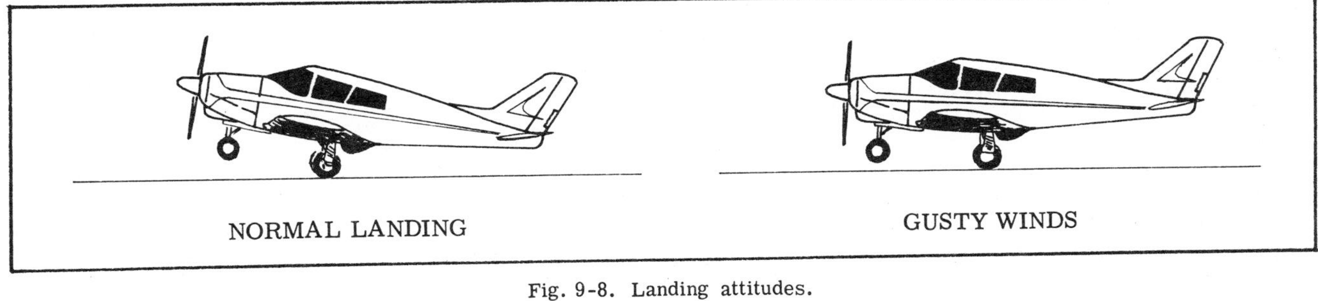

Fig. 9-8. Landing attitudes.

is that in a retractable gear airplane you could inadvertently pull up the *landing gear*. This would preclude a sudden gust picking you up as you would slow down very quickly, but it's more expensive than pulling up the flaps.

After you touch down, lower the nose immediately to decrease the angle of attack, and raise the flaps. If the wind is very strong, don't use flaps for landing.

COMMON ERRORS

1. Overdoing the idea of flying the airplane on and "slamming" it on the ground with the possibility of damaging the nosewheel.
2. Failure to take into consideration the gusty conditions and holding the nosewheel off after landing.

CROSSWIND LANDINGS

As in the gusty wind landings, the approach is the same for the two types of landing gear. You have four choices in making the approach and landing for either type of landing gear.

1. The Wing-Down Method. This was probably the method taught you as it is the simplest for light-to-moderate crosswinds. It's easy because you do not need to raise the wing or kick rudder to straighten the airplane at the last second. You hold the wing down with aileron and use opposite rudder all through the final approach and landing as necessary to keep the nose lined up with the runway. (Fig. 9-9) With strong crosswind components the wing may be down to such a degree that the slipping approach is uncomfortable to the passengers. In extreme cases, the lowered wing may be in danger of striking the ground, particularly in a low wing airplane. Chances are in such cases you wouldn't land at that airport, but would find one having a runway more into the wind.

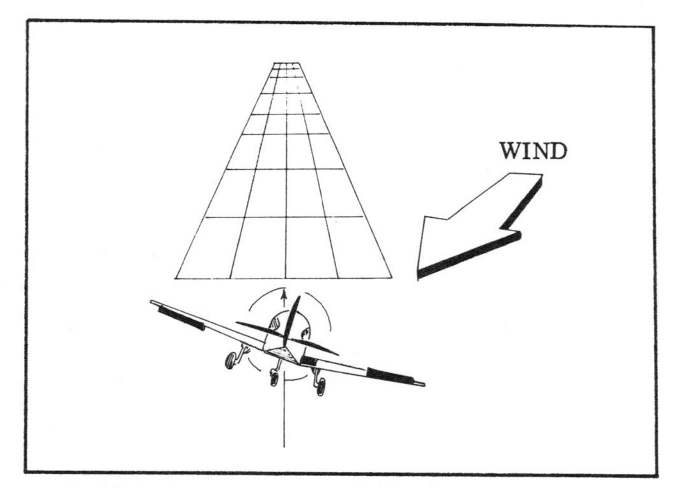

Fig. 9-9. Wing-down method of crosswind correction.

If you are low on fuel and must land, this method may limit your correction for strong crosswind.

Common Errors — Probably the most common error committed by private pilots in this type of correction is using too much top rudder, yawing the nose away from the runway. The nose should be lined up with the runway during the approach and landing. Also, some pilots try to raise the down wing at the last instant. This isn't necessary.

2. Crab Method. This makes for a comfortable approach as the plane is not slipping, but it has the

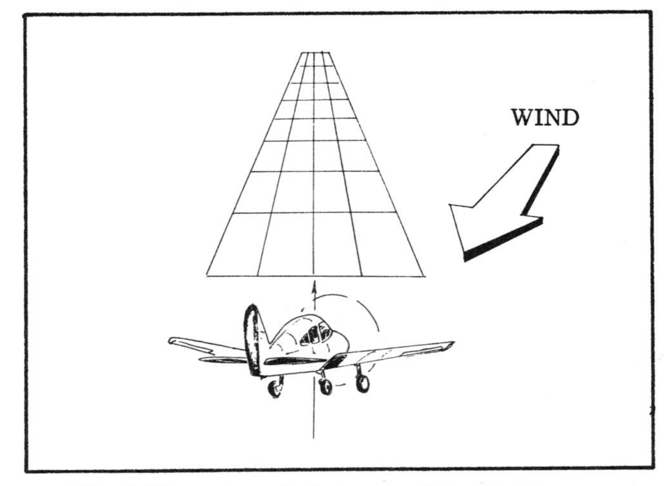

Fig. 9-10. Crab method of crosswind correction.

disadvantage of requiring fine judgment in knowing when to kick the airplane straight. Also, if the crosswind is strong, the crab angle will have to be so great as to make it doubtful that you would have enough rudder effectiveness left to completely straighten the airplane before it touched — and this makes for a possible ground loop. If the airplane hits in a crab, you'll have a weathercocking tendency. In the tailwheel type airplane forces will be set up to aggravate the ground loop once it has started. So, the crab method also is limited to moderate crosswind components. (Fig. 9-10)

Common Errors — Not straightening the airplane at the right time. Gusts may cause the airplane to float and start drifting after you've kicked it straight — or you may touch down before you're ready, still in a crab.

3. Combination Crab and Wing-Down. The limitations of the previous methods may be overcome by combining the two. If you are able to comfortably correct for 15 knots of crosswind component by either

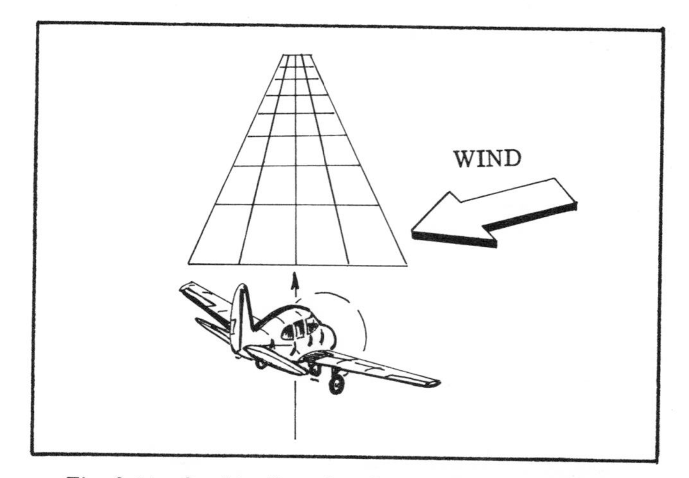

Fig. 9-11. Combination wing-down and crab methods.

method outlined above, chances are you'll find the results additive if they are combined; and you will be able to correct for nearly double the crosswind component. (Fig. 9-11)

You'll be crabbed *and* have the wing down, which means that the wing is not uncomfortably low and the crab is not such that the plane cannot be yawed straight as it touches. The idea is to yaw the plane straight but don't bother trying to raise the wing. Land on one wheel as was done in the wing-down method.

Common Errors — Getting so engrossed in one of the corrections that the other is neglected. Usually the pilot forgets he's in a crab as well as having the wing down.

4. Crab Approach and Wing-Down Landing. This makes a comfortable approach for light to moderate crosswinds. The crab is used during the final approach and, as the landing flare is begun, the nose is straightened and the wing lowered. From this point on it is the wing-down method. This avoids the long slipping approach, but it may require a couple of practice periods to make a smooth transition from the crab to the wing-down attitude.

Common Errors — Poor transition from crab to wing-down attitude, with some frantic scrambling around and poor use of controls.

THE GROUND ROLL

Here's where tricycle gear pays for itself. The center of gravity is ahead of the main wheels, which tends to straighten the airplane out. The tailwheel type reacts just the opposite. The ground roll is the toughest part of the problem in strong crosswinds. Keep that aileron into the wind to help fight wing lifting tendencies and to also utilize the down-aileron Drag to help keep you straight. The aileron is more important for the tailwheel airplane during the ground roll because of the airplane's attitude, but use it for both types.

Tailwheel Type — Keep that wheel back (if you made a three-point landing) because this will allow the tailwheel to get a good grip and help fight the weathercocking tendency.

Tricycle Gear — Ease the nosewheel on and keep it there.

10. AIRPLANE STABILITY AND CONTROL

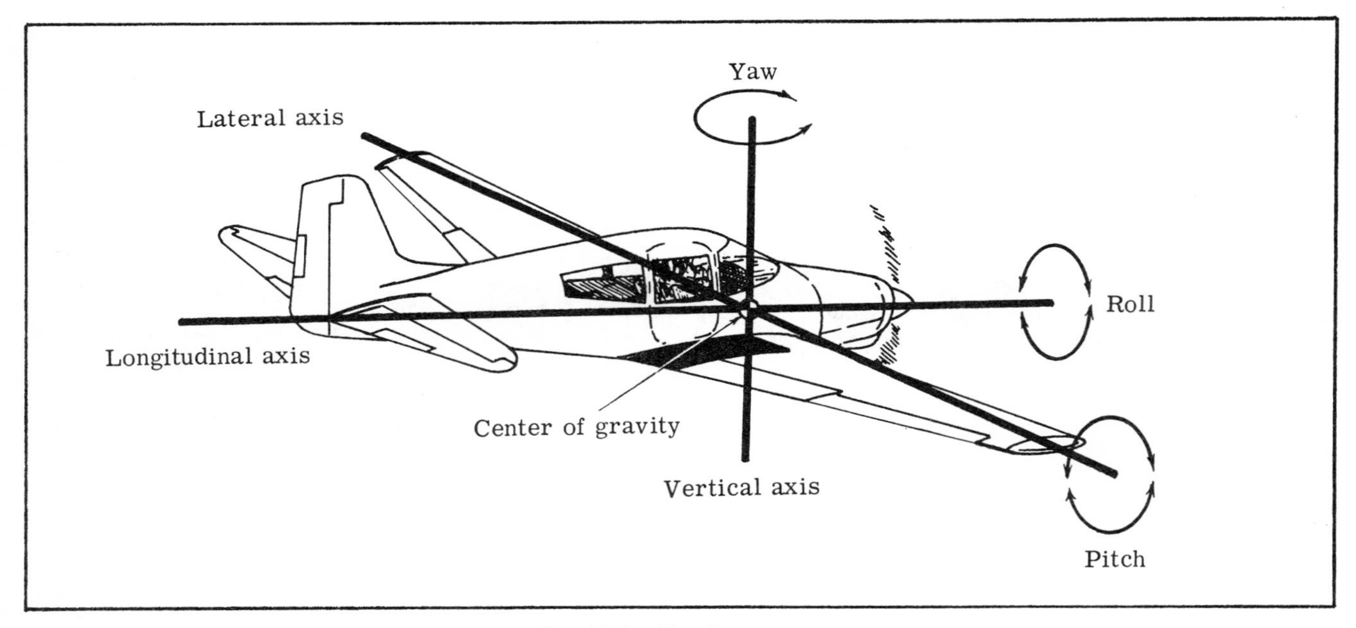

Fig. 10-1. The three axes.

THE THREE AXES

There are three axes around which the airplane moves. These axes pass through the airplane's center of gravity, or the point where the airplane weight is considered to be concentrated.

An airplane that is stable requires little pilot attention after it is trimmed for a certain airspeed and power setting. Airplanes certificated by the FAA for use in private and commercial flying must be stable around all three axes — otherwise the pilot could get into a dangerous situation if he should suffer a momentary lapse of attention. All the airplanes you have flown to date, licensed as "normal" or "utility" category, can be trimmed and flown "hands off" at all speeds from just above a stall to the red line speed.

STABILITY IN GENERAL

Stability, as defined by the dictionary, means "fixedness, steadiness, or equilibrium." An object that is "positively stable" resists any displacement. One that is "negatively stable" does not resist displacement; indeed, it tends to displace itself more and more if acted upon by an outside force. An object that is "neutrally stable" doesn't particularly care what happens to it. If acted upon by a force it will move, but neither tends to return to its position nor move farther after the force is removed.

STATIC STABILITY

Static (meaning "at rest") stability is the initial tendency of a body to return to its original position after being disturbed. An example of positive static stability is given as follows: Imagine a steel ball sitting inside a perfectly smooth hubcap. From Fig. 10-2 you can see that the ball has an initial tendency to return to its original position if displaced.

Fig. 10-3 gives an example of negative static stability. The ball is carefully balanced on the peak of the hubcap and the application of outside force results in its falling. It does *not* tend to return to its original position; on the contrary, it gets farther and farther from the original position as it falls.

Neutral static stability can be likened to a steel

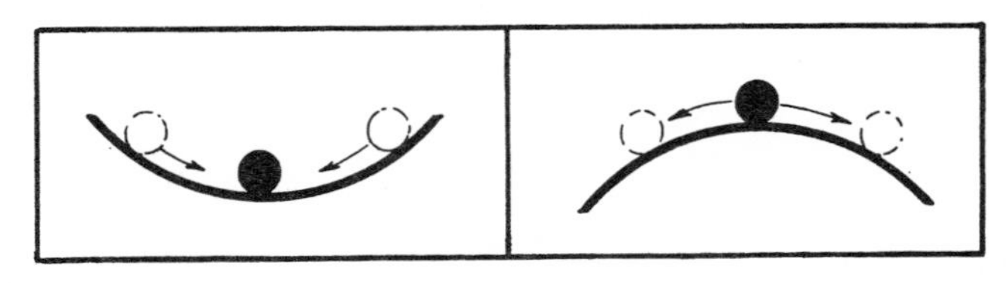

Fig. 10-2. Positive static stability.

Fig. 10-3. Negative static stability.

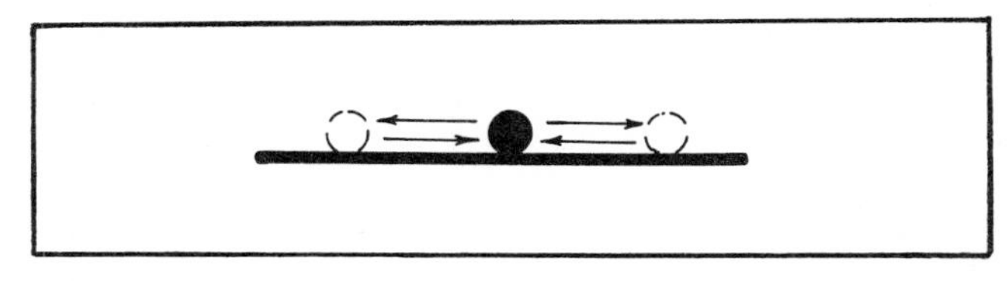

Fig. 10-4. Neutral static stability.

ball on a perfectly flat smooth surface. If a force is exerted on it, the ball will move and stop at some new point after the force is removed. (Fig. 10-4)

DYNAMIC STABILITY

The actions a body takes in response to its static stability properties show its dynamic stability (dynamic = active). This dynamic stability usually is considered to be the time history of a body's response to its inherent static stability. Looking at the earlier examples of static stability and their tie-in with dynamic stability you can note some interesting facts.

Take the example of the steel ball and the hubcap. Fig. 10-2 shows that the ball when inside tends to stay in the center of the hubcap — it has *positive static stability*. It requires force for you to displace it up the side, and it returns immediately to its original position.

Now suppose you push the steel ball well up the side of the hubcap and quickly release it. The ball will roll toward the center position, overshoot, return — keeping this up with ever shortening oscillations until finally it returns to rest in the center. The ball has positive *static* stability because it resists your pushing it up the side and has positive *dynamic* (action) stability because its actions tend to return it to the original position. *That is, the oscillations about its original position become less and less until it stops at the original point.* This would be called "periodic motion" in that the ball would make a complete oscillation in a given interval of time or "period." These periods would remain approximately the same in length (exactly the same under theoretical conditions) even though the "amplitude" (movement) would be less and less.

This, then, is called "periodic motion." You can also see this by suspending a heavy weight on a string or by making a homemade pendulum. The pendulum at rest has positive *static* stability — it resists any attempt to displace it. It has positive *dynamic* stability in that it finally returns to its original position through a series of periodic (equal time) oscillations of decreasing amplitude.

The ball in the hubcap could be given the property of "aperiodic" (non-timed) positive dynamic stability by filling the hubcap with a heavy liquid such as oil. (Fig. 10-5)

The liquid would "damp" the oscillations to such an extent that the ball would probably return directly, though more slowly, to the original position with no overshooting and hence no periodic motion. Through design or manipulation of your system (adding oil) you have caused its motions to be aperiodic.

Taking the steel ball *inside* the hubcap again; we know that it is statically stable, it resists any displacement and we saw in the above example that it had positive dynamic stability as well. The properly designed airplane does not necessarily have positive dynamic stability under all conditions (see LONGITUDINAL DYNAMIC STABILITY OF THE AIRPLANE).

The fact that an airplane has positive static stability does not mean that its dynamic stability is also positive. Outside forces may act on the airplane so that the oscillations may stay the same or even become greater.

Back to the ball inside the hubcap. Suppose you start the ball rolling and then start rocking the hubcap with your hand so that the oscillations do *not* decrease. Because of the outside force set up by you, the ball's oscillations retain the same amplitude. The system has *positive static stability* but *neutral dynamic stability* — the ball's oscillations continue without change. The airplane may also be affected by outside (aerodynamic) forces which may result in undiminishing oscillations, or *neutral dynamic stability*, even though it is properly balanced, or statically stable.

Now suppose you rock the hubcap even more violently. The ball's oscillations get greater and greater until it shoots over the side. You introduced an outside factor that resulted in *negative dynamic stability* — the oscillations increasing in size until "structural damage" (the ball went over the side) occurred.

Thus the system (or airplane) with positive *static* stability may have positive, neutral or negative *dynamic* stability. *A system that is statically stable will have some form of oscillatory behavior.* This tendency may be so heavily damped, as was shown by the oil in the hubcap, that it is not readily evident. The oscillations show that the system is statically stable in that the ball (or airplane) is trying to return to the original position. Outside forces may cause it to continually equally overshoot this position or may be strong enough to cause the oscillations to increase until structural damage occurs.

For a system that has *neutral* static stability such as the ball on a smooth flat plate, there will be no oscillations because the ball isn't trying to return to any particular position. It's displaced and stays displaced.

A system that has *negative* static stability or is statically *unstable* (the terms mean the same thing)

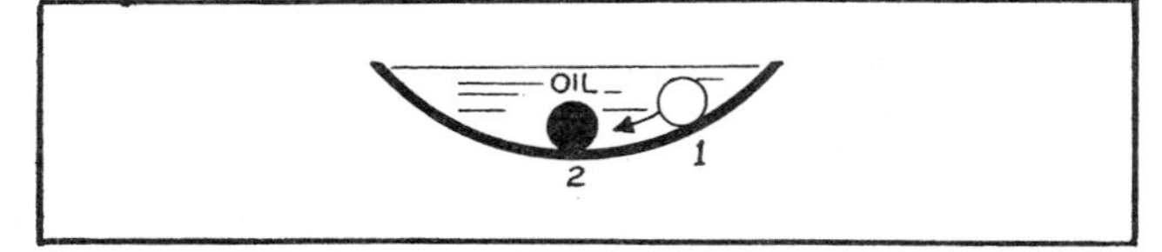

Fig. 10-5. Aperiodic positive dynamic stability.

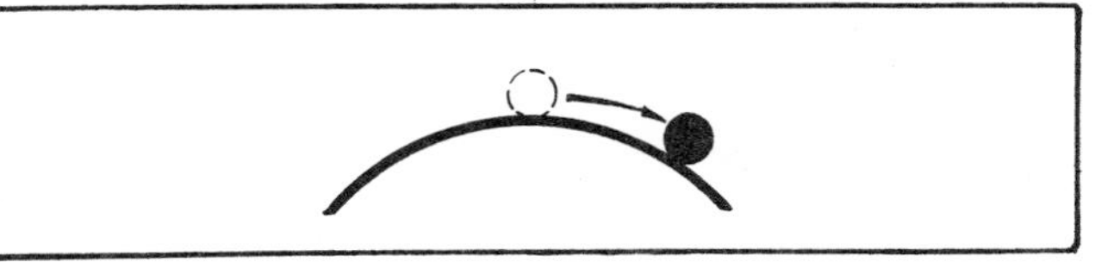

Fig. 10-6. A statically unstable system.

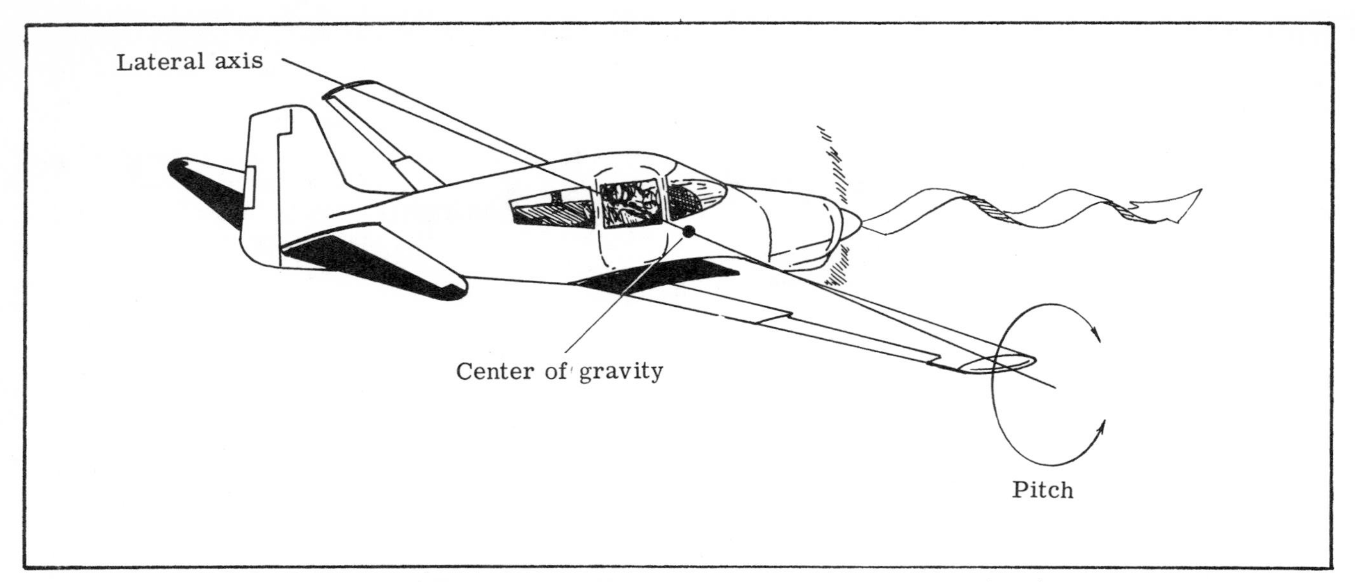

Fig. 10-7. The elevators control movement about the lateral axis (pitch).

will have no oscillations — there will be a steady divergence. Let's use the ball and hubcap again. This time turn the hubcap over and balance the ball carefully on the peak (sure you can) and take another look at the statically unstable system. (Fig. 10-6)

If even a small force is applied, the ball rolls down the side of the hubcap. The ball does not resist any force to offset it from its position — on the contrary, it wants to leave in the first place and when displaced leaves its original position at a faster and faster rate. There will be no oscillations as there is no tendency to return at all. *The statically unstable system has no dynamic* (oscillatory) *characteristics but continually diverges.* The action this system takes in diverging is not always as simple as it might appear, but *that* we'll leave for the slide rule boys. A *statically stable system* (or airplane) *may have either positive, neutral or negative dynamic stability characteristics.*

How this applies to you as a pilot will be shown shortly.

LONGITUDINAL OR PITCH STABILITY OF THE AIRPLANE

The elevators control the movement around the lateral axis (pitch). The pilot's ability to control his airplane about this axis is very important. In designing an airplane a great deal of effort is spent in making it stable around all three axes. But longitudinal stability, or stability about the pitch axis, is considered to be the most affected by variables introduced by the pilot, such as airplane loading.

Take a look at an airplane in balanced, straight and level flight. (Fig. 10-8)

Making calculations from the center of gravity you find the "moment" (force times distance) about the center of gravity caused by the wing's Lift is 5 x 3100 or 15,500 inch-pounds. This is a nose-down moment. To maintain straight and level flight there must be an equal moment in the opposite direction or

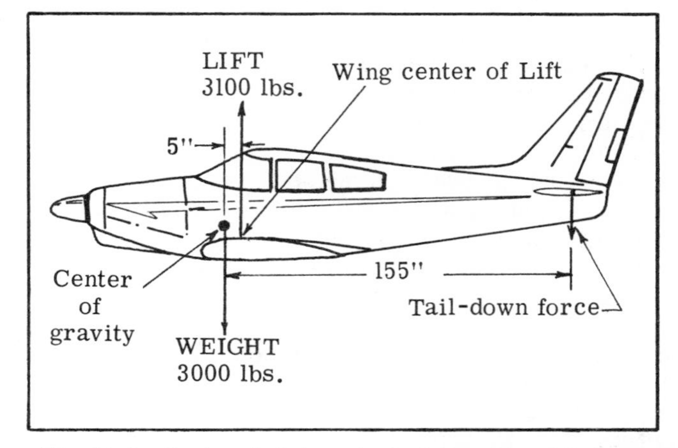

Fig. 10-8. Airplane in balanced, straight and level flight.

the airplane would be attempting to do an outside loop. This opposite moment is furnished by a down force on the tail. Its moment must be 15,500 inch-pounds in a tail-down direction. The distance shown from the center of gravity to the center of tail lift is 155 inches, therefore the down force at the tail must be 100 pounds (force x distance = 100 pounds x 155 inches). The tail-down moment is also 15,500 inch-pounds which balances the nose-down moment. The airplane is statically balanced.

In order for the airplane to maintain level flight the upward forces must balance the downward forces as was covered in Chapter 2. (Fig. 10-9)

The down forces are the airplane's weight (3000 lbs.) and the tail-down force (100 lbs.) for a total of 3100 lbs. In order to balance this, the up force (Lift) must be 3100 lbs. The wing itself contributes some pitching effects; this was also mentioned in Chapter 2.

For airplanes with fixed, or non-adjustable stabilizers, the stabilizer is set by the manufacturer at an angle that furnishes the correct down force at the expected cruising speed and center of gravity position.

The tail-down force is the result of propeller slipstream, downwash from the wing and the free stream velocity (airspeed). (Fig. 10-10)

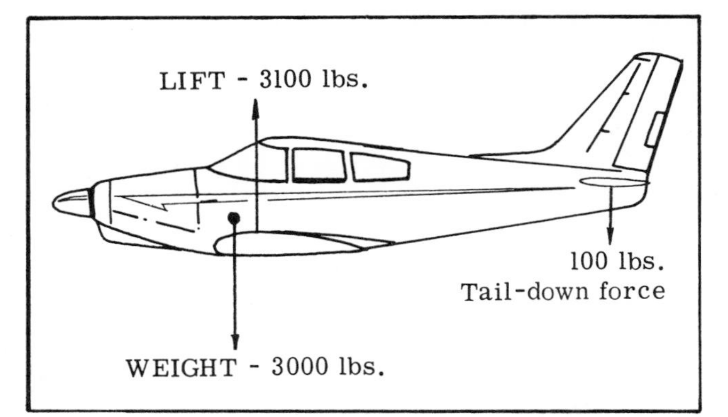

Fig. 10-9. Summation of vertical forces: total up force 3100 pounds; total down force 3100 pounds. Forces balanced.

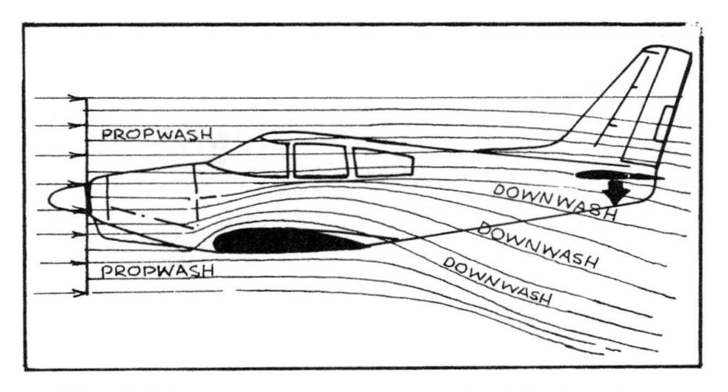

Fig. 10-10. Factors contributing to the tail-down force.

Suppose you're flying straight and level (hands-off) at the design cruise speed and power setting and suddenly close the throttle. The slipstream force suddenly drops to practically nothing; the airplane starts slowing as Thrust is no longer equal to Drag, and the free stream velocity also drops. You've suddenly lost some of the tail-down force. The result is that the nose drops. This is a healthy situation in that the airplane is trying to pick up speed and re-establish the balance. (Fig. 10-11)

Of course, as the airplane slows Lift decreases and the airplane starts to accelerate downward for a very short time, but this is not so noticeable to you as the nosing-down action.

We'll disregard the airplane settling and think only in terms of the rotational movement caused by closing the throttle.

One way of looking at it is to return to the see-saw of your earlier days. When the kid on the other end suddenly jumped off you set up your own "nose down." (The moments were no longer balanced.)

You set the desired tail force for various airspeeds by either holding fore or aft wheel pressure or setting the elevator trim. If you are trimmed for straight and level flight, closing the throttle means more up-elevator trim if you want to glide hands-off at the recommended glide speed. A propeller-driven airplane will always require less up-elevator trim for a given airspeed when using power than under power-off conditions. You can see this for yourself the next pretty day when you're just out flying around. Trim the airplane to fly straight and level at the recommended glide speed and use whatever power is necessary to maintain altitude. Then close the throttle and keep your hands off the wheel. You'll see that the airspeed will be greater in the power-off condition — the airplane's nose will drop until it picks up enough free stream velocity to compensate for the loss of slipstream. This may be up to about cruise speed — depending on the airplane.

You can see that the arrangement of having the center of gravity ahead of the center of Lift, and an aerodynamic tail-down force results in the airplane always trying to return to a safe condition. Pull the nose up, the airplane slows and the tail-down force decreases. The nose will soon drop unless you re-trim it or hold it up with increased back pressure. Push the nose down and it wants to come back up as the airspeed increases the tail-down force. The stable airplane wants to remain in its trimmed conditions and this *inherent* (or built in) *stability* has gotten a lot of pilots out of trouble.

Another arrangement is shown in Fig. 10-12.

A lifting tail is necessary on this airplane in order to maintain balance. From a pure aerodynamic standpoint the two lifting surfaces (wing and tail) are

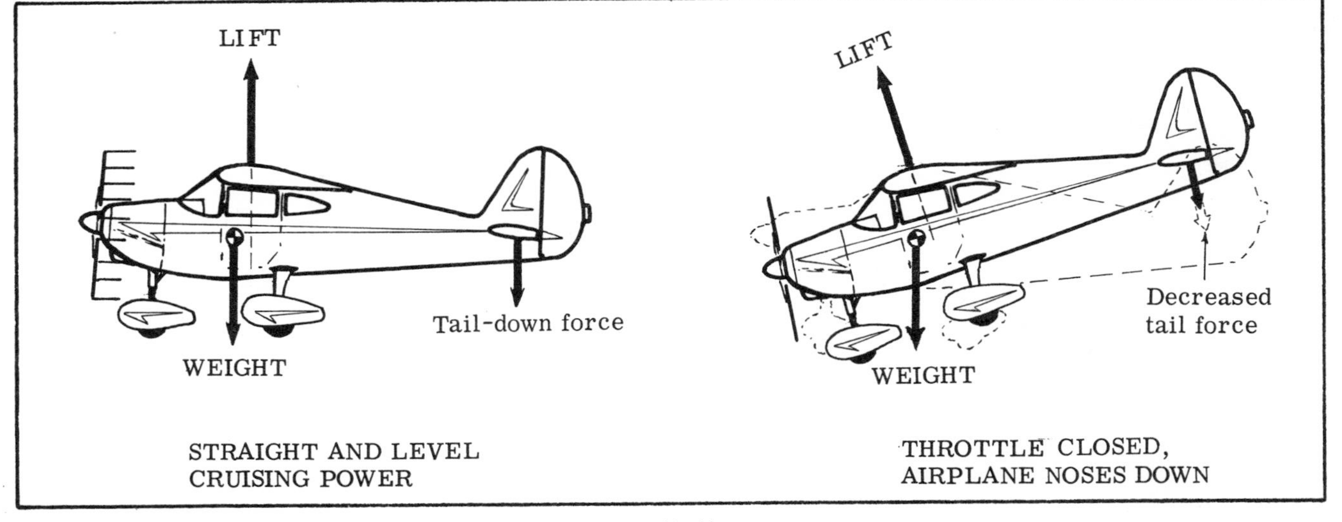

Fig. 10-11.

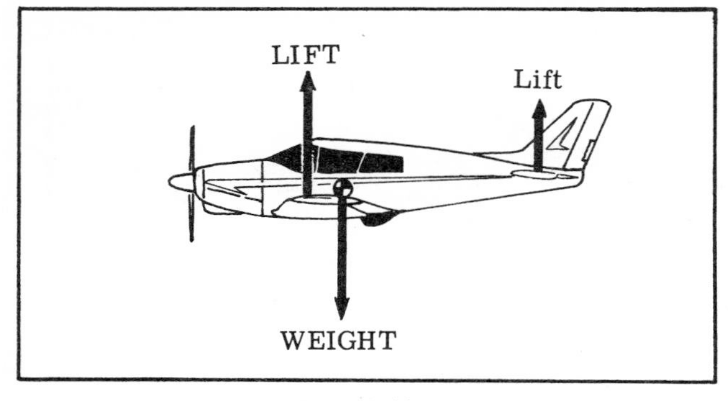

Fig. 10-12.

a good idea; from a stability standpoint this type of configuration is not so good.

When you throttle back, the tail lift would decrease and the nose would tend to go up! This would not be conducive to easy pilot control. The engineers would rather have a little less aerodynamic efficiency and more stability. So this arrangement is avoided—although it is not nearly as critical in a jet airplane. Actually, in some conditions (high C_L), a tail upload may be present, even for the "standard" airplane which has a tail-down force at cruise.

POWER EFFECTS ON STABILITY

Power is considered to be destabilizing. That is, the addition of power tends to make the nose rise. The designer may offset this somewhat by having a "high thrust line." The line through the center of the propeller disk passes above the center of gravity so that as Thrust is increased a moment is produced to counteract slipstream effects on the tail. (Fig. 10-13)

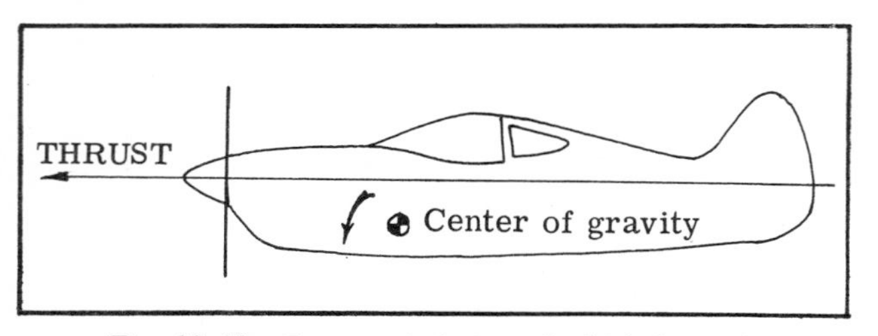

Fig. 10-13. Exaggerated view of a high thrust line as an aid to stability.

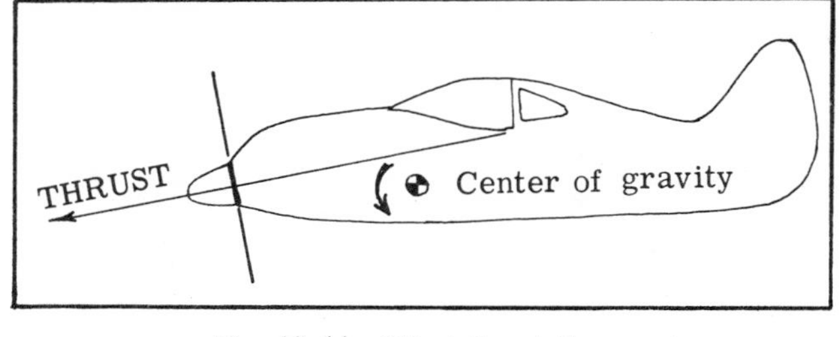

Fig. 10-14. Offset thrust line.

Or the designer may offset the thrust line so that it passes above the center of gravity as shown in Fig. 10-14.

A very low thrust line would be bad as this would tend to add to the nose-up effect of the slipstream on the horizontal tail surfaces. (Fig. 10-15)

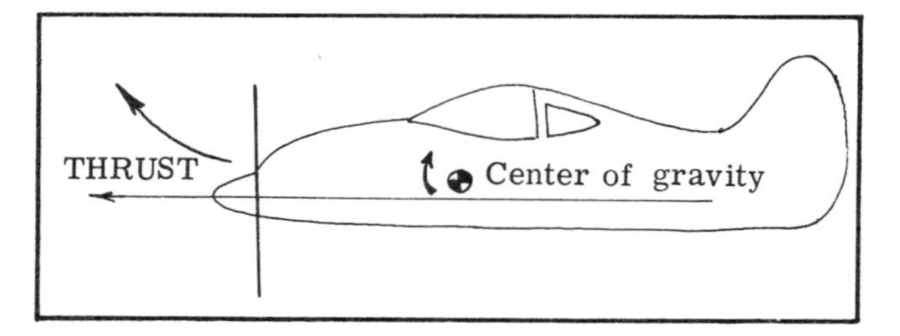

Fig. 10-15 Low thrust line effects with application of power.

The thrust line of your airplane is fixed and there's nothing you can do about it, but this has been presented for your interest. All these factors have been taken into consideration in the certification. No airplane will be certificated in the normal or utility categories if it has dangerous tendencies.

HOW YOU CAN AFFECT THE LONGITUDINAL STATIC STABILITY OF THE AIRPLANE

You can affect the longitudinal static stability of your airplane by the way you load it. If you stay within the loading limitations as given by the Airplane Flight Manual you'll always have a statically stable airplane.

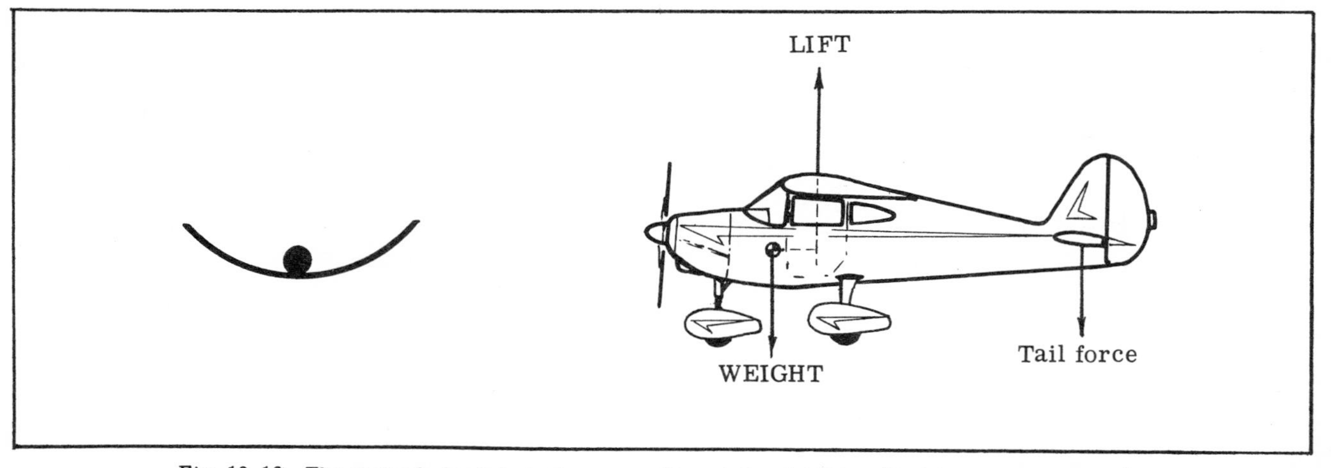

Fig. 10-16. The properly loaded airplane—positive static stability. (Tail force exaggerated.)

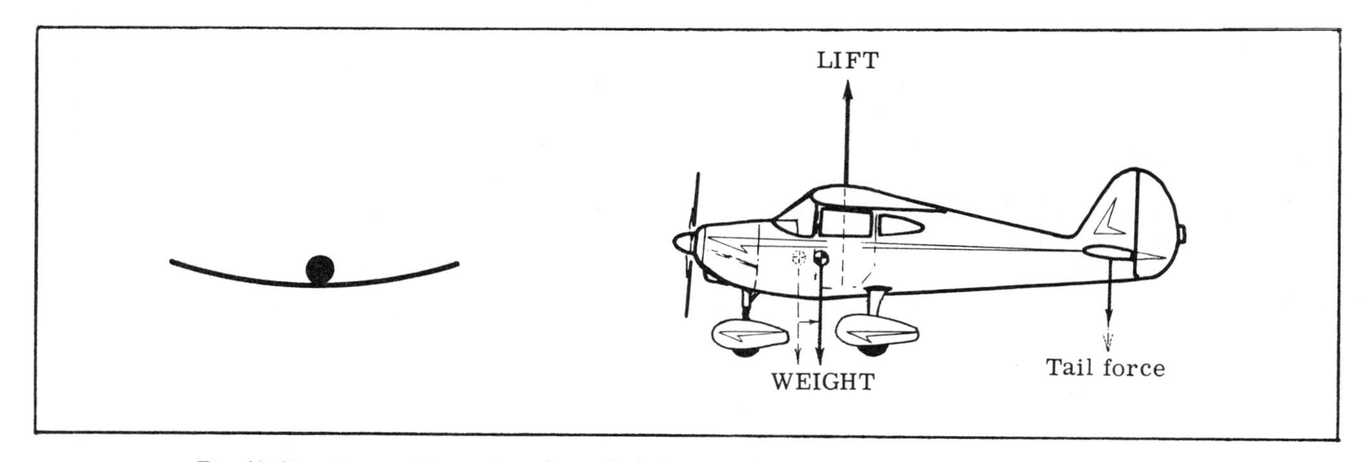

Fig. 10-17. Effects of the center of gravity being moved rearward — less positive static stability.

The properly loaded airplane will be analogous to the steel ball inside the hubcap. It will tend to stay in the attitude and airspeed at which it was trimmed. (Fig. 10-16)

If the C.G. is moved aft, the airplane becomes less statically stable and does not have as strong a tendency to return to its original position. It would be as if our hubcap were made shallower. (Fig. 10-17)

The C.G. can be moved aft to a point where the airplane has *no* tendency to return but would remain offset if displaced. It would be as if the hubcap had been completely flattened. (Fig. 10-18)

By moving the C.G. even farther aft the area of *negative* static stability is encountered. (Fig. 10-19) The hubcap has been turned inside out.

It would seem at first glance that the more statically stable an airplane could be made, the better its flight characteristics. This is true — up to a certain point. If an airplane is so stable that a great deal of force is needed to displace it from a certain attitude, control problems arise. The pilot may not be able to maneuver the airplane and make it do its job. This is more of a problem for fighters than for transports, however.

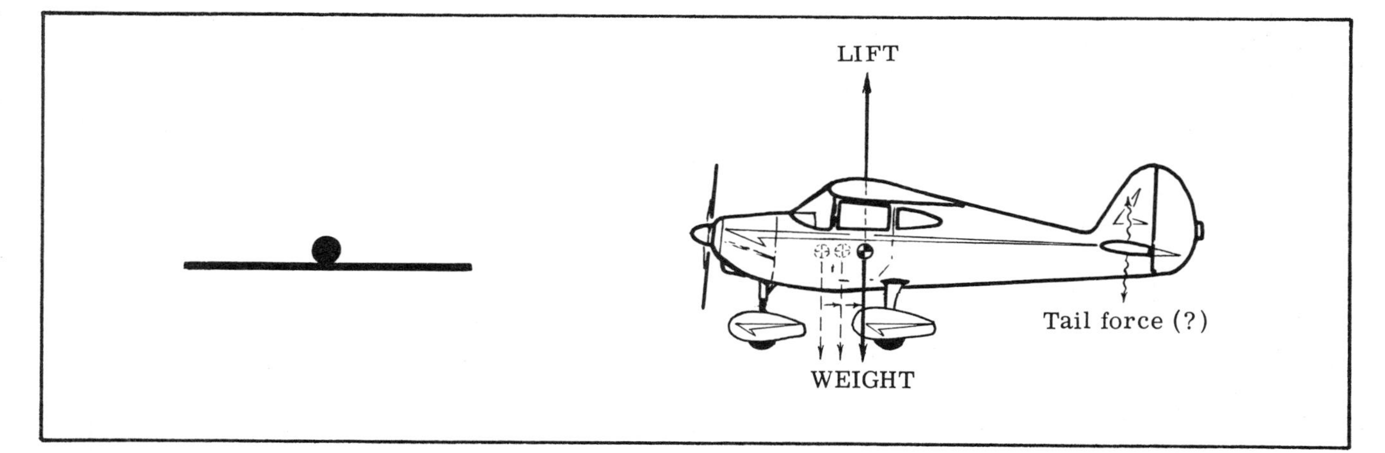

Fig. 10-18. Neutral static stability.

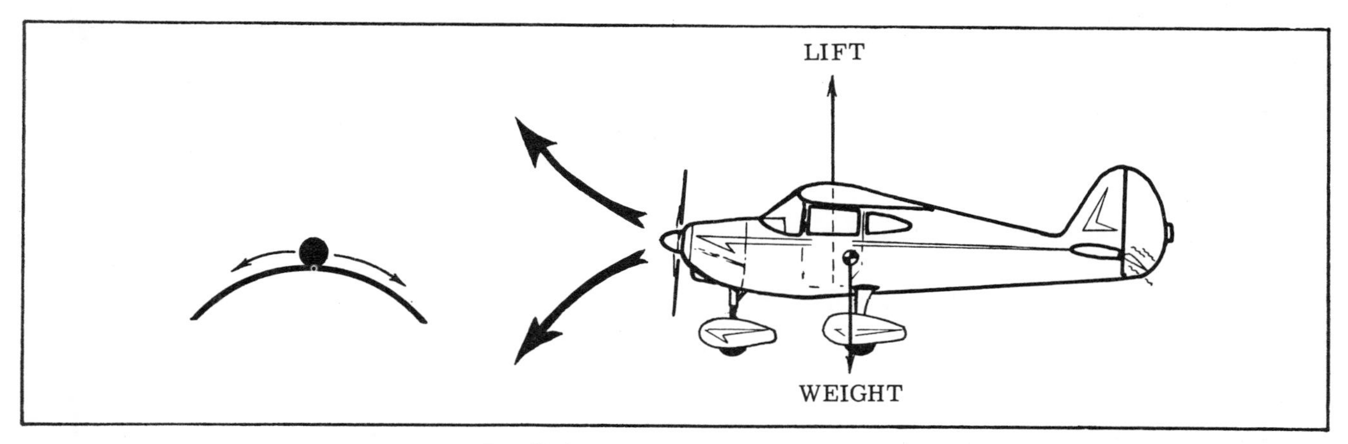

Fig. 10-19. Negative static stability.

The problem with neutral static stability is that the plane does not tend to return to its trimmed state. If you load the airplane to such a condition you might get into trouble. In a plane with neutral static stability the feel is changed considerably. After take-off you may ease the nose up using normal back pressure, and find that the nose attitude has overshot and is too high. You ease it down and again overshoot because you have been used to fighting the airplane's natural stability, and in this case it isn't there. *This is the type of situation that causes accidents.* Particularly dangerous is the fact that the airplane could become unstable during the flight as the fuel is burned. Designers always place the fuel tanks as near to the airplane C.G. as possible. But in a neutral stability condition, a rearward movement of the C.G. could put you into negative static stability. It's possible that the airplane could become uncontrollable or at least be in a very dangerous condition.

LONGITUDINAL DYNAMIC STABILITY OF THE AIRPLANE

Here is a measurement of your plane's actions in response to its *static stability*. As an example, the next time you are flying cross-country, ease the nose up until the airspeed is about 20 knots below cruise and slowly release it. The nose will slowly drop past the cruise position, the airplane will pick up excess speed and slowly rise again. If the airplane has *positive* dynamic stability, it may do this several times, each time the nose moving a lesser distance from the cruise position until finally it is again flying straight and level at cruise. The same thing would have occurred if you had eased the nose down. (Fig. 10-20)

This is like the steel ball in the hubcap as cited earlier. It was dynamically stable and finally resumed its original position.

An airplane which has *neutral* dynamic stability through some design factor would react to being offset as shown in Fig. 10-21. This would be as if some outside unknown force were rocking the hubcap, keeping the ball constantly oscillating.

The airplane having *negative* dynamic stability would have oscillations of increasing magnitude. (Fig. 10-22) You see that the system (hubcap and ball) and the airplane are statically stable, but other factors may be introduced so that it would have neutral or negative dynamic stability.

The oscillations shown are called "phugoid" or "long mode" oscillations.

They are of such length in time that they can be easily controlled by the pilot and are considered to be of relatively little importance. The airplane you are flying may have a neutrally or negatively stable phugoid and still be completely safe. (So don't be disappointed or worried if, in trying the experiment at the beginning of this section, you find that these slow oscillations do *not* decrease in amplitude.) An airplane with neutral or negative dynamic stability in these "long modes" can be flown quite safely with little or no effort as the periods may be many seconds or even minutes in length.

Of primary importance is the "short mode" or rapid oscillation. The periods of the short mode may be in fractions of seconds. You can see that if the short mode is unstable the oscillations could increase dangerously before the pilot realized what had happened. Even if the oscillations were being damped, it's possible that the pilot, in trying to "help" stop the oscillations would get out of phase and rein-

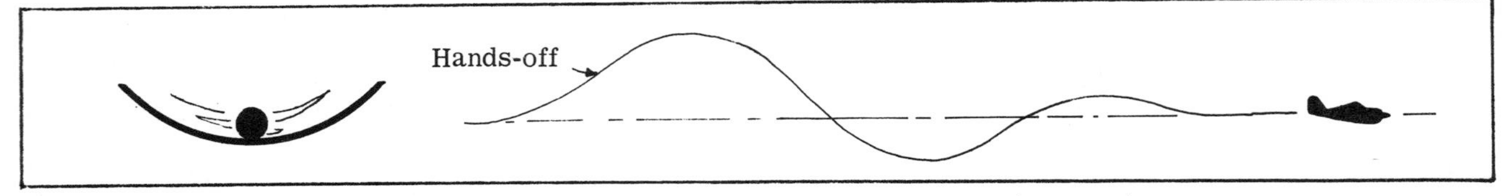

Fig. 10-20. Flight path of a dynamically stable airplane (hands-off).

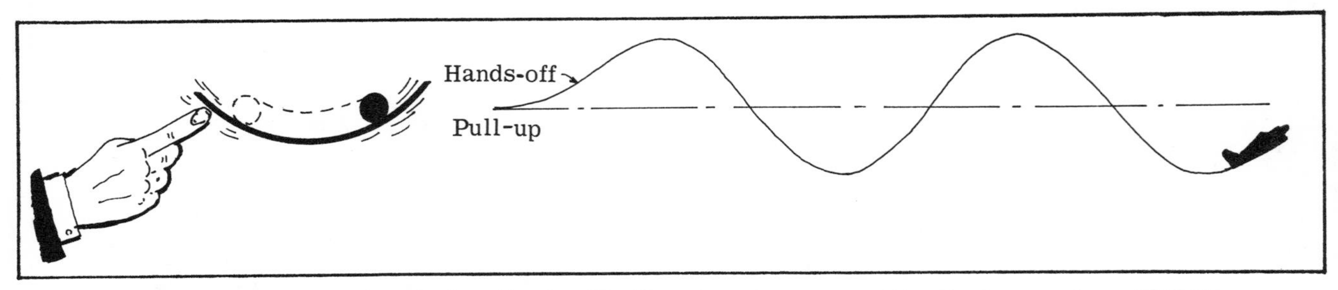

Fig. 10-21. Flight path of an airplane (hands-off) with neutral dynamic stability. No decrease in oscillations.

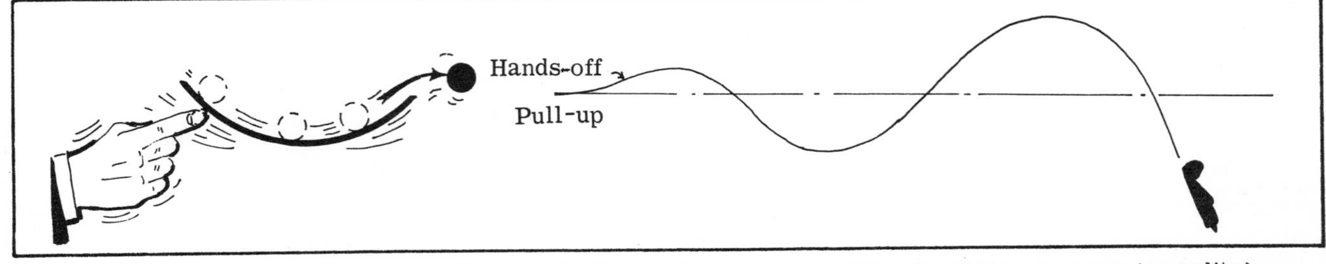

Fig. 10-22. Flight path of an airplane (hands-off) with negative dynamic stability. Oscillations increase in amplitude.

force them to the point where "g" forces could cause structural failure. Usually such problems are caused by poor elevator design or balancing, and are always solved by the manufacturer before the airplane is certificated. Airplanes certificated by the FAA have *positive dynamic stability* in the short mode. If the plane is offset from its path abruptly, it will return in a series of rapid, converging oscillations. (You can see this when flying hands-off on a gusty day.) An airplane that is *statically unstable* would have no oscillations at all but would continually diverge (the hubcap is upside down as in Fig. 10-19).

The point to remember is to keep your airplane statically stable by correct placement of Weight and you won't have any stability problems.

LONGITUDINAL CONTROL

ELEVATOR

You are familiar with the elevators as a means of longitudinal control, but airplanes also have been designed utilizing a "stabilator." You may have trained on such an airplane. First though, let's review the stabilizer-elevator system.

You use the elevator to change the camber of the horizontal tail system, which changes the tail force. For the design cruise airspeed, the elevators are designed to "float" parallel with the stabilizer. Any change from this speed and power setting must be compensated for by elevator deflection. The normal airplane requires that forward pressure must be held for speeds above this, and back pressure held for any speeds below cruise. You rotate the airplane to the desired attitude by exerting fore or aft pressure on the wheel. If an airplane is too stable longitudinally, the elevator control may not be effective enough for good control. One problem that airplane manufacturers face is that of the too-stable airplane, although to be truthful, it's not as much of a problem to them as the unstable type. It has been found through experience that the total horizontal tail area should be 15 to 20 per cent of the effective wing area and the elevator should make up about 35-45 per cent of the total horizontal tail surface area. The farther the horizontal tail is from the C.G., the less area is necessary for the required stability (Tail Moment = distance x force).

The stabilizer-elevator combination is an airfoil, and you vary the tail-force by positioning the elevators with the wheel or elevator trim.

The properly designed airplane will require forward pressure for airspeed above the trim speed you have selected, and back pressure for speeds below the trim speed (this applies whether you trim it for a speed just above stall or at the airplane's maximum speed). This indicates *positive static stability.*

You can see that this is what you've been encountering all along in your flights in FAA certificated airplanes. A happy medium should be found in that the airplane must be stable but not be too hard to displace, or maneuvering problems might arise. The more variation in velocity, the more pressure is required — because the airplane is stable and resists your efforts to vary its airspeed from trim speed.

As you move the C.G. aft you might find that you can't set any particular speed as trim speed.

Suppose you are indicating 120 knots, and have the elevator trim tab set at what you think is the correct position. In the normal airplane this would mean that if you applied fore or aft pressure on the wheel the pressure necessary to hold this nose position would increase as the airspeed changed. Not so the neutrally stable airplane. You are fooling yourself by even trying to trim the airplane.

When you pull the nose up and the speed decreases, the airplane isn't fighting the back pressure and will continue in this attitude without any help from you. Wheel pressure is *not* a function of airspeed in this case and the airplane could continue to a stall, the nose would then drop and it would maintain this nose-down attitude no matter how fast it flew. This is, of course, assuming you are flying "hands-off." Naturally you'll be flying the airplane and will ease the nose back down before the stall occurs. This is not to imply that the airplane is uncontrollable, but does mean that you'll have to make a conscious effort to return the nose to the proper position if it is displaced — and on a bumpy day this could get mighty tiresome. On landing, you will not be fighting the plane's stability. When you flare for the landing you will probably overdo it, because of the very light back pressure required, as opposed to what you've been used to. You'll probably get the nose too high, then consciously have to ease it over and may set up a cycle as is shown in Fig. 10-23.

If you've trimmed the airplane for a glide, you'll normally expect the required back pressure to increase as the plane slows during the landing — that's why you could get into trouble in this situation.

Fig. 10-23. Landing (?) a neutrally stable airplane.

The airplane with *negative* longitudinal stability will aggravate any displacement. If the nose were raised in the neutrally stable airplane, it stayed at that attitude until you (or some other force) lowered it. The airplane that is negatively stable would tend to get an even more nose-high attitude.

If you load the airplane with too much weight in the rear and get into an unstable condition, a serious accident is almost certain to occur. Take an extreme situation: You've loaded the airplane until the C.G. is much too far aft. You realize that there's quite a bit of weight back there and set the elevator tab to what you think is about the right amount of nose-down trim. You go roaring down the runway, ease the nose up — and it just keeps going up. This is neither the time, place or altitude to be practicing stalls. It may require more down elevator than you have available — and that's that.

LONGITUDINAL TRIM

There are several methods of longitudinal trim for the airplane. The most familiar is the elevator trim tab which acts as a control surface on a control surface and has been put to good use by you for many flying hours. When you trim "nose up" the tab goes down and the force of the relative wind moves the elevator up. (Fig. 10-24)

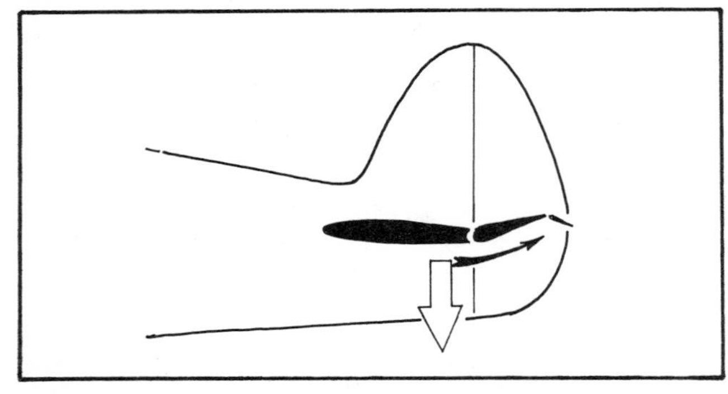

Fig. 10-24. Elevator trim tab; set for "nose-up."

You've found this very handy for nose-heavy conditions and for help during the glide. You may have found that the tab set this way caused trouble, too. Some pilots use almost full up-trim for landing, as this makes for light back pressure during the landing process. If they should suddenly have to take it around, the application of power may result in a severe tendency for the nose to rise. The slipstream hits the trim tab and elevator, which are greatly displaced at the low speed, power-off condition; this is particularly dangerous if the elevator tab control is geared so that it requires many turns to get back to a more nose-down setting. If the airplane were not so close to the ground, it would be amusing to watch the pilot — he can't make up his mind whether to take one hand off the wheel (he's pushing forward with both hands by now) and make a grab for the trim control, or hold the nose down with both hands and try to get altitude first. The usual result is that he does both, frantically moving his right hand from wheel to trim control and gradually getting things under control. Seen from outside the airplane, the maneuvering looks like a whale with a severe case of hiccups. This problem usually is not quite so critical in multi-engine planes, as the elevators are not so much in the slipstream. In any airplane — single, multi, prop or jet — the nose will be harder to keep down as the speed increases, though the single engine high-powered prop plane would give you the most trouble under these conditions. The airplane with an adjustable stabilizer can give you the same problems, so don't think that the trim tab alone is the culprit.

Another method of elevator trimming is the use of bungees. These consist of springs that tend to hold the elevator in the desired position when you set the trim control. (Fig. 10-25)

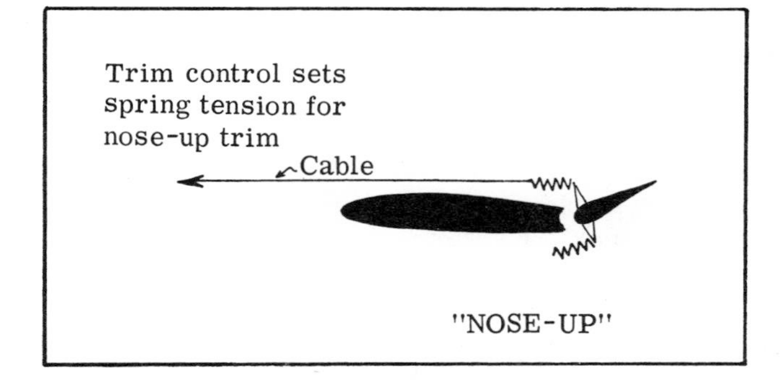

Fig. 10-25. Bungees as a method of elevator trim.

The spring also acts as a damper for any forces that might be working through the system. You set the spring tension with the trim control. Actually you couldn't care less what the spring tension is — you move the trim handle or wheel until you get the desired result. It's doubtful that you would be able to tell by "flight feel" that the airplane had a bungee instead of a trim tab unless you noticed it during the pre-flight check (and you should have).

Another method of longitudinal trimming is the movable or controllable stabilizer. Shown in Fig. 10-26 is the stabilizer trimmed "nose-up." The cockpit control merely turns a jack screw to position the stabilizer.

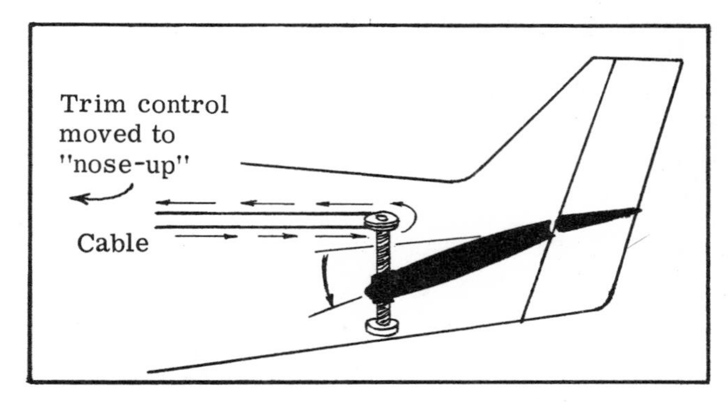

Fig. 10-26. Stabilizer trim — "nose-up."

SPECIAL TYPES OF TABS

Link Balance Tab — If an airplane is found during flight test to have heavy control forces it may use what is known as a "link balance tab." This is a tab mechanically linked so that it moves opposite to the control surface and makes the control forces much lighter. For instance, the link balance tab on an elevator moves down as the elevator moves up — a sort of mutual aid society. These tabs, by the way, also can act as trim tabs when variable length linkage is used. (You vary the linkage length when you move the trim control in the cockpit. This type of arrangement is called a "lagging tab.") (Fig. 10-27)

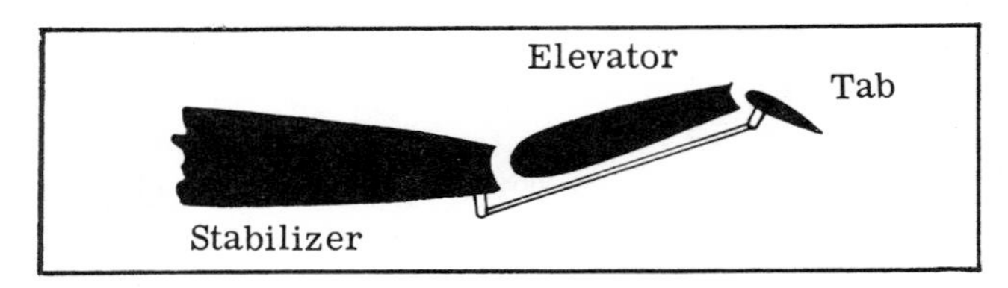

Fig. 10-27. Link balance tab — lagging.

In some instances the control forces may be too light — that is, the elevator may move so easily that in extreme conditions the pilot could inadvertently overstress the airplane. The manufacturer may use a leading link balance tab (Fig. 10-28) in order to increase the control forces necessary to displace the control surface.

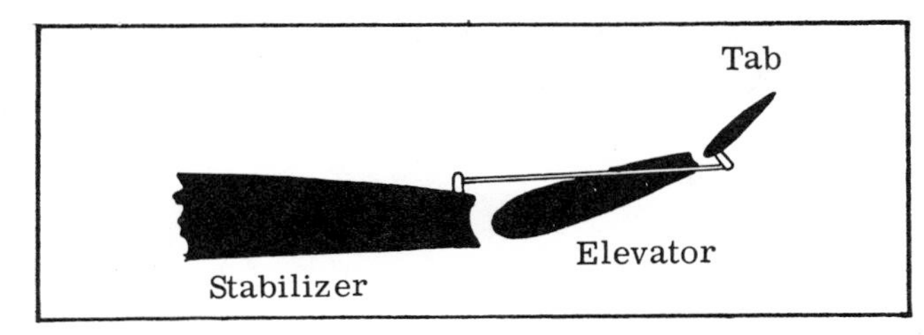

Fig. 10-28. Link balance tab — leading.

Servo Tab — It'll probably be a long time before you use this system as only large planes have used them. The control wheel is connected to the tab, rather than to the elevator itself. When the pilot moves the wheel back the tab is deflected downward. The impact air pressure strikes the tab and the elevator is deflected upward. The elevator is free floating and moves in accordance with the tab deflection.

The principle of the trim tab is simple. It uses a small area, long arm and greater angular deflection to deflect a control surface of greater area to a lesser deflection. (Got it?)

STABILATOR

The stabilator is becoming more popular as a means of longitudinal control for lighter planes. It's, in effect, no more than a symmetrical airfoil whose angle of attack is controlled by the control wheel and trim control. The stabilator is hinged at its aerodynamic center. In computing the tail force you could use the Lift equation for a wing of the stabilator's airfoil and area. You would have a C_L versus Angle of Attack Curve and could use equation $L = C_L S \frac{\rho}{2} V^2$; the problem would be complicated by the fact that the downwash and slipstream effects would be hard to predict. But the principle is exactly the same as that for the wing.

The stabilator was found to be an effective means of control for jets in the transonic region (Mach .8 to 1.2). The elevator system lost effectiveness in this speed range due to shock wave effects on the stabilizer, and severe nose-down tendencies were encountered — with little control to offset these forces. In effect, the elevator was "blanketed" behind the shock wave. It was found that by moving the entire stabilizer-elevator (or stabilator) system that longitudinal control could be maintained in this critical range.

This method of control was found to be effective at all speed ranges, and several manufacturers have introduced several models using this system.

The stabilator control, when properly balanced, is quite sensitive at low speeds. (The pilot who has been flying an elevator-equipped airplane usually tends to slightly overcontrol in moving the nose up or down for the first few minutes.) Because the stabilator has more movable area it usually does not use as much angular travel. Whereas an elevator may have a travel of 30 degrees up and 20 down, the stabilator may move less than half this amount. The trim tab for the stabilator works in the same way — impact pressure on the tab holding the stabilator in the desired position.

FORWARD C.G. CONSIDERATIONS

It would seem, from the discussion of the aft center of gravity position, that the farther forward the C.G., the better off you are. (O.K. now, everybody run to the front of the airplane.) This is true for longitudinal stability but not from a control standpoint. Let's start out with a longitudinally balanced airplane. (Fig. 10-8).

Note in Fig. 10-8 that there is no elevator deflection, the airplane is at design cruise speed and properly balanced weightwise. For our hypothetical situation suppose that during the flight Weight is moved forward, so that the center of gravity also is moved forward another five inches. (Fig. 10-29)

In order to maintain balance the tail force must be increased and this is done by back pressure or use of the trim tab.

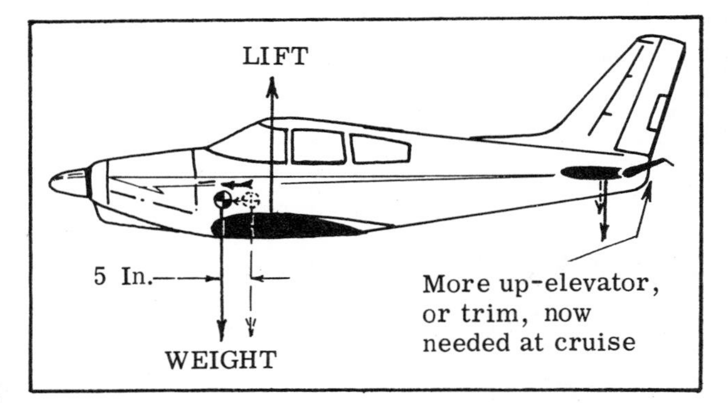

Fig. 10-29. Center of gravity moved forward 5 inches.

Of course, you wouldn't give a hang for the value of the tail force but would trim it until the nose stayed where it belongs. Now move the C.G. forward another five inches. The required tail force would be further increased and more up-elevator would be necessary. By moving the C.G. forward you would soon reach a point where full up-elevator would be required. If the throttle were chopped, the loss in effectiveness of the elevators would result in a definite nosing-down of the airplane. You'd be in the unhappy situation of being unable to slow down because you'd be using full elevator to maintain level flight and would have none left to ease the nose up. Of course, you couldn't chop power because control would be lost. In this exaggerated situation you'd have a tiger by the tail.

The manufacturer will set forward C.G. limitations that will be strict enough so that if you comply you'll never get into a dangerous situation.

You have seen that power nearly always results in a nose-up tendency, so the airplane can have a more forward C.G. with power on. A critical condition could exist at very low airspeeds (approaching a stall) in the power-off condition, where the elevators are relatively inefficient due to lack of slipstream and airspeed. This would limit the most forward C.G. location in flight. It would seem that this would also be the same condition as for the landing, but this is not the case. *The ground effect on landing results in a further decrease in elevator control effectiveness* (as can be seen in Fig. 10-30). *This, then, will be the most limiting factor in establishing the forward C.G. of the airplane.*

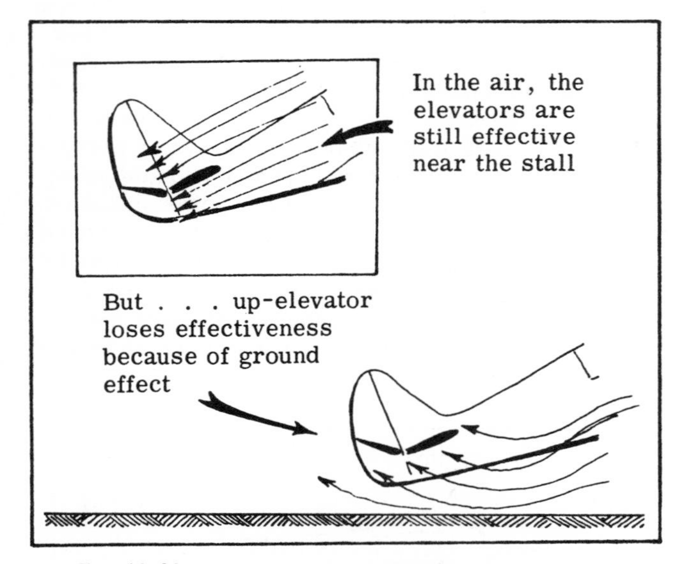

Fig. 10-30. Ground effect and elevator effectiveness.

AIRPLANE WEIGHT AND BALANCE

The center of gravity of an airplane must remain within certain limits for stability and control reasons. These limits are expressed by designers in terms of percentage of the mean aerodynamic chord. (Fig. 10-31)

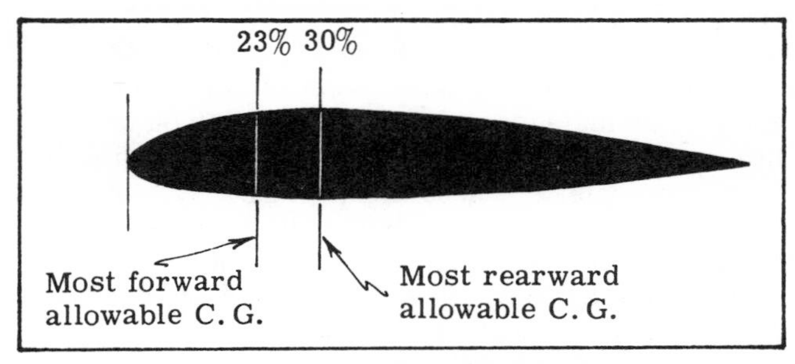

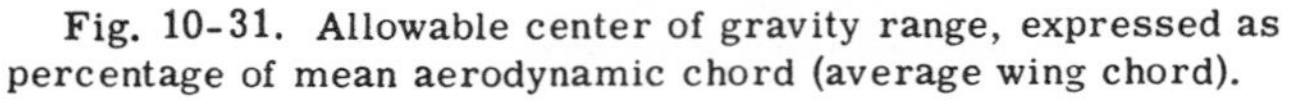

Fig. 10-31. Allowable center of gravity range, expressed as percentage of mean aerodynamic chord (average wing chord).

The airplane's *Weight and Balance Form* will usually express these limits in the form of inches from the datum, or point from which the measurements are taken. This datum is at different points for different airplane makes and models. High-wing, light airplanes usually use the junction of the leading edge of the wing with the fuselage as the datum, and the allowable C.G. range would be expressed in inches aft of this point such as "allowable center of gravity range — from 13.1 to 17.5 inches aft of datum." (Fig. 10-32)

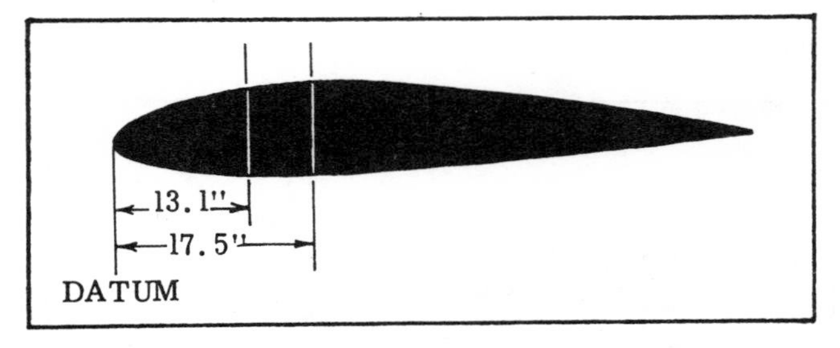

Fig. 10-32. Allowable center of gravity range, expressed as inches aft of datum.

Sometimes the datum is an imaginary point ahead of the airplane. This is easier to compute in that all the moment arms are positive. The datum usually is picked so that it is an even distance ahead of a well defined position such as the junction of the wing's leading edge with the fuselage.

Measurements are taken from a reference point and added to or subtracted from the datum to get the proper arm. In Fig. 10-33 you see that the wing's leading edge is 70 inches aft of the datum. If object A weighing 10 pounds is placed at a point 10 inches behind the junction of the leading edge of the wing and the fuselage, its Weight (10 lbs.) would be multiplied by 70 + 10, or 80 inches. Its moment would be 800 inch-pounds. Object B, also weighing 10 pounds, ten inches ahead of leading edge point would have a moment of 70 - 10 or 60 inches times its Weight or 600 inch-pounds.

The C.G. limits for the airplane in Fig. 10-33 would be expressed as "C.G. allowable range from 80 to 87 inches aft of datum."

Here's how you might run a Weight and Balance for the above mentioned airplane: the manufacturer will give its empty Weight (the Weight at which the airplane is actually ready to fly except for fuel, oil, pilots, passengers and baggage — in other words it's

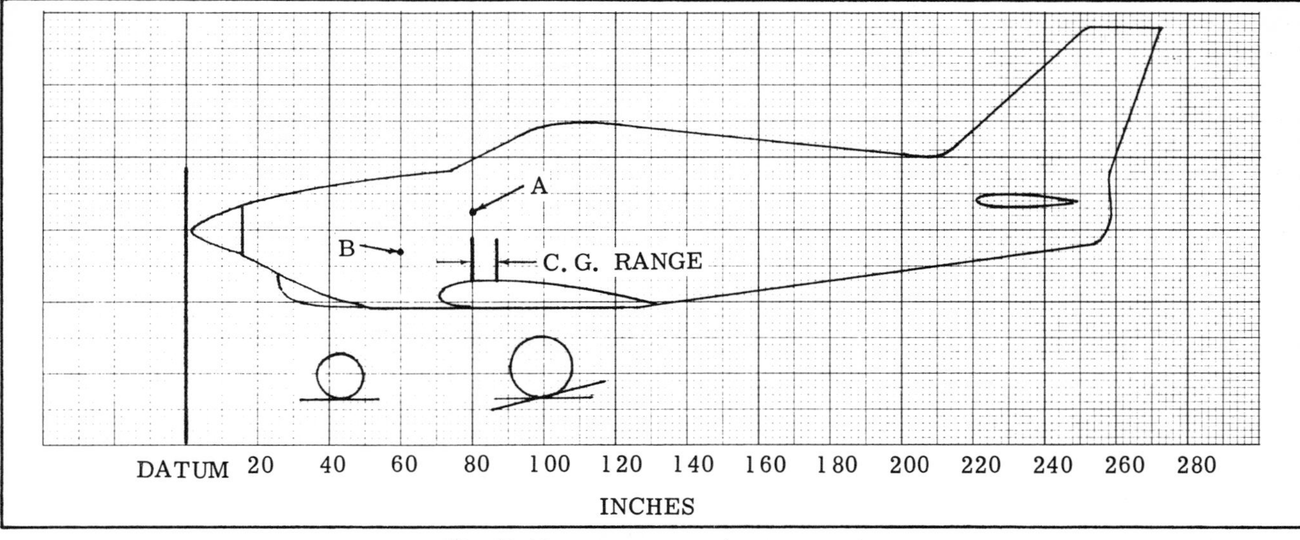

Fig. 10-33. Weight and balance diagram.

not lacking any mechanical parts) and its empty C.G. position.

This hypothetical airplane is a high performance, low-wing type and carries four persons. The empty Weight is 1785 pounds and the empty C.G. position is 81 inches aft of datum. The airplane has an allowable gross Weight of 3000 pounds, leaving a useful load of 1215 pounds (fuel, oil, pilot, passengers, baggage). Calculations could be run as shown in Fig. 10-34.

Item	Weight		Arm (Distance, inches)		Moment
Empty Weight	1785	x	81	=	144585
Oil – 2 gals. @ 7½ lbs.	15	x	45	=	675
Fuel – 70 gals. @ 6 lbs.	420	x	82	=	34440
Pilot	170	x	80	=	13600
Passenger (Front Seat)	170	x	80	=	13600
Passenger (Rear Seat)	170	x	115	=	19550
Passenger (Rear Seat)	170	x	115	=	19550
Baggage	100	x	135	=	13500
	3000 lbs.				259500 inch-lbs.

Fig. 10-34. Weight and balance table with fuel, oil, occupants and baggage located.

The average arm (or C.G. position) then can be found by dividing the total moment (259,500 inch-lbs.) by the total Weight (3000 lbs.), getting an answer of *86.5 inches* which is within the allowable range of 80 to 87 inches. You would use the above information to find the C.G. for various combinations such as half fuel, pilot only, pilot and passengers, no baggage, etc.

Following is an example of an airplane that uses the junction of the wing's leading edge with the fuselage as a datum. The airplane is a two place side-by-side trainer with an empty Weight of 1000 pounds and a gross Weight of 1575 pounds. The manufacturer has found that the empty C.G. is at a distance of 13 inches behind the leading edge. The allowable C.G. travel is from 12 to 16 inches aft of the datum. (Fig. 10-35) Allowable baggage is 100 pounds, fuel twenty gallons, oil two gallons.

$$C.G. = \frac{+22860}{1575} = +14.5 \text{ inches or } 14.5 \text{ inches aft of}$$

Item	Weight		Arm (Distance, inches)		Moment
Empty Weight	1000	x	+13	=	+13000
Oil – 2 gals.	15	x	-24	=	- 360
Fuel (in wing) 20 gals.	120	x	+15	=	+ 1800
Pilot	170	x	+13	=	+ 2210
Passenger	170	x	+13	=	+ 2210
Baggage	100	x	+40	=	+ 4000
	1575 lbs.				+22860 inch-lbs.

Fig. 10-35. Weight and balance table for a two place, side-by-side trainer. For positions ahead of the datum a minus sign is used; for positions behind the datum a plus sign is used.

the datum. This is well within the 12 to 16 inch C.G. range limitation. The moment of 22,860 inch-pounds was found by adding all the positive moments and subtracting the negative moment(s) from the result. You could call all distances behind the datum negative and those forward positive and still arrive at the same answer. Your answer would then be a minus number, meaning that the C.G. is behind the datum under your new set of rules for the calculations. Or you could pick a point 10, 20 or 50 feet behind the tail as a datum and still arrive at a proper answer. The principle applies to any airplane or datum point: Weight times distance equals moment. You can use feet or yards for distance, but inches are the usual measurement so that you get moments in inch-pounds. When heavier components are added to or taken from your airplane (such as radar, radios, etc.) the mechanic will show this on the airplane *Major Alteration and Repair Form* and notes the change in empty Weight and C.G.

The baggage compartment has Weight limitations for two reasons: (1) You might move the C.G. too far aft by overloading it and (2) you could cause structural failure of the compartment floor if you should pull "g's" during the flight. For instance, you are flying a four-place airplane and are the only occupant; the boss asks you to deliver some anvils to another town. You figure that there'll be no sweat on the C.G. and throw 400 lbs. of anvils in the 200-pound limit baggage compartment. For sake of the example let's say there is no C.G. problem and away you go.

Fig. 10-36. "I dunno, do you think we might have put too much stuff in the baggage compartment?"

Enroute, you suddenly see another airplane coming head-on and without thinking, pull up abruptly. You could send 400 pounds of anvils through the bottom of the airplane down through somebody's greenhouse — or worse. (Fig. 10-37)

Fig. 10-37.

The baggage compartment floor is designed to withstand a certain number of "g's" with 200 pounds in it. If you pull this number of "g's" with 400 pounds in there, something will give.

The pilot has a couple of early indications that will help him realize that maybe things are not as they should be and that he may be approaching a dangerous condition: (1) After loading a tricycle gear airplane heavily he will see that the airplane is in an extremely tail-low position on the ground, and that taxiing is very sloppy because the nosewheel doesn't have enough weight on it for effective directional control. The nosewheel "bounces" slowly as you taxi — it doesn't know whether it wants to stay on the ground or not. (2) In the tailwheel type airplane the tail may be extremely hard to raise during the take-off run, even with a farther nose-down trim setting than normal. If this happens, chop the power before you've gone too far to stop the process.

This is all common sense. Even if you ignore rear baggage compartment placards and other rear loading limitations, maybe the fact that the airplane "just doesn't feel right, even taxiing" may give that extra warning. *But — heed those loading limitations and don't depend on "feel" to save your neck.*

You won't always run a Weight and Balance calculation on the airplane every time you fly it, but this discussion will give you an idea of the principles involved. Stay within the limitations on passengers and baggage as given in the Airplane Flight Manual and you'll have no fear of exceeding the C.G. limits or suffering in performance. *Excessive Weight, poorly placed* will result in a dangerous situation both from a performance *and* a stability and control standpoint. (See Fig. 10-38 for a summary.)

DIRECTIONAL STABILITY

Directional stability, unlike longitudinal stability, is not greatly affected by the pilot's placement of Weight in the airplane. An airplane is either designed with good -- or less than good — directional stability. For a simplification of the idea take an airplane with no fin or rudder as is shown in Fig. 10-39. (Page 111.)

You can see that if the center of side area is even with the center of gravity, little or no directional stability is present. If the airplane were displaced in a yaw by turbulence it would not tend to return to its original heading. If the offsetting force were strong enough, the airplane might pivot on around and fly backwards for awhile.

The designer must insure positive directional static stability by making sure that this center of side area is behind the center of gravity. This is done through the addition of a fin. (Fig. 10-40 on page 111.)

You can see that a restoring moment is produced if the airplane is yawed. (Fig. 10-41 on page 111.)

The fin (like the horizontal stabilizer) acts like the feather on an arrow in maintaining stable flight. Naturally the farther aft this fin is placed and the larger its size, the greater the airplane's directional stability.

The rudder, as a part of this area, is furnished to give the pilot control in yaw. If the fin is too large in comparison with the rudder area and deflection limits,

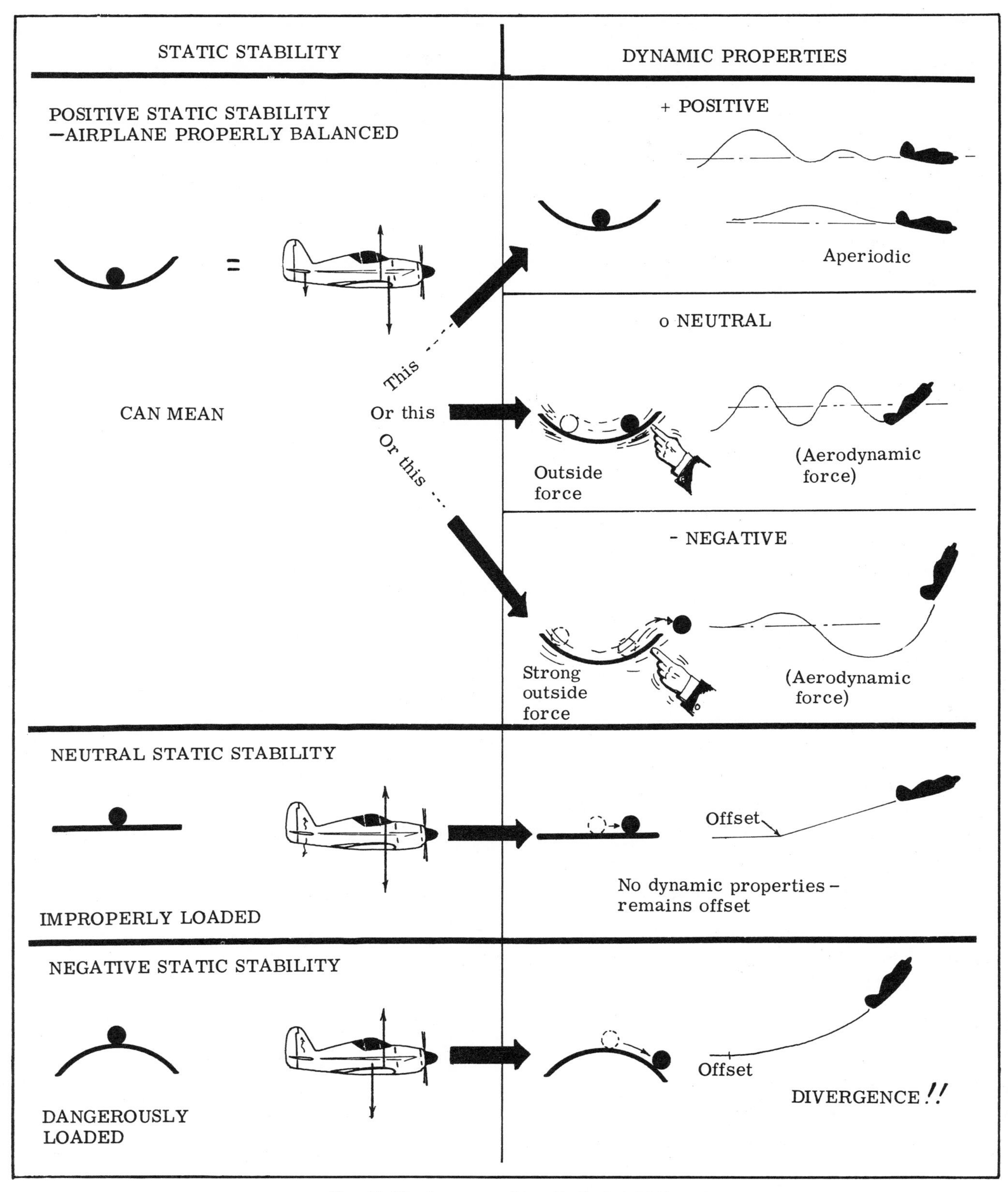

Fig. 10-38. Summary of longitudinal stability.

poor yaw control results. The airplane may be so directionally stable that the pilot is unable to make forward- or sideslips or safe crosswind landings and take-offs.

The rudder effectiveness, or "rudder power" as it is called by engineers, is very important in the event of an engine failure on a multi-engine airplane at low airspeeds. The pilot must be able to offset the asymmetric thrust of the working engine(s) at full power. For multi-engine airplanes this rudder power governs the minimum controllable speed with one engine out, or in the case of a four-engine plane, two engines out on the same side. Lateral control also enters into consideration in establishing the minimum

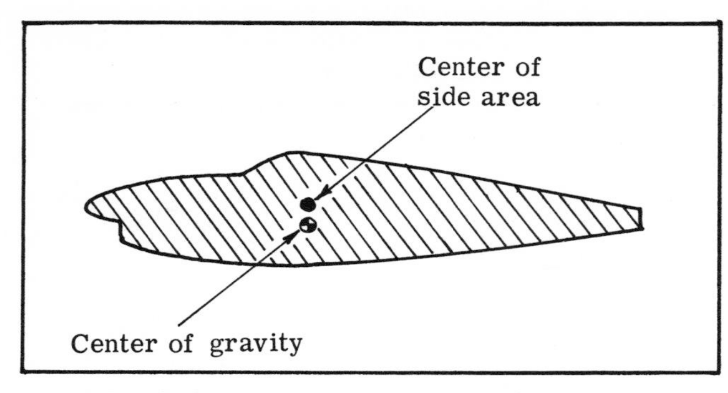

Fig. 10-39. Center of side area — fuselage only.

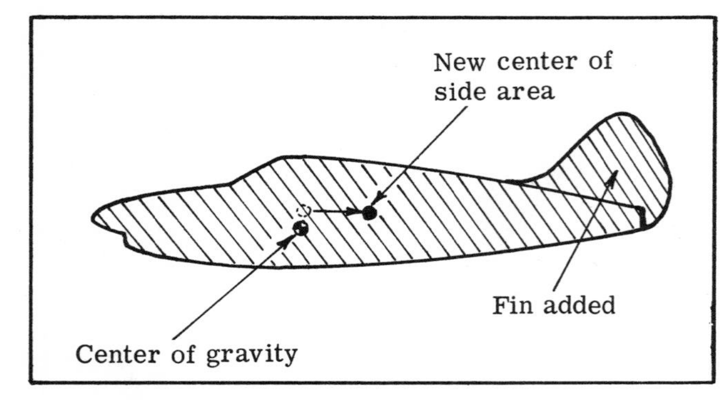

Fig. 10-40. Center of side area — fin added.

controllable speed, but this will be covered more thoroughly in Part II, "Checking Out in Advanced Models."

Rudder deflections usually are held below 30°, as the effectiveness falls off past this amount. Another factor in rudder design is the requirement for spin recovery. However, modern airplanes usually are so spin resistant that this requirement does not hold the importance it once did. It's very seldom these days that deliberate spinning is required (only on the flight test for the instructor's certificate).

A properly designed airplane requires more and more rudder force to be exerted as the yaw angle is increased at any given airspeed. You've found this to be the case when steepening a side or forward slip to land. You found that you were limited in steepness of slip by the rudder more than by the ailerons.

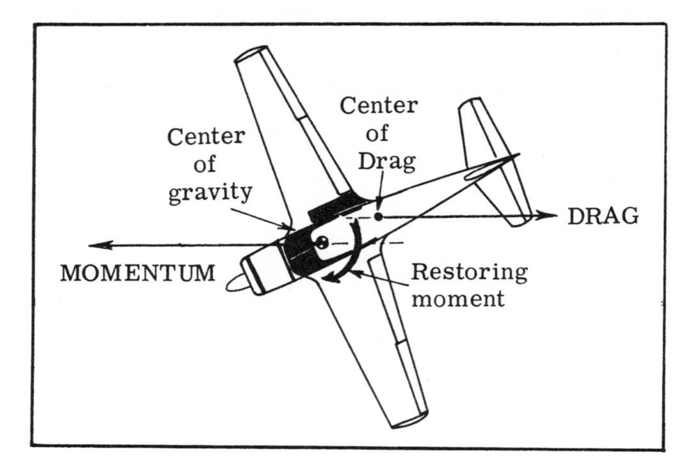

Fig. 10-41. Restoring moment of a properly designed fuselage-fin combination.

The rudder is considered to be an auxiliary control in flight, and in newer airplanes is losing even this value. Primarily its purpose is to overcome adverse aileron yaw. With the advent of differential aileron movement and other means of overcoming adverse yaw its importance is becoming less for normal coordinated flight. It's still mighty handy in slipping and other unbalanced conditions, though!

Steerable tailwheels, nosewheels, and separate wheel brakes have even decreased the rudder's importance for ground control. When airplanes had tail skids or free-swiveling tailwheels, the slipstream and relative wind on the rudder were the only means of turning on the ground. The rudder is still of primary importance for tailwheel type airplanes during take-offs and wheel landings — and particularly so in a crosswind!

A swept-back wing contributes to directional stability but very few light planes use this idea any more. However, a wing with double taper does help, as there is some sweep effect. (Fig. 10-42)

Although there are "cross effects" between lateral and directional stability, for simplicity's sake, sweep-back will be discussed here only as it affects directional stability. Take a look at an airplane in a yaw as shown in Fig. 10-43.

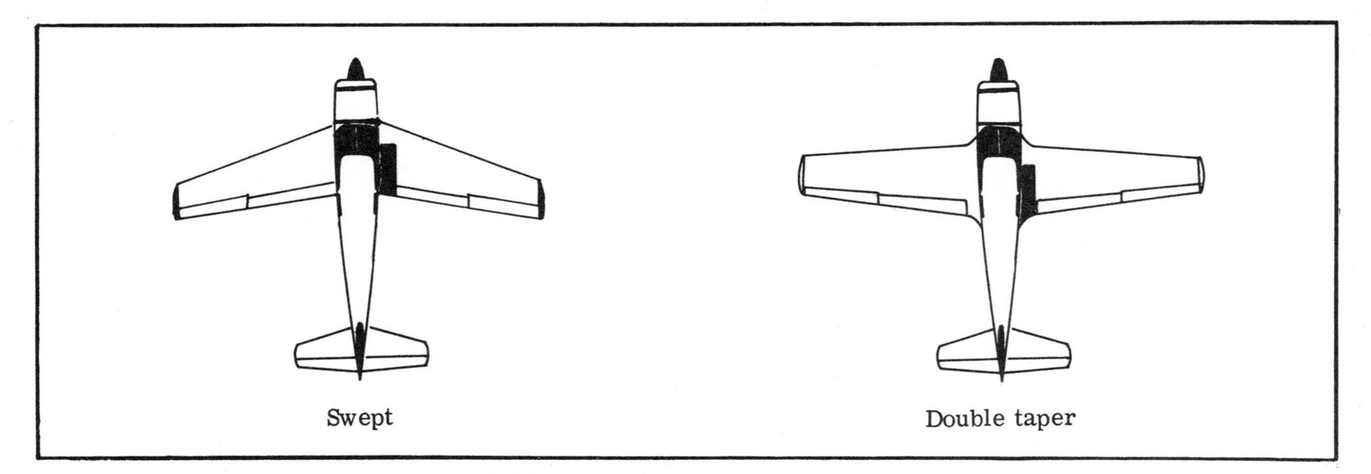

Fig. 10-42. Swept-back and double tapered wing.

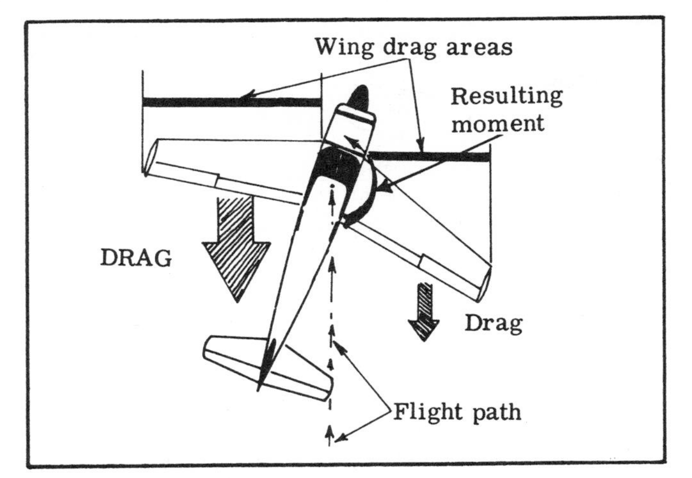

Fig. 10-43. Effects of sweep-back in a yaw.

In Fig. 10-43, the plane is yawed and you can see that there is a difference in Drag that results in a restoring moment. Of course the fin area would be helping, too.

It's hard to separate lateral and directional control effects even though we did it to discuss sweepback. You know yourself that kicking a rudder yaws the airplane, but it also causes reactions in a roll. You also know that abrupt application of aileron alone normally results in adverse yaw.

DIRECTIONAL DYNAMIC STABILITY

With directional dynamic stability, a maneuver which may result when a rudder pedal is pushed and released is called "Dutch Roll."

Suppose you kick right rudder and quickly release it. The nose of the airplane yaws to the right initially. This speeds up the left wing (and slows down the right wing) so that a rolling motion is effected. The airplane, having dynamic or oscillatory properties, (assuming positive static directional stability), will return and overshoot, speeding up the right wing which raises – etc. You get a combination yawing and rolling oscillation which, putting it mildly, is somewhat disconcerting. An airplane that "Dutch Rolls" is miserable to fly in choppy air. (Fig. 10-44)

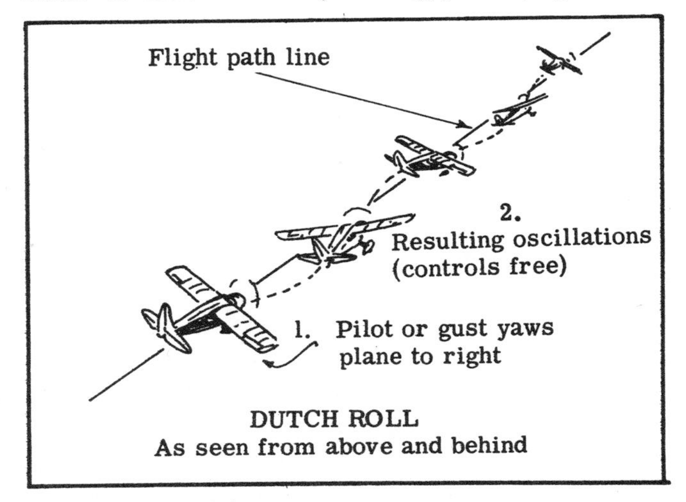

Fig. 10-44. Dutch Roll.

The manufacturer tries to reach a happy medium between too great and too little directional stability (called weathercock stability by some texts).

Such factors as rudder balance and design have strong effects on the dynamic properties of directional stability. If your airplane is certificated by the FAA, it has positive directional stability (both dynamic and static). It's hard to separate lateral and directional control effects. Dihedral, which helps assure positive *lateral* static stability, may result in cross effects and give the airplane "Dutch Roll" problems.

Because it is better to have a situation known as "spiral instability" than Dutch Roll nearly all airplanes are designed this way. You will note in your own airplane that if a wing lowers (controls free) and a spiral is allowed to develop, the bank increases and the spiral tightens if no effort is made by the pilot to stop it. However, the rate of increase of bank is normally slow and causes no problem in VFR conditions. This is the lesser of two evils as compared to the annoyance of Dutch Roll. In situations where visual references are lost by the pilot who is not instrument qualified and/or doesn't have proper instrumentation the tendency of the airplane is to get into a spiral of increasing tightness. (Check it sometime under VFR conditions – at a safe altitude and in an area clear of other airplanes, of course.) Manufacturers are beginning to recognize this problem and automatic wing leveling devices will likely be standard equipment in a few years. One manufacturer has already started such a program.

SUMMARY OF DIRECTIONAL STABILITY

STATIC STABILITY	DYNAMIC PROPERTIES
POSITIVE STATIC STABILITY Airplane with sufficient fin area. (Ball inside hubcap) Three Possibilities....	POSITIVE DYNAMIC STABILITY Airplane returns to straight flight after several decreasing oscillations (yaw). Rudder properly balanced.
	NEUTRAL DYNAMIC STABILITY Airplane oscillates from side to side (yaws) when offset, neither increasing nor decreasing in amplitude.
	NEGATIVE DYNAMIC STABILITY Airplane, when offset, oscillates (yaws) with increasing amplitude. Poor rudder balance.
NEUTRAL STATIC STABILITY Airplane with small fin. (Ball on flat plate)	NO OSCILLATIONS When yawed, the airplane tends to stay in that position. No tendency to return to straight flight.
NEGATIVE STATIC STABILITY Airplane with critical shortage of fin area. (Ball on top of hubcap)	NO OSCILLATIONS When yawed, the airplane tends to increase the yaw. Divergence occurs.

LATERAL STABILITY

DIHEDRAL

The most common design factor for insuring *positive static lateral stability* is dihedral. Dihedral is

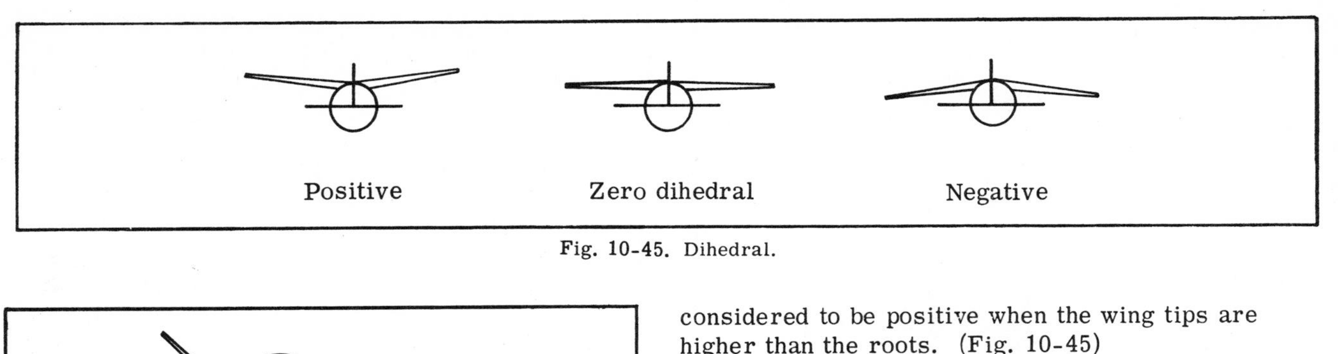

Fig. 10-45. Dihedral.

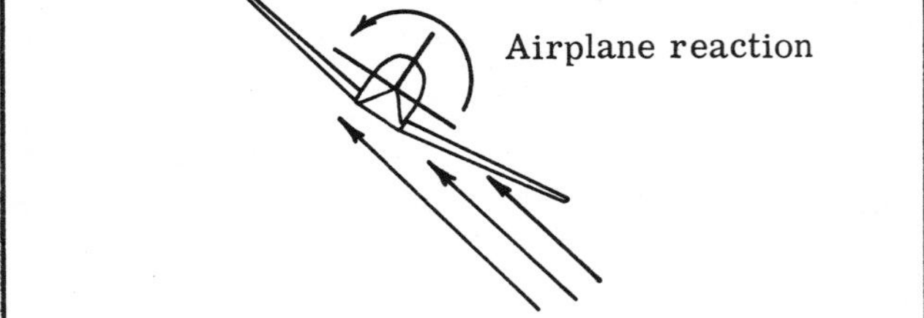

Fig. 10-46. Dihedral effect in a sideslip.

considered to be positive when the wing tips are higher than the roots. (Fig. 10-45)

The effect of dihedral is to produce a rolling moment tending to return the airplane to a balanced flight condition if sideslip occurs.

Fig. 10-46 shows what forces are at work in a slideslip. You can see that a rolling moment is produced which tends to correct the unbalanced condition.

A high-wing airplane, even though it may actually have zero dihedral, has a tendency to return to balanced condition because of its wing position. A low-wing

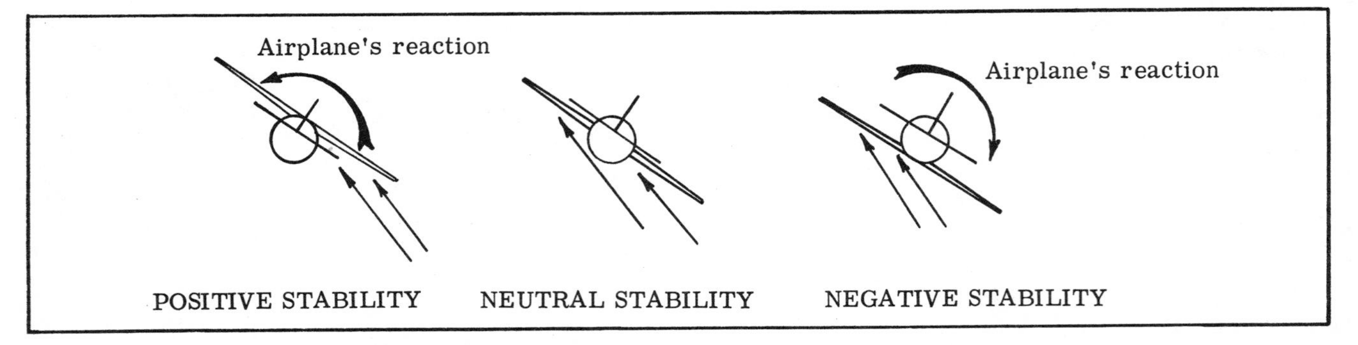

Fig. 10-47. The effects of various wing positions on lateral stability; zero dihedral in each case.

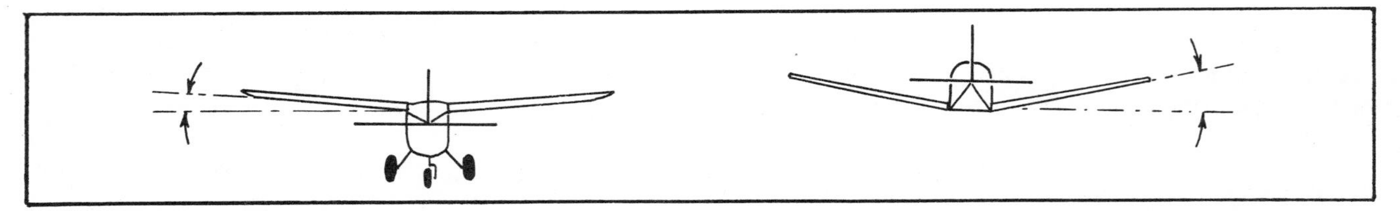

Fig. 10-48. Dihedral requirements for a high- and a low-wing airplane to obtain equivalent lateral stability.

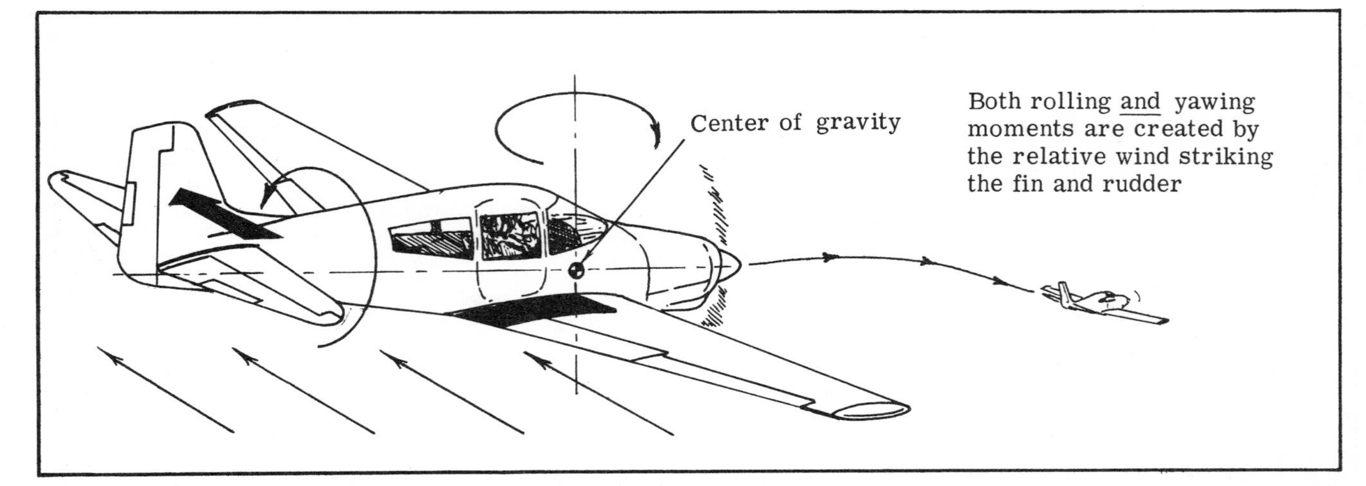

Fig. 10-49. Fin effects in a slip.

airplane with zero dihedral will generally have negative lateral stability because of its C.G. position.

A mid-wing airplane with zero dihedral usually exhibits neutral stability. (Fig. 10-47)

Understanding this idea, you can see that two airplanes of similar design in all other respects, but one having a high wing and the other a low wing, will have different dihedral angles for the same lateral stability requirements. (Fig. 10-48)

Excessive dihedral makes for poor rolling qualities. The airplane is so stable laterally that it is fighting any rolling motion where slipping might be introduced (such as in a slow roll). For this reason, airplanes requiring fast roll characteristics usually have less dihedral than a less maneuverable airplane. A fighter pilot doesn't go around doing slow rolls in combat but he does need to have a high rate of roll in order to turn quickly — and sometimes in the heat of the moment he may not coordinate perfectly. If the airplane has a great deal of dihedral, it may be hard to maneuver laterally, particularly if sideslipping is a factor in the roll.

Lateral and directional stability are hard to separate. For instance, the fin which is primarily designed to aid in directional stability may contribute to lateral stability as well. Fig. 10-49 shows that the fin and rudder contribute a rolling moment as well as a yawing moment in a sideslip.

In a slipping turn the effect of the fin is to stop the slip and balance the turn, which it does through roll *and* yaw effects.

POWER EFFECTS ON LATERAL STABILITY

Power is destabilizing in a sideslip for a propeller-driven airplane. That is, it may tend to counteract the effects of dihedral as is shown by Fig. 10-50.

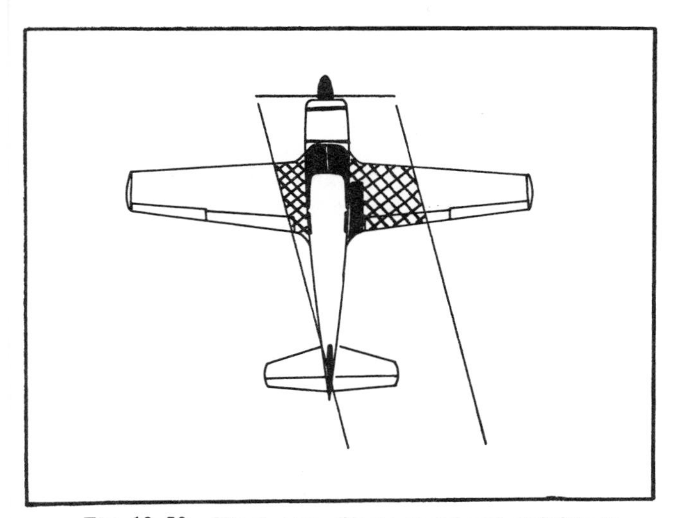

Fig. 10-50. Slipstream effects on lateral stability in a left slip with power. Note that the slipstream tends to increase the Lift of the high wing.

You can visualize that in an accidental slip the slipstream will tend to make the bank steeper, whereas the dihedral effect is to recover from the unnatural position. This effect is aggravated in the flaps-down condition.

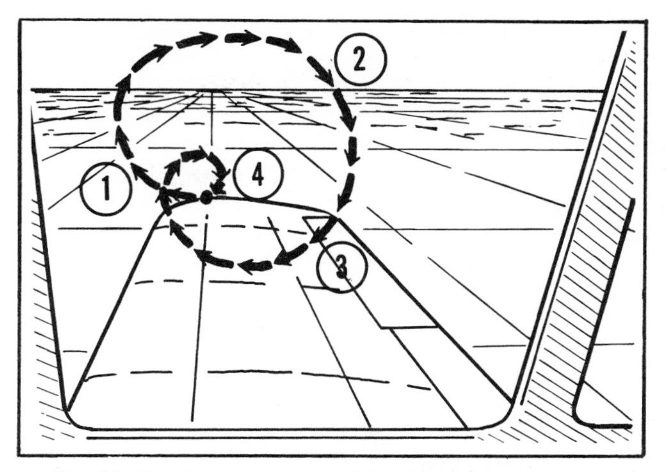

Fig. 10-51. Apparent movement of a point on the wing tip during Dutch Roll (damped oscillations). The pilot pressed left rudder pedal and released it.

LOADING EFFECTS

Usually the airplane loading has no effect on lateral static stability, as the fuselage is too narrow to allow offset loads. However, planes with wing tanks may present slight wing-down tendencies with asymmetric fuel load, but the only result is a tired arm if there's no aileron trim.

LATERAL DYNAMIC STABILITY

Dynamic stability is not a major concern in lateral stability. You recall that control surface balancing and other design factors introduced aerodynamic effects which furnished the "outside force" acting on the system, and the principles apply in the same way. The ailerons, if mass balanced (balanced around the hinge line by Weight) and if comparatively free in movement, usually assure that the pure lateral movements are heavily damped. However, cross effects of yaw displacement may result in lateral oscillations, and also "Dutch Roll."

LATERAL CONTROL

The aileron is the most widely known form of lateral control, although spoiler type controls have been used to some extent.

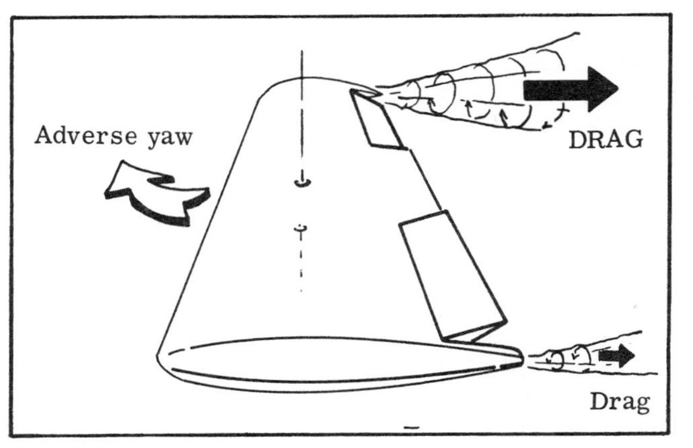

Fig. 10-52. Adverse yaw.

A problem with ailerons is that they introduce cross effects between lateral and directional movement. You were taught from the beginning that aileron and rudder go together 99.99 per cent of the time in the air. You were also shown that the ailerons are the principle banking control, with rudder being used as an auxiliary to overcome adverse yaw. Well, everybody's been busy trying to decrease aileron yaw and no doubt you've flown airplanes with Frise ailerons (Fig. 10-53) and ailerons with differential movement. (Fig. 10-54)

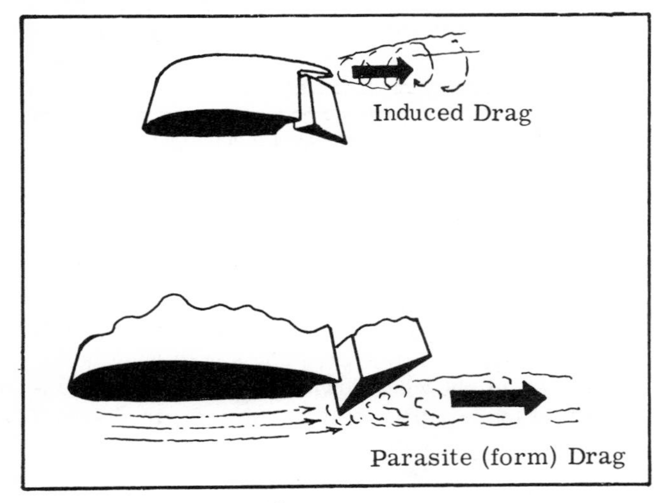

Fig. 10-53. Frise type ailerons.

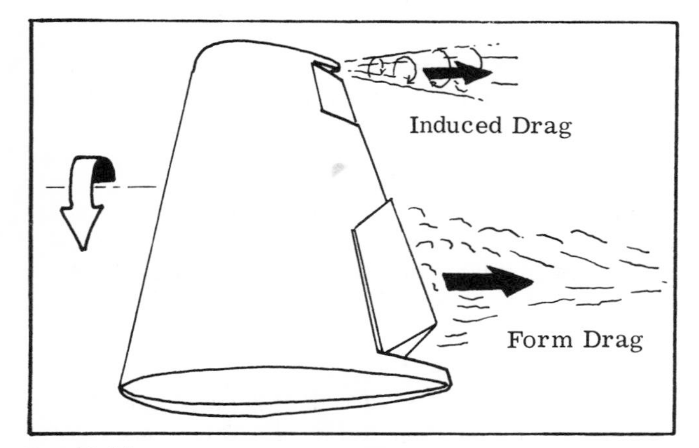

Fig. 10-54. Exaggerated view of differential aileron movement.

The reason for this adverse yaw is simple. You remember from Chapter 1 that induced Drag was caused by Lift. When you deflect the ailerons, the down aileron causes a higher C_L and higher induced Drag which results in yaw.

You can see in Fig. 10-53 that the up-aileron has some area hanging down which causes Drag — and helps overcome the Drag of the down aileron. The design also helps the pilot — as soon as the aileron is deflected up, aerodynamic forces help deflect it so that the pilot's stick force is small. A disadvantage of the Frise type aileron can be seen in the burbling and separation shown in Fig. 10-53. Another disadvantage is that at high speeds the aileron may tend to "overbalance." The pilot's stick force may become too light and the ailerons don't tend to return to neutral, but, if deflected, tend to deflect to the stop. You can imagine the discomfiture of your passengers when you start a nice turn and suddenly do a couple of aileron rolls. The Frise aileron usually is modified to get rid of overbalance or aileron buffet by rounding off the leading edge or making other minor shape changes.

DIFFERENTIAL AILERON MOVEMENT

This is the most popular method today of overcoming adverse yaw, and though the control system is slightly more complicated, there are not as many aerodynamic problems to cope with. The principle involved is to balance induced Drag with flat plate or form Drag. (Fig. 10-54)

SPOILER TYPE CONTROL

This method of control has been used on some jets and heavier prop planes but at the present is not on any light planes. The resulting force lowers the wing as well as adding Drag to that side — a desired combination. These spoilers usually are hydraulically actuated for use at high speeds. (Fig. 10-55)

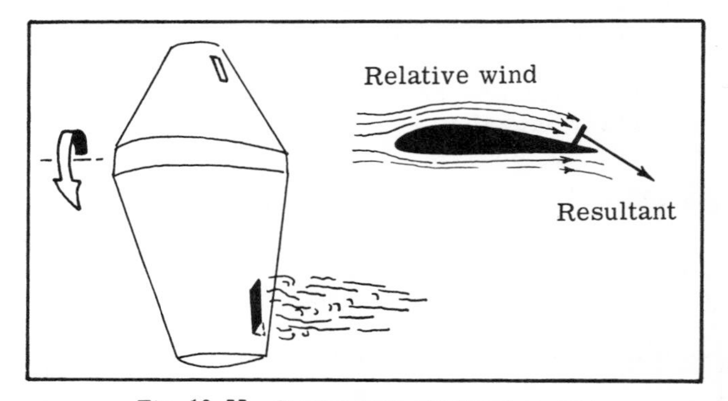

Fig. 10-55. Spoiler type of lateral control.

AILERON REVERSAL

This is presented more for interest than anything else, as you won't be affected by it in the airplane you are flying. But it is a problem for high-speed airplanes with thin wings and hydraulic (or irreversible)

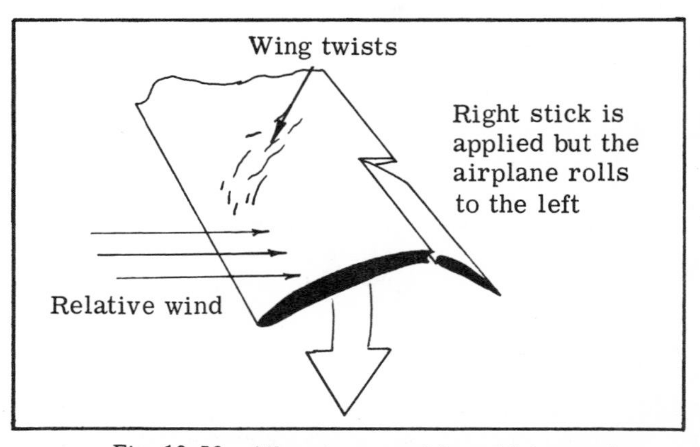

Fig. 10-56. Aileron reversal for a high-speed airplane with irreversible controls.

control systems. If an aileron is deflected by hydraulic power, the result may be that the aileron acts as a trim tab and the wing is twisted the opposite direction by aerodynamic forces. The application of right stick could result in a roll to the left! (Fig. 10-56)

There is another use of the term "aileron reversal" that does apply to light planes. It is the reaction of the airplane to the sudden sharp application of ailerons near the stall. The down aileron causes added induced Drag that results in that wing slowing and dropping — which further increases its angle of attack and stalls it. Aileron application in this case also results in the opposite effect desired. This problem will be covered more thoroughly in the chapter on stalls and slow flight.

SUMMARY OF THE CHAPTER

Static stability is the measure of the *initial tendency* of a body to return to its original position. This initial tendency to return may further be broken down into positive, neutral or negative static stability.

Dynamic stability is the *action* of a body caused by its static stability properties. To have oscillations, the system must have *positive* static stability. A system that has *neutral* static stability has no initial tendency to return and therefore has no oscillatory properties. Likewise, a system that has *negative* static stability diverges — it tends to leave the original position at a faster and faster rate — it cannot have oscillatory properties but diverges if offset.

The ball inside the hubcap is statically stable, but outside forces may act on the system so that it may have neutral or negative dynamic stability. This outside force may be compared to aerodynamic forces set up by improper control balance or design, even though the airplane itself is properly loaded and the over-all design is good.

Longitudinal stability is the most important, as it is most affected by airplane loading. If the airplane is loaded properly, as recommended by the Airplane Flight Manual, it will fall well within safe limits for longitudinal static stability.

Entire texts have been written on airplane stability and this chapter merely hits the high points. For more thorough coverage you are referred to the references at the end of this book.

11. STRESSES ON THE AIRPLANE

BACKGROUND

In Chapter 7 it was illustrated that the pilot may put large loads on the airplane in a steep turn. Not mentioned, but as well known, is the fact that pull-ups may be even more critical. It takes time to roll into a level turn, and during the process the airspeed is decreasing as the back pressure is being increased gradually. Not so the pull-up. Back pressure may be applied almost instantaneously at very high speeds.

High-performance, retractable-gear airplanes are very clean aerodynamically, and unqualified pilots who try to fly under instrument conditions sometimes learn the hard way about load factors. The chain of events might be like this: The pilot flies into instrument conditions without proper training and/or instrumentation. The airplane gets into a "graveyard spiral" and, because of its aerodynamic cleanness, builds up speed very fast. It breaks out of the clouds at a very low altitude and the pilot instinctively pulls back on the wheel — *hard!* The airplane loses a wing (or wings) and private aviation gets another black eye. Unfortunately there are usually passengers aboard. Is this the airplane's fault? Manufacturers do not furnish brains with their airplanes.

Take a look at the strength requirements for various airplane categories. These requirements are expressed in "g's," the term used for load factor. You remember from the discussion of the turn that a *level* 60° banked turn results in a load factor of 2, or 2 "g's." You're pushing down against the seat twice as hard — and the wings are carrying twice their normal load. The back pressure in the 60° banked turn results in a positive load factor, or positive "g's." The force you feel when you apply forward pressure briskly is a negative load factor. In straight and level flight you and the airplane have one positive g working on you. By smooth forward pressure on the wheel you may lower this to one-half g or even to zero g. Pencils and maps may want to float around the cockpit and you are "floating" against the belt — and you can go even farther and get into negative g forces.

Excessive positive g's result in the pilot's "blacking out" as the blood is forced down from his head and upper body. As the positive g's are introduced, the pilot begins to feel heavy and as the force increases, starts to "gray out" or lose vision. Increased positive g's result in blacking out or complete loss of vision and, finally, unconsciousness results. As the force is removed, the process is reversed. Several seconds usually are required for complete recovery from the effects. The average

Fig. 11-1. Positive and negative g effects on the pilot.

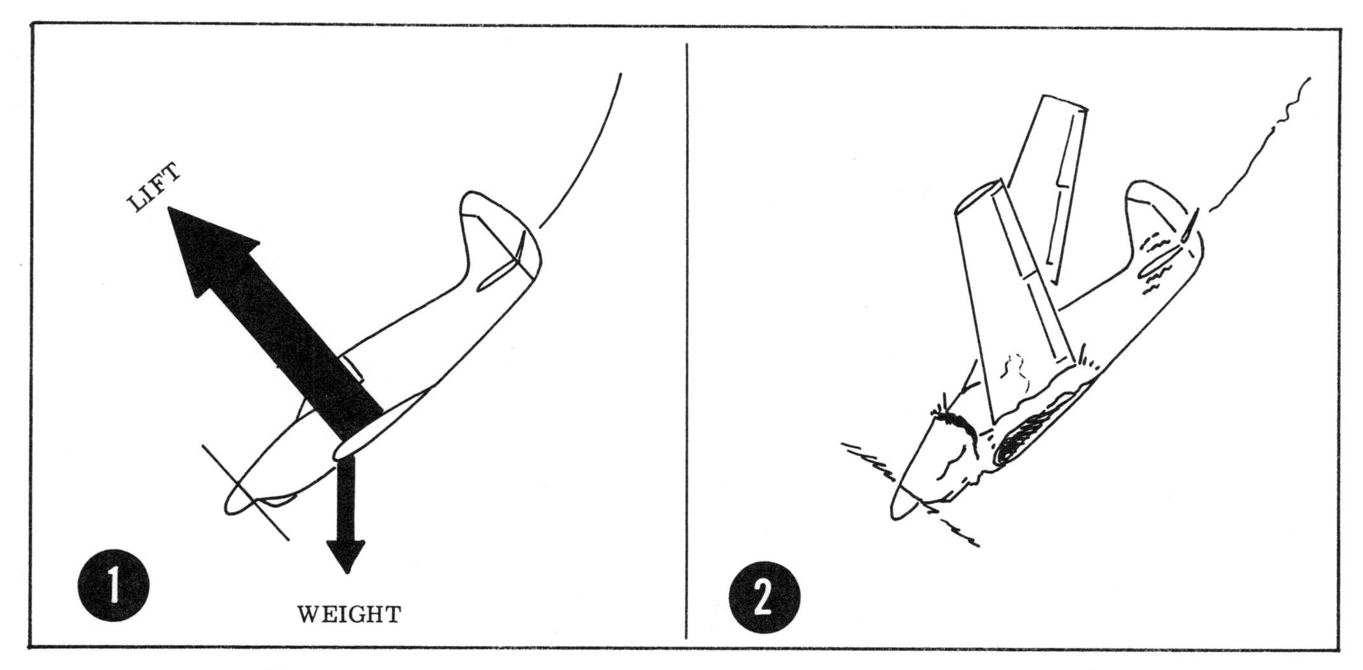

Fig. 11-2. (1) The airplane's Lift may exceed its Weight by several times in a pull-up and (2) sometimes with disappointing results if overdone.

person in good health will gray out at about a positive 4 g's and black out above 6 g's.

You won't be able to take as many negative g's; in fact people get pretty uncomfortable at minus 2 g's, and permanent damage to the eyes or brain could occur at minus 4 g's or higher.

The physical symptoms of excessive negative acceleration (minus g) are called "red out." The blood rushes to the head and small blood vessels in the eyes may hemorrhage, leaving the eyes looking like road maps for several days after the occurrence.

The pilot's positive g tolerance may be raised by wearing a "g-suit," a tight-fitting flight suit having air bladders at the stomach and thighs. The suit is plugged into a pressure source in the airplane and the bladders are automatically inflated when positive g forces are encountered.

The problem of raising the negative g tolerance remains unsolved. Of course, an automatic tourniquet around the pilot's neck might stop the flow of blood to the head but this would result in a high turnover of pilots, to say the least.

The airplane, because of the physical limitations of the pilots, is designed to withstand a higher positive than negative load factor.

Loads may be exerted on the airplane either by the pilot's handling of the controls or by encountering vertical gusts.

PILOT INDUCED (MANEUVERING) LOAD FACTORS

The primary way in which the pilot may cause high load factors is by use of the elevators — although additional loads may be induced by aileron use, as will be mentioned later.

Suppose that you are practicing stalls and that your airplane stalls at *50 knots* at cruise power setting at gross Weight. When you normally practice power-on stalls the procedure is to *ease* the nose up until the airplane stalls (be sure to clear the area before doing stalls). Throughout the entire process you have a load factor of one — there are only normal forces exerted on you and the airplane. The stall recovery also should result in no adverse g forces (unless you get too eager to recover and slam the wheel forward, getting negative g forces).

Now suppose you decide to practice accelerated stalls and through carelessness pull back abruptly so that the wing reaches its critical angle of attack at *100 knots*. You have doubled the speed at stall *but have increased the load factor 4 times!* If you stall this same airplane at 150 knots or at 3 times the normal stall speed, you'll put a load factor of 9 on the airplane (the wings, of course, will be fluttering back behind you somewhere). So . . . the load factor is a function of the speed squared — twice the normal stall speed = $(2)^2$ or 4 g's; 3 times the normal stall speed = $(3)^2$, or 9 g's. These figures are arrived at from the Lift Equation: Lift = $C_L S \frac{\rho}{2} V^2$. The airplane will stall at a certain *maximum* Coefficient of Lift. The $C_{L_{max}}$ is dependent on angle of attack. This means that this part of the equation is fixed (although it has been found that in high speed pull-ups the level flight maximum, or stall, Coefficient of Lift may be slightly exceeded for a brief period, so you might actually pull even more g's than is shown by the Lift Equation).

The density is constant for a given altitude. The wing area is also a fixed quantity unless you pull *too* many g's, then it may be reduced considerably — suddenly. There is nothing left to vary in the Lift Equation except the velocity, which is squared. In the *100 knots* stall you are exerting *4* times the Lift of the 50-knots stall. The wings are subjected to *4* times the strain — or 4 positive g's. A simple measure of the number of g's being exerted is that of $\frac{\text{Lift}}{\text{Weight}}$. The Lift-to-Weight ratio of the airplane was 4 at the 100-knot stall. Remember in the 60° bank (Chapter 7) it was found that the Lift vector had to be doubled to maintain altitude and this gave a load factor of 2.

The FAA minimum strength requirements for general aviation planes are as follows:

Airplane Category	Positive g's Required	Negative g's Required
Normal	2.5 — 3.8	40% of the positive g's or -1.0 to -1.52
Utility	4.4	40% of the positive g's or -1.76
Aerobatic	6.0	50% of the positive g's or -3.0

Notice under normal category that the minimum positive required g's are 2.5 to 3.8. Actually the requirement states that the positive limit maneuvering load factor shall not be less than 2.5 g's or need not be higher than 3.8 g's. Most manufacturers design for the 3.8 figure. This means that the pilot can pull up to 3.8 positive g's with no danger of permanent deformation or structural damage to the airplane. In addition to this, a safety factor of 1.5 is built in.

A typical normal category airplane might be designed for the following load factors:

Limit Load Factor (No deformation)	3.8 +g's and 1.52 -g's
Ultimate Load Factor (1.5 safety factor)	5.7 +g's and 2.28 -g's

About this 1.5 safety factor: this means that when this limit is reached *primary* structure (wings, engines, etc.) will start to leave the airplane. Secondary structures will have long gone. In other words, if you exceed the limit load factor you can expect some damage to occur to the airplane. Just how *much* damage will depend on how far you exceed it. The airplane could have been damaged previously, so you don't have a 1.5 safety factor any more. *As far as you are concerned the limit load factor is not to be exceeded at any time.*

If the airplane in the illustration of the 100 knot accelerated stall had been a normal category airplane, you can see that it was in an unsafe condition at 4 g's. The stall at 150 knots (9 g's) would have resulted in structural failure. You can see that the pilot who comes spiraling out of the clouds at 200 to 250 knots and suddenly pulls back on the wheel will put stress on the airplane that even fighter planes might find hard to handle.

Another thing — a *rolling* sharp pull-up is actually much worse than a straight pull-up. Even if a g meter (accelerometer) indicated that you were within limits, the down-moving wing will have a higher angle of attack and a higher load factor even though the "average" load on the airplane still may not be critical.

In addition to the fact that the wing Lift loads are not equal, the deflection of the ailerons exerts torsional forces on the wing; an illustration of this is shown back in Fig. 10-56. If the wing is close to the limit of its endurance when the torsional forces are introduced, you may witness an exciting event.

THE VG DIAGRAM — MANEUVER ENVELOPE PORTION

The area of operation for the airplane is given in the Vg diagram (or V-G, V-g, or V-n diagram as it is sometimes called). This is merely a chart showing the limits of the airspeed (V) and load factor (G, g or n).

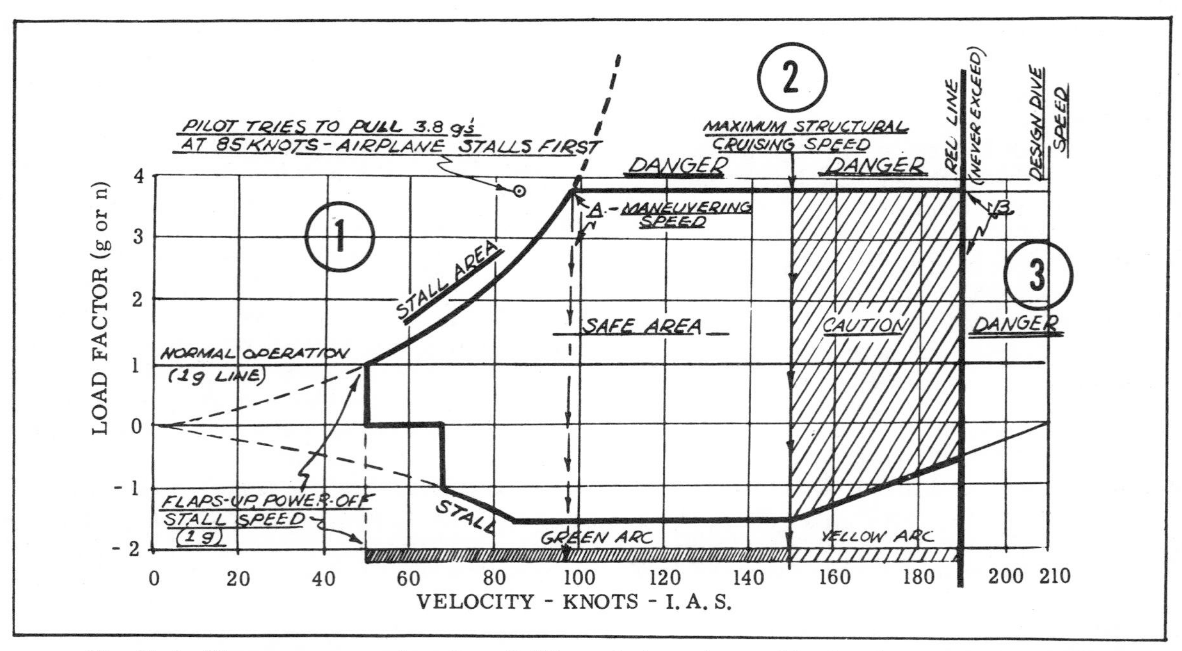

Fig. 11-3. The maneuver envelope for a fictitious airplane at gross Weight. Assume no instrument or position error so that I.A.S. = C.A.S.

Fig. 11-3 shows the maneuver envelope portion of such a diagram for a fictitious normal category airplane.

If you try to operate in Area 1 (for instance you try to pull 3.8 g's at 85 knots), the airplane will stall before you reach the limit load factor of 3.8 g's. This is the least dangerous of the three areas outside the envelope, as the airplane will stall before you can pull it apart. This is assuming that you have enough altitude for the stall recovery.

Point A marks the Maneuvering Speed (which is 97.5 knots in this example), the maximum speed at which the controls can be deflected without overstressing the airplane. This maneuvering speed decreases with decrease in weight. If you are flying below this speed the airplane will stall before you can exceed the limit load factor (positive).

In the speed range from Point A to Point B you can pull up to 3.8 g's without causing undue concern. There is one danger, however. You *can* get into trouble by inadvertently pulling enough g force to get into Area 2 because the airplane won't stall before you begin pulling it apart. Sure, you have a safety factor of 1.5 before things come completely unglued, but as will be shown later, a slight hunk of turbulence during the period you're operating in Area 2 can cause a great deal of disappointment (and *this* is the understatement of the year).

Of course, you don't have an accelerometer like the jet boys so you'll have to more or less play it by ear. For normal flying you'll probably never pull over 2 g's anyway. You might check what the acceleration of 2 g's feels like by making a 60° banked level turn (720° power turn). You actually should have no reason to exceed 2 g's unless you plan to do aerobatics -- and then should fly an aerobatic airplane.

Area 3 is a place to be avoided at any g. The airplane will start getting rid of a few surplus parts like wings even though you're not pulling any forces of acceleration. You may encounter all kinds of flutter problems.

The maneuvering envelope given here is for subsonic airplanes at lower altitudes. The envelope for higher altitudes and speeds is somewhat smaller, as expressions like Mach numbers and buffet boundaries come into use.

The negative side of the maneuvering envelope works in the same way as the positive limits just discussed except, as you can see, the negative g limits are smaller in value.

THE VG DIAGRAM — THE GUST ENVELOPE PORTION

Vertical gusts can put acceleration forces on the airplane, too, so don't get cocky and think that you alone can raise Cain.

Check what happens on a bright sunny day when the plane flies into a 30-feet-per-second upward gust.

Fig. 11-4 shows the wing of the airplane when it's flying straight and level. Lift equals Weight (for illustration purposes, anyway), ceiling and visibility are unlimited and you and the airplane are being worked on by the usual 1 g.

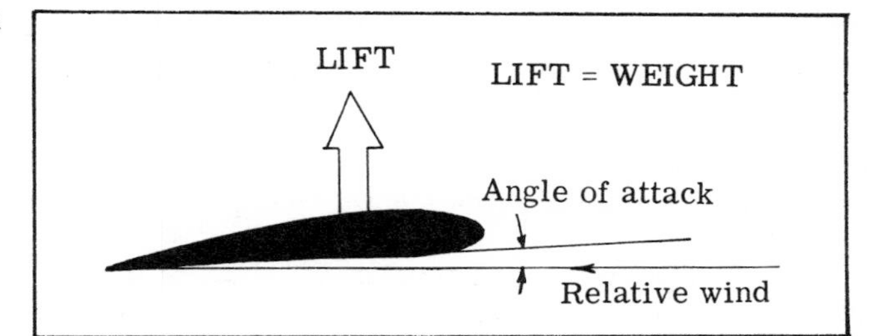

Fig. 11-4. Straight and level flight.

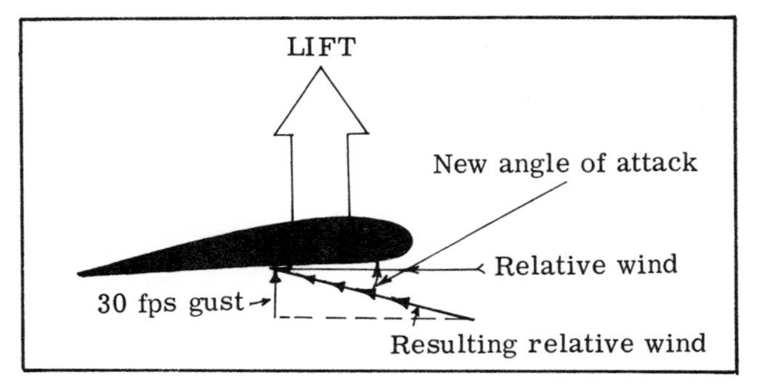

Fig. 11-5. Encountering a 30 fps upward gust.

In Fig. 11-5 you have just flown into a 30-feet-per-second upward vertical gust.

The angle of attack of the wing is suddenly increased. As far as the wing is concerned you suddenly pulled sharply back on the wheel. Because of this sudden increase in Lift there'll be positive g forces exerted on you and the airplane. If you are at extreme ends of the speed range when you fly into the gust you could stall or exceed the g limit.

Upward moving gusts are easier on you and the airplane than the downward variety. You can well remember the occasions you have flown into a downward gust. Those were the times you would have left the imprint of your head in the ceiling if the seat belt weren't snug. It's a good idea to *always* keep that seat belt on, even if the air is smooth, because you could fly into clear air turbulence or a shear area with no warning.

The faster you go, the more stress is put on the airplane when a vertical gust is encountered. Fig. 11-6 shows the gust envelope for the same airplane as used in Fig. 11-3.

You are safer, when encountering moderate to severe turbulence, to slow the airplane. Even if the 30 fps gust at your cruise speed does not cause your airplane to exceed the limit load factor as shown by the gust envelope, this doesn't mean that you won't get gusts higher than this — maybe the gust hasn't read that most vertical gust velocities are 30 fps or less.

The vertical gusts mentioned are considered to be "sharp edged"; that is, there's no relatively gradual transition from smooth air to the 15 or 30 fps velocity. The gusts are considered to be instantaneous because this is a safer line of thinking. If they actually aren't instantaneous, so much the better.

Then it's agreed that the airplane should be slowed but the question is — how much? If the turbulence is light or moderate, just make sure the

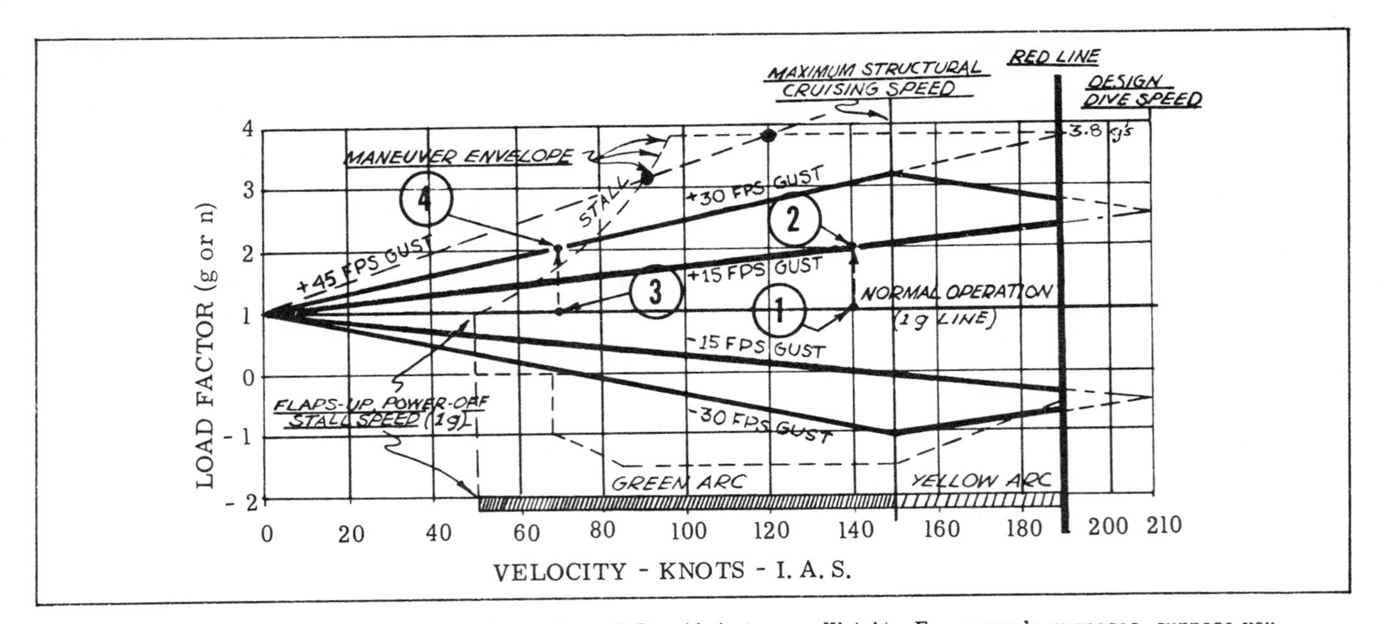

Fig. 11-6. The gust envelope for the airplane of Fig. 11-4 at gross Weight. For example purposes, suppose you are flying at Point (1) (140 knots, 1 g normal flight). If the airplane encounters a 15 fps upward gust, your position in the maneuver envelope is then at Point (2), still well within the envelope. The same airplane slow flying at 70 knots, Point (3), is moved toward Point (4) and stalls momentarily. In this airplane at *gross Weight* you can encounter a 30 fps vertical gust at maximum structural cruising speed and still be well within the maneuver envelope. The dashed line shows a 45 fps upward gust and the two dots show the limits of the very narrow range of operation between stalling and exceeding the limit load factor (approximately 91 knots to 120 knots). It would be wise to keep the airspeed below 120 knots in this case! As Weight decreases, the gust envelope spreads so that at very light Weights even a 30 fps gust could cause you to reach the maneuvering limit load factor at maximum structural cruising speed — a good argument for slowing down in turbulent air.

airspeed is kept in the green range of the airspeed indicator. If it's severe and the airplane starts jumping around, slow it below the maneuvering speed. This is fine except the maneuvering speed isn't put on the airspeed indicator anywhere. The designer finds the value of the maneuvering speed by the equation: Maneuvering Speed = Vstall x $\sqrt{n}$ or Vstall x $\sqrt{g}$, whichever way you prefer it. The Vstall mentioned is the *flaps-up, power-off* stall speed, and at gross Weight is indicated by the lower point of the green arc on the airspeed indicator. The n or g in the equation is the design limit load factor (positive g's) of the airplane. For a normal category airplane having a positive limit load factor of 3.8 g's, that stalls at 50 knots (flaps-up) at gross Weight, the maneuvering speed, Va, at gross Weight may be found as follows:

Maneuvering Speed = 50 x $\sqrt{3.8}$ = 50 x 1.95 = 97.5 knots.

As the figure of 1.95 is very close to 2, use that figure so that the maneuvering speed = 50 x 2 = 100 knots. Just remember that in severe turbulence at gross Weight you would slow the airplane to a speed slightly less than twice the power-off, flaps-up stall speed. Decrease the maneuvering speed by one half of the percentage of Weight decrease; 20 per cent Weight decrease means a 10 per cent decrease in the maneuvering speed, etc. This is simply because that is the approximate effect of Weight decrease on stall speed, and the maneuvering speed equation depends partly on stall speed.

However, as a general all-around figure for all Weights, slow the airplane to a speed of about 1.5 times the flaps-up, power-off stall speed as given by the bottom of the green arc on the airspeed indicator. This should assure that you will be below the maneuvering speed even at lightest Weights. As the maneuvering speed decreases with decrease in Weight, the airplane could be overstressed at lower Weights if you use the gross Weight maneuvering speed — that's why the 1.5 ratio comes in.

For utility category airplanes the equation is fundamentally the same except for the limit load factor being 4.4 for this category. The square root of 4.4 is 2.1, so that the maneuvering speed at gross Weight is slightly over twice the flaps-up, power-off stall speed at gross Weight. Look back at Point A in Fig. 11-3 — that point gives the maneuvering speed at gross Weight. You can see that if you stay below this speed the airplane will always stall before being overstressed, no matter how strong the vertical gust. The stall is no problem as it is momentary and the airplane will have probably recovered even before you realize it was stalled (you'll probably be so busy handing out burp cups to your passengers that a brief stall will go unnoticed). Again, you can use a speed of 1.5 times the flaps-up, power-off stall speed, as given by the bottom of the green arc of the airspeed indicator, for all weights of the utility category airplane. Better a momentary stall than overstressing the airplane. One point, however: a stall can spell trouble at low altitudes such as on approach. You'll have to take this into account.

All the speeds mentioned here are *indicated airspeeds*, because this is a measure of the dynamic pressure "q" working on the airplane structure.

Don't be like the private pilot who, while flying in an easterly direction at normal cruise (indicating in the green range of his airspeed), checked his progress with a computer and found that because of an exceptionally strong tailwind, his *groundspeed* was well over the red line speed as given in the Airplane Flight Manual. He throttled back and landed immediately, thankful for his escape. He knew that he shouldn't exceed the red line speed but didn't realize that the red line speed was *indicated airspeed,* not true airspeed or groundspeed. The airspeed indicator hand tells you if you're in trouble in this sort of situation.

The airplane is limited to lower positive g's (2.0) and less powerful vertical gusts (15 fps) when the flaps are down. You might keep this in mind.

Summed up:

1. *Don't exceed the limit load factors – positive or negative.*

2. *In moderately turbulent air keep the airspeed in the green arc.* You don't know when you may encounter severe turbulence. Stay out of the yellow arc when it's bumpy.

3. *In severe turbulence slow the airplane below the maneuvering speed.* Use 1.5 times the flaps-up, power-off stall speed (bottom of green arc) as a rule of thumb for *all* weights.

4. *If you are letting down in rough air, keep the airspeed in the green arc.* If it's really rough, slow it to the maneuvering speed for the letdown.

5. *If you want to do acrobatics, rent an acrobatic airplane* (chandelles and lazy eights are not considered to be acrobatics).

6. *Read your Airplane Flight Manual* — it may have recommendations for turbulent air penetration.

7. *Check the placards on the instrument panel, and keep the airspeed within reason for the conditions under which you are flying.*

PART II / CHECKING OUT IN ADVANCED MODELS AND TYPES

12. AIRPLANES WITH CONTROLLABLE PITCH PROPELLERS

When you first check out in an airplane with a controllable pitch propeller things can be pretty hectic. In the first place, it's a new airplane to you and the addition of the propeller control further complicates matters. After all, it's only *one* more control you say to yourself, but juggling the throttle and prop to get the right combination of power may cause some consternation for the first few tries. The check pilot who's sitting so calmly in the right seat once had the same problem. If he's grinning at your work-out, it's probably because he's remembering his first struggles with throttle *and* prop controls.

The controllable pitch prop allows you to get more efficiency out of the engine-propeller combination in all speed ranges. That extra control is not just put there to complicate matters, although it does impress the ladies to see you making adjustments with those mysterious knobs. With the girl friend along, pilots have been known to do a lot more adjusting than was strictly necessary, but this is understood as a fair maneuver — all pilots are guilty.

The horsepower developed by a particular engine depends on the manifold pressure and rpm. For instance, 65 per cent power, often used for cruise, may be set up in several combinations of manifold pressure and rpm. The manufacturer furnishes a power setting table which allows the pilot to establish the desired horsepower for his particular altitude. Fig. 12-1 shows the chart for the Lycoming O-540-A1B5, 250 HP (unsupercharged) engine.

Another thing you might look at in Fig. 12-1: the manifold pressure required at a particular rpm becomes less as altitude increases. At sea level, at 2200 rpm, 23.3 inches of mp are required to develop 65 per cent. At 5000 feet, only 22 inches are required for 65 per cent power at 2200 rpm. There are two main reasons for this apparent inconsistency: (1) The exhaust gases have less outside pressure (back pressure) to fight at higher altitudes. Remember that the "explosion" in the cylinder is sealed, and power is used to expel the waste gases. Less back pressure means less power used to eliminate this waste (there's better scavenging); power that can be used in making the airplane go. (2) The air is cooler at higher altitudes. If you use the same mp as you carried at sea level, the mixture density and the horsepower developed would be greater (lower temperature means greater density if the pressure remains the same). Therefore, in order to maintain the same power, less manifold pressure is used for a given rpm with altitude increase as Fig. 12-1 shows.

Power Setting Table — Lycoming Model O-540-A1B5, 250 HP Engine

Press. Alt. 1000 Feet	Std. Alt. Temp. °F.	138 HP — 55% Rated Approx. Fuel 10.3 Gal. Hr. RPM AND MAN. PRESS.				163 HP — 65% Rated Approx. Fuel 12.3 Gal. Hr. RPM AND MAN. PRESS.				188 HP — 75% Rated Approx. Fuel 14.0 Gal. Hr. RPM AND MAN. PRESS.		
		2100	2200	2300	2400	2100	2200	2300	2400	2200	2300	2400
SL	59	21.6	20.8	20.2	19.6	24.2	23.3	22.6	22.0	25.8	25.1	24.3
1	55	21.4	20.6	20.0	19.3	23.9	23.0	22.4	21.8	25.5	24.8	24.1
2	52	21.4	20.4	19.7	19.1	23.7	22.8	22.2	21.5	25.3	24.6	23.8
3	48	20.9	20.1	19.5	18.9	23.4	22.5	21.9	21.3	25.0	24.3	23.6
4	45	20.6	19.9	19.3	18.7	23.1	22.3	21.7	21.0	24.8	24.1	23.3
5	41	20.4	19.7	19.1	18.5	22.9	22.0	21.4	20.8	—	23.8	23.0
6	38	20.1	19.5	18.9	18.3	22.6	21.8	21.2	20.6	—	—	22.8
7	34	19.9	19.2	18.6	18.0	22.3	21.5	21.0	20.4	—	—	—
8	31	19.6	19.0	18.4	17.8	—	—	20.5	19.9			
9	27	19.4	18.8	18.2	17.6	—	21.3	20.7	20.1			
10	23	19.1	18.6	18.0	17.4	—	—	—	19.6			
11	19	18.9	18.3	17.8	17.2	—	—	—	—			
12	16	18.6	18.1	17.5	17.0	—	—	—	—			
13	12	—	17.9	17.3	16.8							
14	9	—	—	17.1	16.5							
15	5	—	—	—	16.3							

To maintain constant power, correct manifold pressure approximately 0.17" Hg for each 10° F. variation in carburetor air temperature from standard altitude temperature. Add manifold pressure for air temperatures above standard; subtract for temperatures below standard.

Fig. 12-1.

The figures given in the table are for standard pressure altitudes. The footnote shows how to correct for deviations from standard temperature, though in actual practice this is seldom done, as the mp gage can't be read that closely.

While we're on the subject of the power table, notice that you are unable to maintain the various percentages of power above certain altitudes. Naturally you won't be able to hold 75 per cent power at as high an altitude as at 65 per cent — the engine can't maintain the required manifold pressure.

For a given rpm, the higher the manifold pressure the more power developed. Of course you can ruin an engine very quickly by thinking that this is the way it should be operated. At low rpm and high manifold pressure the engine could suffer damage — which is an expensive as well as dangerous problem. Manufacturers do not recommend power settings of over 75 per cent for cruise for reciprocating engines, as the fuel consumption and extra engine wear preclude use above this value.

USING THE THROTTLE AND PROPELLER CONTROLS

The manifold pressure gage tells you of the potential power going to the engine, the tachometer tells how much is being used. With the proper throttle and prop setting you have ideal potential *and* use of the power.

A part of the measure of an engine's power output is bmep (brake mean effective pressure) in the cylinders at the instant of combustion. If this internal pressure is too great the engine can be damaged.

The engine can efficiently use a maximum amount

Fig. 12-2. "I guess I should have eased the prop control forward before opening the throttle." (Improper use of the throttle and prop control can cause engine indigestion.)

of fuel-air mixture at a certain rpm. A high manifold pressure means that a lot of fuel and air is available for the engine. When this higher compressed charge is shoved into the cylinders, more power should be produced. But if the prop control is set at too low an rpm, it's like putting the powder load for a cannon into a shotgun (well, maybe not quite).

If you have trouble remembering which goes first in a power change — throttle or prop — remember this. *Keep the propeller control forward more than the throttle.* If you're increasing power, the propeller control is moved forward first; in decreasing power the propeller control is moved back last. In normal usage, if it were timed, the propeller control would be forward more than the throttle (by seconds). Don't slam the throttle or prop control — the engine and propeller have inertia to be overcome in changing speeds and if you get too hasty you could damage the engine this way, too.

The propeller control may be thought of as similar to an automobile (manual) gear shift. However, whereas the car just has a few set positions, you can set any "gear" combination you want with the prop control. Flat pitch, or high rpm, is comparable to low gear in a car. You set it before take-off (prop control full forward) and "step on the gas" (open the throttle all the way). After getting off the ground you throttle back to the climb mp setting and *then* pull the prop control back to the recommended rpm setting for climb.

About decreasing power: if you pull the prop control back *before* the throttle, the manifold pressure will increase because the engine isn't taking the mixture as fast any more — you're giving it more bmep than it can efficiently use — this can be bad if the difference becomes too great.

SOME ITEMS ABOUT TAKE-OFF

Your first take-off in the new airplane almost can be predicted. You'll take off, feeling strange and maybe a little tense as you want to be sure to do a good job. You'll have the prop full forward in high rpm and the throttle wide open. Fine. The airplane lifts off and you're naturally pretty busy. Watching the runway and the area ahead you prepare to throttle back. Glancing at the manifold pressure gage you see it hasn't moved so you pull the power back some more. There's a power loss felt but the mp gage hasn't moved. Is it broken? The screams of the check pilot (and maybe a groan or two from the engine) direct your attention to the tachometer. Oops — it's back below cruising rpm. You pulled the prop control back first, *instead* of the throttle. You hastily shove it back up — the prop overspeeds for a couple of seconds — and you have to start the power reduction process anew for the climb out, this time pulling the throttle back *first* and *then* the prop control, as it should be. You'll feel about six inches high. Well, welcome to the club; you've joined a group of several hundred thousand other pilots who've done the same thing (and this includes the guy in the right seat over there, who's hollering so loudly).

However, just because you didn't mean to do it doesn't lessen punishment to the engine in a deal like this. The proper procedure will come with practice.

Some pilots use their knowledge of the fact that at a constant throttle setting the mp will increase if the rpm decreases, to save themselves the extra manipulation of the prop and throttle on take-off. Suppose your airplane uses 28 inches and 2700 rpm for take-off at sea level, and the manufacturer recommends 24 inches and 2400 rpm for climb-out. The old pro will throttle back to about 23 inches. Then when the prop is pulled back to 2400 rpm, the manifold pressure will be up to 24 inches and no further adjustment will be required. If he had set the throttle to 24 inches, it would have eased up to, say, 25 inches when the prop was pulled back, and would require resetting the manifold pressure. You will soon note the rise in mp with decrease in rpm after take-off for your airplane and will do this automatically. The one-inch rise used here is an illustration, the exact rise will depend on the difference between take-off and climb rpm for your engine.

THE CLIMB

You've set the power to the recommended value of 24 inches and 2400 rpm (or whatever is set up for your particular airplane) and now feel you can relax a little. On your first flight in the new airplane you'll want to get some altitude and just get used to the bird before doing anything exotic like take-offs and landings.

As you climb it seems that after a couple or three thousand feet the airplane has lost quite a bit of its go and if the check pilot hasn't already brought it to your attention by tapping you gently between the eyes with the fire extinguisher, a glance at the manifold pressure gage shows that the mp has dropped a couple (or three) inches. A creeping throttle? No, the atmospheric pressure drops about one inch of mercury per thousand feet at lower altitudes (see the Standard Atmosphere Chart in the Appendix) and the engine just isn't able to get the same amount of manifold pressure at the old throttle setting. You'll have to open the throttle of the unsupercharged engine as altitude is gained (and this is also the case for the more simple supercharged engines to be discussed later.)

Going back to the beginning of the chapter, it was mentioned that the engine of the example needed about one-fourth inch less of mp per thousand feet to maintain the same percentage of power. The mp drops one inch so this puts you about three-fourths of an inch in the hole for each thousand feet. Some engine manufacturers recommend a constant manifold pressure for a particular engine for the climb at *all* altitudes (if you can maintain it) while for engines having limits of continuous power they furnish tables for recommended maximum mp at various altitudes. Maintaining a constant mp to higher altitudes does mean that more power is being developed up there and this power may exceed the manufacturer's recommendation for long-time use for some engines.

CRUISE

On leveling off leave the power at climb setting until the expected cruising speed is reached. This is done for two reasons: (1) The transition from climb to cruise is shorter and (2) You'll only have to set cruise power once. If you throttle back and set the power to say, 23 inches and 2300 rpm for cruise (or whatever the power setting chart recommends for your altitude and chosen power) immediately upon reaching the altitude, you'll find that as the airspeed picks up from climb to cruise the manifold pressure will also increase due to "ram effect" or increased dynamic pressure of the air entering the intake. You'll have to reset the manifold pressure.

LANDING NOTES

Another problem you may have the first few times is remembering to move the prop control forward to a high rpm (low pitch) during the approach to prepare for the possibility of a go-around. Some Airplane Flight Manuals recommend that the rpm be set to high cruise or climb rpm rather than for take-off in this case, as there is a possibility of engine overspeed if throttle is suddenly applied. Others may recommend a full high rpm setting for the landing approach.

For airplanes in which a high cruise or climb rpm is recommended for an approach this is best done on the downwind leg when you have cruise power on. Even the constant speed prop cannot maintain the pre-set rpm when you've throttled back for landing. If you have it set for 2400 rpm at cruise, when you close or nearly close the throttle, the rpm may drop down to 2000, or well below, when you slow up. Of course, as soon as you open the throttle past a certain manifold pressure, the rpm will increase and hold your pre-set value. Actually what happens when you throttle back to idle is this: The prop tries to maintain the pre-set rpm, and as you throttle back the blade angle (pitch) will decrease — trying to maintain the required rpm. Finally you throttle back so far that the blades are as flat as they can go but can't keep up the rpm. Moving the prop controls forward won't help — you'll have to increase power (manifold pressure) before getting a reaction.

On the other hand, if the recommendation for your airplane is to set the prop to full high rpm for the approach, *don't* do it on the downwind leg where you are developing power in the engine. The result here would be a probable rpm overshoot and, at best, a noisy announcement of your presence in the area. The usual practice here is to move the prop control (or controls) forward after turning on final, when you're not using a lot of power. Of the four main items for landing (gear, flaps, mixture and prop) the propeller is set last if a full high rpm setting is recommended for landing.

At any rate, you haven't been setting a propeller for landing before now, and may have to be reminded by the check pilot a couple of times. Remember this: if you have to go around, you may need full power quickly and will be in bad shape if you shove the throttle wide open with the prop control back in a low cruise setting.

PROP CONTROLS

The oil-counterweight prop control is somewhat similar to the throttle. It works the same way and may resemble the throttle, although the handle itself is usually a different shape for quick recognition by feel. The prop control for the single engine airplane usually projects out of the instrument panel; whereas the multiengine airplane throttles, prop controls and mixtures are on a quadrant. The single engine airplane prop control usually can be moved by either of two ways: (1) pressing a manual release (lock) button and moving the control in or out in the same manner

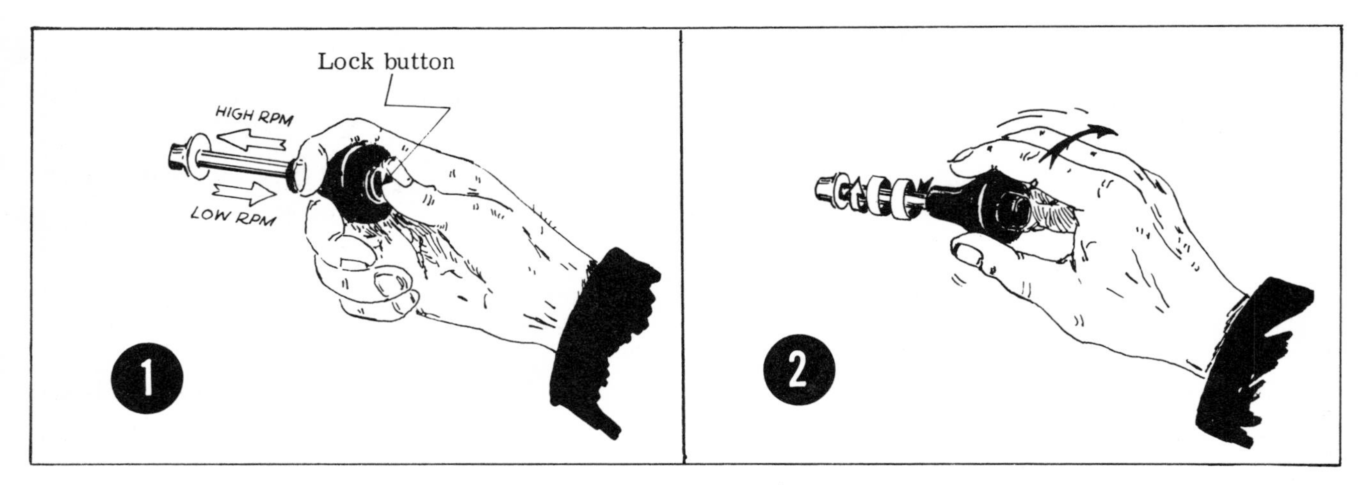

Fig. 12-3. (1) Using the lock button for large rpm adjustments (2) Using the vernier for smoother, minor adjustments. The propeller control is being turned to obtain higher rpm in the illustration.

as the throttle is used, or (2) using the vernier adjustment, screwing the control in (clockwise) to increase rpm and turning it counterclockwise to decrease rpm. The vernier method allows for a finer setting and is used for making adjustments when the engine is developing power such as after take-off, cruise, or setting rpm on the downwind leg. The button lock push-pull method is good for quickly setting full high rpm when the engine is not developing a great deal of power (before opening the throttle for take-off or on final). See Fig. 12-3.

The end result is the same. Naturally, it will take longer, for instance, to go to full high rpm using the vernier, but many pilots prefer it and never use the button release. The *throttles* for some airplanes also may use a combination vernier and button release. *One thing to remember, the prop control moves the same way as the throttle — for more power (rpm) it's forward, for less power (rpm) it's back.*

Some *throttles* for fixed pitch propeller planes are of the vernier type.

ELECTRIC PROPELLER CONTROLS

The electric controllable pitch propeller usually has a spring loaded toggle switch control. Although the toggle switch does not remotely resemble the throttle, pilots still use one when they should be using the other. It still doesn't do the engine any good if you're wrong.

The toggle switch controls an electric motor which sets the blade angle (rpm) desired. Electric propeller controls also may have an "automatic" or constant speed selection.

PRE-FLIGHT CHECKING OF THE CONTROLLABLE PITCH PROPELLER AIRPLANE

Maybe this seems to be a little late in the chapter to discuss the pre-flight check of the propeller since the take-off, cruise and landing procedures have been covered, but you should have an idea of its use so you know what to look for during the check.

It's best *not* to check the magnetos of a constant speed propeller-equipped airplane when the rpm setting is in the constant-speed operating range. If you switch to a bad magneto and the rpm starts to drop, the governor will sense it and automatically flatten pitch to keep the tachometer hand at its old reading. The constant speed prop will, because of its inherent design, tend to mask fouled plugs, bad mags, etc. You will check the magnetos before take-off *below* the governor operating range to get a true picture. Usually this is done somewhere between 1700 and 2000 rpm with the prop control full forward (high rpm).

You will want to check the propeller operation before take-off and will do this by running the prop control through its range, starting with the prop in full high rpm and at the tachometer reading recommended by the manufacturer (usually around 2000 rpm). Pull the prop control aft (or use the electric toggle switch) to reduce rpm. Don't leave it back too long as the manifold pressure will be too high for the rpm. Most pilots pull the prop control back and immediately move it forward before the rpm drops off too far. With practice you can check the response of the propeller the instant the control is moved aft.

A FEW NOTES ON SUPERCHARGED ENGINES

It may be a while before you get a chance to fly an airplane with a supercharged engine, so it will just be covered lightly here. To repeat a little, the power developed by a particular engine depends on the pressure developed in the cylinders during combustion (among other things). This pressure depends on the mass of the fuel-air mixture in the cylinder at the time of combustion. By compressing the mixture before it gets into cylinders, this internal pressure is raised considerably at combustion, resulting in more horsepower being developed. Of course, the engine manufacturer makes sure the engine is structurally capable of taking this increased pressure.

The unsupercharged engine will never get more than the outside atmospheric pressure registered as manifold pressure. The supercharged engine may, theoretically, get any manifold pressure reading. It is, of course, limited by the efficiency of the supercharger and the structural strength of the engine.

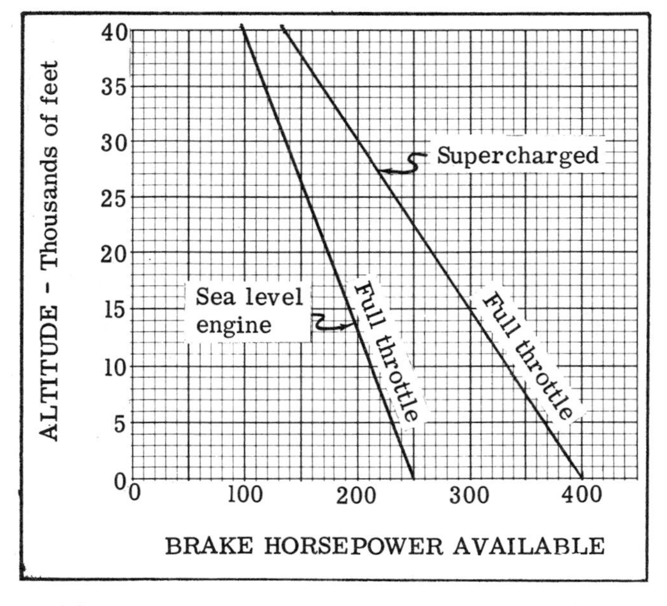

Fig. 12-4. A theoretical comparison of two engines of equal displacement showing greater horsepower developed by the supercharged engine at all altitudes.

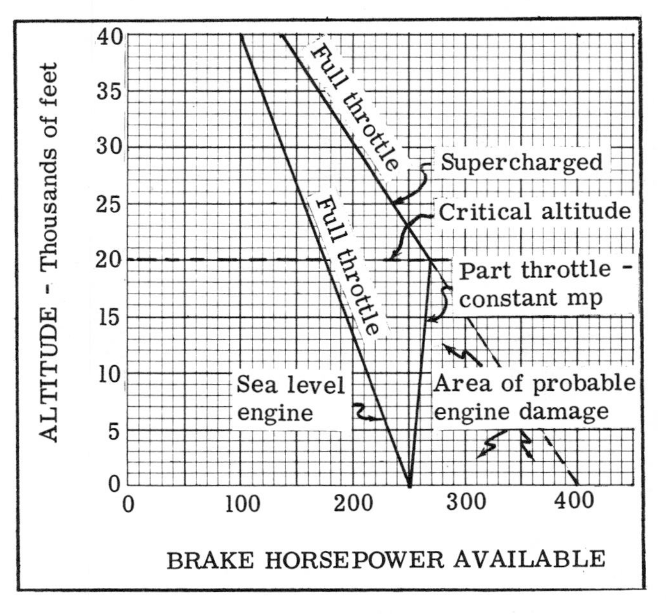

Fig. 12-5. A more realistic comparison of the engines shown in Figure 12-4. It is possible to overboost a supercharged engine at lower altitudes.

Therefore, the throttle of the supercharged engine must be handled with care as it could be "overboosted" very easily at lower rpm and/or at lower altitudes. The principal advantage of the boosted engine is that a great deal more power may be developed at a small cost in Weight. The principal disadvantage is that some care must be used in its operation— and there's some extra cost involved, too.

Most of the superchargers used in light twins are driven by the crankshaft but at a much higher speed through gearing. These superchargers are "single stage, single speed," the single turbine always operates at the same gear ratio and the pilot does not "shift gears" as altitude is gained. Fig. 12-4 shows two engines of equal displacement (cubic inches) with one having a single stage, single speed supercharger.

The supercharged engine has more horsepower available at altitude. Fig. 12-4 shows the theoretical power available, whereas Fig. 12-5 shows the recommended use of the supercharged engine to avoid overboosting at low altitudes. The pilot will start his climb using part throttle, and as altitude is increased will open the throttle to maintain the initial manifold pressure reading. The altitude at which the throttle is wide open to maintain the desired horsepower is called the "critical altitude." Above this the horsepower will drop off as the engine has no more "spare manifold pressure."

When you take off with the unsupercharged engine you can open the throttle wide without danger of overboosting (if the prop control is forward where it should be). The supercharged engine must be watched to insure that it is not overboosted. This means that the manifold pressure gage must be checked as power is increased because these engines are usually restricted to less than full throttle for sea level operation.

Some of the newer light twins have lightweight exhaust-driven superchargers and kits are now available for installation on some unsupercharged twins.

SUMMARY OF THE CHAPTER

When you decide to check out in the airplane with a controllable pitch prop and manifold pressure gage, give yourself a few days. Sit in the cockpit and become familiar with the new controls. Go over in your mind the various steps for take-off, climb, cruise and landing. Ask questions. It's hard to jump cold into a new airplane and do a good job the first time. The professional pilots realize this and spend time in the cockpit before actually flying a new type or model. You might also review Chapters 1 and 3 concerning controllable pitch propellers and the manifold pressure gage if you plan on checking out in an airplane so equipped.

13. NEW AND ADVANCED FUEL SYSTEMS

THE FUEL BOOST PUMP

Probably most of your flying to date has been done in high-wing airplanes using one or, at most, two fuel tanks and a gravity fuel system. When you start flying low-wing airplanes with wing tanks, the need for an engine-driven fuel pump will become evident. Very few cases of airplane engine-driven fuel pump failure are on record. But it *could* happen, so means are furnished to provide fuel pressure in the event of a failure.

In earlier days this was provided by a "wobble pump," a mechanical lever used to pump up fuel pressure for starting and in flight if necessary. The handle usually was placed so that maximum muscle strain could be realized in working it. After a long siege of pumping, the pilot "wobbled" after he got out of the cockpit. (Actually the name is derived from the movement of the handle.)

In multiengine airplanes equipped with a wobble pump, the copilot stands by to use it should one of the engine-driven pumps fail on take-off.

People being what they are — lazy — the electric boost pump came into use. It is turned on to aid in starting and again turned on to standby during the take-off when a loss of fuel pressure is most serious. Of course, it could always be turned on should the engine-driven pump fail, but the delay during take-off could cause serious problems. The boost pump also is turned on before landing, as it would be most embarrassing to lose the engine-driven pump if you were too low to make the runway without power. Your big problem for the first few flights will be remembering to turn it on or off at the right times.

After take-off, turn the boost pump off as soon as a safe altitude has been reached — a lot of pilots use five hundred feet as a minimum — your check pilot may have recommendations. *Always look at the fuel pressure gage as you turn the boost pump off after take-off.* If the pressure starts going down to zero, better get the boost pump back on and get back to the airport because the indications are that the engine-driven pump has failed and the boost pump has been carrying the load. Don't just automatically flick the switch off without looking, as that way the first warning you'll have will be the engine stopping — and it may be hard to get going, even with the boost pump on again.

Assuming you use the boost for starting, turn it off temporarily sometime between starting and take-off to see if the engine-driven pump is operating. In some airplanes it's next to impossible to tell the difference in fuel pressure with the boost pump on or off with the engine running. That is, its pressure is not noticeably additive to the engine-driven pump pressure. On others, it's noticeable right away. Check the operation of the boost pump (or pumps) for the carburetor type engine *before* the engine is started, after turning the master battery switch and boost pump on. The boost pump should give a pressure reading in the normal operation range for the engine-driven pumps. This pre-start check is not recommended for fuel injection engines as it will likely cause flooding if the engine is hot.

Some boost pumps have three-position switches: ON, OFF and PRIME. The PRIME position is a low pressure position for priming and starting. The ON position also runs the pumps at low speed as long as the engine-driven pump is running. If the engine-driven pump fails with the boost pump switch ON, the boost pump will automatically switch to high speed operation. This system, because of its complexity, normally is used on multiengine airplanes.

The Airplane Flight Manuals for some airplanes require the use of boost pumps throughout the start. Others recommend their use to build up pressure *before* the start and are turned off during the actual starting process. Still others don't find it necessary to use the pump for starting.

Some boost pumps are located in the fuel line system where others may be "submerged" boost pumps. The submerged boost pumps are normally only in the main fuel tanks; this being one reason why take-offs and landings are made on the main tanks for most airplanes.

The boost pump may seem to be an additional problem to cope with, but if you remember its purpose — to furnish fuel pressure when the engine isn't running (before the start) or when the engine-driven pump has failed, or may fail — you'll have no problem with its use. Remember, it is a starting aid or a safety standby.

TANK SYSTEMS AND FUEL MANAGEMENT

The airplane may have both main and auxiliary tanks. Naturally the manufacturer would prefer to have all the fuel in one tank as this would simplify the fuel system. Unfortunately, this is not possible from either a structural or a space standpoint for larger airplanes.

When you first look at the fuel system diagram of the new airplane you'll wonder how you'll ever learn

Fig. 13-1. Sometimes the fuel system seems a little complicated at first glance.

which tank should be used at what time. (Fig. 13-1) It may not be clear until you have actually used the tank system. After the check pilot has described the fuel management and you have flown the airplane, go back to the Flight Manual again — it'll be a lot clearer. Ask yourself the question WHY certain tanks are used at certain times and not at others.

Many pilots firmly believe in running tanks dry enroute because they know that it's better to have *all* the remaining fuel in, say, two main tanks than to have a small amount of fuel scattered in several tanks — and having one run dry on approach. This is particularly important if you're approaching the range limit of your airplane.

For some airplanes, running a tank dry presents no problem. As soon as the tank is switched the engine starts right up with no other effort on the pilot's part. With others, the fuel system is such that an air lock might occur, requiring several steps in getting power again. Some airplanes have a fuel line purge button to be used in just this situation.

As a general idea, if this type of difficulty is encountered, close the throttle to about $\frac{1}{4}$ open or less, enrich the mixture, and wait. The windmilling prop will finally purge the line but with some loss of altitude (and a great deal of wear and tear on the nervous system).

Ask your check pilot about running a tank dry on your airplane and *never deliberately run a tank dry below 3000 feet above the surface* — even with an "easy" starting engine.

Fuel management includes the use of crossfeed for multiengine airplanes as well. The use of the crossfeed control will be covered in Chapter 15.

LEANING

Probably the light trainers you've been flying had no means of controlling the mixture, so perhaps you've never encountered a mixture control before. As an all-around figure, gasoline engine mixtures for combustion are about 1 to 15. That is, about 1 pound of fuel to every 15 pounds of air — or about 7 per cent fuel and 93 per cent air by weight.

For richer mixtures an 8 to 10 per cent fuel-air ratio is found.

Although you can set the mixture to a large number of percentages within the mixture operating range, there are actually only two areas in which you are particularly interested: (1) best power and (2) best economy. (Fig. 13-2)

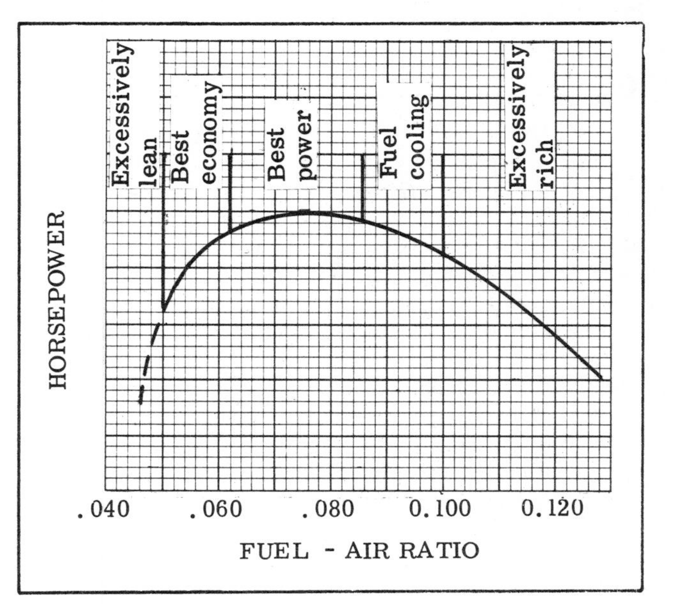

Fig. 13-2. Fuel-air ratio versus power.

For take-off, a mixture setting of "full rich" is used. This setting assures you of the best combination of power *and* cooling. The full rich setting will be slightly richer than best power, but as engine cooling also depends on a richer mixture a compromise must be made.

With the mixture in the full rich position you'll be using a predetermined mixture of fuel and air. As you climb, naturally the air becomes less dense, weighs less per unit volume. On the full rich setting your carburetor is putting out about the same amount of fuel but there's less air to mix with it, so the mixture gets richer and richer. In fact, you may climb so high in full rich that the engine will start to run rough; the fuel-air ratio is too great for smooth operation. So, not only are you losing power, but are using fuel like a madman (as compared to what you should be consuming). After reaching the desired cruise altitude, you'll level off, set cruise power and lean the mixture. Your job with the mixture control is to establish the optimum fuel-air ratio for all conditions.

USE OF THE MIXTURE CONTROL—CARBURETOR

There will be times when your handling of the mixture control will be very important. In Chapter 6 you learned that proper leaning of the mixture is vital for both best range and max endurance.

There are nearly as many techniques for leaning as there are pilots. A couple of the less complicated will be discussed here.

First, it will be assumed that you have neither a fuel mixture indicator (it's very likely that you won't) nor a cylinder head temperature gage. In other words, you'll be leaning the mixture more or less by ear.

After leveling off and establishing the desired power setting, ease the mixture control back until the engine begins to roughen slightly. Ease it forward just until the engine smooths. This is the system most used for airplanes not equipped with an excess of gages for engine information.

A variation of the above technique is to lean the mixture with the engine operating on one magneto, as a too lean mixture will show up more quickly than on BOTH. (Of course, as soon as you've set the mixture you'll go back to BOTH mags.) Most pilots don't bother to do this and it's doubtful if a noticeable advantage is gained by using this technique.

There's not too much leaning precision to be obtained with a single engine airplane because of the way the mixture control juts out of the panel. But for a light twin where the mixture control is on a quadrant it's possible to really improve fuel economy because of added accuracy in using the mixture controls. Let's suppose you are flying a light twin and are ready to lean the mixture. You will lean each engine separately. Leave the ignition on BOTH or one magneto, as you prefer, and retard the chosen mixture control until the first signs of roughness appear. Mark this spot on the quadrant by placing the thumbnail of your other hand against the back of the mixture control. Keep the thumbnail there as you move the mixture control forward brickly to check for a surge in power. If there is an rpm surge the mixture was too lean, so move your marking thumbnail forward about an eighth of an inch and pull the mixture control back against it. Check for another surge. If there is no surge, the mixture is not too lean and you may move the control back about a sixteenth of an inch. You can continue to do this, halving the movement each time until the movement is too slight to notice. Usually the sixteenth of an inch move is good enough.

If a cylinder head temperature is available, many pilots prefer to use this instrument in leaning the mixture, particularly with a constant speed propeller. Method (1) above will work with a constant speed propeller but sometimes requires more experience because the actions of the governor tend to mask the roughness caused by the too lean mixture.

Here is a typical method of leaning a carburetor engine using the cylinder head temperature: Set power at desired rpm and manifold pressure for cruise. Leave the mixture rich and allow cylinder head temperature to stabilize. Begin leaning in increments, observing the cylinder head temperature. When the cylinder head temperature peaks, this is your final mixture setting for that altitude. Don't permit the cylinder head temperature to exceed the limit as given in the Airplane Flight Manual or Engine Manual. If a sudden temperature rise should occur during the process, move the mixture control back to the position it was before the temperature increase. (You overdid it a little.) Let the engine stabilize for at least five minutes before leaning the mixture for a further cylinder head temperature reading. If the cylinder head temperature lead is on a lean cylinder, you may get a drop in temperature as you lean it out (you are decreasing power at that cylinder and decreasing the temperature for the engine as indicated by the gage).

The exhaust gas temperature gage (EGT) is an excellent aid in properly setting the mixture. The system is composed of an instrument on the panel connected to a probe in the exhaust stack(s) so that the temperature of the exhaust gas may be taken and indicated. Since excess fuel or excess air in the mixture produces a cooling effect (lowering the exhaust temperature) the EGT gage may be used to find the optimum for cruise and other conditions.

As an example, to set up a cruise-leaned situation (at 75 per cent power or less) using one of these instruments, lean the mixture until the EGT needle peaks. Then richen the mixture until the needle shows at least a 25°F temperature drop (this puts the mixture on the rich side of the highest exhaust gas temperature). The 25°F drop is best for fuel economy. A drop of 100°F (on the rich side) is in the area for best power. If your airplane has one of these systems you should read the accompanying literature and talk with pilots who've used it before using it on your own.

Leaning is normally effective only above 5000 feet but some engines may be leaned from sea level up. You might check into this for the engine in your airplane. By leaning at too low an altitude you could damage the engine at a high power setting. On the other hand if leaning below 5000 feet is permissible you may save fuel by knowing this. "Don't lean below 5000 feet at power settings above 75 per cent" is how one Engine Manual puts it.

Suppose you are taking off from a field a high altitude and/or temperature (you are at a high density altitude). It is likely that with the mixture control in the full rich setting the weight of fuel going into the engine is too great for the weight of the air mixed with it. That is, the mixture at that setting and altitude is so rich that the engine is not developing full power. (Fig. 13-3)

Under these conditions, particularly if the field is short and power is needed badly, some pilots run the engine up to some point just below prop governor operating speed and move the mixture control back until there is a definite pickup of rpm. The peak indicates the best power mixture setting for the density altitude. The mixture control should be moved slightly forward of this point of maximum power so that the mixture will be a little on the rich side to insure proper cooling. The mixture is normally pre-set to be slightly richer than best power for this reason when in the full rich position, even at sea level

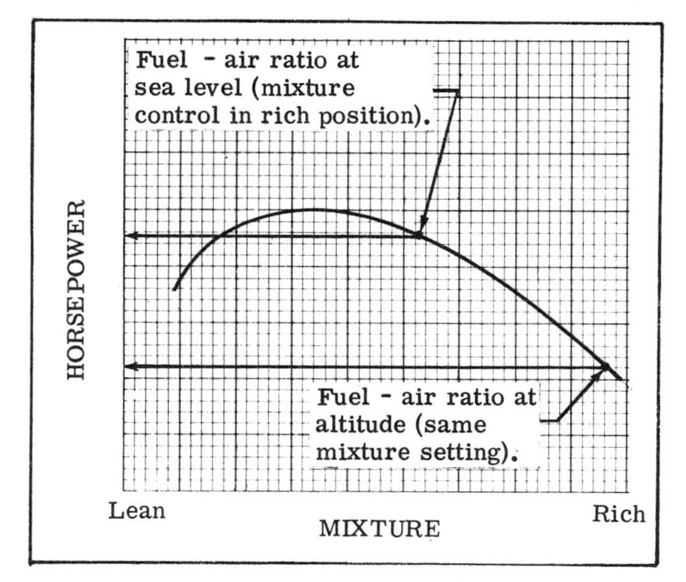

Fig. 13-3. A high density altitude and a full rich setting mean a power loss.

conditions. It is better to be slightly on the rich side to avoid possible engine damage — even if it means a small deviation from best power. The density altitude has to be pretty high before leaning has very much effect in increasing take-off power.

A common error made by the pilot new to the mixture control is, after having set the mixture for best cruise at some fairly high altitude, he starts his descent to the destination airport and forgets about the mixture. As he descends, the air density increases but the carburetor is still putting out the same amount of fuel that worked so well at the higher altitude. Finally the comparable amount of fuel-air becomes so lean that the engine starts running rough and gives every sign of quitting any second. The simple remedy is to enrich the mixture and all is well again. You actually have two choices in the matter of the descent. (1) You can push the mixture control all the way forward to full rich and not have to worry about it or (2) enrich the mixture in increments by guess as you go down. In (1) if you are very high the engine operation may roughen slightly (it's assumed that you plan on using cruise power during the descent) but will soon smooth out as you lose altitude and you won't be as apt to forget to move it into full rich for landing.

VAPORIZATION

It would seem that all other things being equal, the colder the air entering the carburetor the more dense it is, and the more power being developed. However, the fuel particles do not mix as well with cold air, and fuel may be wasted.

For best vaporization a carburetor air inlet temperature of 90-100° F is ideal. This means the use of carburetor heat. If you are interested in maximum economy, *after* setting power, use carburetor heat to establish this temperature and then lean the mixture in one of the ways mentioned earlier. Your airplane may not have a carburetor inlet temperature gage (measuring the air temperature just *before* it goes into the carburetor) so that another technique may be used.

In order to find the probable best heat setting for vaporization, lean the engine until it just starts to run rough — then apply enough carburetor heat to smooth it out (the less dense warm air will result in a comparatively richer mixture hence the smoothness). This is the probable heat control setting for best vaporization. *The addition of carburetor heat usually means higher engine temperatures and could be overdone in warmer weather.*

Some mixture controls have "auto lean" setting positions. The more complex carburetor has an automatic altitude compensator. It contains a bellows that senses the incoming air pressure and controls the fuel metering accordingly. If you set the mixture at best economy at five thousand feet and climb to ten thousand, the mixture will also be at best lean at the higher altitude — even though you haven't touched the mixture control. It is assumed here that you used a *cruise climb* and didn't need to increase the richness for the climb.

FUEL INJECTION

The big advantage of fuel injection is that carburetor icing is no longer present. The fuel is injected into the intake manifold just before going into the cylinder; hence there is no temperature drop in the carburetor due to vaporization. The air temperature drops in the carburetor for two reasons: (1) vaporization and (2) the lowered pressure caused by venturi effect — these two effects being additive.

Of course, you can get impact icing in freezing rain, etc., in either type of fuel system, but this is not the kind caused by invisible moisture.

When fuel injection first came out for light planes many new owners insisted that the airplane have fuel injection. Some of the people who were so in favor of it didn't have the slightest idea of what it was all about. The airplane manufacturers were caught in a race similar to the one in the auto industry when power steering first reared its head. It is now becoming standard for most of the non-trainer type planes.

A particular advantage of fuel injection lies in the pilot's ability to lean the mixture accurately by use of the fuel pressure gage. This gage measures metered fuel pressure or the pressure of the fuel going to the spray nozzles — this being a direct measure of fuel flow. The gage usually is marked with proper fuel pressures for various power settings and/or altitudes. The lower pressure range is for various cruise power settings (45-55-65-75 per cent). The altitude is automatically compensated for by a bellows or diaphragm within the control unit which regulates the fuel flow from the nozzles in proportion to the air pressure (volume) passing through the unit.

In addition, the gage may be marked for best power for take-off and climb for various altitudes (this will be in the higher pressure range). Whereas the fuel pressure gage in the carburetor-equipped airplane remains constant at all power settings (it measures the pressure of the fuel from the pumps to the carburetor), the pressure gage for fuel injection varies with mixture setting. To lean this engine you merely move the mixture control until the fuel pressure or fuel flow indicates that you have the correct mixture for the power setting and/or altitude.

Another advantage of fuel injection is that it theoretically gives a better fuel-air distribution to the cylinders. This does not always occur, but should if the system is properly operating. Here's the reason: the fuel-air is mixed in a carburetor at one place for all cylinders. By the time this mixture reaches each cylinder some variation in mixture may occur between cylinders. This is not as apt to occur in the fuel injection engine because the mixing is done just before entering each cylinder.

LEANING THE MIXTURE

If you've leaned a carburetor type engine, you've noticed that if the mixture is over-leaned the engine will start to get rough. This is because of the initial difference in mixtures in each of the cylinders. You are leaning them all at the same rate, but some cylinders were leaner to begin with and will be too lean, while the other cylinders are still operating smoothly. Naturally the engine will run rough if only part of the cylinders are getting enough fuel. This is the idea you used in manual leaning of the carburetor.

The fuel injection system, having better fuel distribution, will not react this way. As all the cylinders are getting an equal amount of fuel (theoretically), when you lean the mixture excessively there will be no initial roughness but the engine will quietly and smoothly die. Therefore it would be a great deal harder to set the mixture by "feel" so a pressure gage or fuel flow gage is helpful.

Figure 13-4 is a fuel flow indicator used on a current four-place airplane. Most indicators are of this type.

The outside numbers on the meter indicate the fuel flow in gallons per hour. Discussing the various segments of the meter:

STARTING

This is the starting procedure for this particular airplane (cold):

1. Fuel selector — proper tank.
2. Open the throttle approximately 1/2 inch.
3. Turn on the master switch and electric auxiliary fuel pump.
4. Move the mixture control to full rich until an indication of 4 to 6 GPH is indicated on the flow meter, then turn the pump off. (The engine is primed.)
5. Move the mixture control to idle cut-off.
6. Ignition switches ON, and engage starter.
7. When the engine fires, move the mixture control to full rich.

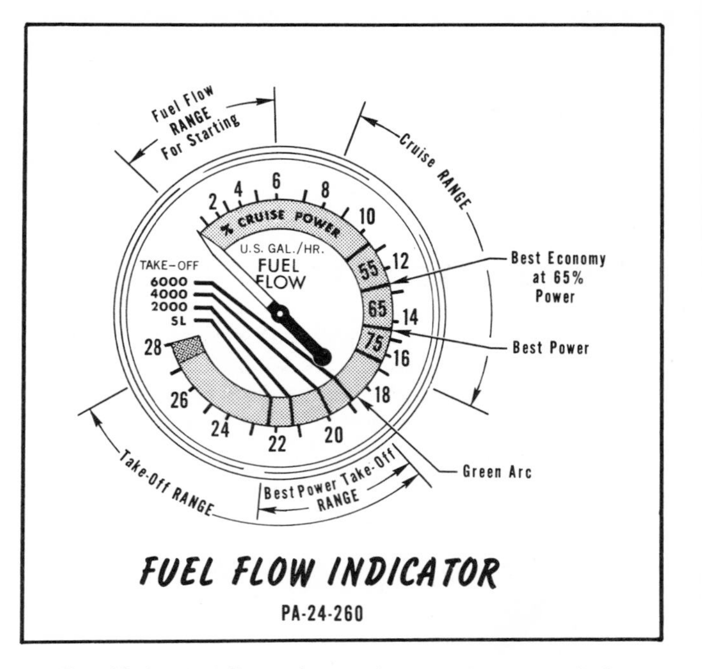

Fig. 13-4. Fuel flow indicator for a single engine, fuel injected airplane. (*Piper Aircraft Corp.*)

For starting hot or flooded engines the flow meter is not used, but the engine is started with the mixture in idle cut-off and fuel pump off. (The throttle is cracked 1/2 inch or open, respectively.) When the engine fires, move the mixture control to full rich (and retard the throttle as necessary).

TAKE-OFF

For this fuel-injected airplane during a normal take-off with full rich mixture the pointer on the fuel flow meter will stabilize between the sea level mark and the red line (see Figure 13-4). This is slightly rich to aid in fuel cooling and is recommended for normal take-offs at sea level.

When taking off from a high altitude field (say 4000 feet density altitude), the mixture should be leaned to maximum power during the pretake-off check. Full throttle is applied and the mixture control is moved toward the lean position until the pointer has stabilized at the 4000 foot mark (between the 19.5 and 20.0 GPH marks). The take-off is made with this mixture. The same technique can be used for obtaining maximum power at sea level, using the sea level mark. (Don't overheat the engine with prolonged climbs; richen it again after clearing the obstacle.)

CRUISE

The flow meter is a good aid in setting up a cruise mixture. The example of 65 per cent power in Figure 13-4 indicates that the two widest variations are (1) 14.5 GPH (best power) and (2) 12.6 GPH (best economy) with any chosen settings between these limits. As you can see, 55 and 75 per cent settings also have a range of possible settings. The

EGT gage is a more precise method of leaning, however.

APPROACH AND LANDING

Set the mixture to full rich. (No reference to fuel flow meter for a *precise* setting.)

PRESSURE CARBURETOR

Your airplane may have a pressure carburetor instead of the float type discussed earlier. The fuel supplied to the carburetor is discharged into the intake air under high pressure. The fuel flow is automatically established within the carburetor through a pressure differential made by the venturi acting on a flexible diaphragm. The pressure carburetor has a lesser tendency to ice because the fuel is injected *beyond* the venturi, thus tending to cancel any additive cooling effects of lower pressure and vaporization.

Figures 13-5, 13-6, and 13-7 show the basic differences between the three fuel metering systems just discussed.

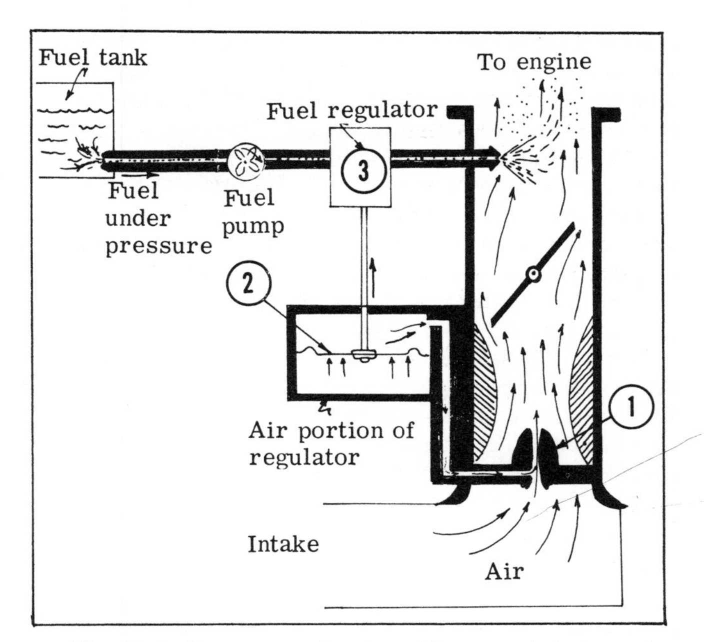

Fig. 13-6. Pressure carburetor. The amount of air moving through the small venturi (1) controls the position of the diaphragm, which in turn (2) regulates the amount of fuel to be added to obtain the correct fuel-air ratio.

FUEL AND OIL FACTS

FUEL

Never use fuel rated below the minimum octane or performance numbers. You may go above the rating as a temporary arrangement, but even then try to stay as close to the recommended octane as you can. If your engine normally uses 100/130 octane and only 80/87, 100/130, and 115/145 fuel are available, use 115/145.

The numbers are the anti-knock quality of the fuel — the higher the number, the better the anti-knock qualities. Take, for instance, 80/87 fuel: The first number (80) is the minimum anti-knock quality of the fuel in a lean mixture; the last (87) is the minimum anti-knock quality in a rich mixture.

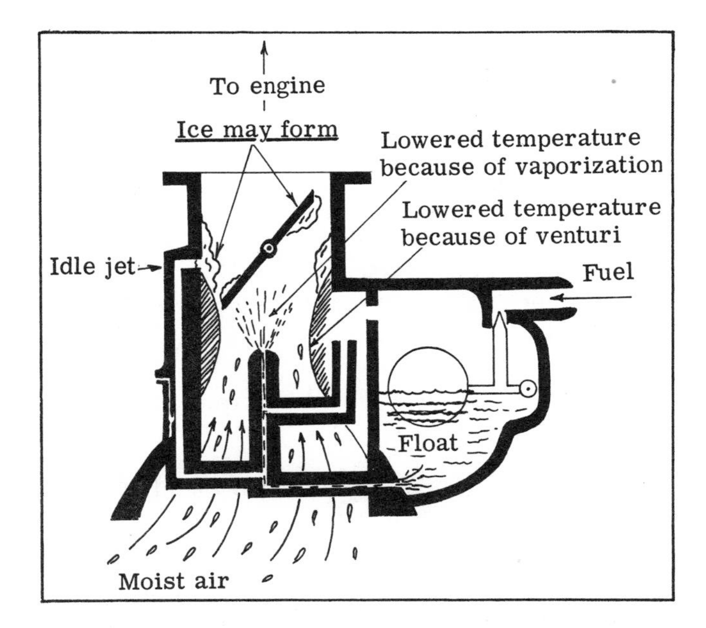

Fig. 13-5. Float carburetor, showing how ice can form.

The anti-knock numbers below 100 are called octane ratings, above 100 they are referred to as performance numbers.

The various octanes and performance numbered fuels are dyed different colors for easy identification.

Octane or Performance Number	Fuel Color
80/87	Red
100/130	Green
115/145	Purple

There is the lady pilot who would never use 80/87 fuel, even if it is recommended, because her airplane is green and the red fuel clashes with the decor. (She pays a little extra for 100/130.)

While the fuel color codes just mentioned are valid in the U. S., some countries use other codes, so if you are in foreign territory don't just assume, for instance, that green fuel is 100/130 — you'd better confirm it.

OIL

This may seem a rather strange place to talk about oil — in a chapter on fuel systems — but you usually check the oil when you check the fuel and should have some information on it.

The viscosity of the oil may be given in one of three ways. You are probably the most familiar with the S.A.E. (Society of Automotive Engineers) number.

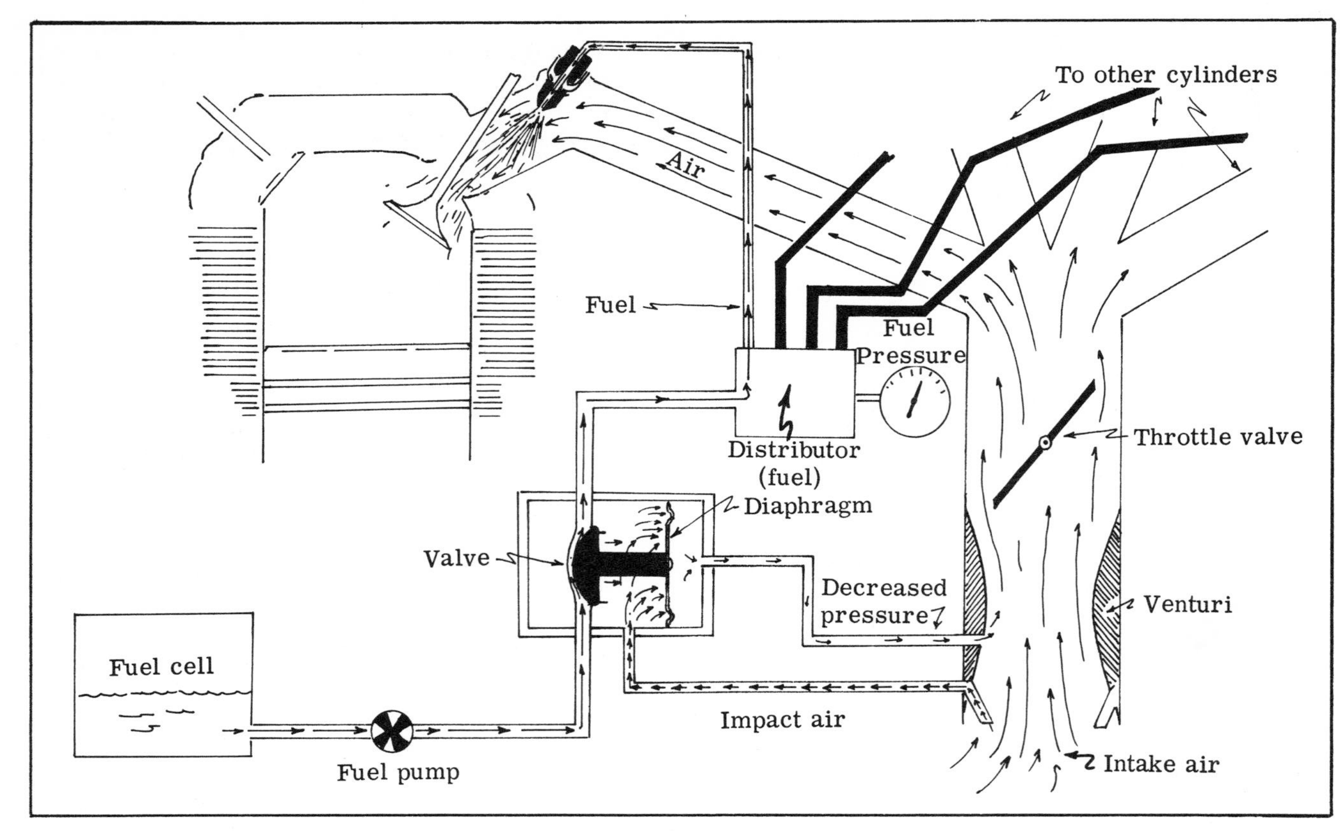

Fig. 13-7. A simplified view of one type of fuel injection system.

Notice that in every case except one, the commercial aviation number is exactly twice the SAE "weight." A rule of thumb to get the SAE equivalent is to divide the aviation number by two and go to the nearest number divisible by 10: $\frac{65}{2} = 32\frac{1}{2}$, the nearest number divisible by 10, is *30*. $\frac{80}{2} = 40$, which is the nearest number divisible by 10, etc.

For conversion from military grade to commercial aviation to SAE:

Military Grade (1120) = *Aviation Grade* (120) = *SAE* (60)
(*1120* minus 1000 = *120;* 120 ÷ 2 = *60*)

SAE Number	Commercial Aviation Number	Military Grade (Army, AF, Navy)
30	65	1065
40	80	1080
50	100	1100
60	120	1120
70	140	--

There are three main types of aviation oil in use today for reciprocating engines:

1. *Straight mineral oil* — This is an oil without any dispersant additives. It is, more or less, an inert lubricating medium.

2. *Detergent oils* — These are "cleaning" oils that remove carbons and other deposits from the interior of the engine as they lubricate. One problem with the detergent oils is that if they are added to high time engines (with heavy carbon build-ups) the additives may "clean" the deposits and re-deposit them in the screens and other places vulnerable to plugging by the released sludge. Wait until the next overhaul before changing to an additive oil.

3. *Ashless dispersant oils* — These are a type of detergent or compound oils (with additive) that keep the foreign particles in solution without the disadvantages of ash-forming detergent additives.

Engine manufacturers recommend generally that new or newly overhauled engines should be operated on straight mineral oil during the first 50 hours of operation, or until oil consumption has stabilized. If an additive oil is used in these engines, high oil consumption might result, since the anti-friction additive of some of these oils will retard the break-in of piston rings and cylinder walls.

Okay, what if you've been using straight mineral oil and decide that it's to your advantage to start using a compounded oil? The least you'd better do is the following according to one major engine manufacturer (make sure the compounded oil is approved for your engine):

1. Don't add the additive oil to straight mineral oil. Drain the straight mineral oil and then fill with additive oil.

2. Don't operate the engine longer than five hours before the first oil change.

3. Check all oil screens for evidence of sludge or plugging. Change oil every ten hours if sludge conditions are evident. Resume normal oil drain periods after the sludge conditions improve.

If you are fairly close to overhaul you might not want to go to all that trouble. The real point is that putting additive oils in an engine which has been using straight mineral oil can cause problems unless you are careful.

Since you are an "advanced pilot" and will be flying more powerful and complex engines, it will behoove you to know *exactly* the type of fuel and oil to be used during servicing. A crew member should be available to oversee the fuel and oil servicing. Jet fuel *has* been mistakenly put in the tanks of airplanes with piston engines — with fatal accidents resulting in some cases.

14. RETRACTABLE GEAR AIRPLANES

SOME POINTS TO THINK ABOUT

The biggest single step forward in decreasing airplane Drag was the designer's installation of retractable landing gear. At first the retracting systems were so complex and heavy that only large airplanes could use them. Now, through the use of electrical motors or very light hydraulic systems, nearly all of the high performance single-engine airplanes and all of the light twins have retractable gear.

The advantages in speed and economy are obvious. You have no doubt already figured out the main disadvantage — the landing gear is sometimes retracted at what might be termed an "inopportune moment." People also forget to put them *down* at the opportune moment. You usually can get away with forgetting to shove the prop control forward, or forgetting to use flaps (you can take care of the prop on the sly as you taxi in) *but* you won't get away with forgetting to put the gear down. Three things are certain about gear-up landings: (1) they are definitely more noisy than the gear-down type; (2) the airplane does not "roll" as far; and (3) expenses are somewhat higher for this type of landing.

At some time in your flying career you will come close to landing gear-up — and you may go all the way if you aren't careful. The purpose of this chapter is to help keep you from going all the way.

PILOT STRESS

Believe it or not, the danger period for the retractable gear plane pilot normally is not the first few hours after check-out. If you are like most new check-outs, you'll spoil the enjoyment of the first few flights by muttering to yourself over and over, "mustn't forget to put the gear down, mustn't forget the gear, mustn't... etc." After a while you'll consider yourself an old pro and the gear check will be important but not the *only* item on the check list as it seemed to be at first.

Back to the idea of stress: One day you'll be going into Chicago Midway or Washington National or some other busy airport. There'll be a lot of traffic and the tower will be giving instructions at a machine gun rate. Suppose you aren't able to finish the approach because of conflicting traffic and are advised to "take it around" by the tower. You pull the gear up and try to work back into the downwind leg. The traffic is heavy and the pressure is on. The tower people may seem unsympathetic but their job is to expedite traffic flow with safety. You are cleared to land again and are very busy, looking for other airplanes and setting up the pattern. *This is a situation in which you could forget to put the gear down again because of stress.* In the daytime the tower operators will probably catch you before you land gear-up. But don't count on their doing a job that is rightfully yours. They're very busy. Many a pilot has been saved from a dangerous or embarrassing situation by an alert tower controller but the controller could be distracted too.

At night you don't even have the possibility of a tower controller spotting the results of your memory lapse. The shower of sparks when you land will show him your problem.

Always have the gear down before turning on base leg under normal conditions. If the tower clears you to enter base leg, have the gear *down* and *locked* before starting the descent on base. *Always check the gear indicators again after turning on final.* Some pilots point to the gear indicators so that they're sure their attention is directed to it. Of course, it's possible that they could point to a gear-*up* light absentmindedly.

CHECK LIST

The check list is a valuable aid if used right. The trouble is that after a while you'll "know" it so thoroughly that using it is just a ritual that must be done at certain times. Some pilots glance at it and don't read it. It's very easy to skip an item this way. A check list is a liability if not used correctly because a quick glance at it may lead you to believe you've done what's necessary and you get a false sense of security. On the other hand, if you use the check list religiously and always put the gear down at the same point, habit may save you embarrassment some time when your conscious mind is out to lunch.

Remember — just because you went through the motions and moved the right lever doesn't mean the gear is down. CHECK IT! Mechanical devices have their off-days too, you know.

EMERGENCY PROCEDURES

Next to the fear that you'll forget to put the gear down will be the thought, "What if it just won't come down?" The newspapers, movies and television have probably milked more drama out of this situation than any other one phase of flying. If it won't come down, you'll probably bend the prop and scrape some paint off the belly. But the cases of the pilots of general aviation planes being physically unable to lower the gear by any means are extremely rare. Manufacturers frown on people belly landing their products. This makes their airplanes look bad, and they try to arrange it so gear-up landings aren't necessary. Actuating arms and other mechanical parts have failed, but 99.999 per cent of the belly landings made by general aviation planes are due to an oversight by the pilot — not plain structural failure.

While in flight suppose you put the gear handle or switch down and can see no green light? You probably got in, started the airplane and went about your business — overlooking the fact that there was no down-light when you taxied out and not noticing that there was no up-light after you pulled the gear up.

Here's where the ball was dropped at the beginning of the flight. *As soon as you get in the airplane, check the position of the gear handle or switch. When the master switch is turned on for start, check for a down-light.* Somebody might have tinkered around in the cockpit and moved the control to the up position. If the safety lock isn't working, the plane could slowly sink to the ground as you start to taxi. This is unlikely, but there's no need of taking a chance.

When the navigation lights are on, the landing gear indicator lights normally are dimmed because the bright lights are disconcerting at night. If the navigation lights are on in the daytime, the gear indicator lights may be so dim as to appear to be off. There have been many cases of newly checked-out pilots who called up on Unicom to state that the gear wasn't down. One of the first things the old pilots in the airport office will ask is, "Are your navigation lights off?" This usually is answered by a long pause and a rather weak, "Uh, Roger." The embarrassed pilot comes in and lands, his gear having been down all the time but he couldn't see the dimmed lights.

The new pilot has a red face but this is far better than taking a chance on bellying it in. In cases like this, new pilots have been known to use the normal *and* emergency means of lowering the gear but still not seeing a down indication (naturally).

Some airplanes with electrically operated landing gear have a three-position switch (UP, OFF, and DOWN), and it's possible for you to stop in the middle or OFF position instead of DOWN.

If you have landing gear problems in flight get yourself some altitude, where you can think — at least get out of the traffic pattern.

The FAA requires that the Airplane Flight Manual or equivalent form be in the airplane at all times — and this is one of the main reasons why. It's funny how blank a usually sharp mind can get sometimes. You no doubt learned the emergency gear-down procedures until you could say them in your sleep, but now the steps have eluded you. *Take your time.* Get the Airplane Flight Manual out and read the emergency procedures if you have to. Some airplanes have the step-by-step instructions printed near or on the cover plate of the emergency gear handle or switch. Follow them carefully.

Slowing the airplane down makes the landing gear extension a lot easier. Don't fly it around just above stall, but have the airspeed well below maximum gear-down speed.

In most airplanes with hydraulically actuated gear, the emergency procedure requires that the gear handle be placed in the *down* position before going on to the extension of the gear. Pilots have forgotten this and, when using a CO_2 bottle emergency extender, have wasted their one shot by having the gear handle up. They got in a hurry and didn't bother to follow the step-by-step procedure, or "thought" they knew the emergency procedure and didn't need to read it.

Getting out of the traffic pattern allows you to analyze the situation. It may be just a popped circuit breaker for electrical gear or a problem requiring a little hand pumping for hydraulic gear.

Don't use the emergency procedure until you are ready to land. This sounds like a rather inane statement but what it means is that the emergency gear *extension* is usually a one way affair. Once the gear is put down by emergency means you have to leave it there.

There's the case of the curious private pilot who suspected after take-off from a strange field that he might have trouble getting the gear down by normal means because it didn't act right coming up — so he did everything wrong. Home field with a good repair station for his airplane was only one hour away (gear up) and he had five hours' fuel. On the way home he started thinking "Will it go down?" until he couldn't stand the suspense any longer and used the emergency procedure. Of course, the gear came down but he had a mighty slow trip and almost got an overheated engine for his impetuous action.

Then there was the private pilot who did the same thing, but, being heavily loaded over mountainous terrain, decided that he had to get the gear back up. By clean living and hard work he managed to get the gear started back up (where it stuck halfway, naturally) and did a fine job of messing up his new Zephyr Six when he landed.

If you have trouble getting the gear up after take-off, don't force the issue, leave it down. *Make sure it's down,* return and land — unless it would be wiser to fly (gear down) to a nearby airport where the trouble can be more easily fixed after you land.

GEAR-UP LANDING

If the emergency procedure doesn't work (you forgot to have the gear handle down when you pulled the CO_2 bottle as a last resort), or there has been a mechanical failure or damage that won't allow the gear to come down by any means, you might remember a few points on gear-up landings. The following applies to the majority of belly landings:

1. Comparatively little damage will be done.
2. The plane's occupants won't even be shaken up (physically that is).

A quick summary of your probable procedure:

1. Tighten seat belts and harness.
2. Make a normal approach, then after the field is made:
3. Battery and generators "OFF."
4. Chop the power and turn off all fuel system switches.
5. Make a normal landing.

If the runway is long enough, don't extend the flaps on the low-wing airplane. This will save a few more dollars, as extended flaps can be damaged. If the terrain is rough it would be better, though, to extend the flaps to further decrease the touch-down speed.

Figure on the prop being damaged. It will still be windmilling when you touch if you cut off the engine after the field is made. If you have any idea of killing the engine at altitude, slowing the plane up until the prop stops and making it horizontal with the starter, forget it unless the runway is extremely long (say, ten or twelve thousand feet). This is no time to be practicing dead stick landings. *A bent prop is a small price to pay for assurance that the field is made.*

SUMMARY OF EMERGENCY PROCEDURES

The exact emergency procedures will vary, as some airplanes use electrical power for gear actuation and others use hydraulic means.

1. Take your time and analyze.
2. Know your emergency procedures.
3. Keep the Airplane Flight Manual handy to help you remember each step.
4. Again, take your time.

SOME ADDED POINTS ABOUT RETRACTABLE GEAR

Some guys get the idea that the sooner they get the gear up on take-off the better they look to the airport crowd. 'Tain't so!

The landing gear has a safety switch (electric gear) or a by-pass valve (hydraulically actuated gear) on one of the main oleos to insure against inadvertent retraction on the ground. As long as the weight is on this gear (the oleo is compressed) the landing gear can't be retracted (oh yeah?). Don't depend on this safety switch — it might not be working that day. Curiosity can cost money, so don't test the anti-retraction safety features.

Even if the safety mechanism is working normally, don't get any ideas of putting the gear handle up during the take-off run to "look sharp" because a gust might lift the plane enough temporarily to extend the oleo and the gear starts up before you're ready.

Don't raise the gear before you are definitely airborne *and* can no longer land gear-down on the runway should the engine quit.

Apply the brakes after take-off before retracting the gear. Otherwise the wheels will be spinning at a good clip when they enter the wheel wells, and can burn rubber that you might want to use later. Most manufacturers have buffer blocks on strips in the main wheel wells to stop the spinning, but you might as well save the tires as much as possible. Of course there's nothing you can do about a nosewheel. On larger airplanes with high take-off speeds this braking is frowned upon as the rapidly spinning heavy wheel can have a great deal of inertia and the sudden stopping of the wheel may cause the tire to slip around the rim.

Know your maximum gear extension speed. (Fig. 14-1)

Fig. 14-1.

If you're taking off through puddles or slush and the temperature is near freezing, leave the gear down for a while after take-off to allow the airflow to dry the landing gear. You may want to cycle the gear a time or two to clear it before leaving it up. If the landing gear has a lot of water on it and this freezes, it might cause problems in extending the gear later.

Some new retractable gear airplanes have "automatic" gear lowering systems designed to help the pilot who *inadvertently forgets* to put the wheels down where they belong. These systems are not intended to replace good headwork. The pilot who flies airplanes so equipped, and automatically relies on the systems to do *his* job, could be unpleasantly surprised sometime when he lands the usual type of retractable gear airplane.

15. CHECKING OUT IN THE LIGHT TWIN

If you're like many single-engine pilots, you may have sold yourself on the idea that twin engine flying is strictly for people with thousands of hours and such skill as seldom is found in lesser mortals.

Remember when you first started flying you sometimes wondered if you'd ever really solo? (Particularly after one of those flights where everything went wrong.) Also, maybe there for a while it looked as though you'd never get the private certificate because you had to take the written again, and then had checkitis for days before the flight test. That's all behind, and now you've found a new subject to worry about — whether you'll be able to fly one of those light twins you've been drooling over.

Under normal conditions the airplane is flown *exactly* as if it were a single-engine airplane. Many new pilots don't believe this even after being told by the check pilot. It *looks* more complicated than the single engine airplane, so you may convince yourself that you'll have to be working a lot harder all the time.

Although you have two of each of the engine controls (throttle, prop and mixture), think of each pair of controls as one handle — at least at the beginning. The check pilot will allow you to get well familiarized with the airplane before he starts into engine-out procedures.

You'll find after a while that you'll be using the controls separately as needed — without any trouble.

Before flying, you and the check pilot will discuss the airplane and its systems in detail, and you'll spend a great deal of time with the Airplane Flight Manual. Following are a few pertinent points:

PREFLIGHT CHECK

There'll be a few more items to check than you've been accustomed to, but take your time. The check pilot will point out pertinent points to look for on that particular model. You'll have to check the oil for two engines and probably will have a couple more fuel tanks to visually check and more strainers to drain to check for water. A typical light twin Preflight Check is given below (make sure switches are off):

1. The tires are satisfactorily inflated and not excessively worn.
2. The landing gear oleos and shock struts are within limits of extension.
3. The propellers are free of detrimental nicks.
4. The ground area under propellers is free of loose stones, cinders, etc.
5. The cowling and inspection opening covers are secure.
6. There is no external damage or operational interference to the control surfaces, wings or fuselage.
7. The windshield is clean and free of defects.
8. There is no snow, ice or frost on the wings, or control surfaces.
9. The tow-bar and control locks are detached and properly stowed.
10. The fuel tanks are full or are at a safe level of proper fuel.
11. The fuel tank caps are tight.
12. The fuel system vents are open.
13. The fuel strainers and fuel lines are free of water and sediment by draining sumps (once a day).
14. The fuel tanks and carburetor bowls are free of water and sediment by draining sumps (once a week).
15. There are no obvious fuel or oil leaks.
16. The engine oil is at proper level.
17. The brakes are working properly.
18. All required papers are in order and in the airplane.
19. Upon entering the plane, ascertain that all controls operate normally, that the landing gear and other controls are in proper positions and that the door is locked.

The check pilot may give you a ground briefing several days before flying which will give you a chance to learn the various control locations and their use. Spend some time in the cockpit by yourself after the ground check, using the Airplane Flight Manual (or Owner's Manual) to mentally review the steps for starting, take-off, etc. This generally makes the first flight a little easier for both you and the check pilot.

STARTING

Normally the left (number one) engine is started first in the light twins. This got its start because many of the earlier light twins had a generator only on the left engine and it could then be working to help start the right engine. Many light twin airplane manufacturers recommend starting the left engine first (or port engine if you really want to be sharp about it), while others leave it up to the pilot. The starboard engine usually is started first on the big twins. Check the boost pumps before starting as was discussed back in Chapter 13 (carburetor engines).

Safety is still the big item in starting. Make sure that the areas around the props are clear before engaging the starter. The tendency is to be so busy with procedures that new pilots sometimes forget to shout "CLEAR!" and get an acknowledgment before starting the engine.

After one engine has started, run it at a high enough rpm to insure the generator is cut in to aid in starting the other engine. It's sometimes more than a weak battery can do to start two engines in close succession. If it's wintertime, you may not want to run the engine at higher rpm right away. If this is the case, don't be in too big a hurry to start the second engine. A short wait will allow the battery to build up again — plus the fact that you can soon run the operating engine up until it's helping.

If the engine you are starting first is cantankerous, you'd better forget it and start the other one. The generator of the second engine can help give the boost needed to start the laggard one.

Leave all unnecessary electrical equipment OFF. This goes for starting any airplane (single or multi-engine) with an electrical system. Radios in particular are power stealers (also, the sudden surge of power required for starting may damage them). Pitot heat also causes very strong current drain and, unlike the radios, is less noticeable as being on (the radio hum can be heard — particularly the ADF — even though the volume is down). Unless you happen to directly check the switch or notice the ammeter gasping at the lower end of the discharge range, the fact that the pitot heat is on may be overlooked.

Fig. 15-1. Joshua Barnslogger, private pilot, sometimes seems to have trouble getting the prop to turn over for starting (lousy electrical system design, he figures).

If you *really* want to give the battery (or batteries) the supreme test, turn on the landing lights also. With the pitot heat, radios and landing lights on, the chances of the engine getting started are very slim indeed.

A lot of people are awed by the idea of starting a twin engine airplane. One way of looking at it is that you are starting a single-engine airplane twice.

Before taxiing you might check your radios for proper functioning (this goes for single- *or* multi-engine).

TAXIING

In earlier times, when nearly all multiengine airplanes had tailwheels, one of the biggest problems was learning to taxi. The new pilot was taught that the use of asymmetric power was helpful in steering the airplane. This is true, but it was sometimes over-emphasized to such an extent that both the check pilot and the pilot checking out became discouraged. It always started about like this: The new man started taxiing and maybe the plane began to turn to the left a little. He applied a touch of power on the left engine to help straighten matters out and, of course, overdid it, requiring use of right engine power. This see-saw usually went on until the airplane was thundering down the taxi-way at an ever increasing speed and in sharper and sharper S-turns. The check pilot finally had to take over and slow the airplane down, the new man was given the controls and went through the same procedure again.

Taxi the airplane as if it were a single-engine type. It's very likely that the twin you are checking out in has a nosewheel and the separate use of throttles will have a much lesser effect. However, you'll soon be subconsciously using extra power on one engine whenever it's needed to make a sharper turn, so don't worry about using it right away as it only complicates matters. One good thing about tricycle gear and nosewheel steering as was stated above — if you do overuse either of the throttles, the plane isn't as apt to get away from you.

PRETAKE-OFF CHECK

A good check list pays off here. The same checks that applied to an advanced single-engine airplane apply here. You'll run the engines at a setting that allows the generators to be charging and, in the airplane with augmenter cooling, gives efficient exhaust venturi action (usually 1400 to 1600 rpm). Check for freedom of controls, check the instruments and other items as required by the check list. A simplified list for a typical light twin is given below:

1. Controls free — This is nothing new to you. Make sure the ailerons, elevators and rudder or rudders move in the right direction (it's hard to check rudder movement in a nosewheel airplane when it's sitting still but you can check the rudder pedal and nosewheel action while taxiing). *In fact, never try to turn the nosewheel while sitting still.*

2. Fuel on proper (main) tank or tanks — Make sure that you are on the main tanks. *Always,* repeat, *always* make your run-up on the tanks you plan to use on take-off. This will give you a chance to discover if that tank is furnishing fuel properly. If you make a run-up on one set of tanks and just before take-off switch to another set, you may find that the last tank

or tanks selected were not working properly. This discovery usually occurs at the most inconvenient point shortly after take-off. It's an old aviation truism that after unknowingly switching to a dry or bad tank, there'll be just enough fuel in the carburetor and fuel lines to get you into an uncompromising position during take-off. Always run the engines for at least a minute at moderate (1400-1600) rpm before take-off if you see the need for changing tanks during or after the run-up.

3. Electric fuel pumps OFF temporarily — to check the action of the engine-driven pumps. *After the check make sure they are both ON for the take-off.*

4. Crossfeed checked and then OFF for take-off — Here's a new control for you. Normally each engine will use fuel from the tanks in its own wing. However, in the event of an engine failure this would mean that there is a great deal of unusable fuel on the dead engine side, limiting single engine range as well as causing lateral trim problems as fuel is used from the operating engine side. The crossfeed valve allows the working engine to draw fuel from the dead engine's tanks. Fig. 15-2 shows a schematic of the normal operation of a typical light twin fuel system.

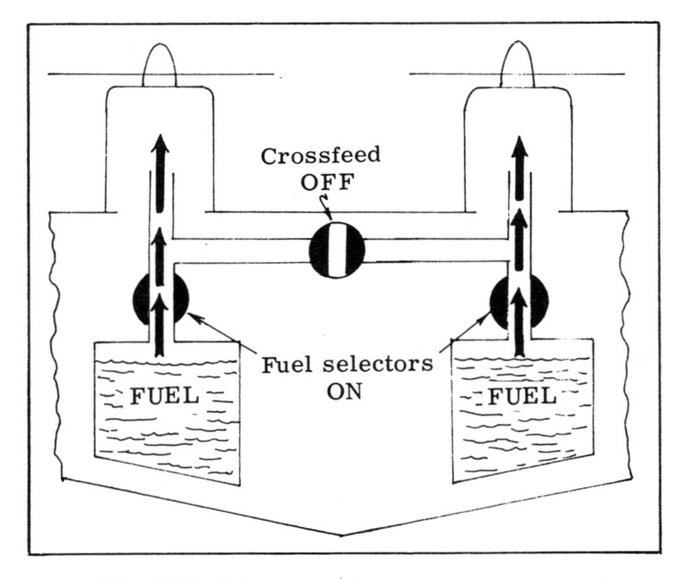

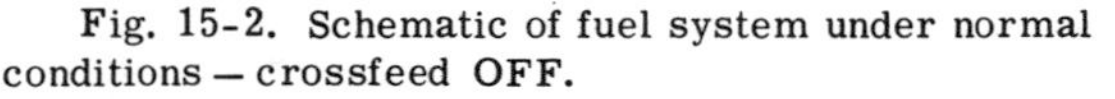
Fig. 15-2. Schematic of fuel system under normal conditions — crossfeed OFF.

Some airplanes do not have a separate valve for crossfeed but have a selector position on each of the two main fuel valves. If you needed to shut down the right engine in flight, for instance, you would "secure it" by throttling back, feathering it, pulling the mixture back to idle cut-off, turning off ignition switches and putting the fuel selector to the OFF position. If you begin to run low on left wing fuel for the good engine, you can select the crossfeed setting which would allow the good engine to draw fuel from the opposite tank. (Fig. 15-3)

Other manufacturers have a set-up whereby the pilot merely selects the tank he wishes to use fuel from — and no particular mention is made of crossfeed — which results in a great deal less confusion.

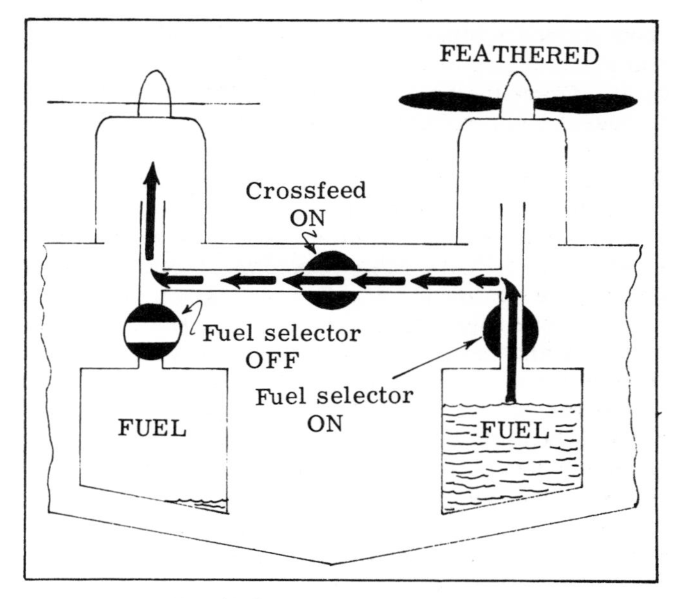

Fig. 15-3. The use of crossfeed.

You could, under normal conditions on some airplanes, run the left engine from the right tank and vice versa. This is frowned upon, just on general principles, as it could cause confusion at a time when instantaneous selection is necessary. On most airplanes the set-up is such that each engine does use its own fuel and only by turning off the fuel valve and selecting crossfeed can you operate an engine from an opposite tank (both engines running from the same wing tank). There are several combinations and you will learn your airplane's particular fuel system.

You might especially check on which tanks the crossfeed can be operated. The crossfeed on some airplanes works only for the main fuel tanks — auxiliary fuel in one wing cannot be used by the other engine. No matter how complicated it may sound, remember that the only purpose crossfeed has is to enable you to use fuel that would otherwise be dead weight.

The use of crossfeed will be covered later as is necessary.

5. Tabs set — You may have aileron tabs to contend with (the airplane will certainly have elevator and rudder trim controls). Make sure there's no wild setting on any of the trim controls.

6. Flaps — check operation — If flaps are required for take-off, or if you plan on using them, it might be better to wait until after the engine run-up before putting them down. The props, being run at high rpm on the ground, may pick up gravel and bat it into the flaps. You might find that for your particular airplane you would prefer setting the flaps just before taxiing onto the runway.

7. Check all instruments — You've been doing this for the single-engine airplane but will now have two of each of the engine instruments to check. Be sure that oil and fuel pressures, cylinder head temperatures and other gages are operating normally.

8. Engine run-up — Make sure the mixtures are full rich and the propellers are full forward (low pitch, high rpm). Run each engine up individually.

The required rpm for prop and mag check varies with each airplane.

(a) Exercise the propellers — This goes for either electrical or oil counterweight type. At a recommended rpm move the propeller controls through the range from high rpm to low rpm several times as covered in Chapter 12.

(b) Check the propeller feathering — Multiengine airplanes have featherable propellers because it was discovered that if the propeller blades of a dead engine could be turned edgewise to the airflow, much better engine-out performance could be realized. Naturally you'll be interested in making sure that you can feather a prop if necessary. A windmilling propeller on a dead engine cuts performance to such a degree that a critical condition could result. Most multiengine pilots would almost as soon skip checking the mags as to not check the feather system. As the check pilot will tell you, don't let the prop stay in the feathered setting too long, as the comparatively high manifold pressure and low rpm are not good for the engine.

Your airplane may have a special feather button for each propeller, or may have a detent on the quadrant that requires you to pull the prop control aft past the normally lowest rpm setting. Either way, the end result is the same.

(c) Check the magnetos — Here's the place where you'll realize that there are two engines instead of your usual one. It seems that to check the four mags is a good day's work. In fact, single-engine pilots have been known to have gotten writer's cramp, or its aeronautical equivalent, checking the mags of a multiengine airplane that first time. The usual maximum allowable drop for most light twins is 125 rpm but you might check to confirm this for your airplane.

(d) Carburetor heat or manifold heat — Use the carburetor air temperature gage if available, or check for a drop in manifold pressure as heat is applied. You remember that with the fixed pitch prop airplane you checked for an rpm drop when the carburetor heat was applied. A drop in rpm showed that the warmer, less dense air was going into the engine, proof that the carburetor heat was working normally.

The constant speed propeller, when in its operating range, will tend to cover any rpm drop. So lacking a carburetor air temperature gage, the manifold pressure would be the most positive indication that the system was working. In fact, the mp gives an *immediate* indication, whereas the carburetor air temperature gage needs a short period to indicate temperature. The mp gage is the primary indicator of the presence of carburetor ice, as you can no longer rely on rpm drop as a warning with a constant speed propeller. You may be able to notice a very brief rpm drop, but it will immediately recover if the rpm indication is in the constant speed prop operating range.

If you apply the carburetor heat with the rpm below the constant speed operating range, you'll get an mp drop *and* an rpm drop that remain as long as the heat is on.

The carburetor air temperature gage is a handy device for controlling the air temperature going through the carburetor. It allows you to keep the optimum temperature — warm enough to stay out of ice problems, yet not so warm as to cause unnecessary power losses. The carburetor heat for most engines of this class is so efficient that the application of full heat may result in power losses of 15 per cent or higher.

Your airplane may have an induction air temperature gage which measures the temperature of the air *before* it goes into the carburetor. Assuming a possible drop of up to 60° F as the air enters the carburetor, you would probably set a temperature at about 100° F, or appropriate Centigrade indication, to make sure the air temperature stays above freezing in the carburetor.

In the light trainers it was "all or nothing at all" as far as carburetor heat was concerned. Here you will use full or part heat as necessary. These larger airplanes usually take pre-warmed air from around the engine. As an added safety feature they have a spring loaded door in the intake so that should intake structural icing occur the airplane will have an alternate air source. The engine suction opens the door as required. Chapter 16 will go into this more thoroughly.

Make sure that the carburetor or manifold heat is OFF for take-off.

(e) Volt-ammeter — As the power is increased for the prop and mag check the ammeter should show a charge as soon as the rpm hits 1200–1300. A good check of both generators is to run both engines up until the ammeters show a charge and then turn each generator off separately to check that the other one is carrying the proper load. If the airplane in which you are checking out has an ammeter for each engine, this is an aid in the check. Multiengine airplanes usually have a paralleling relay that equalizes the load between the generators. Both generators ON for take-off.

To repeat, some *older* model light twins only have one generator (on the left engine) and this certainly shortens the time required to check the electrical system, but is not the best set-up should the duty generator fail, particularly at night or on instruments. The number of light twins with only one generator is small and growing smaller — fortunately.

(f) Suction — Check both sources to see that the engine-driven vacuum pumps are properly operating. This is done with a selector that allows the pressure of each source to register on the single suction gage. Most vacuum-driven instruments operate at about 4 inches of mercury suction, meaning that the air pressure on one side of the instrument is 4 inches lower than the other. A departure from the 4 inches of mercury requirement is the needle-and-ball, which normally uses a suction of 2 inches.

TAKE-OFF

There is very little difference in the taking off of a single- or multiengine airplane because the

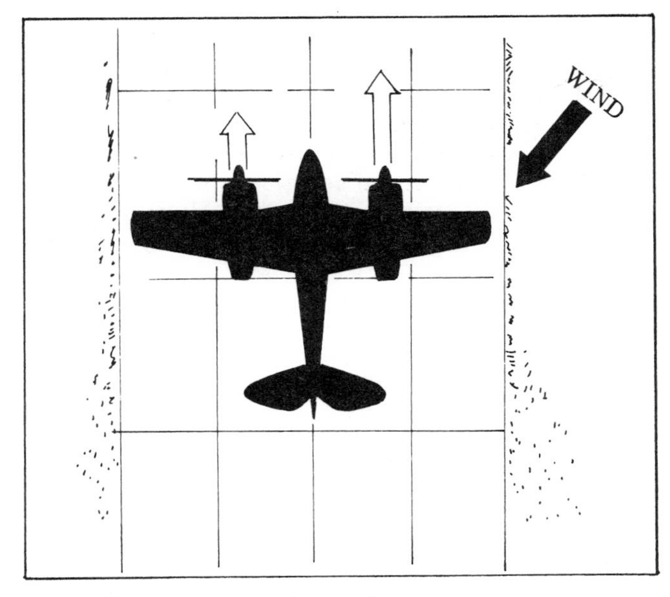

Fig. 15-4. The use of asymmetric power during the beginning of a crosswind take-off.

throttles are normally treated as one control. However, if there is a strong crosswind you may increase the power on the upwind engine first and carry more power on that side during the initial part of the run to help offset weathercocking tendencies. (Fig. 15-4) As the airspeed picks up and steering improves, increase to full power on both engines. This is helpful even for airplanes with a steerable nosewheel.

You'll have to watch your throttle handling if the engines are supercharged — you might overboost them and the check pilot will remind you to check the manifold pressure gage as power is applied.

Shortly after take-off is the point where the fun will begin. You will raise the landing gear and set climb power (throttles back first, then props!). One of your most frustrating experiences will be trying to synchronize the propellers when you are busy getting set up for the climb. (Wait until you have 500 feet altitude before reducing power.)

One tip for synchronization is to use sound as much as possible. After throttling back to the climb mp, move the propeller controls back to the proper rpm setting. Some twins use a single tachometer with two hands which makes the problem a little easier. In addition there may be a "synchronizing wheel," or indicator, which tells if one engine is turning faster than the other. Other twins use two separate tachometers. When moving the prop controls back, try to keep them in the same relative position to each other (don't worry, you won't the first few times). Use one tachometer hand as a "master" and note the relative position of the other. If the other hand is at a higher rpm, ease its prop control back until the "throb" sound has disappeared. This "throb" is your indication of closeness of synchronization. The faster the pulse, the greater the difference between the rpm of the two props. Use common sense, of course; you could pull one prop so far back that it is in feather and would have no pulsating noise at all — performance would suffer though. You'll soon be able to smooth out the props with a flick of the wrist. (Fig. 15-5)

An expression for the maximum climbing power, or maximum continuous power, is METO — Maximum Except Take-Off. Many engines are limited in time for full power operation; this being given in the Airplane Flight Manual and Engine Manual.

Some of the engines in this class have unlimited

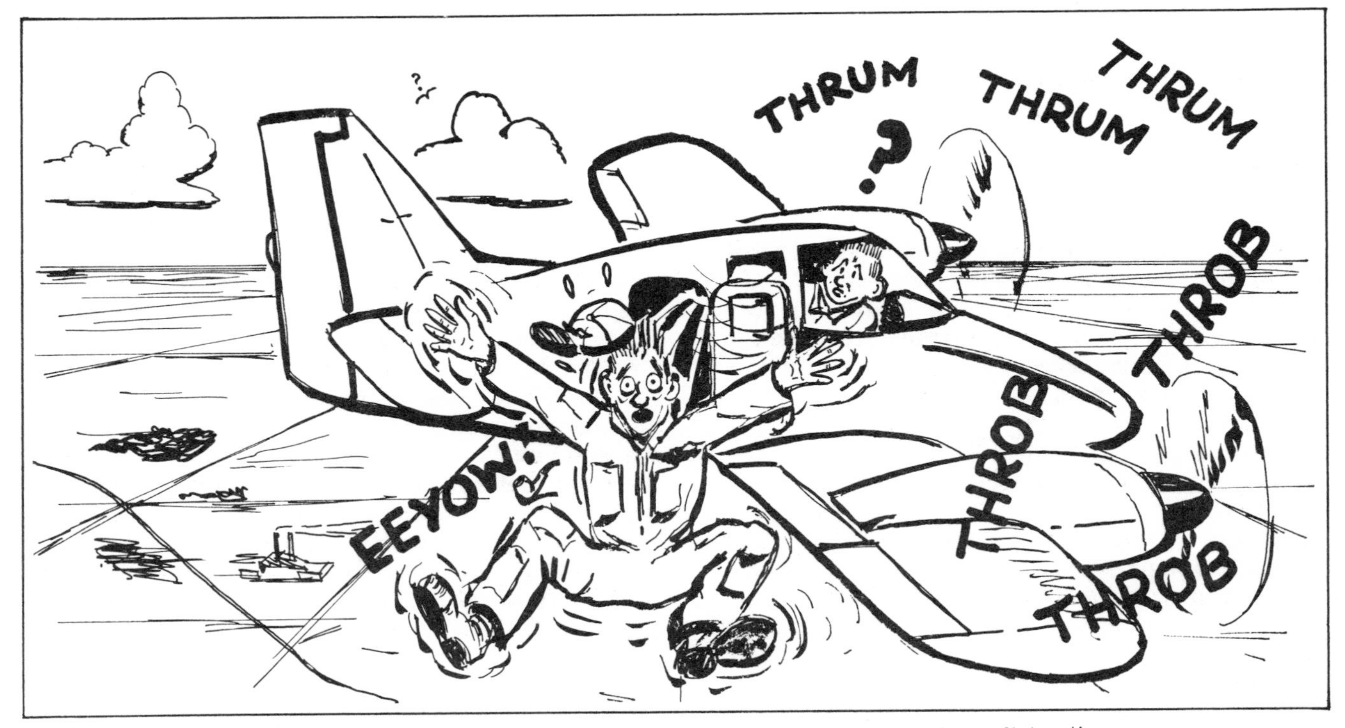

Fig. 15-5. Sometimes poor prop synchronization can drive a check pilot to distraction.

time allowed for full power operation. The engine manufacturer usually states that while there is no danger of failure or immediate damage, the overhaul period will be made shorter by abuse of this privilege.

SINGLE-ENGINE MINIMUM CONTROLLABLE SPEED

The single-engine minimum controllable speed (Vmc) is the minimum speed at which you can maintain directional control of the airplane with one engine inoperative and the other at full power. This speed is arrived at by the manufacturer as follows:

1. Take-off or maximum power on all operating engines (one for the light twin).
2. Rearmost center of gravity.
3. Flaps in take-off position.
4. Landing gear retracted.
5. Dead engine windmilling.

You will normally want to attain this speed before taking off so that should an engine fail you'll have directional control.

On a short field or obstacle take-off you may choose to ignore Vmc for a short while as you want to get off and climb as soon as possible. This means that you may be airborne at a lower speed than the minimum controllable speed and could be in trouble directionally should an engine fail. However, the *best angle* climb speed is usually *above* the Vmc, so that you will be accelerating on past Vmc and won't stay in this region very long. In essence, you are fighting against the definite hazard of the short field and obstacle and taking a chance on the remote possibility of an engine failure during the short period of acceleration. This will be covered more thoroughly in EMERGENCY PROCEDURES.

Know the max rate and max angle climb speeds for your airplane — both multi and engine-out.

CLIMB

The airplane will be cleaned up and power set for proper climb, using the recommended best rate of climb speed. After reaching a safe altitude the flaps will be retracted, if used, and shortly afterward the boost pumps will be turned OFF. Check the fuel pressure as you turn off each pump individually.

CRUISE AND AIRWORK

After reaching the practice altitude, the check pilot will show you the proper cruise power setting and leaning procedure. You may have a little trouble with synchronization again, but this is to be expected. A good procedure for leveling is to ease the nose over to cruise attitude, using trim as necessary, and leaving the power at climb setting to help acceleration. As the cruise airspeed is approached, throttle back to cruise mp and set the props. After getting the airplane trimmed to your satisfaction, switch tanks and lean the mixtures using the technique as described in Chapter 13 or as recommended by the check pilot. He may have a method that he has found to be the most effective for your particular airplane.

The check pilot will have you do shallow, medium and steep turns (up to 45° of bank) to get the feel of the airplane. You will stall the airplane in various attitudes and configurations: gear-up or down, flaps at various settings and at different power settings. He will probably demonstrate the effects of the loss of an engine when you are making a power-on stall. The best thing is to pull the power back on the other engine and lower the nose to pick up Vmc before reapplying full power.

He will throttle back or feather one engine and have you continue to slow up the airplane to below Vmc so that you will have a graphic demonstration of the required rudder force and what can happen when you get too slow on one engine.

Later you will have a chance to feather and unfeather a propeller and to fly around on one engine to check the performance. If possible, you should fly the airplane at gross Weight and, at a safe altitude, simulate or actually feather a propeller to see what effect Weight has on performance. The FAA states that a propeller will be feathered only above 3000 feet above the surface and in a locality within range of a suitable airport in the event the inoperative engine cannot be started. It also makes a note that the airplane should be capable of maintaining at least 1000 feet above the surface of the preselected airport, taking into consideration loading and density altitude.

There have been cases where a plane was damaged or destroyed because the just-mentioned rules weren't followed during a simulated engine failure. It would be mighty embarrassing to clamber from the wreckage and try to explain to an irate operator-owner that you "were just practicing single-engine flight so as to avoid damaging the airplane should the real thing occur." He might not be too understanding.

You'll probably practice slow flight and will slow the airplane to about 10 knots above the stall warning or buffeting point for the configuration used, at a constant altitude. You will also fly at landing configuration to demonstrate your ability to fly the airplane safely, maintaining altitude, speed and a constant direction through proper use of power and the flight controls. You'll fly it long enough in each configuration to demonstrate the acceleration and deceleration characteristics of your airplane. You'll do straight and level flight, level flight turns, climbing and gliding turns at slow flight speeds. The check pilot will be particularly interested in your transitions to and from slow flight. You will want to avoid accidental stalls, of course.

In general, this phase of your transition to multi-engine flying will be quite similar to the check-out in an advanced single-engine airplane (except for the engine-out demonstration). You'll be finding out how the airplane reacts under normal conditions. The full treatment on engine-out procedures will come later.

APPROACH AND LANDING

You'll have a few more items to check than you've been used to and should use the check list religiously, pointing to each item as you check it. Everything mentioned about mixtures, boost pumps, gear, flaps and props for the advanced single-engine airplane will still apply except that you'll have two of some of the controls to move. But, again, under normal conditions the two controls can be handled as one. (Use the main tanks for landing unless the manufacturer recommends differently.)

Note the gear and flap-down speed and give yourself plenty of time and room on the downwind leg and approach, particularly the first few landings. Check the gear again on final.

The approach and landing will be just like a single-engine airplane except that you must keep in mind one thing: it's best to maintain an approach speed above the single engine minimum controllable speed (Vmc). As the airplane has a comparatively high wing loading, you will be making the majority of your approaches with some power. This means that should you have a complete power failure (both engines) you probably wouldn't make the runway. The chances of both engines quitting are practically nonexistent (but not impossible), but one engine *could* quit on you. Suppose you get low and slow (below Vmc) and are dragging it in from 'way back. An engine fails, and as the plane starts sinking you apply full power on the operating engine. You'll find that you made a bad mistake by being too slow — because the directional control is nil with full power on the good engine. You are too low to nose over and pick up Vmc (and then best single-engine climb). You might also find that the only thing to do is to chop the other throttle, turn all the switches off and hit something soft and cheap. You can get caught in a trap of your own making.

Even if you are at or slightly above Vmc, you'll have to accelerate to best single-engine climb speed, so take this into consideration on the approach. Avoid dragged out finals (this goes for any airplane).

For short fields you still should have no reason to get below Vmc as that speed is usually low enough to assure that you won't float before touching down. Again, when you drag it around close to or below Vmc, you're taking a calculated risk as you did on the short field take-off.

You'll be given plenty of chances to shoot normal and short field take-offs and landings before making single-engine approaches or go-arounds.

The landing roll, taxi, and shut down procedures of the light twin follow closely those of advanced single-engine airplanes. The check pilot will cover any peculiarities of the check-out airplane.

EMERGENCY PROCEDURES

The non-pilot may feel that loss of an engine on a multiengine airplane is either a terrifying disaster, or nothing to be concerned about. The experienced pilot knows that the multiengine airplane, if properly flown with an engine out, has a strong safety factor. He also knows that at certain times the airplane must be flown precisely, and in some cases it is safer to chop the other engine(s) than to try to continue. New pilots have been killed by the loss of an engine on take-off or approach when they believed that they could go around. Ironically enough, they might have survived had the engine quit at the same place in a single-engine airplane. They would have had to land straight ahead in the single-engine airplane but attempted the impossible because of overconfidence or ignorance of the single-engine performance of their twin.

Here's where you'll start running into the age-old problem of decisions. You may end up like the orange sorter who finally went berserk because, "although the work was easy, the decisions finally got me down." As one single-engine pilot noted, one advantage to the plane with just one fan is that if the engine quits, you don't have to make the decision whether to go around or not.

So, multiengine flight is safer — you'll feel more comfortable flying over rough terrain and at night, but you must realize that you must earn this increased safety by knowing what to do in an emergency.

One of the first things you'll find is that a windmilling prop can cause a great reduction in performance, and that gear and flaps cause a problem on a single-engine go-around.

Another point: because you've lost half your power with one engine out doesn't mean that you'll have half the performance. *You'll have considerably less than half the performance and must take this into consideration.* For instance, you remember that the rate of climb was dependent on *excess* horsepower. When you cut the power being produced you'll be losing nearly all of that excess horsepower — what you'll have left will be enough to fly the airplane plus some small amount of excess power. So, the excess horsepower is what suffers. (Unfortunately, there is no way for you to lose the power required to fly the airplane and keep the *excess* horsepower.)

For some comparative climb performance with both engines operating as against single-engine climb, you might check Fig. 15-6. The figures are for a sea level standard day at gross Weight.

You can see that at best you might be able to get up close to 20 per cent of the normal twin-engine climb with one engine out.

Your en route performance will not suffer nearly as much as the climb or acceleration characteristics, but all phases will be affected. The single-engine

Airplane	Twin-Engine Climb fpm	Single-Engine Climb fpm	Single-Engine Percentage Rate of Climb
1	1270	195	15.3%
2	1650	250	15.1%
3	1300	180	13.9%
4	1050	180	17.1%

Fig. 15-6. Comparisons of twin- and single-engine climb for four light twin airplanes.

rate of climb is based on a clean airplane with the inoperative prop in the minimum Drag position — feathered if possible, or in high pitch (low rpm). Figure 15-7 is a THP Available and Required versus Airspeed curve for the light twin of Fig. 1-52 with both engines operating and with one feathered at gross Weight and sea level. Notice that even with a prop feathered the THP required is greater than normal because of control deflection, loss of efficiency, etc.

By looking at the excess horsepower in Fig. 15-7 you could see that with both engines operating there is about 240 THP in excess of that required at the best rate-of-climb speed of 100 knots. Using the equation for rate of climb and assuming an airplane Weight of 5000 pounds, the rate of climb is found to be:

$$R/C = \frac{EHP \times 33,000}{Weight} = \frac{240 \times 33,000}{5000} = 1584 \text{ fpm.}$$

Checking the single engine situation in Fig. 15-7, it is found that about 45 excess THP is available at the speed for max rate of climb (90 knots) in that condition. $R/C = \frac{45 \times 33,000}{5000} = 297$ fpm.

You can climb 297 fpm in this airplane at gross Weight *at sea level under ideal conditions.* Turbulent air and/or a higher density altitude can wreak havoc.

You can imagine what percentage of normal rate of climb you'd have at gross Weight on a hot day with an engine out, the gear and flaps down, and a windmilling prop. You would likely end up with a negative rate of climb.

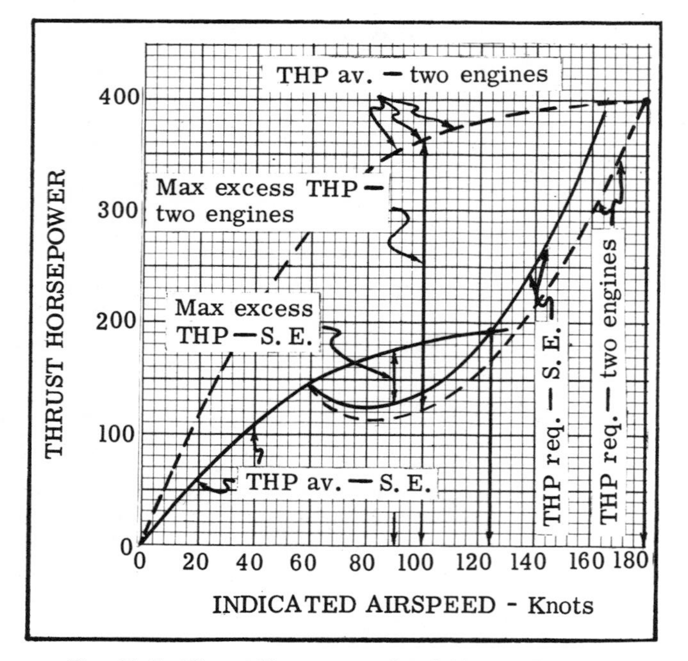

Fig. 15-7. Thrust Horsepower Available and Required versus Airspeed for a light twin in both twin and single engine flight. Gross Weight at sea level.

The lighter the airplane's load, the better the single-engine performance. But even an airplane at light weight doesn't have much get up and go with a lot of garbage hanging out in the slipstream. Of course, you could always lighten ship by throwing your wife's numerous suitcases overboard but this is a last ditch effort because: (1) you'll have to replace all the gear, which costs MONEY and (2) you'll never hear the last of it if you do. One method used by some pilots with success on an occasion of this sort is: after you've thrown her baggage over the side and she is giving you a piece of her tongue in no uncertain terms, eye her speculatively as if estimating her weight and its effect on the airplane's performance. This has been known to act as a great pacifier, and silence will reign supreme. It is assumed that during all this byplay you have a copilot or the autopilot flying the airplane.

ENGINE FAILURE ENROUTE

This is usually the least critical place for engine failure (except during taxiing or warm-up) but can lead to trouble if things are allowed to progress too far.

The check pilot will usually pull an in-flight emergency on you after you've had a chance to get the feel of the airplane. He will demonstrate the procedure to be followed in the event of an engine failure and, after you have had a chance to run through the procedure several times, may quietly turn the fuel off on one engine in order to catch you by surprise. He isn't trying to see how badly he can foul you up but, like the primary instructor who used to give simulated emergencies at unexpected times, is giving realistic training. It's a lot different to watch the check pilot pull back one of the throttles (you'll know immediately which engine is going to be "bad") than to suddenly have one of the engines quit (which one?). You'll find that no matter how hard you've practiced or memorized the procedure, you'll be all thumbs and feet in the cockpit the first time one stops when you aren't ready for it.

The biggest problem at first is knowing which engine is out. In flight you'll have more time to judge and make a decision. If you feather the wrong prop on the check ride . . . well, you can always take the flight test again. What is needed is caution with some speed of action, but make sure that the decision is a good one. Better to be a little slow and be right, than fast but wrong.

To go through a typical case:

An engine fails. You can't tell which one immediately by looking at the tachometers. Remember the constant speed prop will tend to flatten pitch and maintain the chosen rpm.

As long as the prop is windmilling the engine is still acting as a pump and the manifold pressure will tend to stay at the former indication. Although a slight change may immediately occur, it's hard to tell at a quick glance just which manifold pressure hand did the moving. However, as the airplane begins to slow down, the constant speed propeller of the dead engine can no longer maintain rpm. (The governor continually flattens the blades to maintain rpm but the low pitch limit is finally reached.) Because the dead engine is still "making the motions," movement of that throttle will still result in manifold pressure change but no feel of power variation as would normally be expected. One visual indicator is the ball in the turn and slip indicator. The ball, realizing that things are amiss will tend to move *away* from the engine that's causing the problem. But to be on the safe side you should take the following steps before feathering.

Advance the engine controls for both engines as follows: mixtures, props, throttles. You will be needing more power on the good engine and, as you have not definitely ascertained which is good or bad, will move all engine controls forward. Some engines are limited in the time allowed for full power, and you will not want to leave the power up too long. But get in the habit of increasing power on both engines (of course, you will actually only be increasing the power of one engine but you'll be sure this way). Now the problem becomes one of definitely isolating the bad engine.

So, an engine has quit. *Working foot — working engine.*

This cryptic statement means that the airplane will yaw when power is lost on one engine. You will consciously or unconsciously try to hold it straight, which requires the use of rudder — and that foot is the *working foot.* Therefore, that engine is working O.K. You can also use the idea, *loafing foot — loafing engine.*

Let's say for instance that it requires right rudder to keep the airplane straight (it wants to yaw to the left). The right engine is working. Do you feather the left prop as soon as you can get your grubby little hand on the control? You do not! First, you pull the left throttle back. If the left engine *is* dead as you figured, nothing will happen — no change in power effects or sound or feel of the airplane. If somehow you made an error in feel of the rudder and the left engine is the working engine, you'll feel and see the loss of power and discover the mistake before feathering the good engine.

You'll find that holding aileron into the good engine so that a bank of about 3°—5° is established will be a big help in keeping directional control. You are setting up a slipping condition that tends to help keep the airplane straight. If you are banked, the airplane naturally wants to turn, and you want it to try to counteract the turning force set up by the good engine. (Fig. 15-8)

Fig. 15-8. Banking into the good engine to aid in directional control.

FEATHERING

The order of engine control usage for feathering will vary between airplanes but may generally be given as this: (1) Throttle back to idle, (2) Mixture — idle cut-off, (3) Prop control into the feather detent.

In an actual engine failure at cruise don't be in too big a hurry to feather. After you've discovered which engine is the culprit you might turn on the boost pump for that engine (or better still, turn *both* boost pumps on to make sure). You can also switch tanks and check for other problems (carburetor or ram icing will generally hit both engines more or less equally). You enriched the mixture when the engine controls were moved forward.

Okay, so you've checked everything, but the problem still exists and it looks like you'll have to feather it. If you make a thorough check during the simulated failure you'll discover that the check pilot has turned the fuel off, but you'll go ahead with the feathering procedure for practice.

The oil-counterweight propeller must be rotating in order to be feathered. If the engine "freezes up" before the prop is feathered, you'll have some flat blades out there giving lots of Drag and there won't be anything you can do about it. If, under actual conditions, the oil pressure is dropping or has gone to zero and the oil temperature and cylinder head temperature are going up out of sight, you'd better feather while you can.

After the prop has stopped, trim the airplane, secure the dead engine mag and boost pump switches and turn the fuel to that engine off.

Under actual conditions you will want to land at the nearest airport that will safely take your plane. This is no time to be landing at an extremely short field with poor approaches to the runway. On the other hand, don't figure on finishing the last four hundred miles of your trip either.

CARE OF THE OPERATING ENGINE

Now that you have feathered the propeller, you are once again a single-engine pilot. You are interested in taking care of the operating engine and you don't want to be the pilot-in-command of the only twin-engine glider in the area.

There are two ways to combat possible engine abuse: (1) *Airflow* and (2) *Richer Mixture.* If you throttle back and slow down, you're decreasing the airflow and *in some light twins, throttling back automatically leans the mixture as well.*

Watch the cylinder head temperature (if available) and the oil temperature carefully. It's a lot easier to keep the engine temperature within limits than to cool it *after* things have gone too far.

If you are above the single-engine ceiling, you will lose altitude after the failure of one engine. If the engine gets too hot, you also may have to ease some power and make a slight dive to get increased airflow if altitude permits. Manufacturers check their engines for cooling at gross Weight, best rate of climb speed, full power and full rich, so unless you really get wild with the good engine, you'll have no problem with it.

LANDING WITH ONE ENGINE (ACTUAL EMERGENCY)

A twin-engine airplane with an engine out is an airplane in distress, no matter how glowingly the manufacturer describes the single-engine performance of his airplane. You'll certainly let the tower know of your status. They may see it as you enter the pattern but give them a little advance notice so they can do some traffic planning. You'll certainly have the right-of-way — unless somebody else has *both* engines out. At an uncontrolled field you might let Unicom know that you have one out — other pilots in the pattern on that frequency will give you plenty of room. It's a sad fact that many pilots would literally rather die than let anybody know that they think they have a problem. There have been many cases of serious or fatal accidents being caused purely by stubbornness. The pilot is afraid that he might be joshed by his fellow pilots for asking for precedence or preference in an unusual situation. Your passengers have more or less blindly entrusted their lives to YOU and you have no right to risk them to save your pride. The pros will congratulate you for recognizing an unusual situation; the amateurs are the ones who scoff.

Enough of the philosophizing. You are interested in landing on the *first* approach.

Don't fly in such a manner that you get low and slow and have to apply full power to get to the runway. You might find that it will take more power than you have available to drag the airplane up to the landing area — which brings up another point. Don't lower the gear and flaps until you are pretty well assured you'll make the field.

The light twins that use hydraulic pressure for actuating the gear and flaps normally have only one engine-driven hydraulic pump (usually on the left engine). Should this engine be the one that is secured, you'll have to remember to hand pump the gear and flaps down. This may take some time, so give yourself plenty of leeway on final. Forgetting this is one of the most common errors for the new twin pilot during simulated engine-out maneuvers at altitude. The pilot makes a good pattern and uses good headwork until the time comes to lower the gear and flaps. He's got a good final, so he pushes the gear lever down but forgets about the necessity for hand pumping. Valuable seconds go by before he realizes that in an actual approach he would have to start pumping — pronto! Many a new pilot, making a simulated approach at altitude with the propeller feathered on the engine-driven hydraulic pump side has "landed gear-up" at 3,000 feet. This could cause certain inconveniences in an actual landing — so do it right the first time to avoid a potentially dangerous situation.

Keep your approach speed above the single-engine minimum controllable speed (Vmc) until landing is assured. You can get yourself into "Coffin Corner" by slowing it up too soon below Vmc and, if full power is needed for any reason, you may lose control of the airplane.

A good single-engine approach is one that requires gradual throttling back of the good engine as you approach the field. As you throttle back take care of the rudder trim so that when the power is off the airplane will be in trimmed flight. Some pilots neutralize the rudder trim on final and hold the required rudder pressure with their foot. This is good except that should a sudden go-around be required things could get complicated as the pilot will get no help at all on the rudder and must quickly trim the airplane as he executes the required steps.

Another common mistake for the new twin pilot is overshooting the runway on a single-engine approach. He overdoes the idea that he shouldn't undershoot and is so high and fast that when the flaps are extended he balloons to new heights of glory and has to take it around — *on one engine*. (This makes for problems!)

The perfect single-engine approach is one that allows the pilot plenty of time to correct for crosswinds and to get his airplane in the landing configuration. So, an approach that requires a slight amount of power (with gradual reduction) all the way around is much better than a high, hot and overshot one.

This is a time when the check list is most important. To follow the idea that the good engine should be taken care of, make sure that the mixture is rich, boost pumps on and fuel on best tank. A double check of the gear is important. In the stress of the moment you may overlook it if the check list isn't used. Make sure the prop is in high rpm (low pitch) in case you should have to go around.

TAKING IT AROUND ON ONE ENGINE

It may be that after careful planning on your part somebody taxis out on the runway just as you are on final, or for some other reason it is necessary that you must go around. Once you've decided to make the big move — the sooner the better! The sooner that power is applied on final, the more airspeed and altitude you'll have.

Don't ram the throttle open as this will cause directional trim problems. Ease it open and retract the landing gear (you may have to pump it up). Flaps up gradually. Don't try to climb too soon — remember that you must attain and maintain the best single-engine climb speed. If you start the go-around early on final, you may use a small amount of altitude — after opening the throttle and cleaning it up — to help attain the best single-engine climb speed. Remember that flaps require the use of vital horsepower, so don't be *too* slow about getting them up. There are some light twins for which it is recommended that the flaps be retracted *before* the gear in a go-around. Check the recommended sequence of cleaning-up for your airplane. One engine go-arounds are extremely risky, no matter what light twin you are flying. Obstructions ahead might make it better to land anyway, even if you did forget to lower the gear (and realize it at the last second).

ENGINE FAILURE ON TAKE-OFF

You may wonder why we've waited so long to be talking about engine failures on take-off. It might seem more logical to talk about this *first* and then go into the in-flight emergencies. The fact is that you won't cover take-off engine-out procedures until you've had plenty of practice in the air and have a good idea of the principles of single-engine flight.

There's no doubt about it, during take-off is the most critical time to lose an engine. The plane is at its heaviest and the airspeed and altitude are low. This is the time for cautious haste. You won't have a great deal of time but will have enough to make a decision.

The check pilot will give you a single-engine emergency on take-off by throttling one engine back to zero Thrust and you will go through the necessary recovery actions. He will probably give you simulated take-off emergencies and engine-out approaches at altitude, where you can actually feather a prop, before simulating one close to the ground with zero Thrust.

Always pick up the best single-engine climb speed as soon as possible after lift-off, then assume the best twin-engine climb speed. Don't accelerate above the twin-engine climb speed; altitude is much more valuable than *added* airspeed.

One thing that is sometimes overlooked—if an engine quits on take-off you do not always take it around. Most new pilots have drilled themselves so thoroughly on what *to* do that they forget that there are things *not* to do, also.

1. If an engine quits before leaving the ground, chop the power on the good one and taxi back to the hangar, and complain.

2. If an engine quits after you become airborne and there is enough runway left (and your gear is still down) *always* chop the power on the good engine, land and go back to the hangar, and complain. Light twins on a standard sea level day at gross Weight need 2500-4000 feet of runway to accelerate to Vmc and stop (depends on the make and model). Check your Airplane Flight Manual for the Accelerate and Stop Distance if available. If you are taking off from a 2000 foot strip and your airplane requires 3000 feet to accelerate to Vmc and then stop, you could be pretty well committed to go around after getting much above that speed. Remember too, that the Accelerate and Stop Distance will *increase* with an increase in temperature and/or altitude (higher density altitude).

3. If you have lifted off above Vmc but have not attained best single-engine climb speed and the runway is rapidly disappearing:

a. Clean the airplane up.

b. Keep the nose down and keep all engine controls forward, and accelerate to best single-engine climb speed as soon as possible.

c. Remember: "Working foot—working engine."

d. Throttle back to check, and after making sure which engine is the culprit, feather that prop.

e. Maintain the recommended best single-engine climb speed and return and land (no low, tight patterns).

To cover all possibilities for an engine failure on take-off would take a set of encyclopedias. For instance, terrain or obstructions well off the end of the runway might make it better to belly it in, even though you have best single-engine climb speed.

Pilots have been killed when they overrated their ability and the airplane's single-engine performance —and forgot about such things as temperature, turbulence and altitude effects.

UNFEATHERING IN FLIGHT

Generally, if an engine is so rough that it must be shut down, it should remain so. Sometimes restarting an engine that's cutting up is asking for a fire or a situation where the prop may not be feathered again. However, for practice purposes, it would be wise for you to try as much actual feathering and unfeathering as possible. Again, this will be done with the check pilot at a safe altitude. Although you may be leery of the whole idea at first, you'll find that your confidence in single-engine flight will be immeasurably raised if *you* feather and unfeather the propeller several times and do a considerable amount of flying on one engine.

The method of unfeathering will vary between models. Some use normal starting procedures (turning it over with the starter, the oil pressure unfeathering the prop as the engine starts) while others have an accumulator that stores oil pressure for unfeathering.

Whatever method used (which will be outlined in detail in the Airplane Flight Manual or Owner's Manual), remember that the secured engine will be cold because of the airstream passing over it. If properly primed, the engine will make an easier start in the air than on the ground because as the propeller starts turning, engine oil pressure will start to build up and it will move farther and farther out of feather and start windmilling.

Here is a typical unfeathering procedure for a light twin without an accumulator and without fuel injection:

1. Fuel ON.
2. Ignition on BOTH.
3. Boost Pump ON.
4. Mixture RICH.
5. Throttle pumped several times to prime the dead engine (the number of times will depend on the outside temperature and length of time the engine has been secured—this will come with experience). Throttle closed for smooth start.
6. Propeller control moved out of feather into the low rpm range.
7. Engage starter for that engine, as soon as prop starts windmilling, disengage starter.
8. After engine starts, gradually increase power; synchronize the propellers.
9. Check pressures and temperatures for proper operation. This is important if you shut it down because of irregular operation.
10. Retrim as necessary.

Airplanes with accumulators require that the prop control be moved to the full high rpm position for starting.

After you've read the procedure and done it yourself several times it will be quite clear to you.

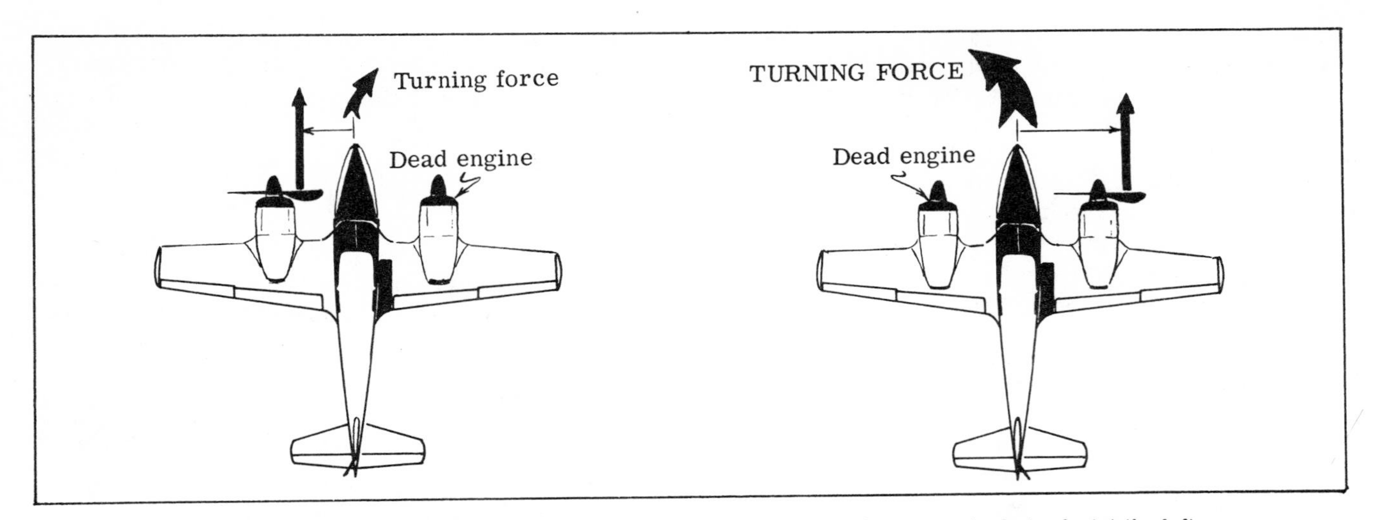

Fig. 15-9. An exaggerated comparison of forces which show that from a control standpoint the left engine is the worst one to lose.

THE CRITICAL ENGINE

You may hear the term "critical engine" used and will probably be asked about it when you take the check ride for the multiengine rating. First, for light twins that have both propellers turning clockwise as seen from the cockpit (which pretty well covers the U.S. light twins) the left engine is the critical engine. This is because of "asymmetric disk loading" which you might review in Chapter 1. Figure 15-9 shows that the yawing force is greater when the left engine is out and, therefore, directional control is more critical with the loss of that engine. Incidentally, asymmetric disk loading is popularly called the "P-Factor."

Don't be like some pilots who, when asked by the check pilot why the left engine is the critical one, answer that "it's because the hydraulic pump is on that engine on some light twins."

One manufacturer is producing a light twin with counter-rotating propellers, a design which will eliminate the problem of a critical engine. (This does not mean that there is no longer a yaw when an engine quits, but single-engine performance will be the same for either engine.) When both engines are operating normally (and equally) the "torque" forces are cancelled with no need for rudder trim change for climbs, cruise or dives.

SUMMARY

Rather than inserting the exact requirements for the multiengine flight test here, it is suggested that you order the latest *Flight Test Guide — Multiengine Airplane Class or Rating* from the U.S. Government Printing Office, Washington D.C. 20402 (15 cents).

PART III / EMERGENCIES AND UNUSUAL SITUATIONS

16. PROBLEMS AND EMERGENCIES AT ALTITUDE

When you became a private pilot you were then "on your own" — no more check rides or sweating out a session with an instructor. You became free to establish as many bad habits as you pleased — and like most of us, probably set about it very soon. If you fly your own airplane and keep up the minimum take-offs and landings in the required period, you may legally fly for years and hundreds of hours without benefit of dual. Your passengers will have blind faith in your ability and they won't know whether you've been up recently with an instructor or haven't had a check ride since your Curtis Robin was new way-back-when. If everything goes normally there's no sweat, *but* if it doesn't, you may have a hard time finding passengers to share the expenses on your trips. If your flying is shaky enough, even the uninitiated will begin to suspect that all is not well. Gone are the days when a pilot boasted of the number of planes he'd pranged during his career.

Sure, you've been shooting landings every chance you get, and this is particularly enjoyable in the late afternoon when the air is calm. Take-offs and landings are fun and you can learn a lot in a good session. But there are other phases of flying, too. For instance, when was the last time you made a power-off approach? Do you have a good idea of the airplane's approximate glide ratio, clean *and* dirty?

Every once in a while you should drag out the Airplane Flight Manual and go over the emergency procedures again. You'll be surprised how much you may have forgotten since the last time you reviewed them.

This chapter will bring up a few points on high altitude emergencies or enroute problems.

ROUGH OPERATION OR LOSS OF POWER

The rough running being talked about is not the "automatic rough" that the engine always jumps into when you're flying over water or rough terrain.

Okay, so the engine really starts to run rough, now what? It's a complicated piece of gear and the trouble could be caused by any one of a thousand things — but a large percentage of problems are caused by a very few items — namely carburetion, fuel management, and ignition. You'll want to analyze the problem and, if possible, correct it.

CARBURETION

This term is meant to cover the fuel-air mixture delivered to the engine cylinders — whether using a carburetor or fuel injection.

CARBURETOR ICE

One of the most common problems for carburetors is plain old-fashioned ice. However, if you've let icing go so far as to cause the engine to start running rough, you've really been asleep. You know that carburetor ice gives warning by (1) a decrease in manifold pressure in an airplane with a constant speed prop or (2) a decrease in rpm for the fixed pitch prop. It's quite possible that carburetor icing in a light trainer can progress to a point that *full* carburetor heat won't undo the damage. It's a vicious cycle: the heat capacity naturally depends on the engine, but the engine is sick because of the ice and the carburetor heat suffers. The engine may quit and there you sit with a windmilling propeller. So you fell asleep and now must pick a field. *Leave the carburetor heat full on as you try for the field.* There may be enough residual heat getting into the carburetor to clear out the ice before you have to land. You pulled it on and nothing happened right away. What may be needed is a little time. This doesn't mean that you won't be picking a field and preparing for an unscheduled landing — because the residual heat may *not* do the trick. Don't count on it and sit up there with your head up and locked.

One thing of importance about carburetor ice if you don't have a carburetor air temperature gage: when you discover you have ice, use *full* carburetor heat to get the garbage cleaned out. After that you can experiment for the best intermediate setting. A little heat can sometimes be worse than none, since a false sense of security might be induced.

On some of the higher performance light planes the carburetor or manifold heat is very effective, and the air going into the carburetor may be raised up to 200° F by application of full heat. *The use of full heat will cause a power loss.*

Remember that the ice will collect around the butterfly and jets of the float carburetor.

Your job will be to use carburetor heat and open the throttle if it looks as though the ice is getting ahead of you. When you open the throttle you've made sure that the butterfly valve is opened so that the fuel-air mixture has a better chance of getting through to the engine. Icing can give you more trouble at part or closed throttle operation — it will take less ice to cut off the fuel-air mixture from the engine. You remember this from your student pilot days when you *always* used carburetor heat before closing the throttle for a glide. More power means more carburetor heat available for use.

Normally, though, you'll apply heat as needed and won't increase the power. You don't want to just ram the throttle wide open without thinking with a constant speed propeller — you might overboost the engine.

As you know, carburetor ice is not as much a function of low temperature as it is high humidity. If the outside air temperature is quite low the air will be so dry that carburetor ice will be no problem.

Any time you are more or less smoothly losing manifold pressure or rpm, use carburetor heat or alternate air (assuming you don't have a creeping throttle). Here's the typical situation: You notice that the manifold pressure is lower than it should be. You haven't climbed and it looks as though you might have ice. Suppose that you were carrying 24 inches for cruise, but it has dropped to 22 inches. You apply heat — now the manifold pressure drops to 21 inches because of the less dense air introduced. After a few seconds the manifold pressure picks up the 2 inches it lost so that now the gage registers 23 inches Hg. When you push the heat off it will again be up to 24 inches. But if it iced once it will do it again, so you'll experiment to get the right setting.

IMPACT ICING

The fuel injection system has the advantage of doing away with carburetor ice, but both types of systems may suffer from impact icing.

Saturated air is the culprit here. Whenever you're flying in rain, clouds, or fog, and if the outside air is near freezing, impact icing may occur. As the air enters the induction system it may be condensed and cooled to the point that ice will form at the ninety-degree bend where the air scoop turns to enter the carburetor or fuel injection control.

Another problem may be in the form of structural icing on the air intake screen. You may run into this when flying instruments later. Structural icing on the airplane may not be serious, but could cause a power loss. Fig. 16-1 shows a simple method of taking care of this problem.

When the intake is iced over, the engine suction will open the spring loaded trap door and the warm air from around the engine is drawn into the intake system. Naturally some power will be lost, as the warmer air is less dense than the cold outside air. But better to lose some power than all of it.

Freezing rain or drizzle may glaze over the intake also, but you will be busy trying to see through a glazed windshield and keep a loggy airplane flying and won't have any time to appreciate the automatic features of your airplane.

MIXTURE

Occasionally an engine will run rough because of a pilot's abuse of the mixture control — but usually he's aware of his misdemeanor. The two main problems are a too-rich mixture at altitude, or descending with the mixture set for a high altitude cruise. Occasionally there will be those who descend and try to apply full power when in the best economy setting.

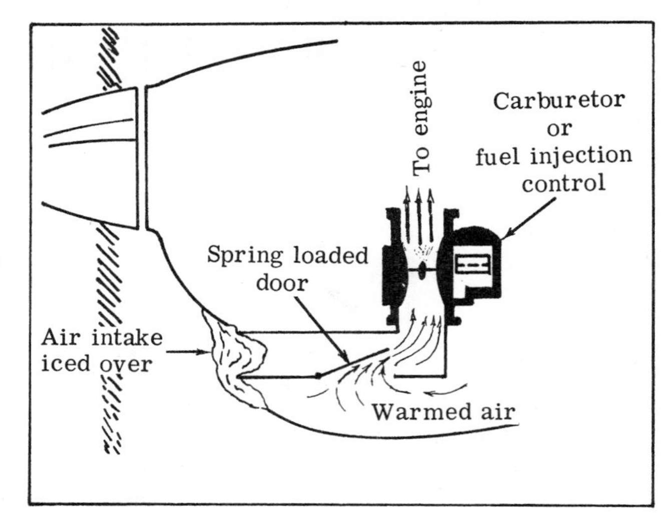

Fig. 16-1. An alternate air source.

You might check to see that the primer is in *and locked.* This is an often overlooked cause of rough running, particularly at lower power settings.

If you have carburetor ice, it is not advisable to enrich the mixture. Carburetor ice tends to cause a richer mixture. You may be heaping coals on the engine's head if you move the mixture control to full rich and throw on full carburetor heat (which further tends to enrich the mixture because the warm air has less weight per unit volume). There for a split second, you may have the richest mixture in town — and the engine may not like it.

For all other occasions of engine roughness, moving the mixture control forward is a good idea. A richer mixture means cooler running and more power, particularly at lower altitudes. You may have over-leaned the engine.

FUEL MANAGEMENT

Breathes there a soul so dead,
Who has not shook his head,
And said,
"Cripes, I forgot to switch tanks!"

There's one thing about running a tank dry — you won't get five minutes of frantic warning beforehand. In fact, if you haven't been watching the fuel pressure, your first warning will be a hiccup from the engine — followed by a loud silence. There is a no more active group than four pilots in a four-place airplane when a tank unexpectedly runs dry. The front seat occupants are both scrambling for the tank selector (getting in each other's way and tripling the required time to switch tanks) while the back seat occupants are shouting advice and maybe trying to get *their* hands into the act. Running a tank dry is a particularly effective attention getter if the three pilot-passengers are dozing when it happens. Funny as it may seem, non-pilots are not as affected by such a practice—they just sit back and quietly tremble. It's best to throttle back to prevent overspeeding of the engine when the switch is made.

Keep up with the flying time on each tank – gages can be wrong. If the engine quits abruptly or starts losing power, you may have had a leak or perhaps there's a stoppage. Don't get the idea that because the tank (or tanks) was full and the consumption time isn't up that it won't do any good to switch tanks—it's worth a try. There have been too many accident reports stating that "in the wreckage it was found that the fuel selector was on an empty tank – the other tank was full." (Admittedly, it's hard to imagine a tank remaining full after the solid impact of a crash, but that's what some reports say.)

IGNITION PROBLEMS

PLUG FOULING

The most common ignition problem on the ground and at very low power settings in the air (extended glides, etc.) is plug fouling. The bottom plugs on a horizontally opposed engine may tend to be fouled after starting, especially if the plane has been sitting for several days. Your indication is an excessive rpm drop during a mag check. This usually can be corrected by leaning the mixture at a fairly high power setting (1500-2000 rpm). Move the mixture back until the first signs of roughness appear and leave it there for 30-60 seconds. You are raising the cylinder head temperature to such an extent that the illegal oil on the plug tip is burned off. Naturally it doesn't do the engine any good to be run on the ground in this way, but it's simpler than going back and getting the plugs cleaned. Normally, fouled plugs will clean themselves out with operation, *but* it may be in the magneto and you just think it's in the plugs. If the just described treatment doesn't work, don't abuse the engine, but enrich the mixture and taxi back to the hangar to find out what's up.

This chapter is about in-flight problems. For plugs to suddenly start fouling in flight is unusual. In-flight fouling may mean serious piston ring problems and it's best to land at the nearest airport.

MAGNETO PROBLEMS

If you start getting a bum mag when you're right over the center of Gitchygoomy Swamp, there's little you can do in the way of repairs. Plug or mag problems usually are characterized by almost instantaneous rough running (or not running). The engine may run smoothly and then abruptly cut in and out. Unfortunately this instantaneous change is sometimes a characteristic of certain carburetion, as well as governor, problems. Generally speaking though, the ignition, by its very principle of operation, is more apt to cause more immediate changes in the engine characteristics.

If the engine starts running rough and you are unable to clear up the problem, start a shallow climb and head for the nearest cleared area or airport – altitude is very comforting on these occasions.

Sometimes the engine will start raising a fuss because the rotor in one of the mags has broken down and is allowing the spark plugs for that mag to fire at random. This is bad because in a sense there is a continual spark, and every time the fuel-air is injected into the cylinders it gets fired – ready or not. In effect the timing has broken down. There's a time for everything and this ain't it, as the young lady said in discouraging an ardent swain during church services.

In this situation (talking about the magneto again) you'd be better off to turn off the one mag completely and run on one – that's why you have two magnetos. If the engine really sounds bad, it might behoove you to check the mags to see if this is the problem.

LOSS OF OIL PRESSURE

The average pilot tends to neglect his scan of the engine instruments during flight. He looks at them before take-off and then averages checking them once an hour, if that often. It's quite a jarring experience to casually look at the oil pressure and find that it's gone – the hand is nestled up against ZERO and you don't know how long it's been there.

Well, take heart; for every instance of actual oil pressure loss, there are several cases where the instrument is the culprit. You won't know whether this time it's a bad instrument or not.

There's one way of telling if you're in trouble: watch the oil temperature gage (and cylinder head temperature if available). Usually a couple of minutes will tell the tale. If the temperature doesn't start up it's just the instrument and you can relax – a little. In other words, just because you have a low (or no) oil pressure doesn't mean that all is lost. But if the temperature starts a rapid rise into the red, you'd better look for a field in a single-engine plane or start feathering that engine in a twin. No oil means engine seizure in a very short while.

HIGH ALTITUDE FORCED LANDINGS

PICKING A FIELD

Here's where you wish you'd practiced more power-off approaches. Set up the max distance glide speed, use carburetor heat, switch tanks and go through the other steps previously mentioned.

Naturally you'll pick the best field available and land into the wind if possible. Maneuver so that you have a "Key Position" at a point opposite to the point of intended landing, comparable to the spot on the downwind leg where you've been starting the 180° side approaches at the airport. You are trying to turn an unusual situation into a more familiar one. The Key Position altitude should be somewhat higher above the ground than the traffic altitude you've been flying. You can "S" or spiral to reach this position, but don't give yourself such a high margin of altitude that you overshoot – you may be so high that flaps and/or slipping won't be enough. You will establish an imaginary box at the Key Position; the bottom of the box will be *at least* traffic pattern altitude and the top no more than 300 feet higher. The center of the Key Position box should be at approximately the abeam position on the downwind leg. (Fig. 16-2)

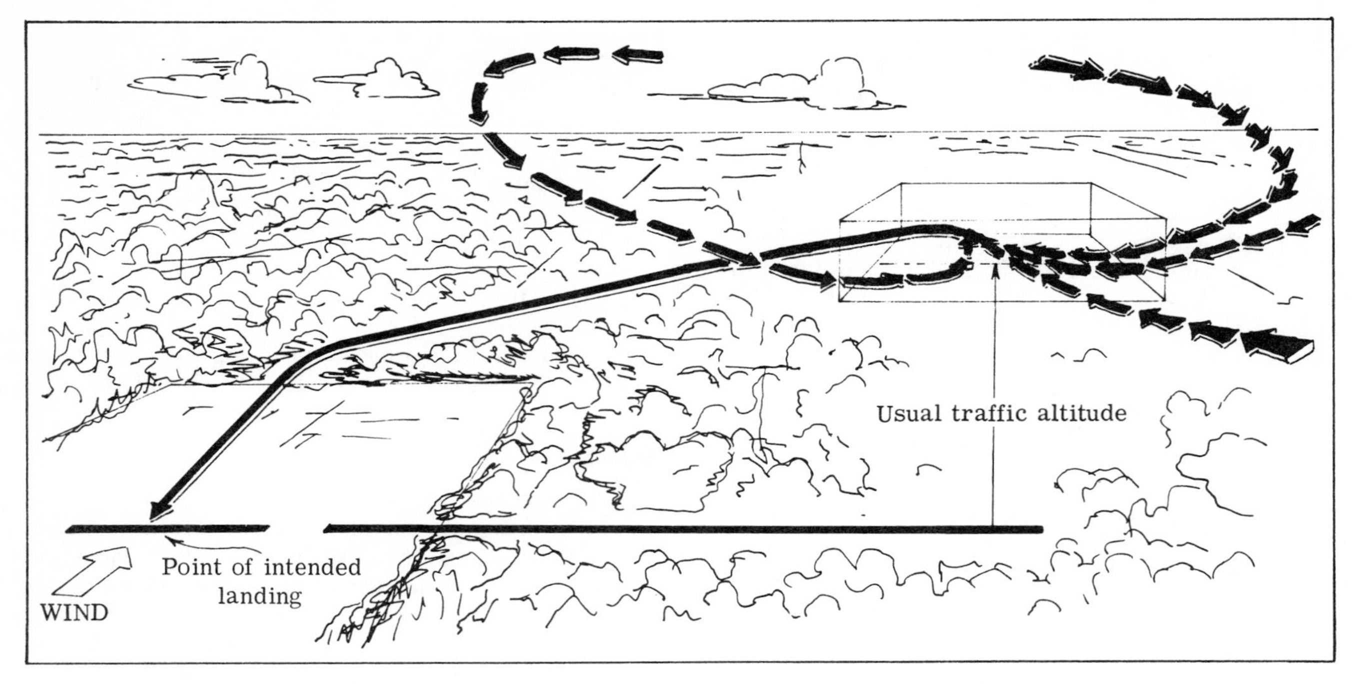

Fig. 16-2. The Key Position box.

You may have been shooting power-on approaches throughout your recent flying career — this means that your downwind and base legs may have been much farther from the field than could be allowed for a power-off approach. *This, and the fact that a windmilling or stopped prop gives more drag than you may have counted on, could result in undershooting.* If you have a controllable pitch prop pull the propeller control full back to high pitch (low rpm).

If the wind is strong the downwind leg or Key Position box should be moved in slightly, as you would do for any power-off approach.

After you've hit the Key Position, the rest is pretty well up to you. Hitting a spot is a matter of experience and practice and there are no printed "crips or gouges" to help you. The greater part of your judgment will be used for the last 90 degrees of turn into the field. If you are low on base, naturally you'll "cut across;" if high, you may "S" past the wind line. (Fig. 16-3)

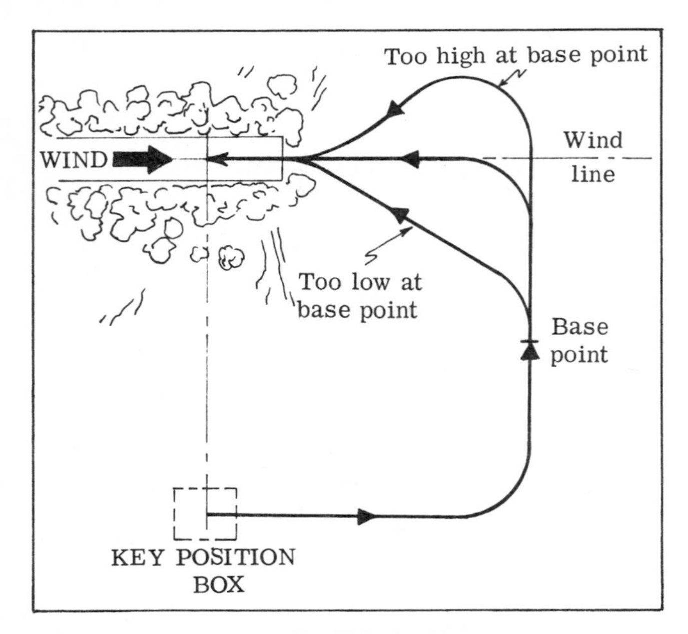

Fig. 16-3. Playing the final turn.

What you definitely do *not* want to do is get the airplane slow and wrapped up. That's the best way in the world to get into serious trouble. It's a lot better to fly into something (a rougher field or bushes) with the plane under some semblance of control than to spin at a low altitude. Many fatal accidents have occurred when a pilot, under the pressure of an actual emergency, got slow and tried to rack the plane around. Some of the higher performance planes you'll be flying are not as forgiving as the trainer you flew earlier.

There'll be times when a right-hand pattern will be better.

If you don't have shoulder harness, have your front seat passenger hold a map case or a folded coat in front of his face — just in case.

GEAR AND FLAPS

GEAR

With a fixed-gear airplane there is no decision to be made as to whether the gear should be up or down. It's down and you can like it or lump it.

For the retractable gear airplane, you'll have to decide whether it'll be gear-up or -down on the forced landing. There's a story that at one of the large military training bases in earlier years, the cadets were admonished to always land gear-up in the event of an engine failure — unless they could land on a "designated military field." This rule was to insure that the cadets did not try to land gear-down in pastures or the short civilian fields in the area. It was felt that there would be less over-all damage done by landing with the gear up. One day on a cross country a cadet had an engine failure right over the center of a busy civilian field (with eight-thousand-foot runways). You guessed it — he made a perfect approach and landed gear-up right in the middle of the busiest runway. Needless to say, the runway wasn't of much use until the airplane was dragged off. The cadet's story? "But sir, the rules: This wasn't a designated *military* field." This left a very frustrated accident board chewing their nails and trying to broaden the regulations.

You won't have any hard and fast rules to go by. If the surface is either firm and smooth or extremely rough (large rocks, stumps, etc.) you'd probably be better off to put the gear *down*. If there are stumps and large rocks, the down gear will take a large amount of the shock before being torn off. You remember that the kinetic energy of your airplane is $\frac{1}{2}MV^2$, or one-half the mass (Weight divided by 32.2) times the square of the velocity. The longer you take to stop the better off you are, even if it's a matter of a split second's difference. If you land gear-up on the rocky or stump strewn field, it's possible that the obstacles may start ripping through the belly while you're at a high speed. In such a situation forget the rest of the airplane, you'll be wanting to keep the cabin intact — leave the gear down. As has been stated about jumping off a roof, "it's not the fall that hurts, it's the sudden stop."

If you think the chosen field will be soft, keep the gear up. And naturally for water landings (ditching) the gear should be up.

Your decision for gear-up or -down will be based on what you see at the time.

If in doubt about the firmness of a field the usual decision is to leave gear up. If it is soft and the wheels sink in, high deceleration forces will result.

All the above considerations will depend on whether you are *able* to get the gear down. You should have enough residual hydraulic pressure or electric power for one cycle anyway (and you may change your mind on the way down).

FLAPS

Proper use of flaps can mean the difference between success or failure. A common error in practice emergencies is for the pilot to put down the gear and use full flaps right away — resulting in an undershoot. Taking off some of the flaps when undershooting is usually too late to do any good — plus the fact that the sudden upping of flaps causes a temporarily increased sink rate, especially at low airspeeds.

Use the flaps in this situation in increments; put a little down as you feel they are needed. If it looks as though the flaps will have to be put down by emergency means because of the no-power condition, you might be better off forgetting about them — this is no time to be fumbling around in the cockpit.

SLIPS

For airplanes without flaps, the slip is the big aid in hitting the field. If you are on a close base leg and are high, waiting until after turning final to slip may be too late — you may miss the field because the final leg will be too short to give you time to get set up for a slip.

If it looks as though you'll be crowded, a slipping turn comes in handy. It's a good way to lose altitude in the turn without picking up excessive airspeed. You're holding a touch of top rudder and a little aileron into the turn. Don't hold so much top rudder that the turn is stopped — you still want to land in the field you picked originally. (Fig. 16-4)

Fig. 16-4. The slipping turn.

Flaps and slips don't always mix well. Some airplanes are unforgiving this way as the flaps may blanket the tail surfaces in a slip. You might go to altitude and try the reaction of your airplane to a flaps-down slip and also talk with some of the local pros.

DITCHING

Plan your approach into the wind if the wind is high and seas are heavy. If the swells are heavy but the wind is light, land parallel to the swells. *If you are uncertain of wind direction remember that the white caps appear to move into the wind.* Land the way the white caps are moving (unless the swells are heavy).

Obviously the best ditching can be made with landing gear up. A fixed gear airplane is not good for ditching, but it is too late to switch airplanes. In this case you will make the impact as light as possible under the conditions by the following steps:

1. Use full flaps.
2. Set up a minimum descent glide (power-on or -off as your situation dictates).

3. *Don't* try to second-guess and flare when you think it's time. You may level off too high and drop in at a nose-down attitude — this will generally insure an unsuccessful ditching. Also if the tail is too low on impact, the result may be a pitching forward and digging in. It is very hard to judge altitude over water, particularly in a slick sea.

4. Make sure all the passengers' (and your) seat belts are snug. If the airplane is equipped with shoulder harness so much the better. It would be wise to have your front seat passenger cover his face with a folded coat if there are no shoulder harnesses available — and you might do the same at the last second.

Expect more than one impact shock. The airplane may skip once or more before the final hard shock. You will swear that the airplane has gone straight to the bottom as there will be nothing but spray visible for several seconds. In witnessing a ditching from the air, the airplane may be completely hidden in spray for a few seconds (you, of course, won't be particularly interested in how it looks from the air at this time).

As soon as all forward motion has stopped, leave the airplane (it is assumed that you planned on flying over water and have flotation gear — or that you are a strong swimmer). The length of time that the airplane will float depends on how empty the tanks are and how much the airplane was damaged.

It's important that the wings be level when you hit the water — this is no time to be cartwheeling.

It is very important that you get on the radio as soon as you know you'll have to ditch. Broadcast "MAYDAY" and your position on "Emergency" (121.5 MHz). You might also transmit on communications frequencies (122.1, 122.6 or 123.6 MHz) because you seldom transmit on 121.5mc and who knows if it's working or not. A wise move, if you plan on flying an extended overwater trip, is to make sure that "Emergency" and other important transmitter frequencies are working, and above all — FILE A FLIGHT PLAN.

The best time to consider emergency procedures is *before* the flight.

PRECAUTIONARY LANDINGS

In the event of imminent fuel exhaustion, fast deteriorating weather or engine problems it may be necessary to make an off-airport landing. It's always best, naturally, to be able to pick your spot while the picking is good. You should know what to do when you'll have to land in the hinterlands and the technique of "dragging the area" is valuable for this situation.

Basically, it consists of picking a field that appears to be able to take your airplane (gear down, if possible) and flying a normal pattern and approach to the area of the field on which you plan to land. Add power and level off at about 100 feet altitude on final and fly to the right side and parallel to the landing area. Look over the landing area for holes, ditches, field condition (soft, high grass, etc.), and in general look for items that couldn't have been seen from altitude. If everything looks O.K. then open the throttle, climb out and make a pattern with a short field approach and landing. If you see that the field isn't for you on the first pass, then repeat it at another one.

During the approach and pass watch for wires or other obstructions of that nature. You can't always see the wires but if there are poles you'll know which way they are running. In the case of wires look for other poles well to both sides of the pole line; there may be wire take-offs that would cause you trouble if you were landing parallel to the known wires.

Your decision whether to land gear-up or down will depend on the points mentioned earlier in the chapter. You'll have to decide whether things have gone so far that you may be better off to belly it into a fairly bad field rather than take the time to find another one. (You may figure you have only a couple of minutes fuel left or it looks as if it may only be that long before the weather goes to zero-zero and you're practically at rock bottom altitude now.)

If you're not going to be able to make it to an airport it's better to land at a field of *your* choice.

After you're down, get to a telephone and let the proper offices or people know of your plight.

17. LOW-ALTITUDE PROBLEMS

ENGINE FAILURE ON TAKE-OFF (SINGLE-ENGINE AIRPLANE)

This has been discussed in many ways throughout this book and during your flying career, but one thing will be repeated. *Don't try to turn back!* You can make a minor turn but keep it shallow. If you have time, cut the mixture, ignition and fuel. Have the airplane as slow as possible when you touch down. If you are going into trees, don't stall it out above them but fly into them at the lowest possible speed, still having control. Fly between two trees if possible so that the wings will help take the impact forces as they are carried away.

If the engine starts running rough after passing the airport boundary (or any point where it's too late to try to land again) fly straight ahead (if the terrain and populated area allow) and try to gain altitude without getting too slow. The engine may quit at any time and at least you're headed more or less into the wind. *Ease* your way back to the airport.

THE OPEN DOOR

A door opening suddenly during take-off or in flight is quite an experience. There is a loud bang as it opens and the noise of the air moving past the crack is enough to set up a good case of combat fatigue in a short while. In addition to the sundry noises associated with such a problem the airplane may have tail buffet or the wing on that side may tend to drop because of the disturbed airflow.

The usual situation is that the pilot has forgotten to lock the door (it should be on the checklist) and no problem occurs until the airspeed is such that the drop in pressure caused by the air moving past the door is enough to pull it from the latching mechanism – then the fun begins. It seems that the airspeed required for this is that amount found just after lift-off on a short airport in turbulent air with obstacles ahead when the airplane is at gross Weight and your nervous maiden aunt is sitting by the door; this seems to be the usual setting for a door opening episode.

If the door opens on the take-off run and there is room to stop, naturally this is the thing to do.

If it opens after lift-off and there isn't room left to land keep *full* power on climb at the normal climb airspeed. In most cases performance will suffer but the airplane can be flown. Fly a normal pattern (don't wrack it around), make your approach at a slightly faster airspeed and carry the airplane closer to the ground before starting the transition – more or less fly it on. As was mentioned, the wing on that side may tend to drop out early. It would help if your passenger would hold the door as closely shut as possible to minimize the effects.

Your biggest problem will be plain old-fashioned fear. It's a nerve-shattering occurrence and between the noise and the buffeting, plus possibly yells from distraught passengers you could be fatally distracted. Fatalities have occurred in airplanes that were perfectly capable of flying with the door open, when the pilot tried to cut corners and spun in.

If you're at altitude and your Aunt Minerva catches her knitting bag on the handle and suddenly presents you with fresh air in copious quantities, you may be able to shut the door in flight.

Generally, the procedure will be to throttle back to idle, slow and trim the airplane to a speed just above the stall. Open the small storm window on your (the pilot's) side to help equalize the pressure and shut the door – and lock it. On some airplanes the storm window can be held open with your left elbow while holding the control wheel with your left hand. You can reach over and shut the door with your right hand if you're by yourself. Needless to say, these gymnastics could be hairy at lower altitudes.

There may be different techniques for your airplane but the main idea of this section is to bring this possible problem to your attention. Check on your airplane now with other pilots and/or the Airplane Flight Manual and go over in your mind what to do in such an event. Better to set up a plan now than to have to work something out in all that noise and confusion.

GUSTY AIR, TURBULENCE AND GRADIENT WINDS

Wind velocities and directions will vary with altitude – this is expected – but what's not always expected is the fact that the wind may change velocity and/or direction with only a small altitude change. On a gusty day the wind velocity may change almost instantly and you might be in a tight spot if you're flying the airplane too close to the stall.

A problem that should be watched closely is the effect on the wind of obstacles such as trees or buildings on the windward side of the runway. (Fig. 17-1)

Even if the wind is steady and you've been holding the same crosswind correction all the way down on final, things can go to pot pretty fast when you get close to the ground. Not only may obstacles cause trouble, but the wind itself could change velocity

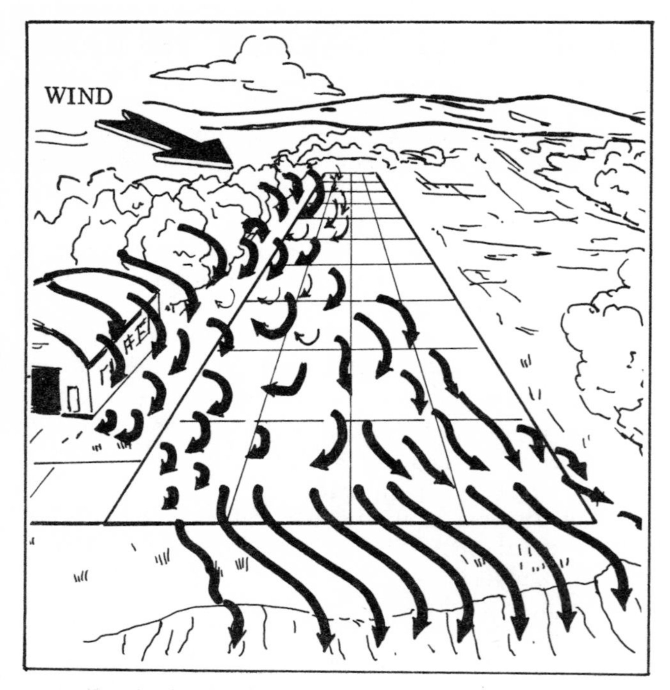

Fig. 17-1. The effects of obstacles on the wind.

abruptly. The airplane feels as if the bottom were dropping out and the air may be extremely choppy. Added power (and fast!) usually is necessary to keep things under control.

There may be a time when you are racing a thunderstorm to the field so remember something here, too. The wind may shift abruptly; it can be very strong one way and when you get all set up for landing, you find that you are now trying to land cross-downwind. Keep a close eye on the wind indicator on final when there are thunderstorms in the vicinity or a cold front passage is expected momentarily.

WAKE TURBULENCE

You may not have had the dubious pleasure of encountering the above-named menace but chances are good that sometime you will.

Naturally, the most critical place to fly into wake turbulence is when you are at a low altitude and the most likely place for trouble is on take-off or landing. Give the big planes plenty of room; take it around again or ask for another runway if you think there might be wake turbulence hanging around.

The turbulence may be around for several minutes after the instigator has made its get-away.

If there is a crosswind, always take off or land on the upwind side of the runway. After taking off, stay off to one side of the big airplane that's causing the problem. (Fig. 17-2)

If the wind is straight down the runway or calm and you *must* take off, then the following steps would be best: (1) don't lift the airplane off until you have plenty of flying speed — hold it on a little longer than usual but plan to get off before the point of the big plane's lift off, since its vortices will begin at the point of rotation for lift-off and (2) after lifting

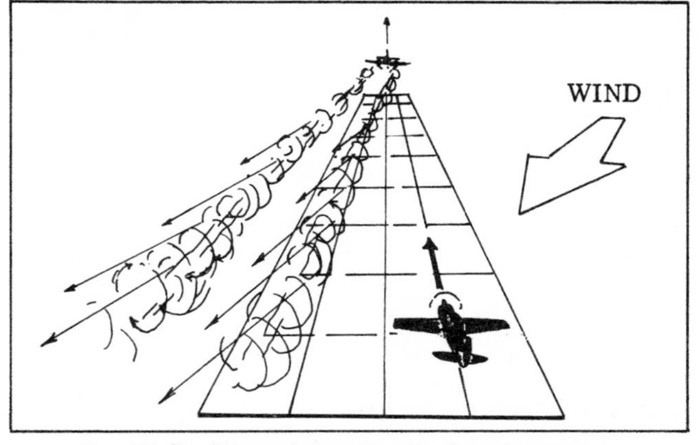

Fig. 17-2. Always take off or land on the upwind side of the runway if there is a crosswind and you are close behind another airplane.

off make a shallow turn to one side and then turn back to parallel the runway. (Fig. 17-3)

Even this idea of clearing the runway will be dangerous if you take off directly behind the other plane with no waiting time.

Though you will get out of his way as soon as possible there will be a period when you will be in the turbulence. Whether you'll have airspeed enough to get through it safely is a matter for conjecture — so it's best to wait.

If you are landing behind another plane and the wind is directly down the runway or calm, you'd be better off to land past the spot he touched down, making a slightly steeper approach than usual.

Of course, if he overshoots and lands so far down the runway that reversible pitch props were the only thing that saved him, you'd better take it around rather than land long.

The best thing is to avoid wake turbulence but

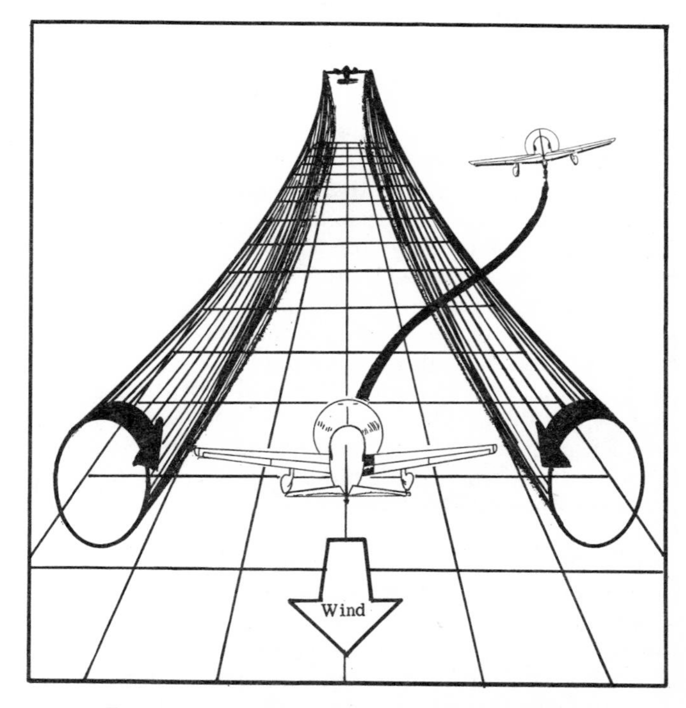

Fig. 17-3. Clearing wake turbulence on take-off.

you may be caught unprepared. The most exciting place to get caught is during take-off or landing; you're low and slow and don't have ideal control response.

WAKE TURBULENCE ON TAKE-OFF OR LANDING

You've just lifted off and suddenly turbulence makes the airplane go berserk; it rolls violently, and opposite aileron and rudder have little effect in stopping it. In addition to the roll the turbulence may try to pick the airplane up or slam it back on the ground.

Keep full power on and do not try to climb up through it. Keep the nose down. You need airspeed. If you are slow when hitting the disturbance you'll make more trouble by hauling back on the wheel. Don't let it put you into the ground (if you can help it) but don't get to thinking that you can pop up through it either — you might end up on your back at twenty feet and get socked with a violation of Federal Aviation Regulations for putting on an airshow without permission. Besides you wouldn't want to land this way as there are no wheels on the top of the airplane.

When the airplane rolls, try to stop the bank from becoming too steep but don't attempt to completely level the wings. A turn is needed right now and some bank will help you get out of the choppy area. Turn in the direction that the turbulence tends to roll you.

You may cut some fancy capers at a low altitude and end up flying 45° from your take-off heading for a few seconds but you want to get into smooth air — and stay there. After doing all sorts of graceful and not so graceful rolling and maneuvering right off the ground because of wake turbulence one day, a private pilot was asked by the tower if he were in trouble. His answer remains a classic in radio communication and resulted in a violation filed by the FCC for "improper language over the air." Tower operators realize that wake turbulence is a definite menace to light planes and warn pilots if there is a chance of it during take-off or landing.

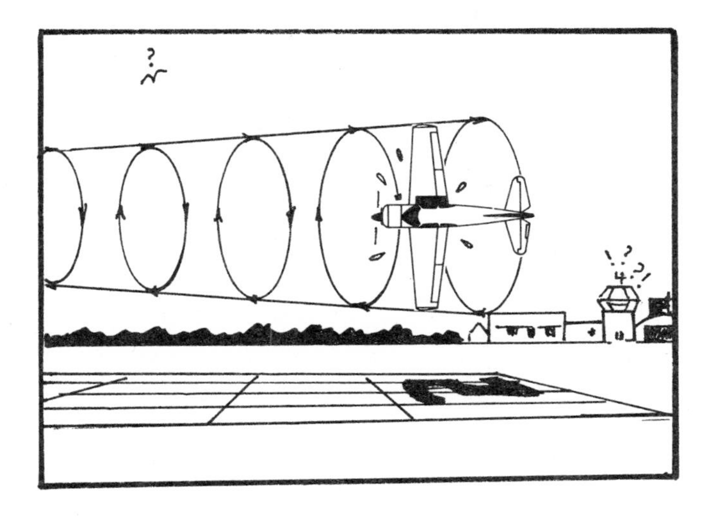

Fig. 17-4. Putting on an impromptu airshow can be a dangerous procedure.

On take-off or landing, airspeed is insurance when wake turbulence is encountered. You may have to deliberately fly a few feet off the ground for a few seconds until clear of the disturbance.

Wake turbulence is unpredictable; the airplane's reaction will depend on how you fly into it. As seen from behind, its motion is rotational but you may fly into only one side of the vortex.

If turbulence is encountered on approach, the best answer is full power. Keep the nose down, turn out of the area and go around.

If you suspect there will be wake turbulence during an approach and landing, keep the airspeed up and literally fly the airplane on. *Better yet, if you suspect that there'll be turbulence on approach or landing, take it around — always.* When you take off or land into a known area of turbulence you're betting a large repair bill against a couple of minutes saved. Anytime you think there may be wake turbulence on landing you can request another runway — a crosswind usually is much easier to cope with.

Wingtip vortices have a direct tie-in with induced Drag. You remember that induced Drag was a function of a high Coefficient of Lift. This means that induced Drag is greatest at low airspeeds (high angles of attack). Wingtip vortices are the worst for an airplane with a high span loading (pounds of weight per foot of span) and at low speeds with flaps down. Because Fowler flaps give a large increase in Coefficient of Lift when down, wingtip vortices are strongest for airplanes so equipped; all other things being equal.

Avoid flying into wake turbulence at high speeds -- stresses may be imposed on the airplane to such an extent that structural failure could occur. Fly so as to avoid crossing directly behind a large airplane. If you should encounter wake turbulence at altitude, *don't pull up sharply to climb out of it.* Throttle back but don't try to slow up too quickly as you will be adding stress to the airplane. The most aggressive maneuver you should try would be a *shallow* climbing turn. If you are crossing the turbulence at a 90° angle, you'll probably be through it before you have a chance to do anything.

You can see that at altitude the greatest danger is found at high speeds where the sudden encountering of turbulence would result in overstressing the airplane. On take-off and landing the problem is that the airspeed is low and control may be marginal.

If you see that you are going to cross directly behind a large airplane, if possible go above him (this is assuming that you have time to make up your mind) because (1) the downwash of the wing tends to carry the disturbances downward and (2) you'll be slowing up as you start to climb. (Fig. 17-5)

There's no need of your adding further stress on the airplane by sharp pull-ups or other radical departures from the norm.

PROPERTIES OF WAKE TURBULENCE

Wake turbulence moves downward. You may encounter flows in that direction at rates up to 1500 feet per minute. If your plane has a maximum rate of climb of 1000 feet per minute, you can expect to be descending at a minimum rate of 500 feet per minute. Some of the flow in a vortex can produce a roll rate of approximately 80 degrees per second, which is twice as much as the *maximum* roll rate of some general aviation airplanes.

Vortices generated more than 100 feet above the surface will drop nearly vertically for about one-half wing span of the generating aircraft and then spread out.

With a light crosswind, it's quite possible that one of the vortices is kept over the same position relative to the runway. (The downwind vortex will be moved even faster laterally.)

Wake turbulence can be a menace to your health.

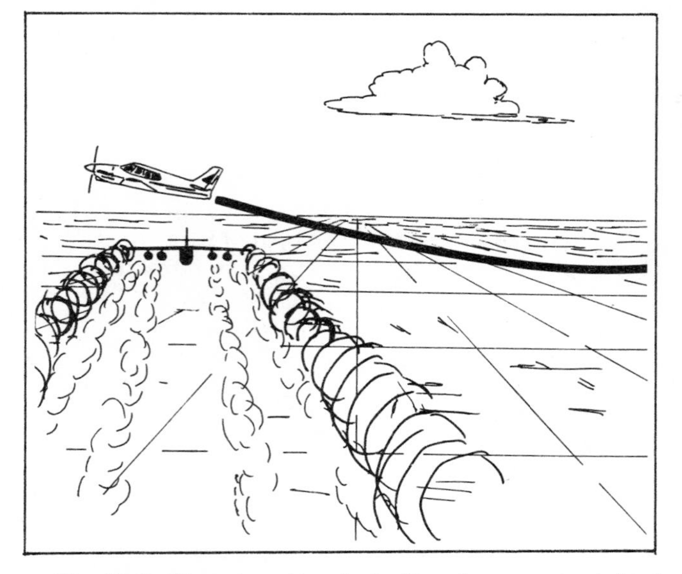

Fig. 17-5. If you have time to decide, when crossing behind a large airplane always go above him.

18. EMERGENCY FLYING BY REFERENCE TO INSTRUMENTS

INTRODUCTION

This may be your introduction to the art of instrument flying. As you know, the Federal Aviation Agency requires that applicants for a private or commercial certificate must demonstrate the ability to recover from unusual situations caused by inadvertently flying into instrument conditions (notice the *inadvertently*). The best tactic is a 180° turn before getting into weather.

One of the arguments against the ruling was that there was a possibility that pilots might become overconfident of their ability because of this training. Whereas the pilot without any knowledge of instrument flying might turn back immediately, it was felt that some pilots with a *little* knowledge might push on "just a little farther to see if it won't get better."

If you received your private certificate after mid-1960, much of this chapter may be a repeat for you. It covers the instrument requirements for a commercial certificate and serves as an aid in getting a Blue Seal Certificate. Even if you have no plans of getting a commercial or a Blue Seal, it might be wise to get some dual under the hood on these maneuvers -- it's life insurance.

The whole principle of flying under the hood is for you to see what the airplane is doing by understanding the indications of the instruments.

You are required to have ten hours of instrument instruction and to demonstrate on the commercial flight test your ability to fly the airplane by sole reference to the instruments.

It's been found that pilots learn to fly by reference to instruments more easily if the training is started early. In fact, the FAA recommends that the student pilot be given instrument training beginning with the Four Fundamentals. Instructors are finding that this is working very well. No pilot can safely fly under instrument conditions without proper instrumentation *and* training. That word "safely" is an understatement. Your chances of keeping the airplane in the air by "feel" are about as good as those of a revenuer being greeted with open arms by moonshiners.

Vertigo occurs when you don't have visual references. Your body gives false signals and you are sure that the airplane is doing differently than is indicated by the instruments. You will be using the training in this chapter as a life-saving procedure, and shouldn't be in the clouds or fog for more than a very few minutes. *But,* you *can* get vertigo. It will take a great deal of courage and thought, but *trust those instruments -- not your own "feel" of what's going on.*

Pressure on the pilot is the real hazard to safety in flying when the weather is marginal. Probably your biggest problem if you do charter flying will be pressure from the passengers. They're paying you and by gad they have to be in Jonesville by two o'clock. You'll have to make up *your* mind about the weather. The guy who lets the passengers decide for him has absolutely no business in the pilot's seat of an airplane. Every year's accident reports have a long list of fatalities caused by being "pushed" into flying bad weather with no training.

Practice of the following maneuvers is aimed at saving your neck and in no way should be construed as a means of pushing farther into bad weather before acting. If you go busting into a control zone or control area in instrument conditions without clearance, you may have several months of sitting on the ground to consider your sins.

The only additional instruments required for the airplane will be a turn and slip and sensitive altimeter but most planes have these instruments anyway. It's most likely that your plane also has a gyro horizon and a directional gyro.

As far as the use of the turn and slip is concerned, the idea is to use coordinated controls. If for some reason you have to think of the needle and ball as being separate, control the ball with the rudder. Considering the needle displacement (the airplane's rate of turn) to be a function of the angle of bank, the ailerons are the primary control for varying the bank. So, if you must think of needle and ball as being separate, this method is most logical, as the ball moves parallel to a plane passing through the rudder pedals and the needle moves just as the stick or wheel is moved. (More stick or wheel means more needle after any initial adverse yaw.)

Going back to the earlier discussion of the climb, the airplane's altitude, rate of climb or descent is controlled by the throttle, and the airspeed by the elevators. Actually, the throttles and elevators should always be thought of as working together -- just as you use them under visual conditions.

The following maneuvers must be demonstrated on the commercial flight test to the FAA inspector or designated flight examiner to show him that you are able to manually control an airplane depending solely on the flight instrument indications. If you can do these, you'll certainly have no trouble getting a Blue Seal for your private certificate.

1. *Recovery from a well-developed power-on moderate turn spiral in a medium-banked attitude.*

2. *Recovery from a high-angle climb in a turn.* (A high-angle climb is one that if allowed to continue another 30 seconds at cruising power would result in stalling the aircarft.)

3. *Standard rate turns* of 180^{0} and 360^{0} duration to within $\pm 10^{0}$ and $\pm 20^{0}$ respectively, of proper heading, and within ± 150 feet of altitude.

4. *Maximum safe performance climbing turns of 180^{0} duration followed by continued straight climb to predetermined altitude* requiring not less than 1 minute straight climb performed within ± 10 knots of airspeed and $\pm 10^{0}$ of proper heading.

5. *Two consecutive descending 90^{0} turns using normal approach power for reducing altitude* performed within ± 10 knots of airspeed and $\pm 10^{0}$ of proper heading. At completion of first 90^{0} turn continue straight descent for 1 minute. Complete second 90^{0} descending turn and continue straight descent for $1\frac{1}{2}$ minutes.

6. *Straight and level flight* performed within $\pm 10^{0}$ of proper heading, 100 feet of altitude and 10 knots of airspeed.

Test programs have shown that it's usually harder for the experienced pilot to get the instrument idea right away because he's suddenly in a completely new element. He may have been flying cross country so long that he's forgotten that in a diving spiral it doesn't do any good to pull back on the wheel — the wings have to be leveled first. (Remember the 720^{0} power turns you did as a student?)

The best way to approach the problem is to start out flying *both* visually and by reference to instruments. The instructor will probably have you fly a brief period of the Four Fundamentals (climbs, glides, turns and straight and level) and you can note the reactions of the instruments to the various attitudes and airspeeds of the airplane. Later you'll try the required maneuvers in the same manner; finally you'll use a hood and do all your practicing "on the gages."

The main reason for emphasis on the turn and bank is that this instrument will be working long after the gyro horizon has been tumbled by your erratic maneuvering. If your airplane has a gyro horizon, by all means use it, but it's possible that the check pilot will be interested in seeing how you can operate with it tumbled. Figures 18-1 through 18-6 are a review of the Four Fundamentals, plus climbing and gliding turns, as would be indicated by the instruments. You know from using a magnetic compass that it is erratic and inaccurate at steep angles of bank, speed changes and at different nose attitudes. Therefore the compass will be used in the illustrations only to show the idea of the turn. You might review Chapter 3.

THE POWER-ON SPIRAL

This is the maneuver that pilots who cannot fly instruments end up doing after losing visual references. Some of them postpone the inevitable by doing

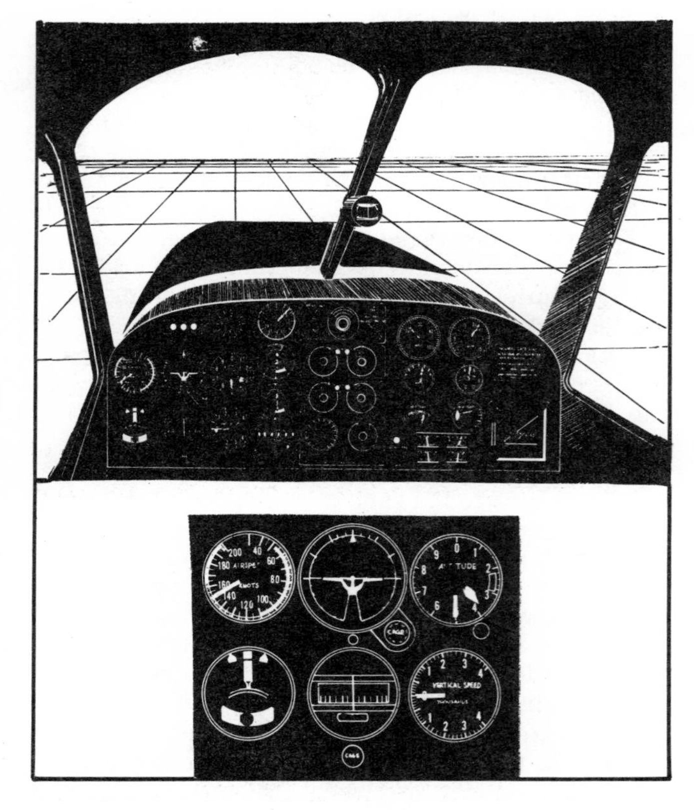

Fig. 18-1. Straight and level.

Fig. 18-2. Level turn.

climbing stalls and other colorful maneuvers first. But nearly every case of flying into the soup without proper instrument training or instrumentation will lead (sooner or later) to a power-on spiral. Fig. 18-7 shows the reactions of the instruments to a power-on spiral.

The basic instruments would indicate as follows in the power-on spiral: (Fig. 18-7)

1. Turn and slip — needle shows a great rate of turn, the ball may or may not be centered.
2. Airspeed -- high and increasing.
3. Altimeter -- showing a rapid loss in altitude.

RECOVERY

1. *Throttle* retarded immediately to keep the prop from overspeeding (fixed pitch prop only).
2. *Center the needle and ball* through coordinated use of ailerons and rudder. If the needle is to the right, apply left aileron and rudder until the needle is centered, then neutralize. This coordination is necessary. If you think of using rudder alone to center the needle as is sometimes advocated, stresses may be put on the airplane and recovery actually delayed as compared to the use of coordinated controls.
3. *Check the airspeed.* As you roll out of the bank apply back pressure to stop the airspeed increase. As you can see by doing this under visual conditions, the airspeed will start to decrease if at a steady value, or will *stop increasing* when the airplane nose is approximately in the level attitude. When this is indicated, *ease* on forward pressure so that the nose is not pulled up into a climb or stall attitude. The nose will tend to continue upward past the level flight position because of the excess airspeed and the forward pressure is needed to stop this. If you hold back pressure until the cruise airspeed is reached you may be in a steep climb. (Fig. 18-8)
4. *Stop the altimeter.* This is the finer adjustment to be made immediately after getting the airplane to the approximately level flight attitude by the airspeed indicator change. Pick some altitude close to the one being indicated and fly the altimeter with the elevators. As the altitude starts to settle down, increase the power to cruise. Continue to keep the needle and ball centered and fly the altimeter. If the needle and ball are centered and you are maintaining a constant altitude and the power is set for cruise, the airspeed will soon settle down to cruise value.
5. Maintain the instrument scan throughout the recovery. Keep that needle and ball centered!
6. When the airplane is under control make adjustments in power, altitude and heading as needed. (Timed turns will be covered later.)

RECOVERY FROM A HIGH-ANGLE CLIMB IN A TURN

Although the spiral dive is the most common result of the loss of visual references, some pilots may be individualistic and decide to stall the airplane first — or the high-angle climb may be a result of doping off and letting the nose rise too high in a spiral recovery. (Fig. 18-9)

Fig. 18-3. Normal climb.

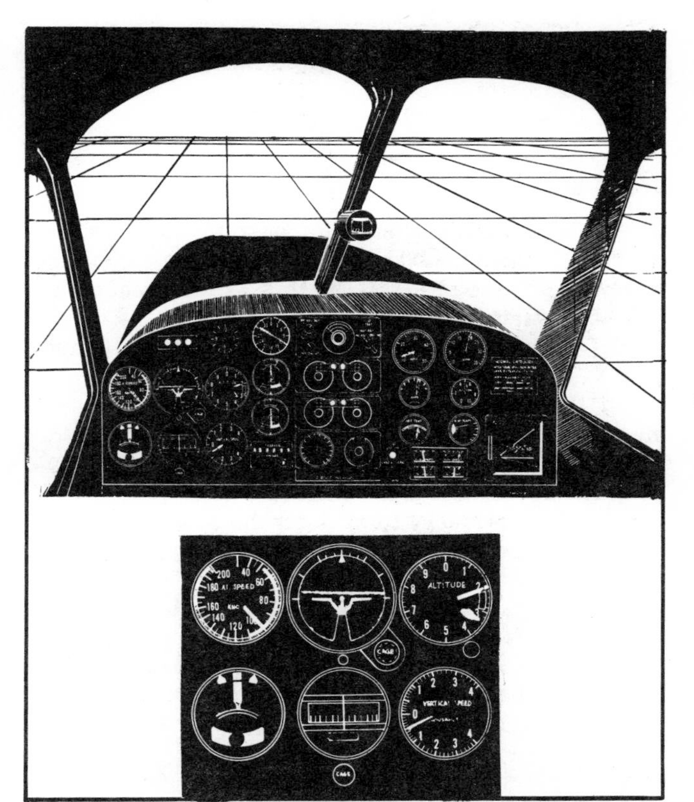

Fig. 18-4. Normal glide or descent.

Fig. 18-5. Climbing turn.

Fig. 18-6. Gliding turn.

The indications of a high-angle climb in a turn are as follows:

1. Needle and ball — needle shows a turn, the ball is apt to be to the right because of torque effects as the climb progresses.
2. Airspeed — decreasing (probably rapidly).
3. Altimeter — increasing altitude, or stationary as stall is approached.

RECOVERY

1. Relax back pressure (or ease forward if necessary) until the airspeed starts to increase. Increase power if close to the stall, as this will decrease altitude loss during the recovery.
2. As the nose is lowered, center the needle and ball.
3. Use the altimeter to level off. Lose a minimum of altitude but don't get a secondary stall.
4. Maintain a constant scan during the recovery. Keep the needle and ball centered.
5. After recovery to cruising flight is made, make power, altitude, and heading adjustments as needed.

Make sure that the airspeed is definitely increasing before trying to stop the altimeter — this is no time to be practicing secondary stalls. The first and most important thing is to get the nose down; you can worry about the heading later.

STANDARD RATE TURNS OF 180 AND 360 DEGREES

The standard rate turn is 3° per second for your type of airplane. This is noted as either a one needle-width turn or a two needle-width turn on the turn and slip indicator. You can find out which type of turn and slip your airplane has by setting up a one needle-width turn and holding it for one minute. If the turn and bank is calibrated to give 3° of turn per second at a one needle-width deflection, naturally you'll have turned 180° (60 seconds x 3 = 180°). If you've just turned about 90° in that time after having religiously kept the needle-width deflection, your turn and slip will require a double needle-width deflection in order to get the required 3° per second turn. Of course, the time to find out which type of needle you have is on a bright sunny day over the airport — not after you've flown into the murk and lost visual references.

Another way of checking your needle is by knowing that as a rule of thumb, dividing your airspeed (mph) by 10 and adding 5 to it will give the required bank for a standard rate (3° per second) turn. For instance you're indicating 160 mph, the required angle of bank would be $\frac{160}{10} = 16$; $16 + 5 = 21^{\circ}$ of bank if the turn is balanced. You would set up about 20° of bank visually or on the gyro horizon and look at the needle. You'll be able to readily check whether one or two needle widths will be required for a standard rate turn. Usually the instruments requiring a double needle for a standard rate turn are marked "Four minute turn," meaning that if you

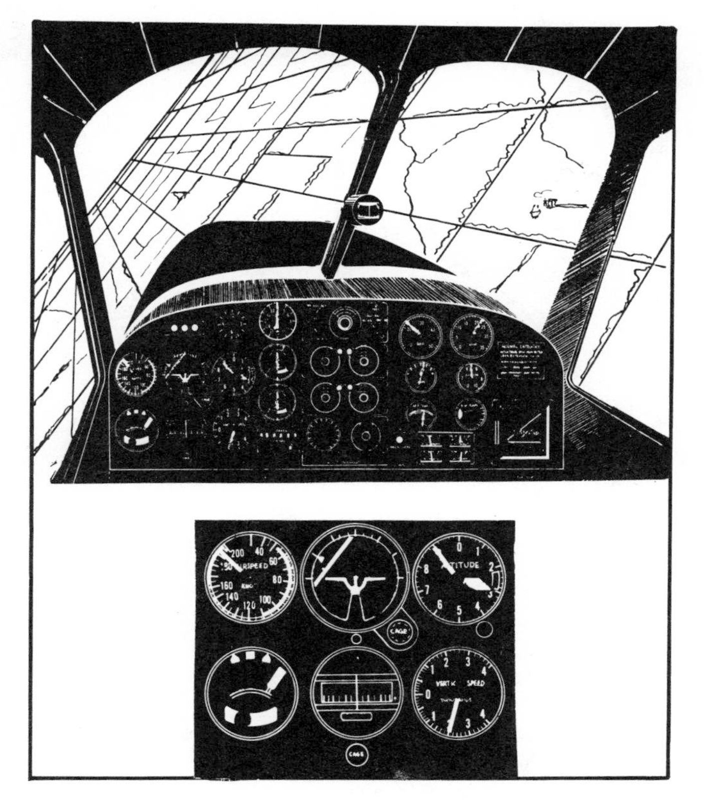

Fig. 18-7. Power-on spiral.

hold a single needle, four minutes will be required to complete a 360° turn — instead of the usual two minutes. You might review Chapter 3 if some confusion results from these last few paragraphs.

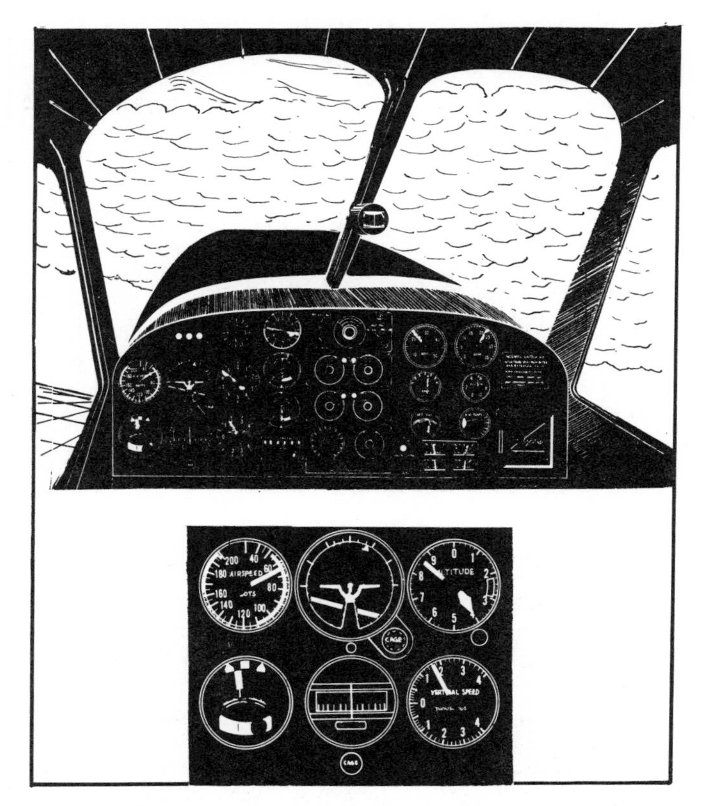

Fig. 18-9. The high-angle climb in a turn.

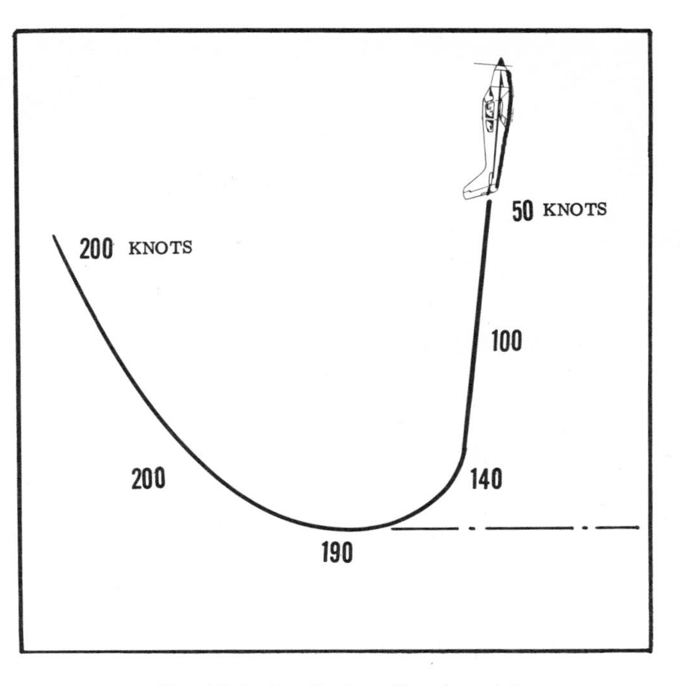

Fig. 18-8. Don't chase the airspeed.

TIMED TURN—RECOMMENDED METHOD

Knowing that the standard rate turn is 3° per second you can turn any number of degrees by dividing the degrees to be turned by 3 and holding a standard rate turn on the needle for that number of seconds.

For instance, 60 seconds is required for a 180° turn. You could glance at the position of the sweep second hand on the clock as you set up a standard rate turn. Hold this until the second hand is about three seconds short of this point again and center the needle with coordinated controls. You'll be very close to being 180° from the original heading. A word of caution: instrument trainees have been known to go into a non-permeable daze and let the second hand go around *twice* before rolling out. While the *360°* turn is sometimes a useful maneuver, it is hardly a means of reversing course. Include the clock in your scan of the instruments — a minute can either be very short or very long, depending on your outlook.

If your airplane doesn't have a clock with a sweep second hand, you may have to use your watch — which won't help your instrument scan.

The perfect answer is to have a directional gyro, then you can set up a standard rate turn and start your roll-out about 10° before reaching the desired heading — then a timepiece is not needed. The next section will discuss what can be done if you have no clock and only a magnetic compass as a heading indicator.

Back to the use of the needle and ball (or turn and slip — whichever name you prefer): in bumpy air the needle will oscillate and you'll have to fly the average position. Don't let the turn get over double standard-rate at any time. As the bank increases, your chances of entering a spiral are increased. If you start losing altitude rapidly your only choice is to center the needle, regain the lost footage and then resume the turn. You'll make points with the check pilot if you use the spiral recovery technique when it's needed. Some instrument trainees do perfect jobs when asked to perform a recovery from a diving spiral. But when in the course of other maneuvers they accidentally get into one, they just sit there fat, dumb and happy. This leads the check pilot to believe that the trainee might do the same thing under actual instrument conditions and it would be impossible to pass an applicant if it was felt that he might be unsafe.

Under actual conditions, if you accidentally fly into bad weather the important heading is the reciprocal of your original course. Don't just automatically set up a 180° turn because you may have turned after entering the clouds. Fig. 18-10 shows what might happen if the pilot thinks only in terms of a 180° turn.

As shown by the diagram, the pilot ends up flying farther into the weather.

USING THE FOUR MAIN DIRECTIONS

It was stated in Chapter 3 that in a shallow banked turn the compass is fairly accurate on the headings of East or West and lags by about 30° as the plane passes the heading of North and leads by about 30° as the plane passes the heading of South. The exact lag and lead will have to be checked for your situation which will include latitude and other variables. For illustration purposes here it is assumed to be 30°. This "northerly turning error" affects the compass while the plane is turning.

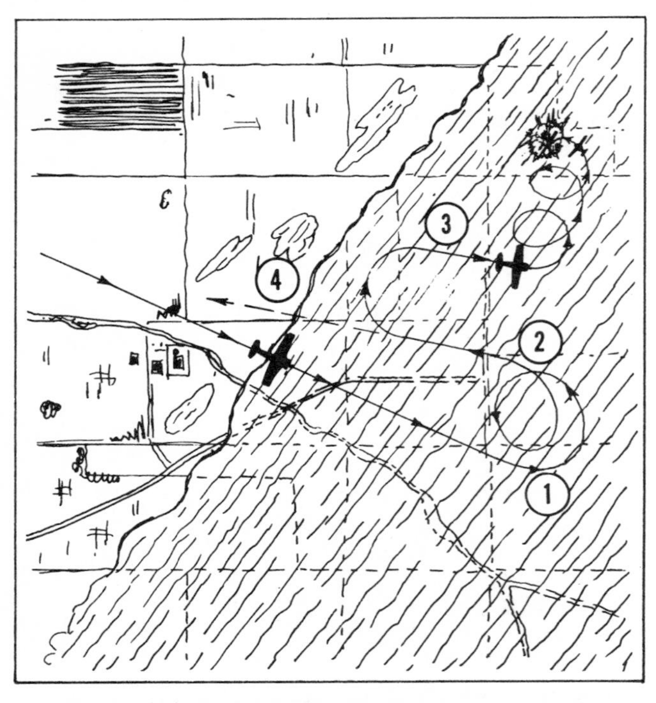

Fig. 18-10. The pilot inadvertently flies into bad weather and gets into a power-on spiral (1). His recovery is effected on a heading that would get him out of the weather in a short while (2). He automatically sets up a 180° turn without checking the compass (3). If he had held his recovery heading he would have been back in good weather in a short while (4).

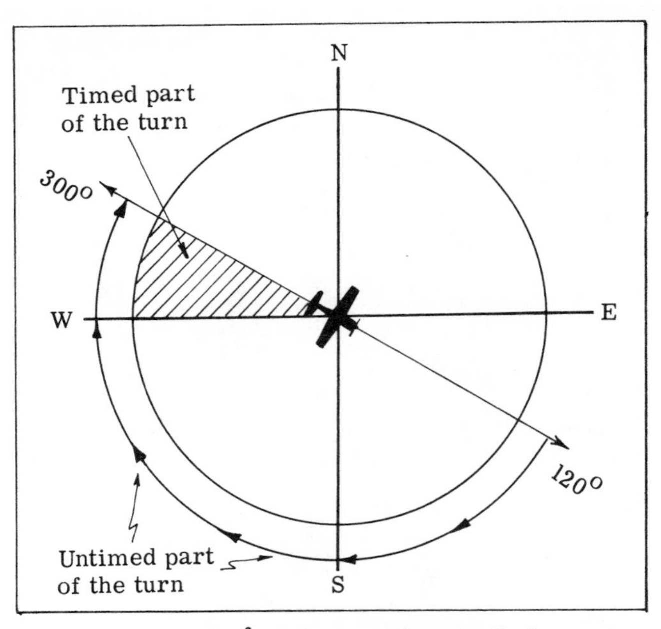

Fig. 18-11. A 180° turn to the right using the four main directions method.

Assume that you are flying on a heading of 120° and want to make a 180° turn. The desired new heading will be 300°. By turning to the right you will have a cardinal heading (West) reasonably close to the new heading. A standard rate turn is made to the right but no timing is attempted. You will be watching for West on the compass, and as the plane reaches that heading you will start timing. The desired heading is 30° (300° minus 270°) past this, and will require ten seconds at the standard rate. The timing can be done by counting "one thousand and one, one thousand and two" and so forth, up to ten, at which time the needle and ball are centered. (Fig. 18-11)

A turn to the left could have been made realizing that the nearest major compass heading (North) will be 60° or twenty seconds short of the desired heading of 300°. In that case you will set up a standard rate turn to the left and will not start timing until the compass indicates 030°. Remember that the compass will lag on a turn through North and will be behind the actual heading by about 30°. (Fig. 18-12)

When the 030° indication is given, the twenty second timing begins, either by the sweep second hand of the aircraft clock, your watch, or by counting.

There are several disadvantages to this system, the major one being that it requires a visual picture of the airplane's present and proposed heading and mental calculations are required. You may not have time or may be too excited for a mental exercise at this point. Also, in bumpy weather the compass may not give accurate readings on the four main headings.

The main advantage is that reasonably accurate turns may be made without a timepiece as the new heading will never be more than 45°, or fifteen seconds, away from a major heading. This method then

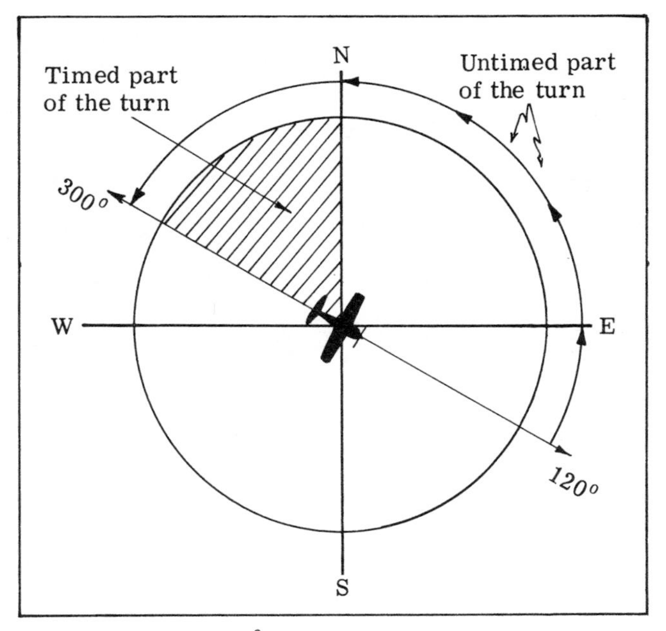

Fig. 18-12. A 180° turn to the left using the four main directions method.

is presented as something to be filed away and used if necessary.

MAXIMUM SAFE PERFORMANCE CLIMBING TURNS

This is a valuable maneuver when you've suddenly lost visual contact at a low altitude and know that the area ahead has high terrain or towers and you must gain altitude – and quickly – as you turn. The point is that the maximum *safe* performance is required. You don't want to stall but still must get altitude as you turn. After the 180° turn is completed you'll be required to perform a straight climb for one minute and fly the airplane within 10° of the proper heading and within 10 knots of the proper airspeed.

It is assumed that you are at cruise airspeed when the need to climb and turn is discovered. This means that there'll be a great deal of "zoom effect" (sometimes called energy altitude) available. Your main problem will be to not overdo the zoom as you turn. You remember that in the recovery from the power-on spiral, one of the big problems was the tendency to overshoot the proper nose position and get too slow. The same can happen here. *There is no reason for you to get below the best rate of climb speed for your airplane,* because: (1) there is a possibility of loss of control at speeds approaching the stall and (2) you will have to continue the climb for one minute after rolling out. If the airplane is too slow, the rate of climb will suffer severely as you struggle to get back to the best climb speed.

PROCEDURE

1. Note your heading and know its reciprocal.
2. As you ease the nose up, increase the power to the maximum (mixture, prop and throttle forward).
3. Establish a standard rate turn in either direction – note the position of the sweep second hand as the turn is started.
4. As the airspeed drops to a few knots above the best rate of climb speed the back pressure is relaxed slightly so that the airspeed does not fall below the best rate of climb speed.

The airspeed value at which you will stop the nose rise will naturally depend on the amount of back pressure exerted and the airplane you're flying. You might be alert as soon as the airspeed drops below a point halfway between cruise and best rate of climb speed. The best way to find out is to get plenty of practice both visually and hooded (with a safety pilot). You know the nose position visually and should be able to develop a technique to get the safest maximum climb. You'll be flying the airspeed with fore or aft pressure on the wheel. Using visual practice you may find that you can ease the nose up slightly past the best rate of climb position to take advantage of the zoom and then lower it as a certain airspeed is reached so that the best rate of climb airspeed is attained. This is a tricky proposition, however, and the best method is to apply full power and don't get too radical with the nose attitude. It's better to come up to a selected airspeed a little less flamboyantly than to overshoot wildly and have to correct. One of the major problems of instrument trainees is trying to obtain a particular airspeed too quickly. Invariably they will overshoot and start chasing the airspeed until, in some cases, control of the airplane is lost. Don't fall into this trap.

You may be so engrossed in catching the airspeed that the turn may be forgotten and the roll-out point overshot, or the standard rate turn not maintained. Things could get rough if you suddenly notice a triple needle-width turn and the airspeed building up quickly again. Center the needle, apply back pressure to stop the airspeed increase and stop the altimeter – then climb and turn to the desired heading – you know the routine.

Stick to a standard rate turn. You're used to it and the bank will be shallow enough so that the climb doesn't suffer, plus the fact that if you don't have a directional gyro the timed turn will be the only accurate way of making a 180° turn. In a fairly steep climb the magnetic compass tends to act like a prima donna.

After the 180° turn is completed, continue a straight ahead climb to the predetermined altitude. You'll climb for no less than a minute. Keep full power on unless the check pilot specifically tells you to use less power throughout the maneuver.

The straight climb allows a brief period in which you can pull parts of TV towers out of the empennage.

It will be important that the ball be kept centered throughout. Torque will be a problem and will probably give you the most trouble during the one minute straight climb. You're only allowed 10° of heading deviation and a couple of seconds lapse of attention can allow torque to pull you off that much or more.

The needle and ball will be the greatest

directional aid if you don't have a directional gyro. The magnetic compass may be used as a reference to see if you have turned. It won't be indicating the actual heading in the climb as you will note during practice of this maneuver. You might review MAGNETIC COMPASS, Acceleration Errors, in Chapter 3.

You can do a pretty good job of holding a heading using only the needle and ball. If the needle has deviated one way for a period, turn the airplane the same amount in the opposite direction for what you estimate to be the same length of time. That is, deflect the needle in the opposite direction through use of coordinated controls. You'll be surprised how well you can hold a heading this way.

Keep your scan going.

TWO DESCENDING 90 DEGREE TURNS

This maneuver is meant to simulate a safe, but not precise, radar controlled low approach (1000 feet) to an airport. (Fig. 18-13)

You'll use normal approach power during the descent and will have to stay within limits of ±10 knots of airspeed and 10° of heading.

It would be better to practice the simulated approach with the landing gear down if you're using a retractable gear airplane. The airspeed is not as apt to get away from you, and in the actual situation you'd want to be ready to land when breaking out.

Don't use flaps during the approach but use an airspeed low enough so that the flaps may be extended at any time. An approach speed of somewhere in the vicinity of 1.5 times the flaps-up, power-off stall speed should be fast enough to allow good control, yet not too fast to exceed the maximum speed for flaps extension. Count on a rate of descent of about 500 feet per minute. Your instructor will show you the approximate power setting for a gear-down descent of 500 fpm at the proper airspeed. If you set the power correctly and hold the right airspeed, the 500 fpm descent will take care of itself. Don't juggle with the throttle. Set the power as was found in practice (it will vary slightly with outside temperature, Weight and altitude) and fly the airplane. If it appears that the rate of descent is too high, increase the power slightly – then leave it alone.

The 500 fpm descent is considered standard for an approach of this type. This is not intended to be a precise maneuver, and if you stay in an area of 400 to 600 fpm descent, things will go all right. A vertical speed indicator will be useful as a reference, *but don't chase it.*

Your main problem will be in the turns. Because you are descending, there may be a slight tendency to get "wrapped up" in the turns.

It would be wise to practice patterns in both directions as many airports have right-hand radar approach patterns – and the check pilot can ask for turns either way on the flight test.

Keep up with the time on the straight legs. If the airplane is performing correctly don't feel that you have to do something anyway. Keep the scan up to make sure it doesn't start to wander. Catch deviations early.

Under actual conditions the straightaway periods will give you a chance to think up a plausible story as to how you managed to get on actual instruments and need a letdown when you were neither on an instrument flight plan nor instrument qualified.

You'd better make it a good one.

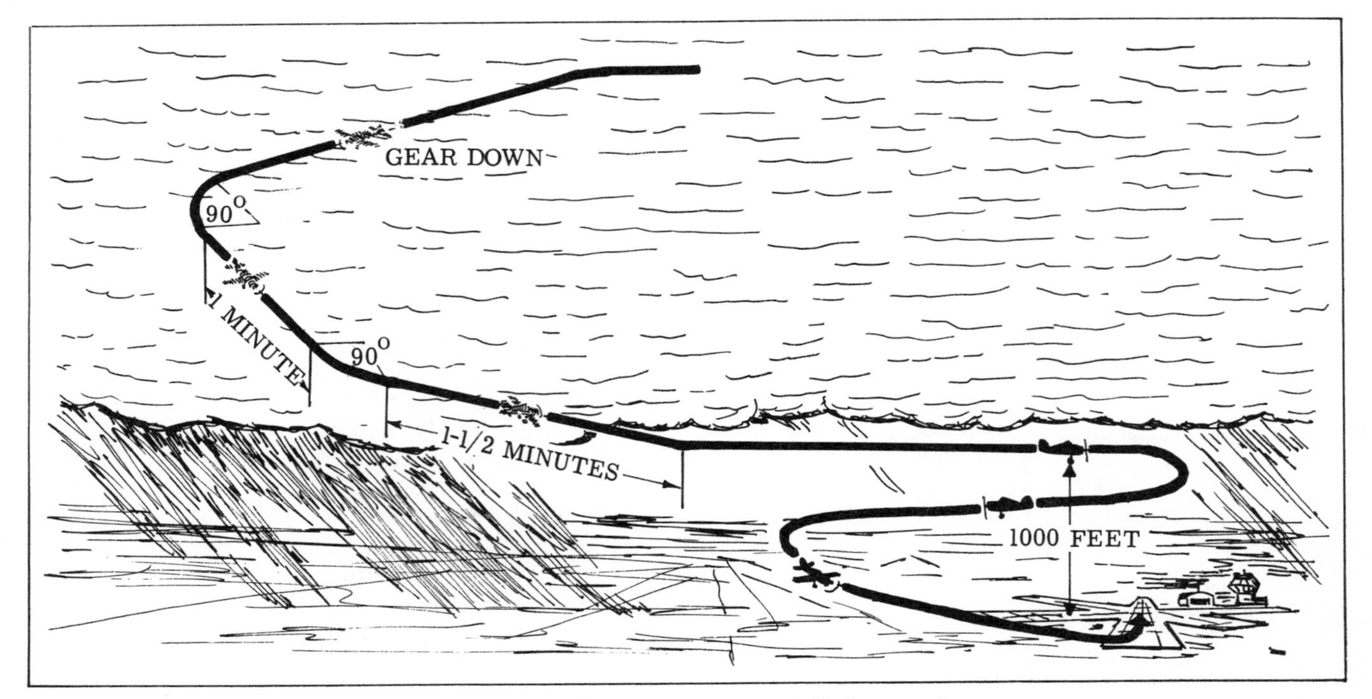

Fig. 18-13. A surveillance radar controlled approach.

STRAIGHT AND LEVEL FLYING

After you've lost visual references and have made a 180° turn by reference to the instruments it's very important that you be able to fly straight and level to get back to the clear area. It won't do any good to make a perfect 180° turn and then roam all over the country trying to get back to visual conditions.

If there's no directional gyro available, then the needle is the primary direction indicator. If the air is not too bumpy, the magnetic compass may be used as a check. Because you are in straight and level, unaccelerated flight, the magnetic compass will be of more use than it was in the climb or descent. Once straight and level is established the natural stability of the airplane will help you, although some minor corrections will be necessary, particularly in bumpy air.

If the compass has settled down and shows that you are off heading, correct with a balanced turn in the proper direction. For minor variations from heading (up to 15°) a half standard rate turn correction is preferred. If you are 10° off, make a half standard rate turn ($1\frac{1}{2}°$ per second) for 7 seconds. The check pilot will undoubtedly give you more leeway if your corrections are smooth and definite. Don't try to turn back to heading using the magnetic compass. Make your correction and, after the compass has settled down, check the results.

In straight and level flying the trainee sometimes has a tendency to stare at the needle and ball, neglecting the rest of his scan. It's important that the instruments be continually cross-checked.

Your limits are to fly within 10 degrees of heading, 100 feet of altitude and 10 knots of airspeed.

SUMMARY

The instruments give information on the airplane's actions and attitudes and take the place of the horizon. Continual practice is necessary to be able to fly safely under instrument conditions. The maneuvers discussed in this chapter are for self-preservation only. If you have to use them under actual conditions it would be wise to consider that it became necessary because of either poor planning or lack of headwork in flight on your part. Then, perhaps the next time you'll check the weather better and cancel, or at least make a 180° turn *before* flying into a condition where visual references are lost. You are encouraged to continue your training and become instrument rated. Then, with proper clearance you may fly the airplane without reference to the ground for as long as necessary.

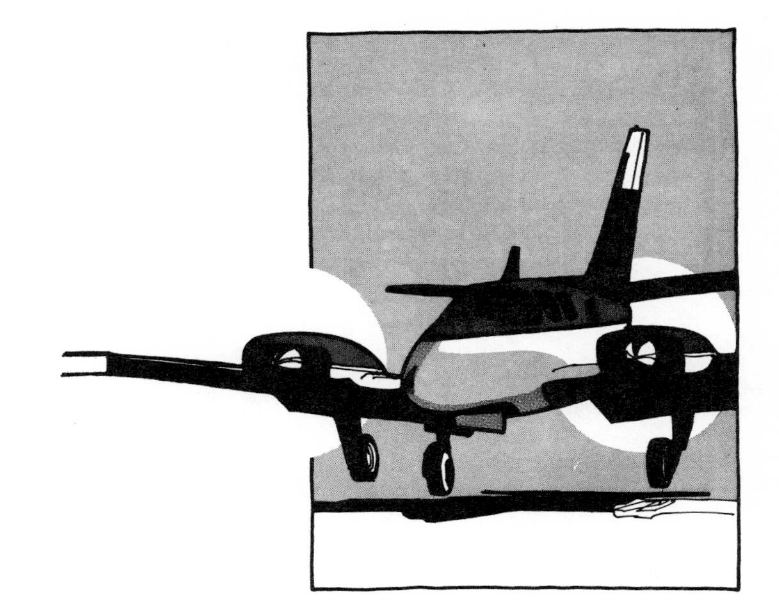

PART IV / CROSS-COUNTRY AND NIGHT FLYING

19. NAVIGATION AND RADIO

NAVIGATION

If you are like most private pilots, the greatest part of your flying lately has been cross country. This is the way it should be – local flying is fun, but the main appeal of the airplane is its use as a fast, enjoyable means of transportation.

This chapter will not attempt to repeat a great deal of basic information on navigation. You know the procedure of laying out a course and flying by pilotage navigation. Instead, safety tips and information of interest to the private or commercial pilot will be covered.

FLIGHT PLANNING

You've made cross country flights where everything seemed to come unglued. Sometimes flights having the benefit of the most careful pre-flight planning go wrong, but the odds are heavy against it. Nine times out of ten the problem is a lack of pre-flight planning. Maybe you were rushed or just careless, but sooner or later your sins of omission catch up with you.

1. *Always* carry up-to-date charts (preferably sectional charts) of the area you plan to fly over – *no matter how many times you have made the trip.* You may never look at them, but should trouble arise you'll be able to break them out and locate the nearest usable airport in a hurry. You may know the area along the course line like the back of your hand, but how well do you know the area and which airports are usable, say, twenty miles to either side of the course? Even though you are using radios as the means of navigation, those sectional charts should be available. Airplane radios fail and VOR's may be shut down for maintenance, and sometimes destination airports are hard to find the first time without a sectional chart. Many pilots use radio facility charts or a flight log for navigation, and may not have to look at a sectional chart for hours – but they have them along just in case. The ceiling and visibility may get pretty low, and even though you are still legal VFR, may be too low to get much help from the radio. The weather may start deteriorating fast and you, not being instrument qualified at present, will have to find the nearest good airport quickly. The best way to do this is to have a sectional chart available. Radio navigation charts only show major airports and *no* geographic features.

2. *Make a flight log for long flights to new territory.* Pilots using a prepared flight log can fly by radio navigation from Washington, D. C., to Miami (assuming the airplane has the range) without using a sectional or radio chart. In your pre-flight planning for a long flight check the sectional, WAC or radio charts for courses and distances. Use the *Airman's Information Manual* (AIM) for correct frequencies of the enroute radio facilities. Don't depend on the frequencies given on sectional or WAC charts at all. Only a little more dependence should be put on the radio charts, even if they are the latest issue.

You can check the latest navigational and tower frequencies in the AIM and note them on the log. You may use the same flight log for the trip many times – if you always check the AIM for correct frequencies and other pertinent information each time. Some pilots keep a small pocket notebook containing flight logs of long trips made fairly often by VFR radio navigation. They check the AIM each time they go, and in between times make sure they have the latest required sectionals and radio charts. The log saves the problem of unfolding and folding charts in a rather confined area such as the cockpit (or cabin, if you prefer).

Even if you are using a flight log, mark the course lines on the sectionals. If you are flying airways they will be shown on the charts. But even here a heavy line to accentuate the course will make it easier to locate on the chart in a hurry. Arrange the sectional charts in the order of use and put them in a place in the airplane where they are out of the way but easily reached if needed.

Some pilots of faster airplanes use only WAC charts, but you'll be better off having a more detailed map in an emergency.

If the course line drawn on the sectionals falls fairly close to the edge of the map take the adjoining chart as well. You could get a few miles off course and have no map of that area.

Marking the course with a red pencil is fine for daylight flying, but at night under the red cockpit lighting a red line is just about impossible to see. For an all-purpose operation a heavy black, or dark blue, ink line is best. An advantage of night flying is that the rotating beacon on a lighted airport can be seen for many miles in good weather. This will be covered in Chapter 21.

Naturally if your trip is a short one, a flight log would be a waste of time. Go ahead and use a sectional. But it would still be wise to check NOTAMS at the nearest Flight Service Station or other facilities.

3. *Always plan on at least 45 minutes of fuel in reserve,* even if it means an extra fuel stop. There are few more uncomfortable feelings than the one of wondering if you have enough fuel to make an airport. Unexpected headwinds or going around weather in the latter part of the trip may put you at a point away from available airports. It seems that you never get cut short in areas with plenty of airports, but always in the hinterlands.

4. *If you are going into questionable weather* near the destination and have to refuel enroute, plan to do so at a point well along the route (but still in the good weather area). (Fig. 19-1)

Although you would in theory have enough fuel to make the trip by refueling at either airport, the best course for safety would be to refuel at Airport 2, even if it doesn't have a restaurant like Airport 1. Or if you feel that you have to stop early along the way, you might plan on making another stop enroute. This can be summed up as follows: *Never fly into questionable weather, particularly in a strange area, without a good reserve of fuel.*

5. *Carefully check the weather.* If the present weather at the destination is marginal but forecast to get better, do your flight planning as if it will stay marginal or even get worse. Forecasts aren't 100 per cent accurate, by a long shot. Chapter 20 will cover the things to look for during the pre-flight weather check.

6. *Always leave yourself a back door.* Keep airports in mind along the route that can be used as alternates.

7. *Make use of all the FAA facilities available.* This includes weather service and Flight Service Stations.

THE FLIGHT PLAN

The FAA's Flight Service Stations are an aid to all pilots. By filing a VFR flight plan you'll have interested people on the ground helping you look out for your neck. To get this service you'll file a VFR flight plan as is shown in Fig. 19-2.

The Flight Service Station personnel and you will get together in checking weather, NOTAMS and other information vital to flight safety. You can call the Flight Service Stations while flying and get information on weather, or NOTAMS enroute or at the destination. Guard the nearest VOR frequency or 122.6 or 123.6 MHz while flying on this type of flight plan.

About filing flight plans in the air: If you've ever been in a bind and wanted to talk to the nearest Flight Service Station to get needed weather information, but couldn't because some guy was filing a VFR flight plan in the air, you'll understand this point. You may need the weather *now* as you must decide *now* whether to turn back, fuel up or make other decisions. Frankly, you would cheerfully strangle the other pilot if you could get a hand on him. (Fig. 19-3)

If at all possible, file a flight plan in person or by phone. Even a long distance phone call won't hurt you. Some pilots think nothing of spending 100-150 dollars for a trip but absolutely will not make a 35-cent phone call to file a flight plan.

Filing an IFR flight plan in the air is a little more forgivable -- but not much. The weather may have deteriorated more than predicted so that the only way for the pilot to proceed would be by instrument flight rules. Still, this is abused much too often, and some

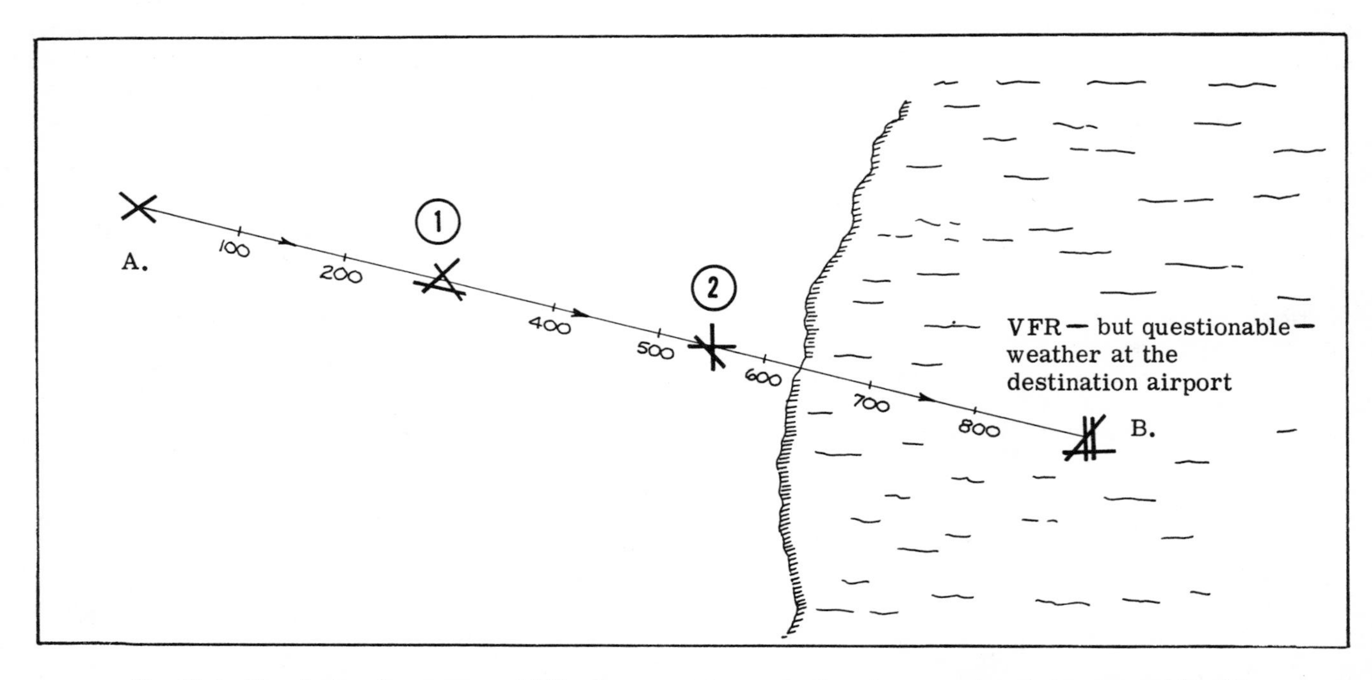

Fig. 19-1. The airplane has a range of 800 miles, requiring a refueling stop somewhere between A and B. The pilot should refuel at Airport (2) rather than at Airport (1). This is assuming that both airports are fairly comparable in servicing facilities.

FEDERAL AVIATION AGENCY

FLIGHT PLAN

Form Approved.
Budget Bureau No. 04-R072.3

1. TYPE OF FLIGHT PLAN	2. AIRCRAFT IDENTIFICATION
IFR / VFR X / DVFR	N8183P

3. AIRCRAFT TYPE/SPECIAL EQUIPMENT 1/	4. TRUE AIRSPEED	5. POINT OF DEPARTURE	6. DEPARTURE TIME PROPOSED (Z)	ACTUAL (Z)	7. INITIAL CRUISING ALTITUDE
PA-24-250	150 KNOTS	DCA	1400		4500

8. ROUTE OF FLIGHT

DIRECT BROOKE, V-3, V-3E, V-3 TO SAV

9. DESTINATION (Name of airport and city)	10. REMARKS
TRAVIS AP, SAVANNAH, GA	~

11. ESTIMATED TIME EN ROUTE HOURS	MINUTES	12. FUEL ON BOARD HOURS	MINUTES	13. ALTERNATE AIRPORT(S)	14. PILOT'S NAME
3	20	4	45	~	STEWART G. DICKSON

15. PILOT'S ADDRESS AND TELEPHONE NO. OR AIRCRAFT HOME BASE	16. NO. OF PERSONS ABOARD	17. COLOR OF AIRCRAFT	18. FLIGHT WATCH STATIONS
TRAVIS AIRPORT-SAVANNAH, GA	3	RED & WHITE	X

CLOSE FLIGHT PLAN UPON ARRIVAL

1/ SPECIAL EQUIPMENT SUFFIX
A — DME & 4096 Code transponder
B — DME & 64 Code transponder
D — DME
L — DME & transponder—no code
T — 64 Code transponder
U — 4096 Code transponder
X — Transponder—no code

FAA Form 7233—1 (4-66) FORMERLY FAA 398

0052-027-8000

Fig. 19-2. A sample VFR flight plan. The marked out portions of Items 1 and 18 are for the FVFR flight plan which is no longer in use.

pilots wait to file IFR until after getting into the air "to save time" even when they know before leaving the hangar that it will be necessary.

When you've filed a flight plan make doubly sure that you close it with the nearest FAA communications facility when you reach your destination.

If you don't report within one-half hour of your ETA (estimated time of arrival), a ramp and communications search will be started. They will check all the flight operators at the destination airport and call the tower to see if they remember your calling in for landing (don't think that the tower will automatically file an arrival report, they are not required nor expected to do this unless requested to do so by pilots; they have a lot of airplanes to handle, many of them not on flight plans). The next probable step would be to check with each Flight Service Station along your route to see if there's been any contact with you.

Within $1\frac{1}{2}$ hours past the ETA, if there's no word from the pilot, the full facilities of the Search and Rescue Service will be activated. When this happens *you'd better be really lost* and not just doping off at

Fig. 19-3. *You* need quick weather information and some so-and-so just decided to file a VFR flight plan over the air.

some little airport where you suddenly decided to stop because it has a good restaurant.

So, if it looks as if you'll have to land at some other airport, or find that you'll be delayed a half-hour or more in arriving at the destination, you'd better contact the nearest Flight Service Station and have them pass this on to your destination.

ENROUTE TIPS

1. When flying cross country in an airplane with a controllable pitch prop (unsupercharged engine) fly the following altitudes:

65 per cent power being carried	8000 – 10,000 feet
75 per cent power being carried	6000 – 8,000 feet

The above suggestions are based on the fact that the altitudes are in the vicinity of full throttle operation for the power settings given. You'll get better true airspeeds at these altitudes for the power used and will be operating more efficiently. Naturally you'll be taking the winds aloft and ceiling into account, and will always fly the correct altitude for your magnetic course (0-179°, Odd thousands plus 500 feet; 180-359°, Even plus 500 feet). But, if the wind at altitude is light and variable and the ceiling is high or unlimited you'll be getting more for your money at these recommended altitude ranges.

2. For better cruising efficiency, in a variable or controllable pitch propeller equipped airplane, choose the setting for a certain per cent of power (say 65 per cent) on the power chart that requires the higher manifold pressure and lower rpm. You might review Fig. 12-1.

3. It's better to have one fuel tank one-fourth full and three empty, than four tanks, each one-sixteenth full.

4. If in doubt about your geographical position, don't circle looking for check points but continue on the original heading. When you see an outstanding terrain feature that you cannot recognize as being on the course line, check the chart on *both* sides of the course for the feature.

Sometime you may be lost and running low on fuel but with all radios working. A VHF/DF (VHF Direction Finding) steer may be obtained by calling Flight Service Stations or towers and requesting such service. They'll alert the nearest point having VHF/DF facilities. You will be instructed in the procedure which will generally be your keying the microphone on the instructed frequency for ten-second periods, interspersed with your call sign so that the station(s) can "home in" on you. Once you've contacted the FAA they'll tell you what to do, so just remember that the service is available and don't try to memorize the exact procedure now. A lot of pilots don't like to admit they're lost and sometimes ask for a "practice steer." If you need help badly don't hesitate to say that you want an emergency steer. The procedure and the FAA VHF/DF stations are listed in the *Airman's Information Manual.*

THE AIRMAN'S INFORMATION MANUAL

The *Airman's Information Manual* is an FAA publication containing information of value to pilots and you should have access to a copy as you do more cross-country flying. The manual is designed to present in one volume the information necessary for the planning and conducting of a flight in the National Airspace System.

The *Airman's Information Manual* is available on annual subscription from the U.S. Government Printing Office; Washington, D.C. 20402 and is divided into four basic parts, each of which may be purchased separately:

PART 1—BASIC FLIGHT MANUAL AND ATC PROCEDURES

This part contains instructional, educational and training material — things that are basic and not often changed, including:

1. *Glossary of Aeronautical Terms* — Definition of control zones and areas and other terms used in aviation.

2. *Aeronautical Publications* — A list of publications of general operational interest to pilots is provided (Advisory Circular Checklist, written and flight test guides, etc.).

3. *Air Navigation Radio Aids* — Theory and operations of such aids as LF/MF ranges, VOR's, radio beacons, Distance Measuring Equipment, Instrument Landing Systems (ILS) and marker beacons. The Frequency Utilization Plan (what frequencies are used where) and VHF/DF (VHF Direction Finding) are included.

4. *Good Operating Practices* — Some hints on operating the airplane in a safe manner, and suggestions on use of the aircraft radio to avoid frequency congestion.

5. *Airport, Air Navigation Lighting and Marking Aids* — Information on airport beacons and runway lighting and marking. Information on enroute beacons and landmark lighting, including obstructions hazardous to flight, is in this part also.

6. *Radar* — General information on FAA and military radar plus specifics on precision and surveillance radar approaches.

7. *Radiotelephone Phraseology and Techniques* -- Background on microphone technique and procedure words and phrases. Includes the phonetic alphabet and Morse code.

8. *Altimetry* — Background and use of the altimeter in the airways system.

9. *Weather* — The weather reporting aids available, in-flight weather safety advisories, weather radar, pilot reports (PIREPS). Information on thunderstorms, such as at which altitudes the most turbulent conditions exist, etc.

10. *Safety of Flight*— Information on Airplane Wake Turbulence (how it's developed, the places where it's most likely to be found, and suggested pilot action if it is encountered). There's also a section on how to report a near midair collision.

11. *Medical Facts for Pilots* -- Discussions of the effects of hypoxia (lack of oxygen), use of alcohol with tips on how long to wait before flying after its use, plus sections on use of drugs, effects of vertigo and carbon monoxide on the pilot, and the danger of flying shortly after scuba diving.

12. *Preflight*— Weather briefing tips and how to file and cancel a flight plan.

13. *Departure* — Communications information (ground control and tower). Light signals. Clearances for taxi and take-off. Departure control procedures.

14. *Enroute* — The VOR and LF/MF airways systems and special control areas. Communications and operating procedures enroute (Instrument Flight Plan). Radar Assistance to VFR (Visual Flight Rules) aircraft. Diagrams of cruising altitude requirements (VFR and IFR).

15. *Arrival* — VFR Advisory Information Procedures. Radar Traffic Information Service and Terminal Radar Service. Airport Advisory Service (Flight Service Stations). Unicom and Multicom information. Approach Control. Instrument approach procedures.

16. *Landing*— Traffic pattern standard procedures and traffic indicators (wind sock, tee and segmented circle). Hand signals for taxi directors.

17. *General*— Airports of entry and departure and procedures. ADIZ (Air Defense Identification Zone) procedures.

18. *Emergency Procedures* — What to do under various emergency situations. Search and rescue procedures and visual emergency signals and codes. Radio communications failure procedures.

Part I is issued quarterly.

PART 2—AIRPORT DIRECTORY

This part is issued semiannually and contains a Directory of all Airports, Seaplane Bases in the conterminous United States, Puerto Rico, and the Virgin Islands which are open to the general public. It includes all of their facilities and services, *except communications*, in codified form. (These are listed in alphabetical order, by states.) Airports with communications are also listed in Part 3, which covers their radio facilities.

A list of selected Commercial Broadcast Stations (having a power of 100 watts or more) is included in this part.

This part contains a listing of Flight Service Station and Weather Bureau telephone numbers.

PART 3—OPERATIONAL DATA

Part 3 is issued every 28 days and contains such information as the following:

1. *Airport/Facility Directory* — Airport data and services with navaid and communications facilities, alphabetically by state and city. (A legend is included for deciphering the coded information.)

2. *Preferred Routes* — Information on preferred routing for IFR flights.

3. *Standard Instrument Departures* — Standard routing for IFR departures from various terminal points.

4. *Standard Terminal Arrival Routes* (STARS).

5. *Substitute Route Structure* — A listing of alternate routes to be used while VOR's or other enroute navigation aids are shut down for maintenance or repair.

6. *Sectional Chart Bulletin*— A tabulation of changes to sectional charts which have occurred since the last publication date of the particular charts listed.

7. *Special Notices* — Information on existing military climb corridors. Civil use of military fields. Special information on terminal areas such as Washington, Atlanta, Los Angeles, etc. A listing of the date of issuance of the latest sectional charts and Instrument Approach Procedure Charts. Also cites changes in FAR's and other important notices. Keep up with this section.

PART 3A—NOTICES TO AIRMEN

This section lists NOTAM information, Airman Advisories, new or revised Oil Burner Routes, hazardous airspace activities and other items considered essential to flight safety. Listed alphabetically by state and issued *every 14 days.*

PART 4—GRAPHIC NOTICES AND SUPPLEMENTAL DATA

Part 4 is issued semiannually and contains such information as:

1. *Abbreviations* used in AIM
2. *Parachute Jump Areas*
3. *VOR Receiver Check Points* — Ground and airborne
4. *Oil Burner Routes* -- Military flight exercise routes.

Part 4 will later contain additional Terminal Area Graphics and other data not requiring frequent change.

Become familiar with the *Airman's Information Manual;* it will become more important to you as you gain flying experience.

RADIO

The radio has become a necessary piece of aircraft equipment. You'll be working with various frequency bands and it would be a good idea to review some background. Your home radio dial is marked with figures from about 550 to 1650. This is in kilohertz ("hertz" means "cycles per second"), and you dial the frequency of the station wanted. This is in the MF band as you can see by checking the following listing:

Very low frequencies (VLF) . . . . 10-30 kilohertz
Low frequencies (LF) . . . . 30-300 kHz
Medium frequencies (MF) . . . . . 300-3000 kHz
High frequencies (HF) . . . . 3-30 megahertz
Very high frequencies (VHF) . . . 30-300 MHz
Ultra high frequencies (UHF) . . . 300-3000 MHz

At the HF band the term megahertz is introduced (1 megahertz = 1000 kilohertz). This is done to keep things from getting too cumbersome as would be done by going on with the kilohertz idea.

In your radio work in flying you'll be working mostly with VHF, LF and MF and maybe later will use UHF.

COMMUNICATIONS

The private pilot who does most of his flying from a small airport sometimes considers with a great deal of nervousness the thought of going into a controlled field. Even the airline pilots with their casual "radio growl" had a first time to use the radio. The chances are good right now that you, as a private pilot, are sharper with a radio than the average commercial charter pilot of fifteen years ago.

Modern aircraft communications equipment has practically eliminated the original purpose of the "formal" phraseology used several years back. However, due to the increasing number of radio equipped aircraft it is still a good idea to be familiar with the use of some of the so-called formal terms since this will shorten transmission time -- an important factor at busy airports. Proper mike technique and a normal conversational voice will result in satisfactory contacts. Also it's better to be a little slow and only have to say it once.

Airport traffic areas enclose an area of a five statute mile radius from the center of airports having an operative control tower (FAA, Military, or privately controlled) and extend from the surface to 3000 feet above the surface. If you are operating into (or out of) an FAA or Military controlled airport you must maintain two-way communications with the tower unless your radio has failed in flight -- then you'll get light signals. Landing at other uncontrolled airports in the area is permissible without contact with the tower, but a listening watch on the tower frequency is recommended. Privately controlled towers, however, still allow operations if you have no radio *but* it's not recommended. *All* airports having government operated control towers have airport traffic areas, require two-way radio and have speed limitations. The following maximum indicated airspeeds apply: (1) Reciprocating engine aircraft -- 156 knots (180 mph) (2) turbine-powered aircraft -- 200 knots (230 mph).

You can't operate within an airport traffic area except for the purpose of landing and taking off at airports located within such airport traffic area, or unless authorized by air traffic control. If you are going to fly past an FAA or Military tower-controlled airport within five miles you have two choices: (1) make sure that you're at an altitude above 3000 feet above the surface or (2) call the tower and get clearance to fly through the airport traffic area if you don't want to, or can't, fly above 3000 feet above the surface.

You can consider this as the airport traffic pattern area of the controlled airport where aircraft are converging and definite VFR control is needed.

Remember that you can still be controlled by light signals in a situation such as loss of radio; however, if traffic is heavy at the controlled airport it would be wiser to land at an uncontrolled field.

The term "airport traffic area" is considered more in terms of VFR flying whereas "control zone" is more an IFR term. The control zone is normally of a five-mile radius, but in addition may have one or more approach leg extensions for aiding IFR operation.

A control zone, as defined by the Administrator, extends from the surface upward to the base of the Continental Control Area (14,500 feet MSL or 1500 feet above the surface in the continental U.S.).

THE CONTROL TOWER

The functions of the tower are broken down into three main divisions or "positions"; that is, Approach Control, Local Control and Ground Control. Following is a condensed coverage of each:

1. Approach Control -- This may be a radar or non-radar control. If it is a non-radar setup, the approach control location is in the tower (the same man may talk to you as approach control and tower). This is the usual system for less busy controlled airports. *The approach controller's primary job is that of coordinating IFR traffic approaching the control zone.* However, he may also be coordinating both VFR *and* IFR traffic during marginal conditions, as well as VFR traffic at busy terminals even in CAVU conditions.

The radar-equipped approach controllers usually are in an IFR room which is located in the tower building but not necessarily in the glassed-in portion. As they will be working strictly with radar they probably won't see the light of day (or night) except during a coffee break. Approach controllers work directly with the tower and their duties are rotated between both tower and IFR room positions.

At congested airports approach control should be contacted by the VFR pilot some distance out (25 miles, or as given in the AIM). This gives them a picture of the traffic approaching the airport and they can coordinate with the local control (tower) to expedite landings. You will be given traffic information if pertinent, as well as runway, wind, altimeter settings and NOTAMS and will be told to switch to the tower frequency at a particular time or place. You would call "LaGuardia approach control" (in the case of two or more controlled airports at a city) or "Memphis approach control" where there is only one airport having approach control. A sample call might be (after choosing the proper frequency):

You: MEMPHIS APPROACH CONTROL (THIS IS)

ZEPHYR SIX SEVEN EIGHT NINER ZULU, OVER.

Memphis Approach Control: ZEPHYR SIX SEVEN EIGHT NINER ZULU, (THIS IS) MEMPHIS APPROACH (CONTROL), (OVER).

You: (MEMPHIS APPROACH) (THIS IS) ZEPHYR EIGHT NINER ZULU, TWENTY-FIVE MILES NORTHEAST, VFR, (AT) TWO THOUSAND FEET, LANDING (AT) MEMPHIS, OVER.

Memphis Approach: EIGHT NINER ZULU, TWENTY-FIVE MILES NORTHEAST, CONTACT MEMPHIS TOWER ON ONE ONE EIGHT POINT THREE (118.3) (MEGAHERTZ) FIVE MILES NORTHEAST (OF THE AIRPORT). (He'll give you wind, altimeter and traffic info unless you have already received it on ATIS.)

(ATIS, or Automatic Terminal Information Service, is the continuous broadcast at selected high activity airports, of recorded noncontrol information such as weather, wind, instrument approach and runway in use, and altimeter setting. As information changes it is given a new phonetic alphabet listing such as "Information Bravo, charlie, etc." On initial contact with approach control or the tower, let them know you have "Information Bravo,"--assuming that you do have Information Bravo. This saves a lot of repetitive chatter on the crowded frequencies.)

You: EIGHT NINER ZULU (You acknowledge all transmissions after the initial contact by the last three numbers or letters of your airplane. No more of this "Roger and Out" routine. After contact has definitely been established, transmissions may be continued without further call-up or identification.)

Suppose, for example purposes, that immediately after you made the last transmission you discovered that you need information repeated right away such as traffic, etc.

You: (THIS IS) EIGHT NINER ZULU, SAY AGAIN TRAFFIC, OVER. (Say "Over" so the controller will know you're through.)

Memphis Approach: TRAFFIC ONE CESSNA ON FOUR MILE STRAIGHT-IN.

You: EIGHT NINER ZULU.

The phrases in parentheses can be left out under abbreviated procedures.

2. Local Control – This is what pilots commonly call the tower. This position controls air traffic within the control zone itself and in the airport traffic pattern – both take-offs and landings.

The local controllers are in the glassed-in part of the tower building because their control is dependent on visual recognition of the aircraft. To follow our sample further: Upon reaching a point about five miles northeast of the airport, you contact Memphis tower on the proper frequency after listening to make sure nobody else is talking.

You: MEMPHIS TOWER (THIS IS) ZEPHYR EIGHT NINER ZULU, FIVE MILES NORTHEAST AT ONE THOUSAND FIVE HUNDRED (FEET) FOR LANDING, INFORMATION BRAVO, OVER.

Memphis Tower: EIGHT NINER ZULU, FIVE MILES NORTHEAST, CLEARED TO ENTER RIGHT DOWNWIND FOR RUNWAY THREE. WIND NORTH-NORTHEAST AT ONE FIVE (15) KNOTS. ALTIMETER THREE ZERO ZERO SEVEN. CALL ENTERING DOWNWIND.

You: EIGHT NINER ZULU. (Don't repeat their instructions back to towers unless there is some question of understanding.) The "EIGHT NINER ZULU," acknowledges that you understand all of the instructions. If, for instance, you did not understand the runway you'd say: THIS IS EIGHT NINER ZULU, SAY AGAIN RUNWAY, OVER.

After getting on the downwind leg you'd call:

You: (MEMPHIS TOWER) ZEPHYR EIGHT NINER ZULU (ON) RIGHT DOWNWIND (LEG) FOR (RUNWAY) THREE (OVER).

Memphis Tower: EIGHT NINER ZULU IN SIGHT, CONTINUE, NUMBER TWO TO FOLLOW A CONSTELLATION NOW TURNING FINAL FOR (RUNWAY) THREE.

You: EIGHT NINER ZULU. You continue your approach, giving the Connie plenty of room.

Memphis Tower (after the Connie has cleared the runway): EIGHT NINER ZULU, WIND NORTH-NORTHEAST ONE FIVE, CLEARED TO LAND.

You: EIGHT NINER ZULU.

If there is no other traffic, towers will normally give you a "CLEARED TO LAND" on the downwind leg. If there is other traffic they'll say "CONTINUE" or "CONTINUE APPROACH." Be sure that you get a "CLEARED TO LAND" before landing, even if you have to ask for it.

Although you are under positive control, you are the person responsible for the safety of your airplane. When the tower says "cleared to land" *it doesn't mean "ordered to land."* You may request another runway, or not land at all, if the situation appears unsafe. Let the tower know of your plans, though.

During the landing roll the tower may give further instructions.

Memphis Tower: EIGHT NINER ZULU, TURN RIGHT NEXT INTERSECTION, CONTACT GROUND CONTROL ONE TWO ONE POINT NINER (121.9 MHz) AFTER CLEARING. (The tower will judge your roll-out before telling you where to turn off.)

You may be too busy to acknowledge but will follow the instructions. *Stay on the tower frequency as long as you are on the "hot" or duty, runway. Do not switch to the ground control frequency until clear of the runway.* The tower may want to contact you in a hurry (there may be an emergency) and they don't want you in the process of switching frequencies while sitting in the middle of the active runway.

3. Ground Control – After you've cleared the active runway but before getting too far into the ground traffic (just after crossing the yellow "hold" lines) contact ground control for taxi clearance to the desired point on the airport.

You: MEMPHIS GROUND (CONTROL) EIGHT NINER ZULU, TAXI TO BLANK FLYING SERVICE, OVER.

Memphis Ground Control: CLEARED TO BLANK FLYING SERVICE (followed by instructions if necessary). Don't hesitate to ask how to get there if you are uncertain – that's what ground control is for.

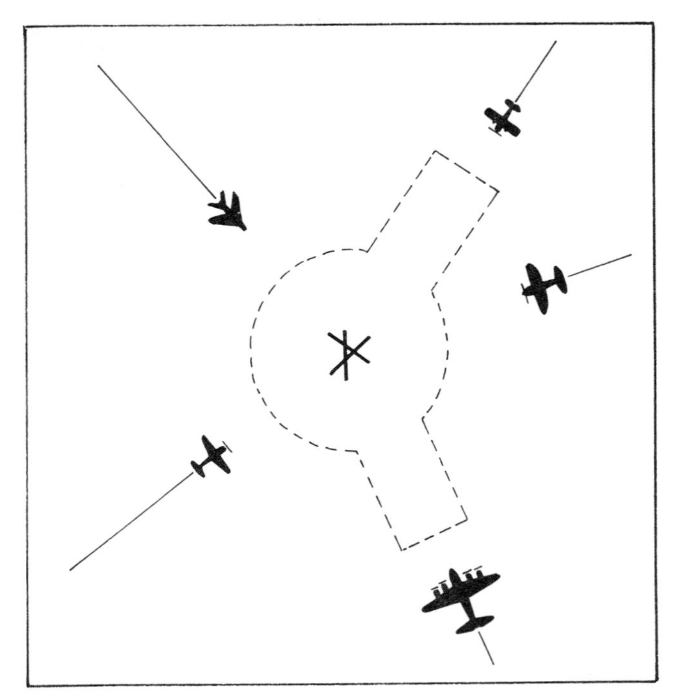

Fig. 19-4. Approach control at the busier airports also is used to coordinate the VFR traffic approaching the control zone.

You: EIGHT NINER ZULU.

After you've completed your business and decided to continue your trip, you'd call ground control again after starting and after listening to ATIS, but before taxiing.

You: MEMPHIS GROUND CONTROL, (THIS IS) ZEPHYR SIX SEVEN EIGHT NINER ZULU, BLANK FLYING SERVICE, TAXI FOR TAKE-OFF, INFORMATION CHARLIE, OVER.

Memphis Ground Control: ZEPHYR EIGHT NINER ZULU CLEARED TO RUNWAY THREE. (Gives wind, altimeter setting and time.) TAXI STRAIGHT AHEAD, MAKE LEFT TURN FIRST TAXIWAY.

You: EIGHT NINER ZULU. (Again, don't hesitate to ask directions if necessary.)

When you have reached the warm-up spot by the end of the active runway and have stopped to make the pre-flight check, switch to the tower frequency as you are now about to enter its domain (the hot runway). When you're ready, call the tower:

You: MEMPHIS TOWER, ZEPHYR EIGHT NINER ZULU READY TO GO (OR READY FOR TAKE-OFF) ON RUNWAY THREE.

Memphis Tower: EIGHT NINER ZULU, HOLD. (For example purposes we'll assume that you will be required to hold short of the runway because of incoming traffic.)

You: (Holding position on warm-up area) EIGHT NINER ZULU HOLDING.

Memphis Tower: (After traffic has landed but hasn't cleared the runway) EIGHT NINER ZULU, TAXI INTO POSITION AND HOLD.

You: (TAXI INTO) POSITION AND HOLD, EIGHT NINER ZULU.

Memphis Tower: (After the other airplane has

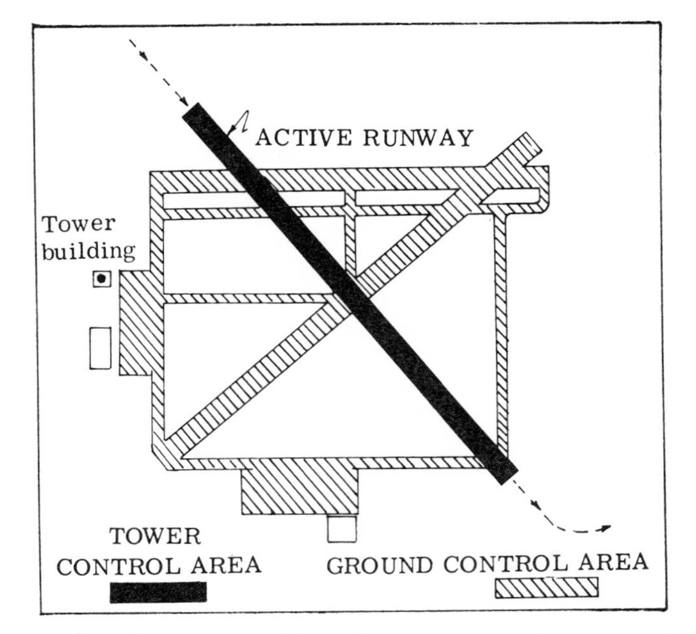

Fig. 19-5. A general idea of tower and ground control areas of jurisdiction when a particular runway is in use.

cleared the runway) EIGHT NINER ZULU CLEARED FOR TAKE-OFF.

You: EIGHT NINER ZULU ROLLING.

At exceptionally busy places like Washington National Airport you would clutter up the air with acknowledgments if you're sitting on the runway and are cleared for take-off. You'd just open it up and go. If there is any doubt in your mind as to whether instructions or clearances are directed to you, ask again. At busy airports two or even three different runways may be in constant use. If you're on the runway waiting to go, and hear, "(Static) . . . CLEARED FOR TAKE-OFF," call in and ask, "IS (your number) CLEARED FOR TAKE-OFF?" Make sure that the clearance is for you before roaring off, otherwise you and another plane could meet at the intersection, with disastrous results.

If, after take-off, you want to make a turn in a direction contrary to normal traffic flow, ask the tower. You can request this before take-off or during climb-out if you like.

Stay on tower frequency until out of the control zone, or you may leave sooner if so cleared by the tower.

As an added note: Use your airplane model, rather than the manufacturer (APACHE 1234P — not PIPER 1234P) as this will help the tower pick you out. The manufacturer may make several models from trainers to twin engines.

SPECIAL VFR

There will be times when the destination airport weather is less than the 1000-foot ceiling and 3 miles visibility required for VFR operation in a control zone, but the weather is good enough for reasonably safe operation such as a 1 to 2 mile visibility caused by smoke or haze. This puts the airport below VFR minimums even though there may not be a cloud in the sky. Remember that you cannot enter any control zone without prior clearance if the weather is below VFR minimums. Control zones are sometimes designated at airports that have only a Flight Service Station (no approach control or tower). You should maintain contact with the Flight Service Station at *any time* when operating to or from or within 5 miles of any airport that has an FSS but no tower, because they will have some information of interest to you (surface winds, altimeter setting, reported traffic, etc.). This is recommended for any weather conditions.

Suppose that you are approaching the destination airport and find that the smoke or haze condition has not dissipated as forecast, and the visibility is still $1\frac{1}{2}$ miles in the control zone. (1 mile visibility is the minimum for special VFR operations for general aviation airplanes.) You must get a special type of clearance to enter the control zone and this must be coordinated with instrument traffic in the area.

Remain clear of *control zones and control areas*, which means that if your path and altitude would take you through an airway you'd better let down to below the control area if obstructions permit. (Remember that a control area extends upward from a given distance above the surface (either 700 or 1200 feet) – a control zone extends from the surface upward to the base of the Continental Control Area.) Don't get so engrossed in avoiding the airport control zone that you fly through an airway – which has the same 3-mile visibility minimum.

The thing to do is to get close to the control zone without entering and call the approach control, tower (some airports have a tower but no approach control) or Flight Service Station as applicable. (Fig. 19-6)

You: HARRISVILLE TOWER (approach control, or radio) (THIS IS) ZEPHYR SIX SEVEN EIGHT NINER ZULU, OVER.

Harrisville Tower: ZEPHYR SIX SEVEN EIGHT NINER ZULU (THIS IS) HARRISVILLE TOWER, OVER.

You: (THIS IS) EIGHT NINER ZULU, SIX MILES WEST AT TWO TWO, ONE THOUSAND THREE HUNDRED MSL, VFR, REQUEST CONTROLLED VFR (INTO CONTROL ZONE), OVER.

Harrisville Tower: EIGHT NINER ZULU IS CLEARED TO HOLD AT PRESENT POSITION. IF UNABLE TO MAINTAIN VFR, ADVISE. EXPECT APPROACH CLEARANCE AT THREE ONE. (Then gives current weather and altimeter setting.)

You: EIGHT NINER ZULU IS CLEARED TO HOLD AT PRESENT POSITION. IF UNABLE TO MAINTAIN VFR, ADVISE. EXPECT APPROACH CLEARANCE AT THREE ONE, OVER.

Harrisville Tower: CLEARANCE CORRECT, OUT. Notice that you repeated the clearance so that there would be no mistake – this is for your own protection. In the above illustration you would set up maximum endurance and hold over an easily recognizable landmark so that you won't go drifting into the control zone unknowlingly. (Fig. 19-6)

Why bother to set up maximum endurance? You reported in at twenty-two minutes past the hour and will "expect approach clearance at three one" (thirty-one minutes past the hour). This is only a wait of nine minutes – so why all the concern? You'll be lucky if you are cleared to enter the control zone at the time first stated. Instrument traffic may cause the approach clearance to be revised several times or the airport visibility may go below one mile, which is the minimum for a controlled VFR approach or departure. Assuming that you are not prepared to make an instrument approach, if the visibility goes below a mile you'll have to go elsewhere, and this takes fuel. We'll assume that this doesn't happen, so the following conversation will take place at the predicted time.

Harrisville Tower: ZEPHYR EIGHT NINER ZULU IS CLEARED INTO THE CONTROL ZONE WEST. CALL TWO MILES WEST FOR POSSIBLE STRAIGHT-IN TO RUNWAY NINER.

You: (Read back clearance – and follow it.)

You may also get a controlled VFR departure under conditions of less than VFR (but one mile or

Fig. 19-6. Hold over some well-defined landmark while awaiting clearance into the control zone when on a controlled VFR clearance.

greater visibility). The tower or FSS may give you instructions on altitude and direction to fly going out of the control zone, or you may be cleared "on course" out of the control zone. The tower usually will request that you report leaving the control zone.

If, as you approach the control zone, the weather looks a little suspicious and you think that it may be below VFR minimums (even though the last report was good, be sure to call the tower before entering, and ask the latest weather. If it's above VFR minimums, follow the usual procedure for getting landing clearance.

You are responsible for maintaining proper terrain clearance and remaining clear of clouds on a special VFR clearance. Controlled VFR means just that; you are being controlled as far as the time and place of entry are concerned and are to conduct your flight by visual reference with the ground, but are being provided with separation from IFR traffic.

Within some control zones, the volume of IFR traffic is such that special VFR flight can't be permitted. There is currently a list of 33 control zones (Newark, Atlanta, Washington National, etc.) in Part 3 of AIM. Keep checking AIM for for additions (probable) or deletions (not likely).

Incidentally, you won't get a special VFR clearance at *any* control zone if IFR traffic will be delayed.

DEPARTURE CONTROL

You may get a chance to use departure control at the larger airports. This is a radar-controlled departure used mainly for instrument traffic, but also is used for VFR departures in marginal VFR weather. The tower normally will have you switch to the departure control frequency shortly after takeoff.

Tower: EIGHT NINER ZULU, CONTACT HARRISVILLE DEPARTURE CONTROL ONE ONE NINER (POINT) SEVEN (119.7 MHz) NOW.

You: EIGHT NINER ZULU, ONE ONE NINER SEVEN.

You: (After switching to the proper frequency) HARRISVILLE DEPARTURE (CONTROL) (THIS IS) ZEPHYR SIX SEVEN EIGHT NINER ZULU, OVER.

Departure control will acknowledge your call and give you heading and/or altitude instructions out of the congested area.

ENROUTE COMMUNICATIONS

You may need information while flying enroute and the best place to get it is from the Flight Service Stations. As may be seen at the end of this section, you can call a Flight Service Station on 122.1, 122.6 or 123.6 MHz and listen on one or more of the following: (1) The VOR frequency (2) the low/medium frequency (if an LF/MF facility is available) (3) 122.2 MHz (4) 122.6 MHz or (5) 123.6 MHz. (At some selected locations you can both transmit and receive on 122.2 and 122.3 MHz — check the AIM.)

As a suggestion, always call a Flight Service Station on 122.1 MHz and listen on the VOR frequency, instead of transmitting and receiving on 122.6 or 123.6 MHz as is sometimes done. You'll have less interference this way, as you or the FSS may be cut out by some pilot using 122.6 or 123.6 MHz many miles away. It's true that other pilots will be transmitting on 122.1 MHz, but the odds on your getting interference are halved, as that frequency is a one-way channel.

Because the VOR frequencies are set up to avoid interference between stations (facilities with the same frequency are established hundreds of miles apart), you'll have no problem of this nature unless you are at a very high altitude. You may still be delayed by the fact that the station you are calling is talking to someone else in the area. It's a good idea to let the Flight Service Station know on what frequency you would like the reply.

You: (Calling on 122.1 MHz) DOTHAN RADIO, (THIS IS) ZEPHYR SIX SEVEN EIGHT NINER ZULU, LISTENING OMNI, OVER.

Dothan Radio: (Transmitting on omni frequency) ZEPHYR SIX SEVEN EIGHT NINER ZULU, (THIS IS) DOTHAN RADIO, OVER.

You: (THIS IS) EIGHT NINER ZULU, I'D LIKE THE LATEST MOBILE WEATHER, OVER.

Dothan Radio: (Reads latest Mobile sequence and gives you Dothan altimeter.)

You: EIGHT NINER ZULU, THANK YOU AND OUT.

Avoid calling Flight Service Stations at fifteen and forty-five minutes past the hour, as these are scheduled broadcast times.

Just because you are listening to a VOR or low frequency identification doesn't mean that voice facilities are available on that frequency. The sectional chart will note "no voice" facilities but may be out of date, so refer to the *Airman's Information Manual.*

Fig. 19-7 shows the navigation and communications facilities and controlled airspace in the vicinity of Chattanooga, Tennessee as indicated on a sectional chart.

When you are calling a Flight Service Station always use the term "radio" ("Jonesville radio"), whether it's a VOR or low frequency facility.

Following are frequencies you'll be using in contacting towers or Flight Service Stations:

Flight Service Station Frequencies

1. *Aircraft Transmitting Frequencies —*
   Normal — 122.1, 122.6 and 123.6 MHz
   Emergency — 121.5 MHz
2. *Aircraft Receiving Frequencies —*
   Normal — The voice facilities of LF/MF aids and VOR (when voice is available), *plus* 122.2, 122.6 and 123.6 MHz
   Emergency — 121.5 MHz

Check the latest *Airman's Information Manual* for frequencies as sectional chart data may be out of date. Remember: You can both transmit and receive on 122.6, 123.6 and 121.5 MHz and on 122.2 and 122.3 MHz at selected locations.

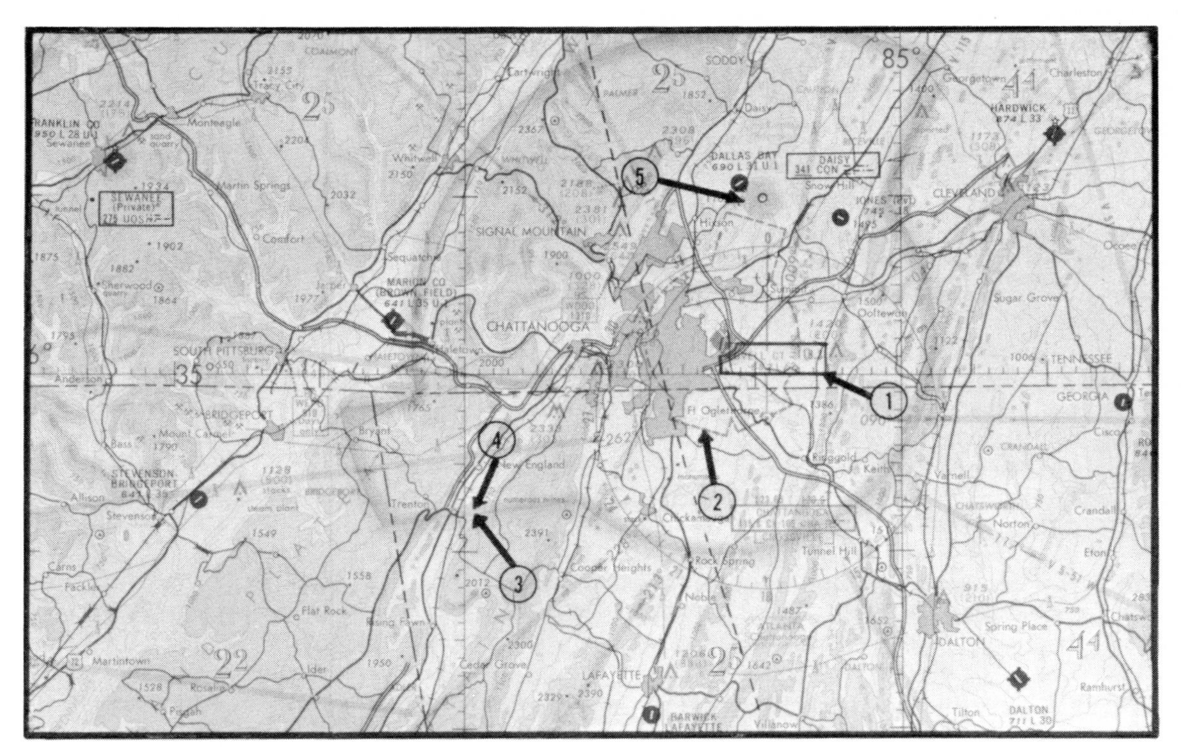

Fig. 19-7. Navigation and communications facilities in the general area of Chattanooga, Tennessee. (1) Tower and airport information. (2) Control zone boundary. Boundary between control area (3) and transition area (4). The base of the control area is 1200 feet above the surface and the transition area is 700 feet.) (5) Radiobeacon (Daisy 341 kHz CQN.)

Tower, Approach Control and Ground Control Frequencies

1. *Aircraft Transmitting and Receiving Frequencies* —
   Normal — 118.0 — 121.4 MHz; 123.6 — 128.8 MHz (towers and approach control)
   121.6 — 121.95 MHz (airport utility, ground control). The frequency most used for ground control is 121.9 MHz, with 121.7 MHz being the next most popular. If you are unable to contact the tower on its assigned frequency, ground control may be used.
   Emergency — 121.5 MHz
2. *Aircraft Transmitting Frequencies Only* — 122.4 122.5, 122.7 MHz. For airplanes with a small number of transmitting frequencies the pilot can call on one of the above frequencies and listen on the assigned frequency (118.0 — 121.4 MHz, etc. for towers and approach control, and 121.6 — 121.95 MHz for ground control). Every tower stands by on one of the above three frequencies as assigned to it. The largest number of towers in the U.S. receive on 122.5 MHz. In the case of two or more controlled fields in close proximity, each is assigned a different one of the above listed three frequencies (the next most popular is 122.7 MHz) in order to avoid confusion. If your plane has the equipment, *transmit* to the tower as well as receive on the tower's assigned frequency of, for instance, 119.1 MHz, rather than the 122.5, etc. Normally the three frequencies are used only by aircraft with limited transmitting facilities. *Keep in mind that some towers can transmit on the localizer frequency also.*

UNICOM

Unicom is intended as an aid to the pilot operating from smaller airports. It's mighty handy to call in and get the wind direction and velocity at the airport equipped with Unicom (but no tower) or maybe have transportation ready when you arrive.

Unicom is the "private pilots' frequency" and is for your convenience. Unfortunately, the privilege is sometimes abused. The technical term for Unicom is "aeronautical advisory service." Get the word *advisory;* that's exactly what it means. If an airport operator gives you the wind and gets carried away and gives the pilot "landing clearance" or "take-off clearance" he is sticking everybody's neck out a mile. *You are still on your own, even though talking to Unicom.*

The smaller airports transmit and receive on 122.8 MHz. Those airports having this service are noted by U-1 in the AIM and U on sectional charts.

Controlled airports having Unicom at one of the flying services on the field are designated by U-2 in the AIM and transmit and receive on 123.0 MHz.

MULTICOM

This is a frequency (122.9 MHz) used to provide communications essential to conduct activities being performed by, or directed from, private aircraft. It

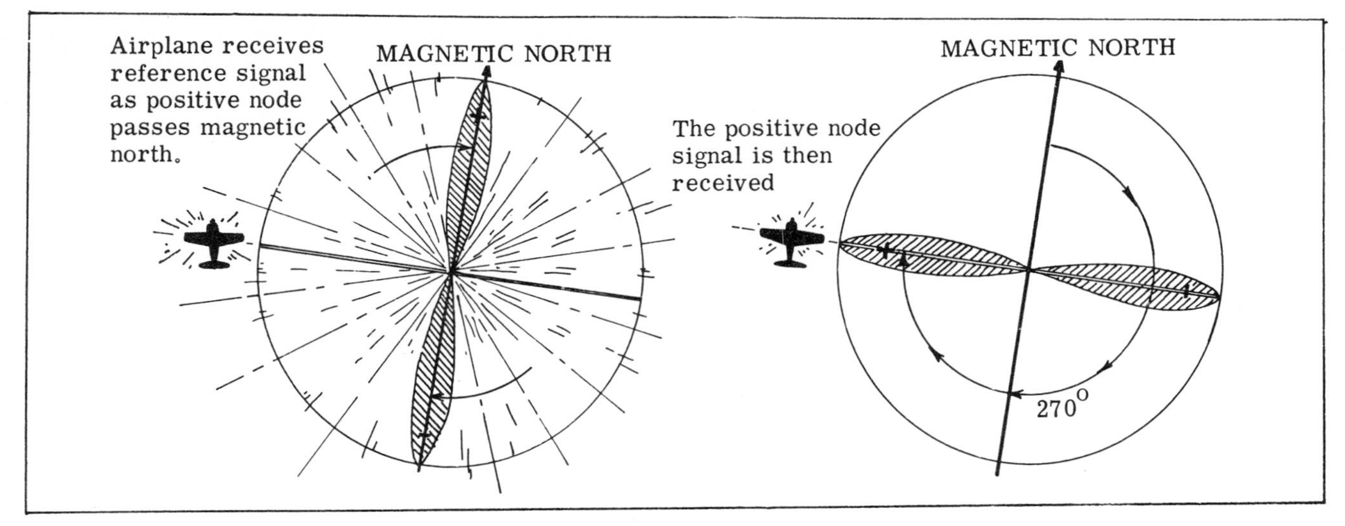

Fig. 19-8.

is sometimes considered the "private pilot's air-to-air communication," and has taken some of the pressure from the Unicom frequencies. Multicom, like Unicom, is not to be used for purposes of traffic control. Both are too often used as a private line for chatting—which was not the purpose in setting up these frequencies.

RADIO NAVIGATION

THE VHF OMNIRANGE OR VOR (108.2 – 117.9 MHz)

It would be well to review the workings of the VOR briefly at this point.

Your omni receiver in the airplane is able to measure your position relative to the station by electronic timing.

The omni station puts out two signals. One is omni-directional; this signal is transmitted in all directions simultaneously at the rate of 30 times a second.

The other is a rotating signal which is turning at the rate of 30 revolutions per second (clockwise). The rotating signal has a positive and a negative side.

The all-directional or reference signal is timed to transmit at the instant the positive side of the rotating signal passes *magnetic* north.

So, let's suppose that instead of 30 times a second, the rotating signal turned one time a minute (and the reference signal also flashed once a minute as the positive side of the rotating signal passed north). Suppose that at your geographic position your omni set receives the reference signal and 45 seconds later receives the rotating signal. Your position is 45/60 or 3/4 of the way around (clockwise); you are somewhere on the 270° radial. (Fig. 19-8)

The airplane's VOR receiver is composed of four main parts (Fig. 19-9):

1. Frequency selector.
2. Azimuth or bearing selector, calibrated from 0 – 360. (Called also OBS or omni bearing selector.)
3. Deviation indicator or left-right needle.
4. TO-FROM indicator.

You are in the vicinity of a VOR and want to fly directly to it. You would: (1) Tune in the correct frequency and identify the station. (2) Make sure that the VHF receiver is selected on VOR. Some older sets may have VOR, *LOC*alizer and *COM*munications selections. You'll hear the identification on LOC or COM but won't be getting an omni bearing. On crystal controlled sets this is taken care of automatically.
(3) Turn the omni bearing selector until the TO-FROM indicator says "TO" and the needle is centered. The omni bearing giving you this result is your *magnetic course TO the station.*

You'll most likely have to correct for wind and will take up a heading that will keep the needle centered. If you are going TO the station and the indicator says "TO," fly the needle. If the needle is to the left, turn left and fly until the needle is centered.

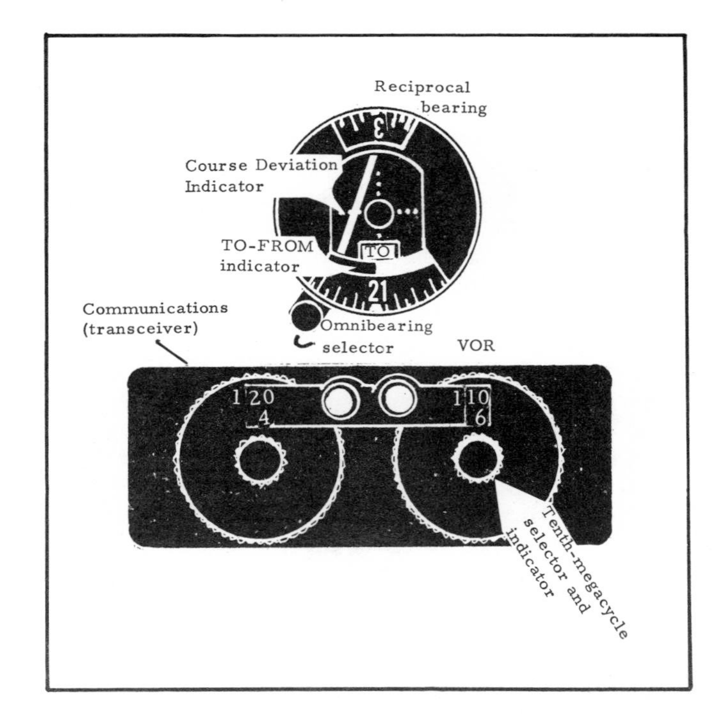

Fig. 19-9. The VOR receiver.

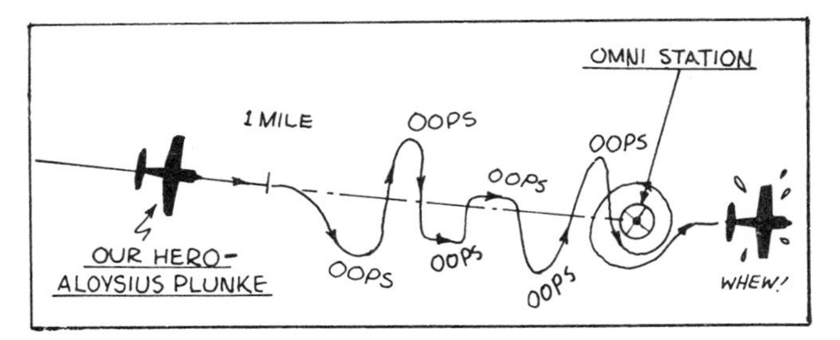

Fig. 19-10. Aloysius Plunke always does pretty well until he gets close to the station, at which point he has a slight tendency to chase the omni needle.

You are then back on your pre-chosen course and will set up a wind correction to keep it centered. The left-right needle is set up so that a full deflection to either side means that you are off 10° or more. A half deflection to either side would mean that you are about 5° from your selected bearing. If the needle is pegged to one side you won't immediately know whether you are 11° or 50° off, without re-centering the needle with the bearing selector and reading your bearing. As you approach the station don't chase the left-right needle. This is the most common problem with the pilot new to flying VOR. If the needle is deflected halfway from center (5°) at a distance of 60 miles from the station, you'll probably be about 5 miles off course. At a distance of 2 miles this would mean a deviation of about 1/6 of a mile or about 900 feet. At one mile this distance would be about 450 feet. If you can consistently fly within even a half a mile of a ground check point on a cross country, you're satisfied, so don't make a lot of work for yourself using omni. (Fig. 19-10)

All of the above assumed that the VOR and your receiver are perfectly calibrated which is certainly not likely to be the case, so don't bother to be too precise close-in. You shouldn't become a wastrel and ne'er-do-well with the omni but don't overwork yourself, either.

As you cross the station the left-right needle will quiver and may peg to either side and the TO-FROM indicator will switch to "FROM." If you continue on the same course, the needle will be flown normally as before.

Always have the bearing selector set to the course to be followed — then the needle will sense correctly.

The VOR has the great advantage of being usable from all directions and is not affected by thunderstorms as are low frequency radios.

One disadvantage to the VOR: being VHF, it's line of sight, so if you're low you may not be able to pick it up at any great distance.

One misunderstanding pilots have when introduced to the idea of the VOR is that the airplane's heading has an effect on the OBS and TO-FROM needle indications at a particular time.

Suppose you tune in a VOR (identify it) and the OBS indicates 250 degrees when the needle is centered and the TO-FROM indicator says "FROM." The receiver merely tells your position relative to the VOR. If you made a tight 360 degree turn the indicators would remain the same because you would still be at a position of 250 degrees FROM the VOR. (The needle may flicker as the plane blankets out the antenna in a certain part of the turn.) If you made a very wide 360, entailing covering of a pretty large area the left-right needle likely would not stay centered because the airplane would be moving to one side of the selected radial. To go to the station you would rotate the OBS until the needle was centered and the TO-FROM indicator said "TO." Fine, now the set tells you that the bearing TO the station is 070 degrees. What you do with this information the VOR and your receiver couldn't care less. *You* are the one who's being paid to think. If you decide to track into the VOR with the OBS still set at 250° and fly opposite to the left-right needle on an inbound track of 070° -- fine, that's your business. The VOR has done its job of giving your geographical bearing and the rest is up to you.

THE RADIO BEACON

The radio beacon is a ground station operating in the LF/MF band. There are UHF (ultra-high frequency) radio beacons in use by the military but you won't have the equipment to use them.

The majority of the low frequency radio beacons are found in the 190-415 kilohertz band (the AM home radio runs from about 550-1650 kilohertz) and a few fall in the 415-544 kHz area.

An advantage of the radio beacon is that it is nondirectional, or can be approached from any direction.

You may use either an MDF (Manual Direction Finder) or an ADF (Automatic Direction Finder) in the airplane to navigate with the radio beacon.

The principle of operation for either piece of equipment is the same, the major difference being that the ADF has a sensing antenna that automatically turns the loop to the null position. Perhaps a little background should be given here.

You have found that reception of a portable radio is directional. You can pick up a station more clearly by turning the radio to a certain direction in relation to the station. One property of a loop antenna is that the maximum reception is found when the plane of the loop is in line with the station and the minimum when the loop is broadside to the station. (Fig. 19-11)

One way to remember the loop principle is that a lot of the signal slips through the hole (this is hardly specific but will do for a memory aid, anyway).

The weakest reception (null) is the most sensitive to change. In other words, the minimum reception position is a very narrow and pronounced arc, whereas the maximum or strongest reception is found over a comparatively large arc. (Fig. 19-12)

Therefore, the null position, being a more accurate measure of direction, is used. Okay, so you have the loop turned so that a minimum signal is received, the signal is "leaking through the hole" but from which direction? This is one of the problems with a manual loop — the problem of ambiguity. One method of solving it is to orient the loop so that the weak signal position is toward each wingtip (or the plane of the loop is parallel to the fuselage) and turn

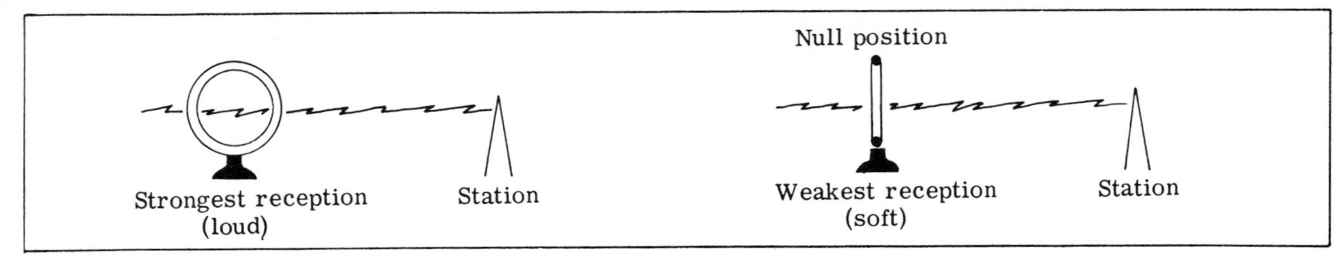

Fig. 19-11.

the airplane until the null is found again. An example is in order:

Suppose that you are temporarily disoriented on a cross country and want to home in on a radio beacon. You tune in the correct frequency and identify it. Assuming that your loop is manual, you find the null when the loop is oriented 30 degrees in front of the right wing and 30 degrees behind the left wing. The station is somewhere along that line – but which way? (Fig. 19-13)

disappears as you fly and you rotate the loop until you find it again – which happens to be at a loop indicator position of 110°-290°. The station is off to the *right*, at a relative bearing of 110°, because the right half of the needle had to be moved backward to find the null. (Fig. 19-15)

So, if the right side of the position indicator has to be moved back, the station is to the right. Notice that your present magnetic heading is 020°, for example purposes, so that your magnetic heading to the

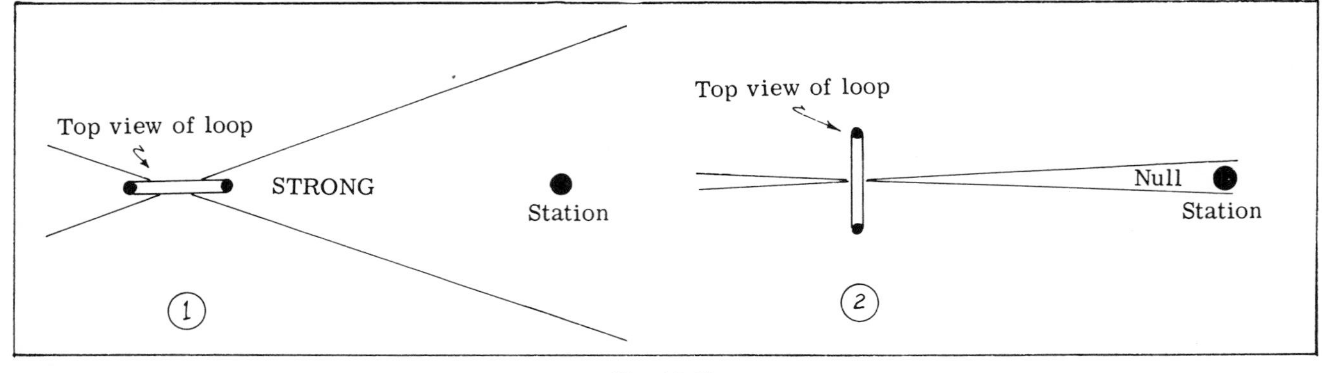

Fig. 19-12.

You would know from the cockpit loop-position indicator that to turn to the left would be the quickest way to get the station off a wingtip. You would manually turn the loop until the plane of the loop was parallel to the fuselage (the null line was parallel to the wingspan line) and then turn to the left until you picked up the null again. Note the airplane heading and maintain it. You are now flying at a heading of approximately 90° to the line of the station – the station is either 90° to the left or 90° to the right – and you don't know which way yet. (Fig. 19-14)

As you fly, the station will gradually move behind the wing (which wing?). Let's say that the null

station is 020° + 110° = 130°. You would start your turn to the right to go to the station and in the meanwhile would turn the loop indicator to the 0°-180° relative position. As long as you have the null, the airplane's nose is pointed toward the station. (Fig. 19-16)

You can readily understand that should there be a wind from the side, in keeping the nose pointed at the station your path would be like that shown in Fig. 19-17.

One solution would be to set up the wind correction. Suppose that the correction is to be 10° to the right; the problem would look like Fig. 19-18. You

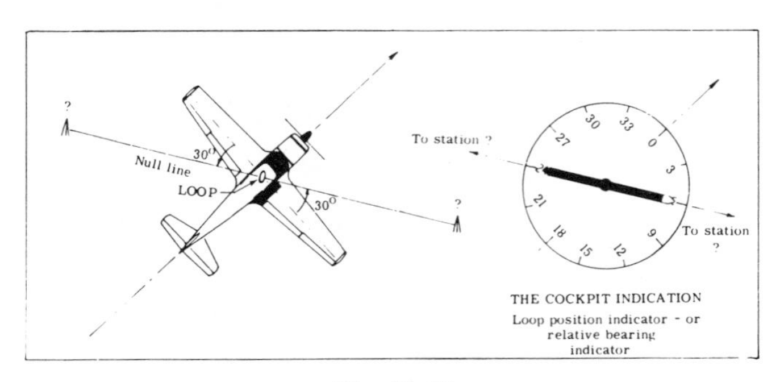

Fig. 19-13.

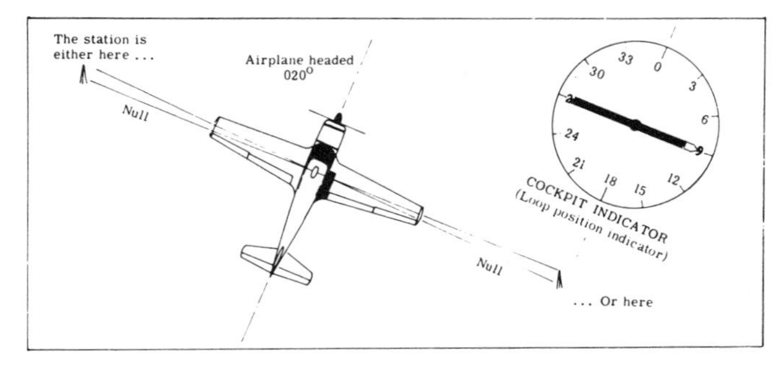

Fig. 19-14.

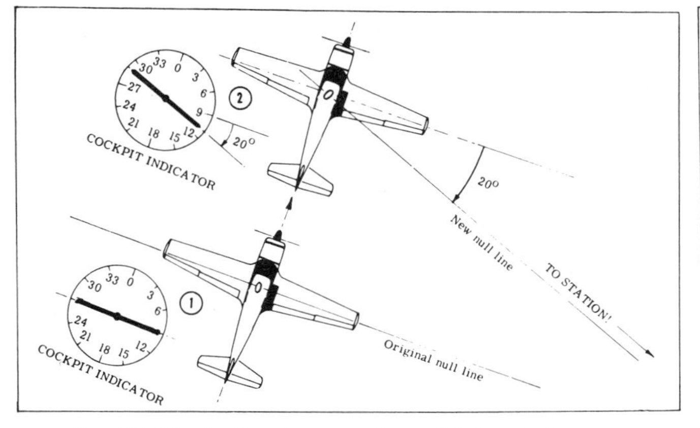

Fig. 19-15. As the plane continues (1) the station moves behind the wing(2).

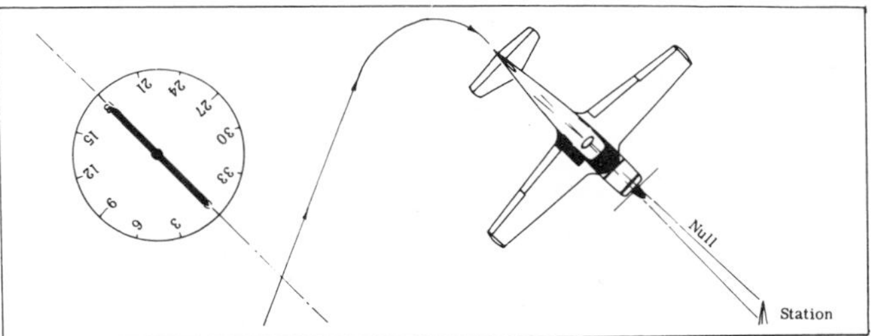

Fig. 19-16.

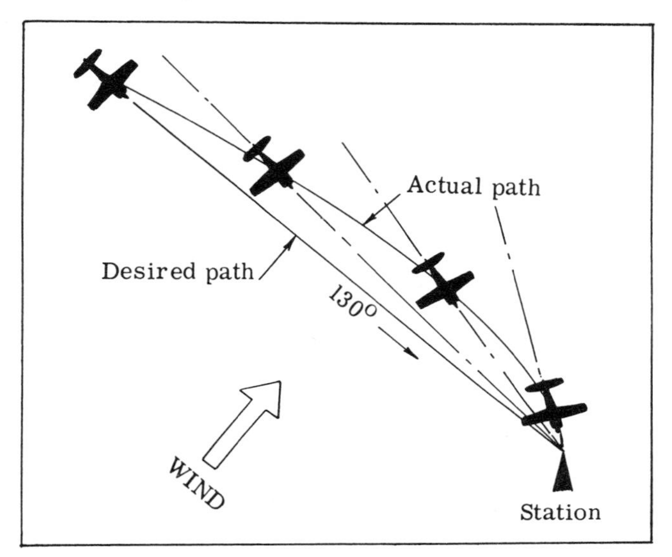

Fig. 19-17. The airplane's path with a crosswind if the nose is kept pointed at the station (homing).

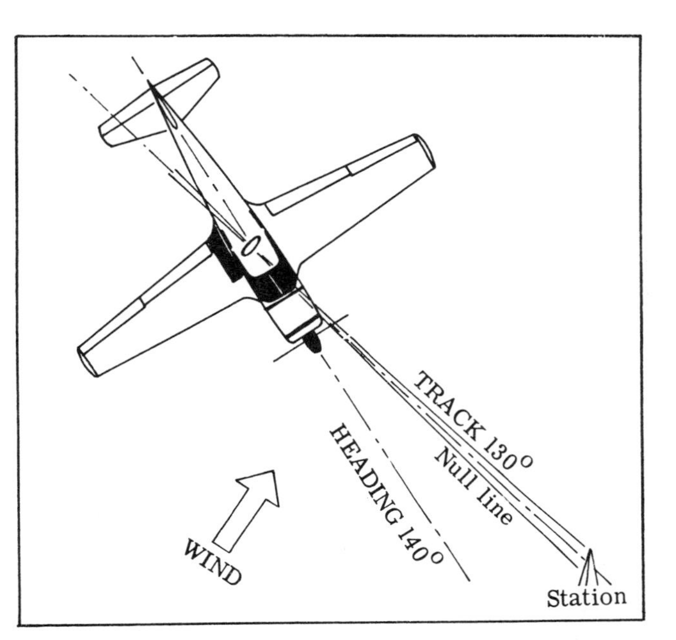

Fig. 19-18. Correcting for wind (tracking).

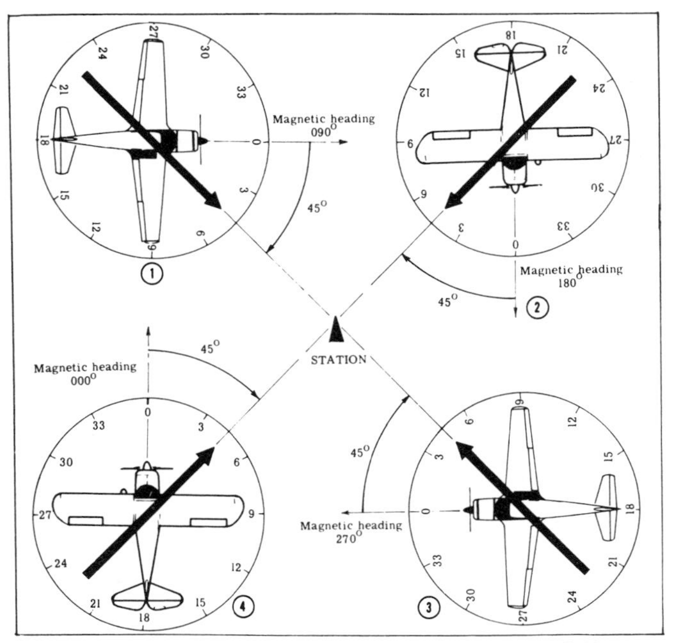

Fig. 19-19. All four airplanes have the same relative bearing indication, even though they are headed in different directions.

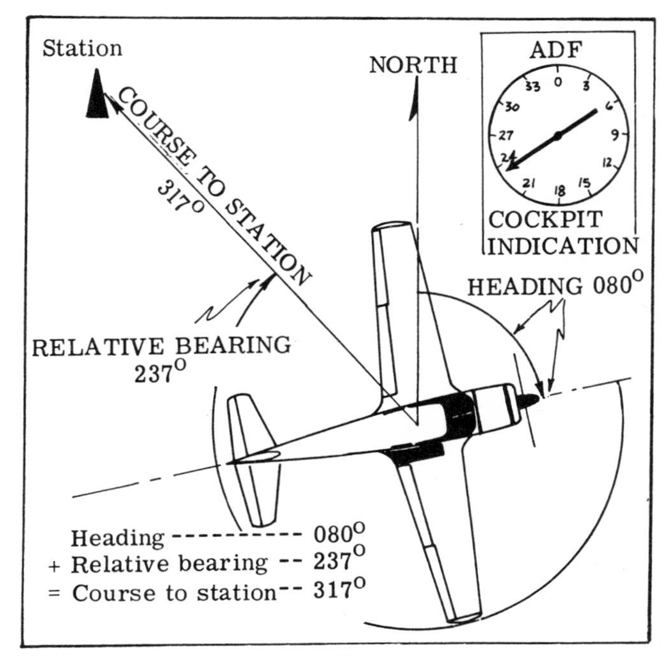

Fig. 19-20.

would make the 10° correction to the right and would turn the loop 10° to the left so that it would still be pointing down your magnetic course of 130°.

Note that *magnetic* heading or course is used here, which means that you would correct your magnetic compass indications for deviation. In practice, deviation is normally assumed to be negligible and the compass is considered to have no error.

The steps just described seem like a great deal of trouble -- and they are. The manual direction finder has pretty well been replaced by the ADF. MDF still has its place in the vicinity of thunderstorms as it is not as affected by atmospheric conditions as the ADF.

The big advantage of the ADF is that the needle points to the station -- the problem of ambiguity is solved electronically. The loop principle is basically the same for both types.

Assuming that you have the ADF set properly tuned and there are no thunderstorms in the near vicinity, the ADF will give you the immediate *relative* bearing of the station from you. (Fig. 19-19)

All four of the airplanes would have an indication as shown on the ADF indicator. The station would be the same *relative* bearing from each airplane but each would have to fly a different course to the station. *Assuming no wind:*

Airplane 1 -- Heading (090°) + Relative Bearing (045°) = 135° heading and course to station.
Airplane 2 -- $180^{\circ} + 045^{\circ} = 225^{\circ}$ course and heading to station.
Airplane 3 -- $270^{\circ} + 045^{\circ} = 315^{\circ}$ course and heading to station.
Airplane 4 -- $360^{\circ} + 045^{\circ} = 045^{\circ}$ course and heading to station.

Suppose that you were headed 080° and the needle showed a relative bearing of 237°. What is your magnetic course to the station? ($080^{\circ} + 237^{\circ} = 317^{\circ}$.)

Fig. 19-20 shows why this is so.

If you are headed 270° and the relative bearing to the station is 270°, obviously you'll fly a magnetic course of 540°. Maybe *your* compass rose doesn't go any higher than 360° -- none of them do. Now what do you do? Why, you subtract 360° from 540° and come up with an answer of 180° heading to the station. Another way to look at it is that you're headed 270° and the needle is pointed 90° to your left (the 270° position) so the heading to the station must be $270^{\circ} - 90^{\circ} = 180^{\circ}$. (Fig. 19-21) (Subtract degrees left or add degrees right.)

(The cockpit indicators have been turned for easier reading of Figures 19-20 and 19-21.)

For VFR purposes, if you want to go to a station tune it in, identify it and turn the airplane until the ADF needle is at the 0° position on the indicator. The station is dead ahead and you can read your heading from the directional gyro or the compass. Under no-wind conditions that heading will take you to the station.

Take the earlier problem of the MDF: Under no-wind conditions the heading (and course) was 130°. It was found that a 10° wind correction to the

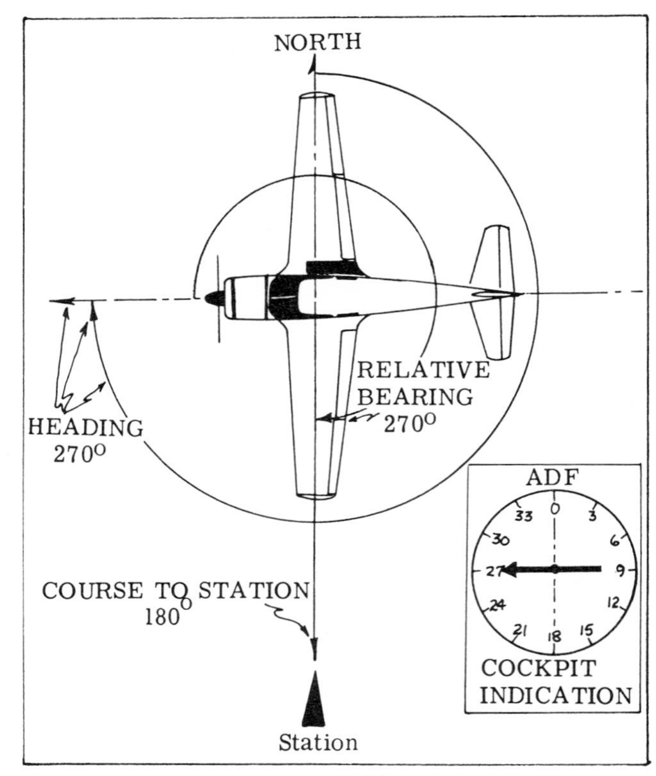

Fig. 19-21.

right was needed. Fig. 19-22 shows how the ADF indicator and compass would look. (Assume no deviation.)

If, as you hold the same heading the relative bearing doesn't change, the heading chosen is correct.

Suppose that you are still holding a 140° heading and notice that the needle is now indicating 20° to the left, or 340° relative, instead of 10° left, or 350° relative, as at the start. It means that you aren't on course anymore -- you've either under- or overcorrected for the wind. To find out where you stand:

Heading $140^{\circ} + 340^{\circ} = 480^{\circ}$; $480^{\circ} - 360^{\circ} = 120^{\circ}$ to the station or:

Heading $140^{\circ} - 20^{\circ}$ (left indication) = 120° to the station. You have overcorrected for wind. (Fig. 19-23)

The ADF is usually the piece of radio equipment that is hardest for the new pilot to understand. Whereas the VOR and low frequency range operations are based on fixed *geographic* bearings, the ADF needle indicates only the *relative* bearing to the station and this is sometimes hard to visualize. After you have used the ADF more you'll see in your mind's eye the relationship between you and the station.

Fig. 19-24 shows a radio compass (ADF/MDF) console.

You'll be able to use commercial broadcast stations as well as radio beacons for homing. The commercial broadcast stations will only give an identification during station break but the radio beacon gives a continuous identification in Morse code.

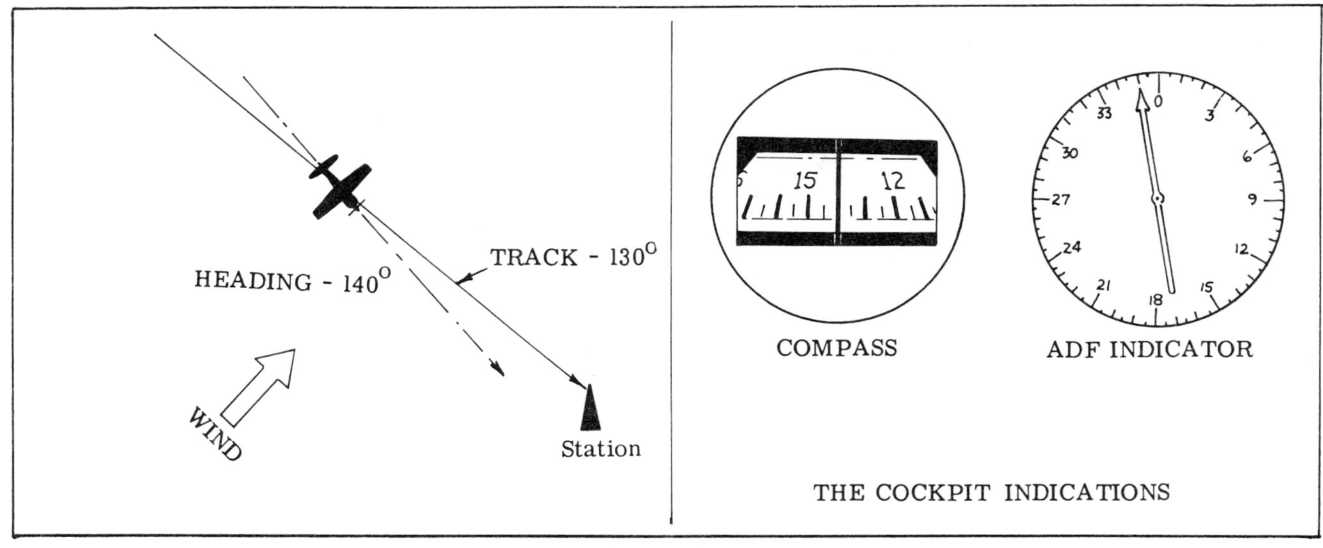

Fig. 19-22.

COMPASS LOCATORS

While it's not the purpose of this book to get you ready for an instrument rating, it would be well to cover the compass locators, or lower powered radio beacons associated with the ILS (Instrument Landing System). Even as a VFR pilot you can help locate a strange large airport by turning in a compass locator and homing into the airport. It's best to use the LMM (the compass locator at the middle marker) if available, because it is only about one-half mile from the approach end of the ILS runway; the LOM (compass locator) at the outer marker is anywhere from $3\frac{1}{2}$ to 7 miles out from the runway — which wouldn't be a lot of help if the visibility is 3 miles.

Figure 19-25 shows an instrument approach chart to Bangor International Airport, Bangor, Maine. The frequency for the compass locator at the outer marker (LOM) is 227 kilohertz (Identification "BG"). This, like most ILS approaches now, uses no LMM (middle marker compass locator, which would normally have the identification in Morse of "GR"). In other words, the three letter localizer identification of I-BGR (all ILS designators are preceded by the letter I) is used for identifying the LOM and/or LMM.

The minima section of the approach plate will likely be of little interest to you as a VFR pilot but you might look over the plate to pick out some pertinent points. (Don't try to use Fig. 19-25 for navigation purposes, it'll likely be out of date.)

One advantage that the radio beacon has over the VOR is that it is not "line of sight"; you can receive it at low altitudes, as it is low frequency, just below commercial broadcast bands.

Most of the less expensive sets only have ADF as this is the setting most used, and the airplanes using this type normally do not fly in weather conditions requiring the use of MDF. The LF/MF range (4 legs) is being phased out and will not be covered in this book.

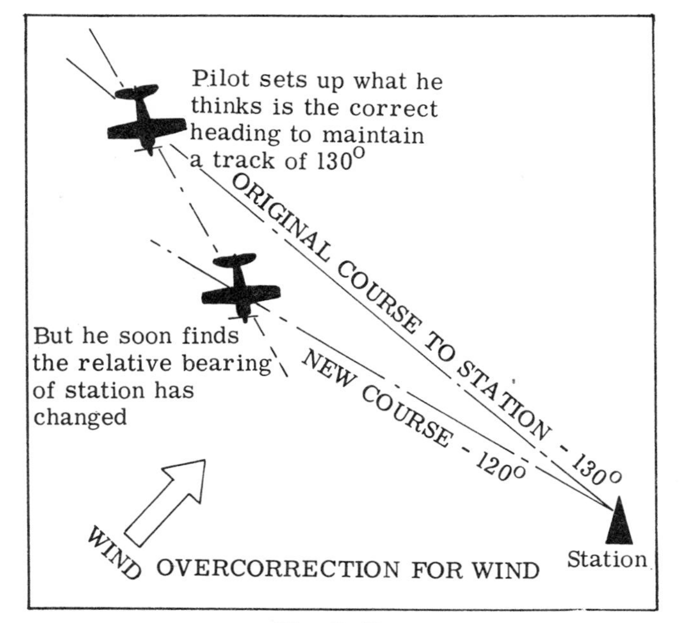

Fig. 19-23.

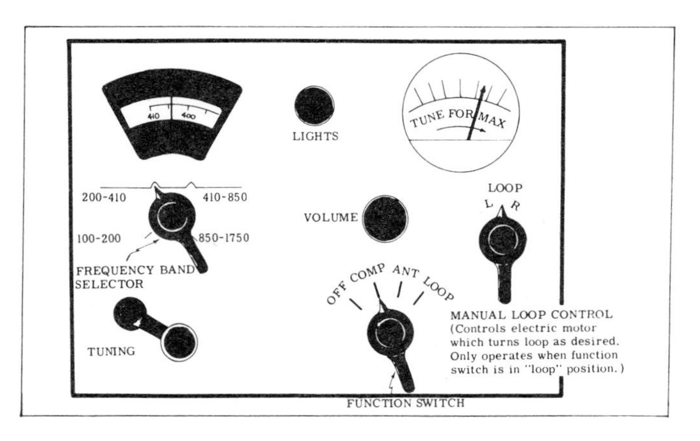

Fig. 19-24. A radio compass console (AN/ARN-7). The function switch is in the ADF(COMP) position. For MDF operation the function switch is set on "Loop" and the loop can then be manually controlled. (Okay, it's an old type ADF, *but* it shows everything you will have to deal with concerning this equipment.)

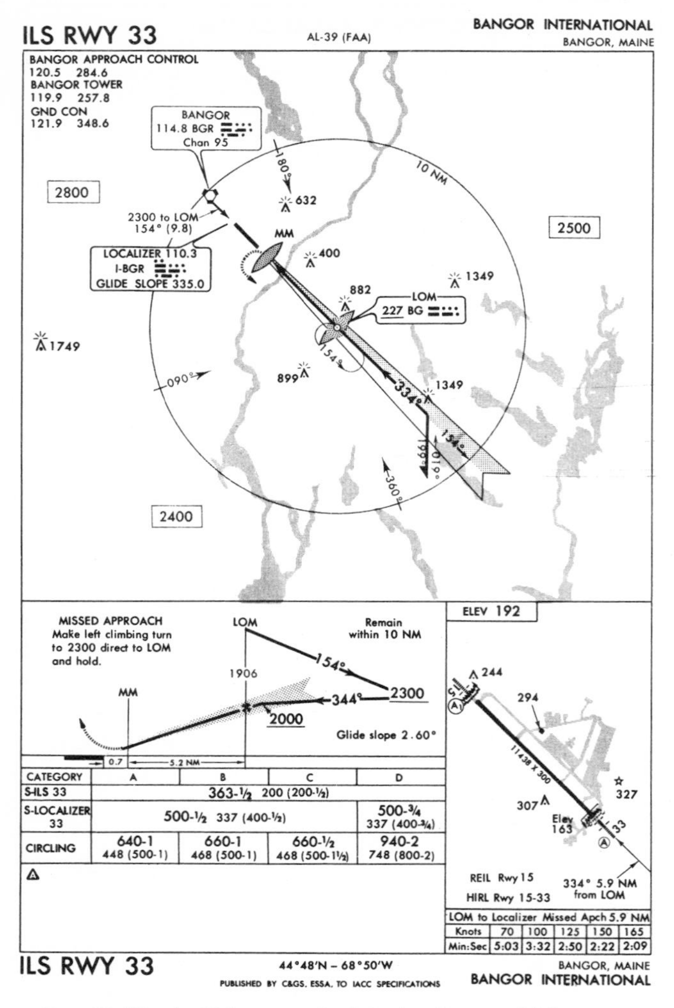

Fig. 19-25. An ILS approach plate for Bangor, Maine.

DISTANCE MEASURING EQUIPMENT — DME

The DME set in the aircraft sends out interrogating pulses at a specific spacing which are received by the ground station. The ground station then transmits a paired pulse back to the aircraft at a different pulse spacing and on a different frequency. The time required for the round trip of this signal exchange is measured by the DME unit in the aircraft and is indicated on a counter or dial as nautical miles from the aircraft to the station.

DME operates on frequencies between 962 and 1213 MHz (UHF) but, because each DME frequency is always paired with an associated VOR frequency the selector of the set may be marked in terms of the VOR frequency band (108.2-117.9 MHz). All you have to do is to know the frequency of the associated VOR and set up the selector of the DME on this number, rather than having to remember the various DME frequencies or "channels."

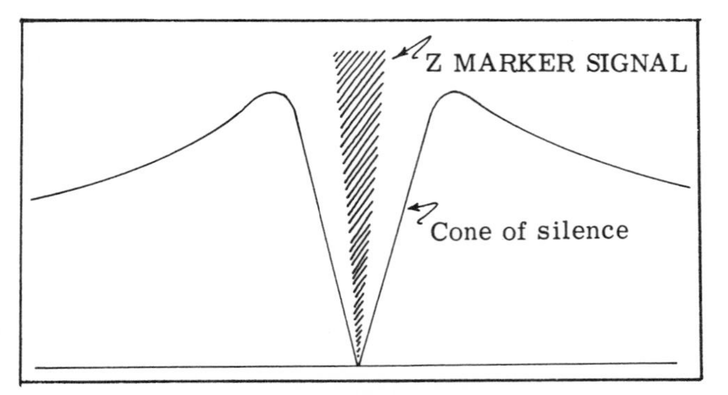

Fig. 19-26. The Z-marker.

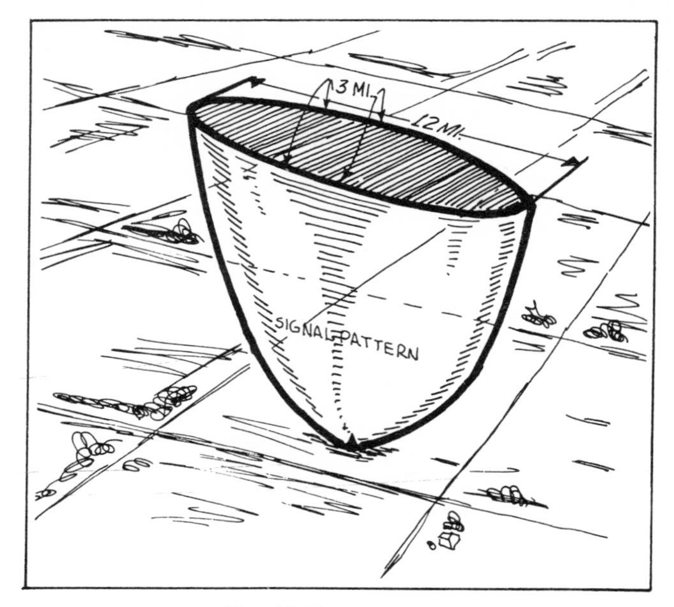

Fig. 19-27. Fan marker.

Like the VOR, the DME depends on line-of-sight reception. Because it is a transmitter (it sends signals, however, rather than words), airplanes so equipped must have the DME listed on the Aircraft Radio Station License (see Chapter 22).

OTHER NAVIGATION AIDS

Z-Marker — A VHF (75 MHz) facility located at the old LF/MF range stations that emits a cone-shaped signal providing a positive identification of the cone of silence position. A receiver in the airplane is triggered by the signal and gives an indication by a tone or a tone and light. The Z-marker is going the way of the LF/MF range (out).

Fan Marker — A VHF (75 MHz) fan-shaped marker which defines a specific area along an airway for navigation and approach purposes. (Fig. 19-27)

Dumbbell or Bone-Shaped Marker — An improved version of the fan marker in that it is "pinched" in the middle for more concise navigation information. (Fig. 19-28)

The airway fan marker is being done away with as it was a part of the old LF/MF range system and only a few of these remain in the U.S. The airway fan marker had a Morse code identifier and transmitted a 3000-cps (cycles per second) signal, as did the Z-marker. (The Z-marker transmitted a continuous tone rather than definite coded signals.)

The fan markers you will be more interested in are the lower powered markers located as a part of the ILS system as was covered earlier. You might take another look at Fig. 19-25.

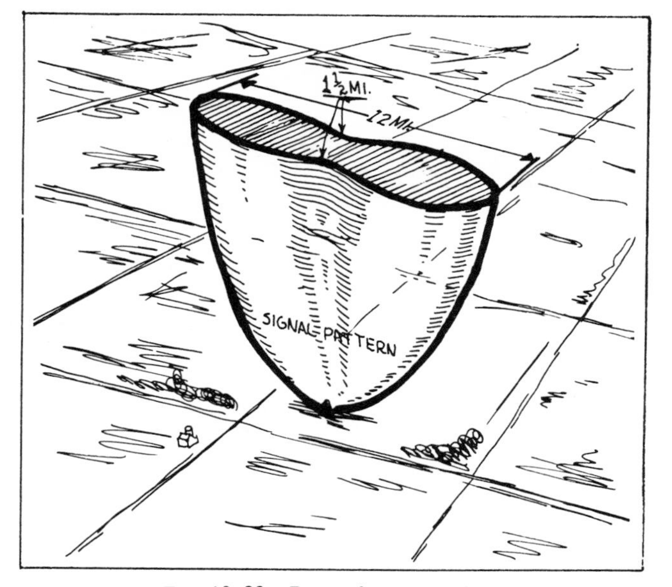

Fig. 19-28. Bone-shaped marker.

The outer marker, as mentioned, is located anywhere from $3\frac{1}{2}$ to 7 miles from the approach end of the runway and operates on a frequency of 75 MHz. The OM is modulated at 400 cps and transmits a continuous series of dashes (about two per second).

The middle marker (MM) is located about one-half mile from the approach end of the runway and like the other markers has a frequency of 75 MHz but is modulated at 1300 cps and transmits a continuous series of alternating dots and dashes.

Airplanes equipped for instrument flying have indicator lights for each type of marker (airways, OM and MM).

SUMMARY OF THE CHAPTER

For more information on the radio it is suggested that you study the *Airman's Information Manual* and the *Pilot's Radio Handbook.*

The FAA navigation aids with voice facilities broadcast current weather, NOTAMS and weather advisories at 15 past the hour.

Use the *Airman's Information Manual* to keep up to date on frequency changes and other navigation and communications data.

Figure 19-29 is the Frequency Utilization Plan as published in AIM.

FREQUENCY UTILIZATION PLAN

AIR NAVIGATION AIDS

108.1–111.9 MHz: ILS localizer with or without simultaneous radio-telephone channel operating on odd-tenth decimal frequencies (108.1, 108.3 etc)

108.2–111.8 MHz: VOR's operating on even-tenth decimal frequencies (108.2, 108.4 etc.).

112.0–117.9 MHz: Airway track guidance. (VORs)

COMMUNICATIONS

118.0–121.4 MHz: Air Traffic Control Communications
121.5 MHz: Emergency (World-Wide)
121.6–121.9 MHz: Airport Utility (Ground Control)
121.95 MHz: Flight Test
122.0 MHz: FSS's, Weather, Selected Locations, Private Aircraft and Air Carriers
122.1 MHz: Private Aircraft to Flight Service Stations
122.2, 122.3 MHz: FSS's, Private Aircraft, Selected Locations
122.4, 122.5, 122.7 MHz: Private Aircraft to Towers
122.6 MHz: FSS's, Private Aircraft
122.8, 123.0, 122.85, 122.95 MHz: Aeronautical Advisory Stations (UNICOM)
122.9 MHz: Aeronautical Multicom Stations
123.1 MHz: Search and Rescue (SAR) Scene of Action
123.05 MHz: Aeronautical Advisory Stations (UNICOM) Heliports
123.15–123.55 MHz: Flight Test
123.3, 123.5 MHz: Flying School
123.6 MHz: FSS's, Airport Advisory Service
123.6–128.8 MHz: Air Traffic Control Communications
128.85–132.0 MHz: Aeronautical Enroute Stations (Air Carrier)
132.05–135.95 MHz: Air Traffic Control Communications

Fig. 19-29. Frequency Utilization Plan. (*Airman's Information Manual.*)

20. WEATHER INFORMATION

WEATHER MAP

Points of equal sea level barometric pressure (actual pressure) on the observed weather map are joined by lines called isobars. The developed system of isobars on the map shows the High and Low pressure areas. The isobar lines are calibrated in millibars rather than inches of mercury; standard sea level pressure being 1013.2 mb (millibars). A conversion from inches of mercury to millibars can be made as shown:

Pressure in inches of mercury x 33.9 = pressure in millibars.

Pressure in millibars ÷ 33.9 = pressure in inches of mercury.

You'll soon get used to thinking of 1013.2 mb as being normal and note deviations above or below this figure. For simplicity, one inch of mercury equals about 34 millibars. Some countries use millibars in the altimeter setting window instead of inches of mercury.

PRESSURE AREAS

High pressure *usually* means good weather. The circulation around a High is clockwise and slightly outward. Low pressure *usually* means bad weather. The circulation around a Low is counterclockwise and slightly inward.

Even though the weather in the High area itself may be good, it can help to bring in bad weather behind it; particularly if a High is followed by a Low moving across the U. S. A strong northerly flow may be induced, bringing warm moist air from the Gulf up to the central or northern part of the country where it is cooled, forming widespread fog and stratus. (Fig. 20-1) This may continue to form until the pressure areas have moved eastward. This condition is most noticeable in the fall or winter.

Another thing that becomes evident from looking at the weather map is the probable wind direction at the surface and lower altitudes, and you can make plans accordingly. A check will show you where a

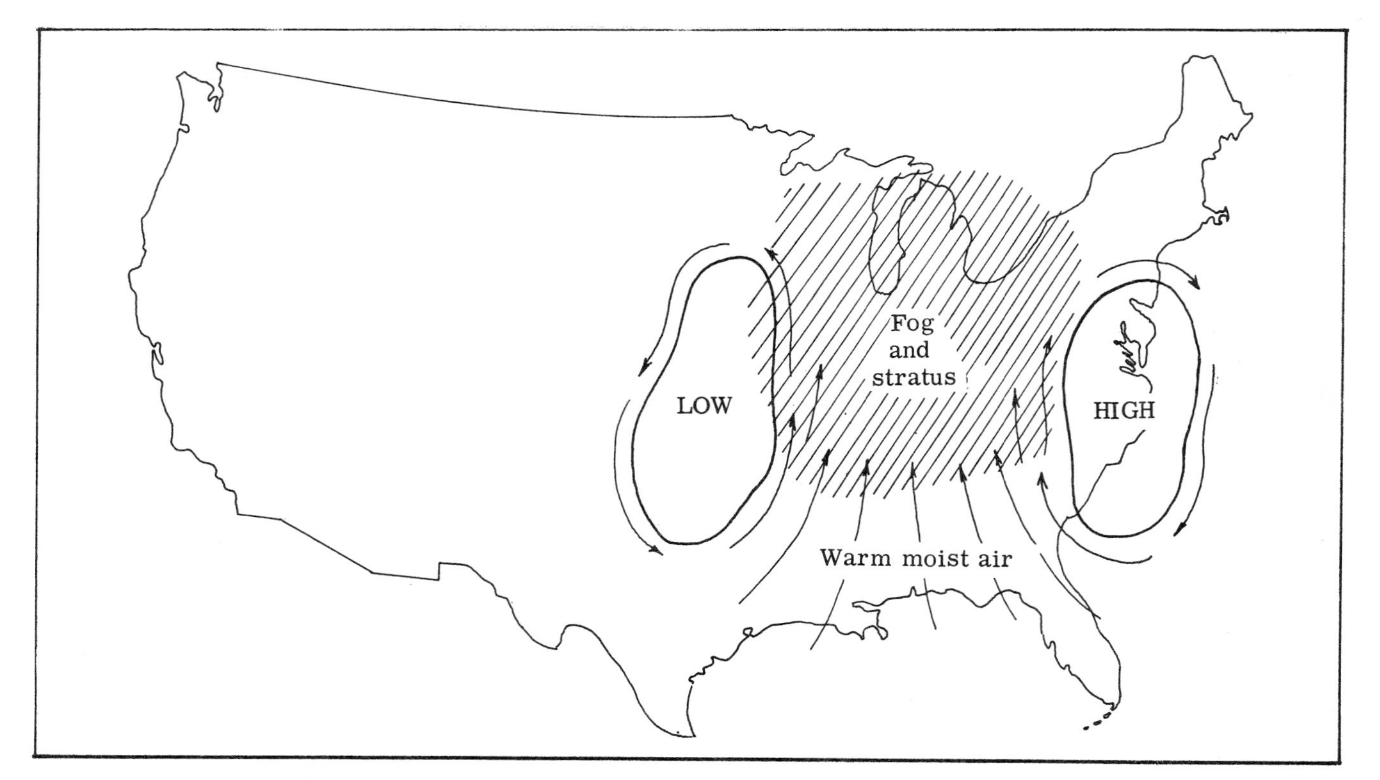

Fig. 20-1.

headwind or tailwind probably will become a crosswind.

If there is a Low in or near your path, be especially careful to check the sequence reports and forecasts for that area.

Elongated High pressure areas are called ridges; the equivalent Low pressure shapes are called troughs.

FRONTAL SYSTEMS

The observed weather map shows the types and positions of the frontal systems in the U. S. at the time of the map release. Fig. 20-2 is a review of the symbols used on the weather map.

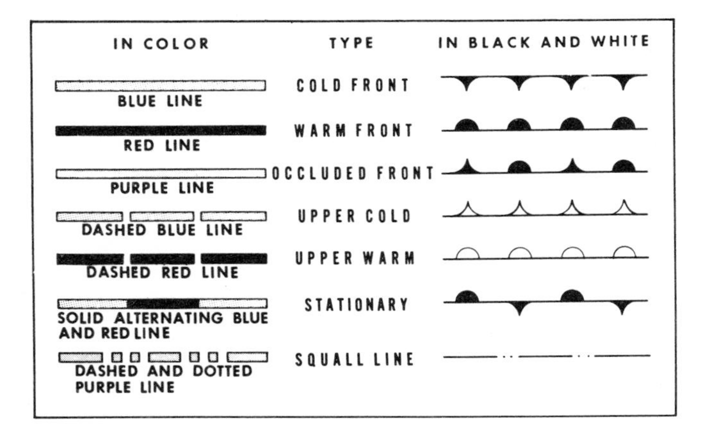

IN COLOR	TYPE	IN BLACK AND WHITE
BLUE LINE	COLD FRONT	
RED LINE	WARM FRONT	
PURPLE LINE	OCCLUDED FRONT	
DASHED BLUE LINE	UPPER COLD	
DASHED RED LINE	UPPER WARM	
SOLID ALTERNATING BLUE AND RED LINE	STATIONARY	
DASHED AND DOTTED PURPLE LINE	SQUALL LINE	

Fig. 20-2. Front symbols used on a weather map *(Pilots' Weather Handbook)*.

The recommendation is for you to glance at the map first before moving on to the sequence reports and terminal forecasts. The map gives a quick *general* picture. The type of frontal systems shown will give you an idea of the weather to be expected in that area.

COLD FRONT

The weather band of the cold front is comparatively narrow; the worst part of the front normally being less than 100 miles from beginning to end. Cumulus, or vertically developed clouds, heavy rain and turbulence usually are associated with it. If the cold front is fast moving, squall lines are likely to form, especially during the summer months, and you may find a solid line of thunderstorms across your flight path if you try to penetrate the front. These squall lines usually are just ahead of the front.

After a cold front passage, the winds will shift from west or southwest to north or northwest. The cold front moves faster than the warm front associated with the frontal system.

Fig. 20-3 shows the circulation about a frontal system.

If you were to slice through a cold front as indicated by A-A in Fig. 20-3 the weather might appear as shown in Fig. 20-4. (You are looking the way the arrows are pointing.)

You can see that the indications of an approaching cold front would likely be cirrus followed by a fairly

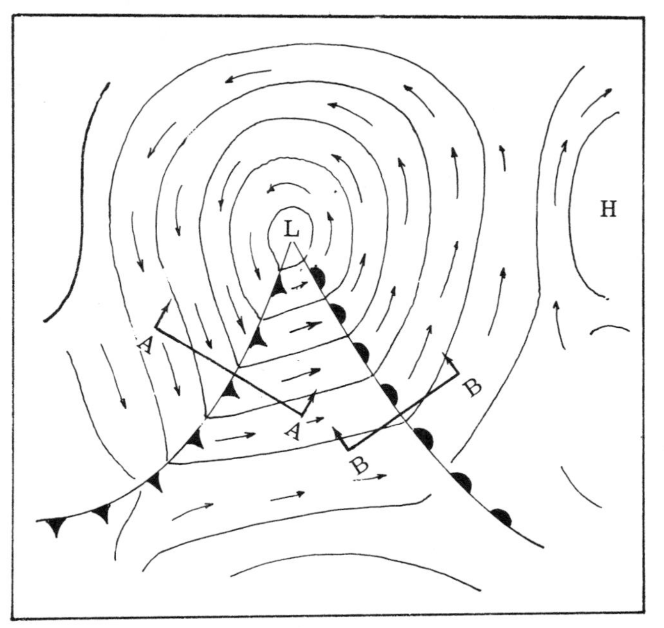

Fig. 20-3. Circulation in the vicinity of a frontal system.

rapid lowering of ceilings followed by comparatively violent weather as the cumulonimbus clouds enter the area. The ceiling raises and breaks up fairly rapidly behind the "average" cold front. The cold front may vary in its characteristics depending on its speed, stability and moisture content of the cold air. The "average" cold front moves from 20 to 30 miles per hour but under extreme conditions may have a speed of 60 mph or more.

Behind the rapidly moving cold front you may expect rapid clearing, gusty, turbulent surface winds and colder temperatures. Flying through a cold front, even if you are instrument rated, can be quite "interesting."

WARM FRONT

The warm front, while normally lacking the violence of the cold front, produces a much wider band

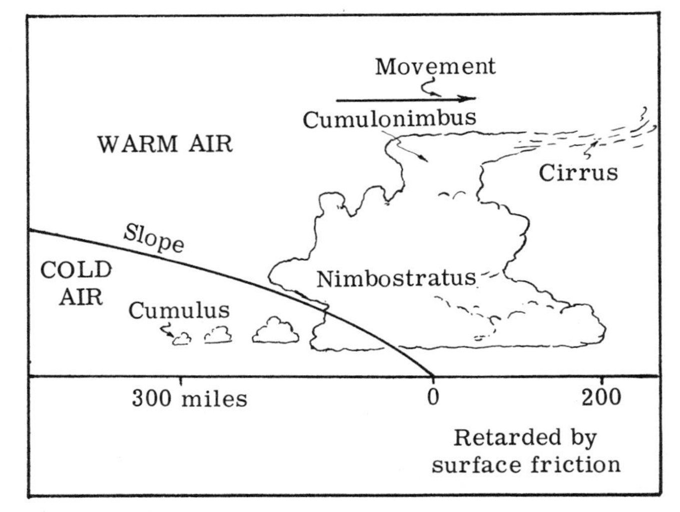

Fig. 20-4. Cross-section of an "average" cold front. The slope varies from 1:50 to 1:150.

of weather. The weather associated with a warm front may be expected to consist of stratus (layer) type clouds with fog and rain. One of the popular misconceptions is that warm fronts *always* consist of stratus type clouds only. If the warm air is unstable, you may find thunderstorms in the frontal area.

The warm front usually moves at about one-half the velocity of a cold front.

Fig. 20-5 is the cross-section of an "average" warm front as indicated by B-B in Fig. 20-3.

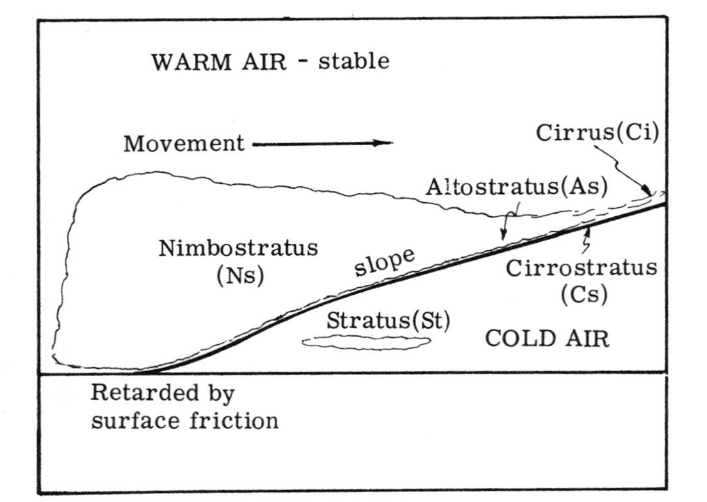

Fig. 20-5. An "average" warm front with the warm air stable. The slope of the warm front may vary from 1:50 to 1:200, the average being about 1:100.

In the fall and winter, check the sequence reports in the area ahead of the front for signs of freezing rain. If the lower altitudes and surface temperatures are below freezing, the rain falling from the warm air above will freeze. (Fig. 20-6)

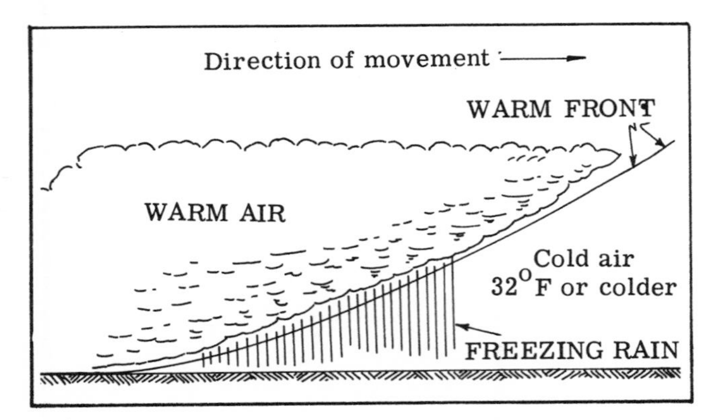

Fig. 20-6. Freezing rain associated with a warm front.

OCCLUDED FRONT

The occluded front contains weather of both warm *and* cold fronts. The cold front, moving faster than the warm front, has caught up with it. You may expect all the disadvantages of both types, with widespread stratus type clouds and build-ups within them. You'll do better to think twice before starting to fly through the occluded portion of the system. There's an old saying that the roughest weather is found in an

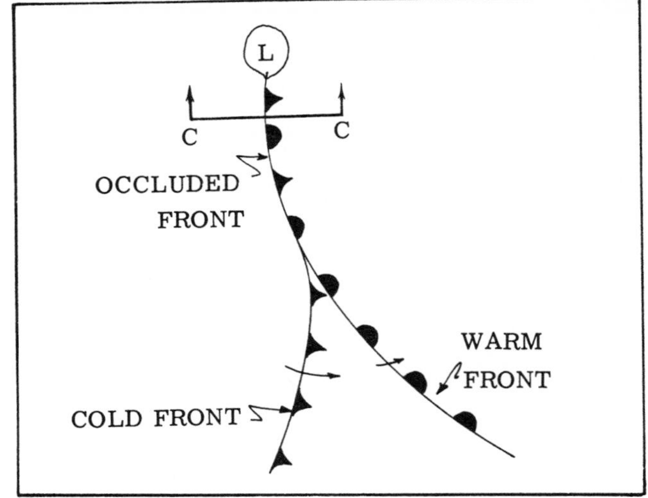

Fig. 20-7. Occluded front.

occluded front within fifty miles of the Low. (Fig. 20-7)

Fig. 20-8 shows the cross-section of an occluded front as indicated by C-C in Fig. 20-7. There are two basic types of occluded fronts and Fig. 20-8 shows the warm type occluded front. If the air in the front of the system is colder (and more dense) than the air behind the cold front overtaking the warm front, it will move up over the cold air and give the cloud cover as shown. Most of the weather will be found ahead of the surface front.

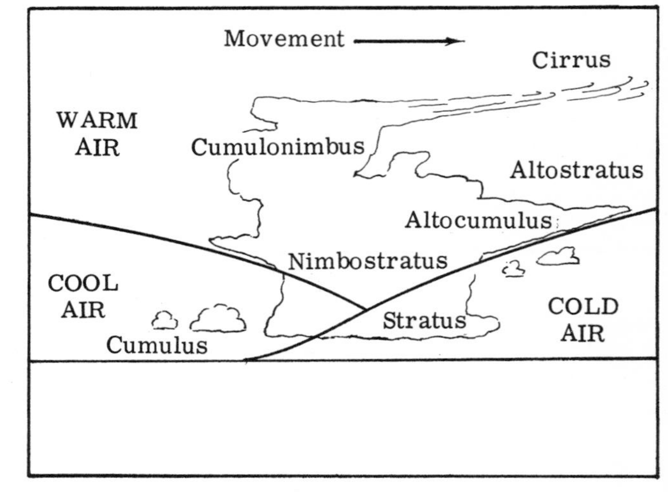

Fig. 20-8. A warm type occluded front.

The cold type occluded front is a situation in which the air behind the cold front is colder than that ahead of the system. It slides under the cool air as shown by Fig. 20-9.

Most of the weather for this type of occlusion will be found near the surface front.

STATIONARY FRONT

Strangely enough, the stationary front is so named because it is stationary. The weather associated with the stationary front is similar to that found in a warm

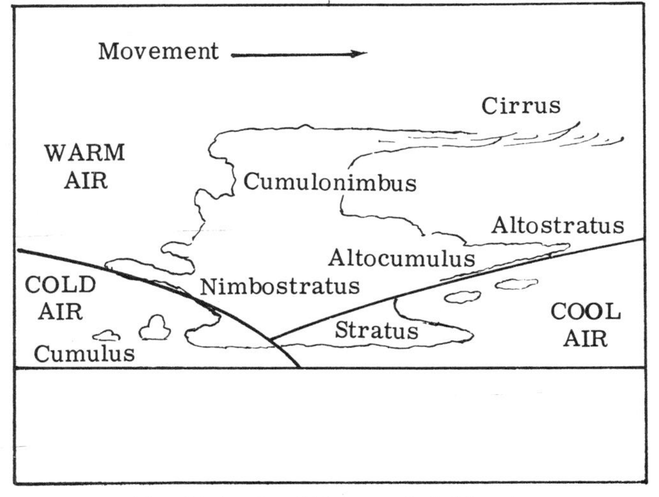

Fig. 20-9. A cold type occluded front.

front but usually not as intense. The problem is that the bad weather hangs around until the front moves out.

CLOUD TYPES

In discussing the weather indicated on the weather map it would be well to take a general look at cloud types you'll expect to encounter in flying.

Cloud types are broken down into four families: High, middle, low clouds, and clouds with large vertical development. Clouds are further described as to their form and appearance. The puffy or billowy type clouds are "cumulus," the layered types are "stratus."

The term "nimbo" (raincloud) is added to clouds that would be expected to produce precipitation.

Fig. 20-10 shows some representative types of clouds.

Normally, flying near clouds of stratus type formation you would expect fairly smooth air. Cumulus clouds by their very nature are the product of air conditions that indicate the presence of vertical currents.

Clouds are composed of minute ice crystals or water droplets and are the result of moist air being cooled to the point of condensation. The high clouds (cirrus, cirrostratus and cirrocumulus) are composed of extremely fine ice crystals. (The biggest puzzle to the layman is that if they are composed of ice – why don't they fall?) For that matter, as the lower clouds are composed of water droplets, why don't they fall also? The answer is, of course, in the comparative lesser density of the moisture as compared to that of the ambient air. When the water droplets become a certain size, rain, snow or ice pellets result, depending on the conditions). Hail is a form of precipitation associated with cumulonimbus type clouds and is the result of rain being lifted by vertical currents until it

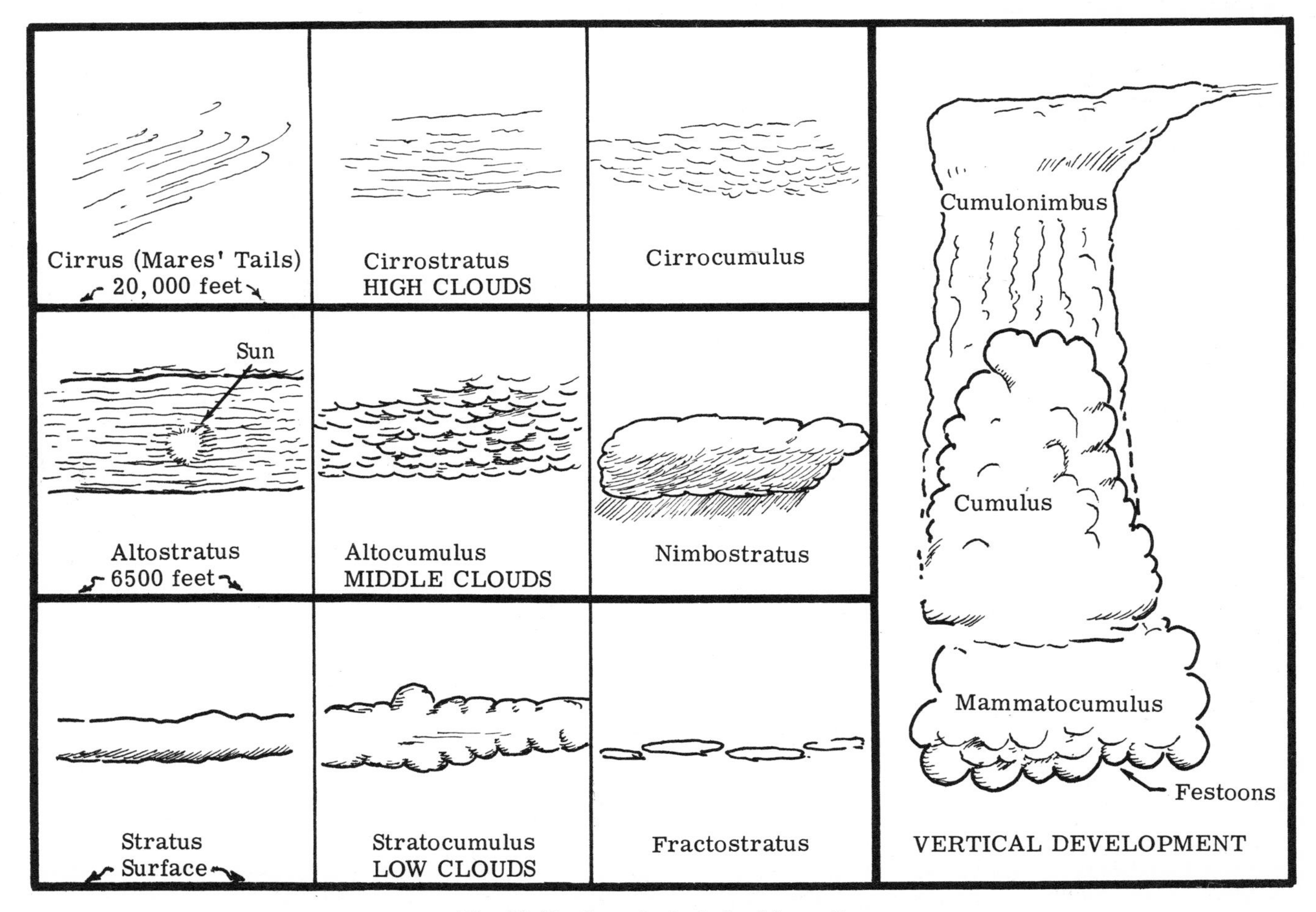

Fig. 20-10. Some typical cloud formations.

reaches an altitude where it freezes and is carried downward again to gain more moisture; the cycle may be repeated several times giving the larger hailstones their characteristic "layers" or strata.

Of course, you wouldn't think of flying into a cumulonimbus because of the extreme turbulence and hail, but remember that turbulence can be found *near* these clouds and hail can fall from the "anvil head" into what could appear to be a clear area.

Clouds may be composed of supercooled moisture and the impact of your airplane on these particles causes them to immediately freeze on the airplane. (Stay out of clouds until you get that instrument rating.)

As was discussed, clouds are formed by moist air being cooled to the point of condensation and this leads to the subject of lapse rates.

For air, the dry adiabatic lapse rate is $5\frac{1}{2}^{\circ}$ F per thousand feet. (Adiabatic is a process during which no heat is withdrawn or added to the system or body concerned.) The normal lapse rate for "average" air is $3\frac{1}{2}^{\circ}$ F, or 2° C. The moist adiabatic lapse rate is the lapse rate produced by convection in a saturated atmosphere such as within a cumulus cloud. At high temperatures it will be 2-3° F per thousand feet and at low temperatures it will be in the vicinity of 4-5° F. The dew point lapse rate is about 1° F per thousand feet.

For cumulus type clouds that are formed by surface heating the base of the clouds may be estimated by the rate at which the dry lapse rate "catches" the dew point. (Dry lapse rate $5\frac{1}{2}^{\circ}$ F, dew point drop is 1° F per thousand so the temperature is dropping $4\frac{1}{2}^{\circ}$ F faster than the dew point, per thousand feet.) Assume the surface temperature is 76° F and the dew point 58° F, a difference of 18° F: dividing this number by $4\frac{1}{2}$ it is found that the temperature and dew point make connections at 4000 feet — the approximate base of the clouds. This only works for the type of cloud formed by surface heating.

SURFACE ANALYSIS

Four times a day a surface analysis is transmitted on the teletype giving locations of high and low pressure areas and fronts. Some smaller airports with teletype service have a plastic overlay on a map of the U.S. and the systems are drawn in by pilots with a grease pencil by referring to an analysis.

Looking at Fig. 20-11 you can see the following: AS WBC 14 Ø75ØZ — The analysis is sent from Weather Bureau Central on the 14th of the month at Ø75ØZ (Greenwich time). The map is valid at Ø6ØØZ on that date.

There is a High (center) of 1024 millibars pressure located 43 degrees (north latitude), 89 degrees (west longitude). (This would put it in the vicinity of Madison, Wisconsin.) There is a Low (center) of 994 mb located at 57 degrees N, 108 degrees W, and another of 994 mb at 55° N, 122° W, etc.

There is a weak cold front which can be located by connecting the points 55° N, 112° W with 50° N, 112° W and the points given by the coordinates

```
 AS WBC 14Ø75ØZ
VALID 14Ø6ØØZ

HIGHS 1Ø24 4389
LOWS 994 571Ø8 994 55112 1ØØ6 29111
2688
COLD WK 55112 5Ø112 46112 43117 4Ø12Ø
STNRY WK 291Ø7 261Ø2 2397
COLD WK 2397 2492 2688
STNRY WK 2688 2481 2776 3Ø7Ø
STNRY WK 571Ø8 5699 5585
WARM WK 5585 5182 4585
STNRY WK 4585 428Ø 4274 417Ø 4165
```

Fig. 20-11. The Surface Analysis is issued every six hours.

following. The other fronts on the analysis would be located by the same method.

At approximately 6-hour intervals, a 12-hour surface prognosis (FS-1) is transmitted, forecasting the future positions of these systems.

SURFACE CHART

Prognostic surface charts are forecast charts and indicate the expected surface positions of fronts and pressure areas 12, 24, 30, 36, 48 or 72 hours from the time of preparation. They are an extrapolation of previous charts and are based on data available at the time of preparation. The system may stop or move faster, leaving a particular prognostic map high and dry; nevertheless it is a good indicator of what you might expect.

Look at several previous observed surface charts to get a trend of movements of the system. You can make your own estimate of the system's probable movement. Fig. 20-12 shows three observed surface charts 12 hours apart (the charts are issued every three hours but a better picture of the changes can be seen here by using intervals of 12 hours).

Taking a look at the isobars on the surface chart you can get an idea of comparable wind velocities; the closer the isobars, the greater the pressure gradient and the stronger the wind. (Isobars are lines drawn between stations of equal pressure.)

You'll notice that nothing has been said about the station model, that cluster of figures and symbols around various stations on the weather map. This is information of primary interest to the meteorologist, not the pilot. If you've had dealings with it, you know that the symbols are very confusing. A late sequence report is of much more value, and you'll find that the station model has little or no practical value to the pilot. Use the time you might spend on memorizing the station model symbols for learning more about sequence reports and forecasts. Charts with station models are found in Weather Bureau Offices so you can always ask questions if necessary.

Take another look at Fig. 20-3. A favorite question on written exams is one that says, "You have been correcting for a left wind, suddenly you have to hold a right correction, what has likely occurred?" Notice that this occurs in crossing perpendicular to

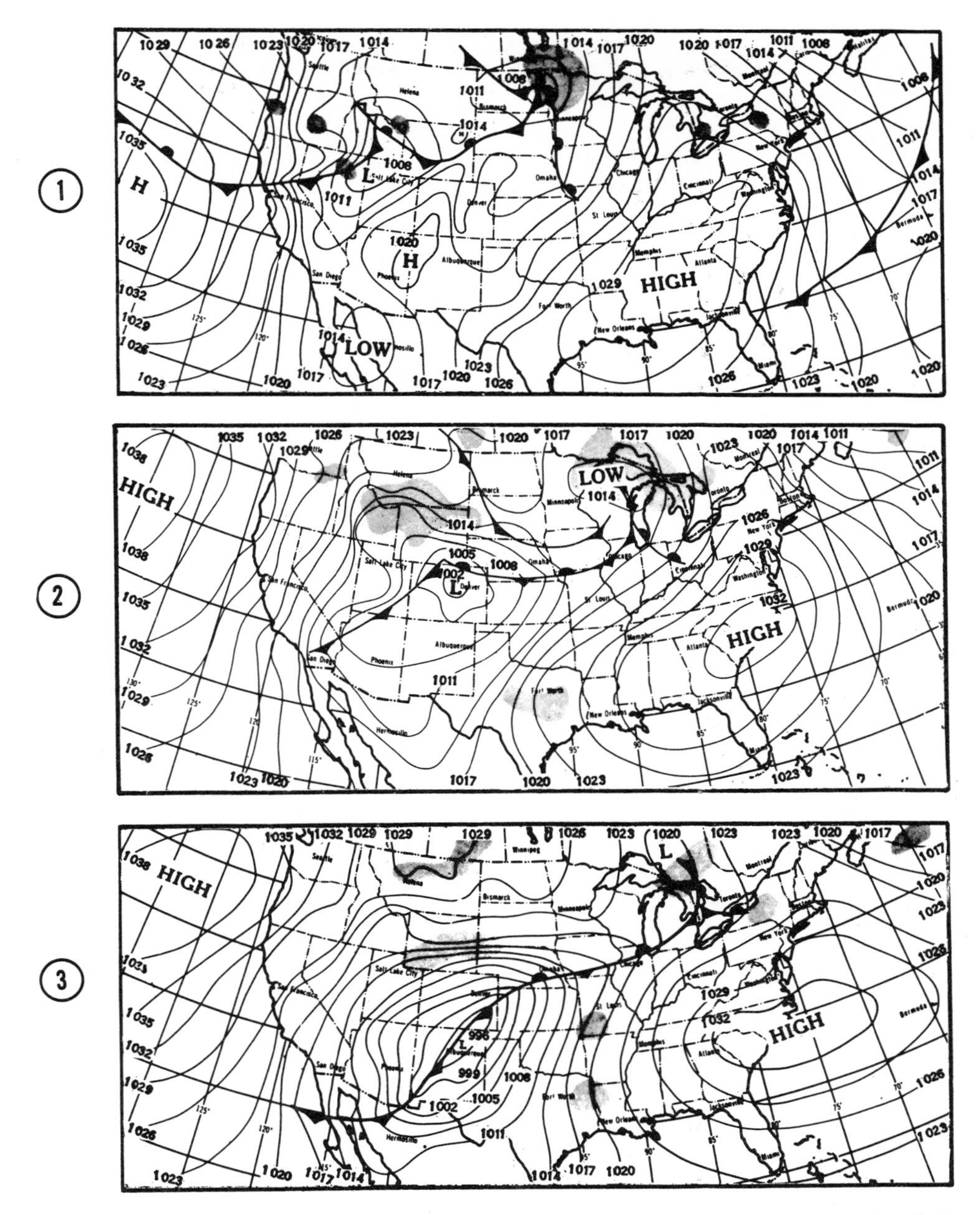

Fig. 20-12. Three weather maps 12 hours apart showing frontal movements. Later charts use 4 millibar intervals between isobars. *(Pilots' Weather Handbook).*

a cold front from either direction. With a warm front it's not such an abrupt shift.

SUMMARY OF THE WEATHER CHART

So far as the surface weather chart is concerned, you should remember the following main points:

1. Low Pressure Area – the circulation is counterclockwise and bad weather normally is associated with it.
2. High Pressure Area – the circulation is clockwise and good weather normally is associated with it.
3. The circulation of the pressure areas may bring warm moist air into your area from large water areas. If the warm moist air is cooled as it moves, it may condense into low clouds, fog, drizzle or other types of visible moisture over a wide area.
4. Cold fronts normally move about twice as fast as warm fronts, and the weather band is narrower and more violent than the warm front. The greater the velocity, the more violent the associated weather is apt to be just ahead of the cold front.
5. Warm fronts move slowly and the associated weather covers a wide area. You may expect low ceilings, fog, and generally poor visibility.

SEQUENCE REPORT

After you have gotten a general picture of the weather by looking at the weather map, check the sequence reports for stations along your route. The front on the map that looks as though it will give you trouble may be a "dry front" with no associated weather. The sequence report and forecasts will be your final criterions as to whether the trip can be made safely or not. Some pilots look at the sequences first and only refer to the weather map if the weather is bad and they want to know why. (Is it a local condition or the result of a large-scale system?) This works quite well and, of course, you will use the method of checking that you prefer. But in any event you should cross-check the various presentations (sequences, maps, forecasts). Of all the information, the sequence reports are the most valuable for pilots. By looking at the sequences for the past few hours you can get a weather trend if no forecasts are available. (Although it's likely that if sequences are available forecast information will be also.)

Sequence reports are issued hourly at Flight Service Stations and Weather Bureau Offices across the U.S. Fig. 20-13 gives a key to the sequence report. Fig. 20-14 shows symbols used in sequence reports and terminal forecasts. (Tables 1 through 4 are mentioned in Fig. 20-13.) Taking a look at Fig. 20-15, an actual hourly report for a few stations, you might note some points: The heading for this particular sequence indicates that it is on Circuit 22 (the U.S. reporting stations are divided into "circuit" areas) and was released on the 3rd of the month (03) at 2000 Zulu time. Zulu time is Greenwich time and for local time you would subtract 5 hours for EST, 6 for CST, etc.

Williamsport, Pennsylvania (IPT) has scattered clouds at 4000 feet with 15 (statute) miles visibility.

The barometric pressure is 1023.3 millibars. (If the first two numbers are less than 50 — 23 is the number here — add a 10 to the numbers; if over 50,

DECODING AVIATION WEATHER REPORTS
Based on Instructions in Federal Meteorological Handbook No. 1, Surface Observations

STANDARD AVIATION REPORT FORMAT FOR MANNED STATIONS

EXAMPLE OF AN OBSERVATION AS FOUND ON HOURLY SEQUENCES

PIT -XM11V⊕38⊕ 7/8VL-FK 146/ 66/65/2713/967/ R10LVR26V55 FK3 /⊕27/CIG 9V12 VSBY1/2V1

VSBY1/2V1	REMARKS: Visibility variable between 1/2 and 1 mile.
CIG 9V12	REMARKS: Ceiling variable between 900 to 1200 feet.
/⊕27/	BASES AND TOPS OF CLOUDS: Tops broken layer 2700 ft. msl. Height of bases not visible at the station precede sky cover symbol. "U" indicates layer amount unknown. If the report is more than 20 minutes old, the time (GMT) precedes the entry.
FK3	REMARKS: Fog and Smoke hiding 3/10 of sky.
R10LVR26V55	RUNWAY VISUAL RANGE: Runway 10L, Visual Range variable between 2600 and 5500 ft. in past 10 minutes. When visual range is constant for past 10 minutes, only the constant value is reported, e.g., R10LVR60+.
967	ALTIMETER SETTING: 29.67 inches. Three figures, representing units, tenths and hundredths of inches, indicate the altimeter setting. "Low" is used preceding figures to indicate values below 29.00 inches.
2713	WIND: 270° true, 13 kts. To decode direction, multiply first 2 digits by 10. If product is >500, subtract 500 and add 100 to speed. Gusts and squalls are indicated by "G" or "Q" following speed and peak speed following the letter.
65	DEWPOINT: 65°F. A minus sign indicates temperatures below zero.
66	TEMPERATURE: 66°F.
146	SEA LEVEL PRESSURE: 1014.6 millibars. Only the tens, units and tenths digits are reported.
L-FK	WEATHER AND OBSTRUCTIONS TO VISION: Light Drizzle, Fog & Smoke. Symbols used in reporting weather and obstructions to vision are in Table 1. Algebraic signs (Table 1) following symbols indicate intensity.
7/8V	PREVAILING VISIBILITY: Seven eighths statute mile and variable by the amount given in REMARKS.
-XM11V⊕38⊕	SKY & CEILING: Partly obscured sky, ceiling measured 1100 ft., variable broken, 3800 ft. overcast. Figures are height of each layer in 100s of feet above ground. A number preceding an X indicates vertical visibility into phenomena. A "V" indicates height varying by amount given in REMARKS. Symbol after height is amount of sky cover (Table 2). The letter preceding height indicates that height to be the ceiling and the method used to determine the height (Table 3).
(blank)	TYPE OF REPORT (Table 4): "R" omitted when observation is in hourly sequence.
PIT	STATION IDENTIFICATION: Identifies report for Pittsburgh by using FAA identifier.

TA B-0-1

Fig. 20-13. See Fig.20-14 for Tables 1 through 4 mentioned here.

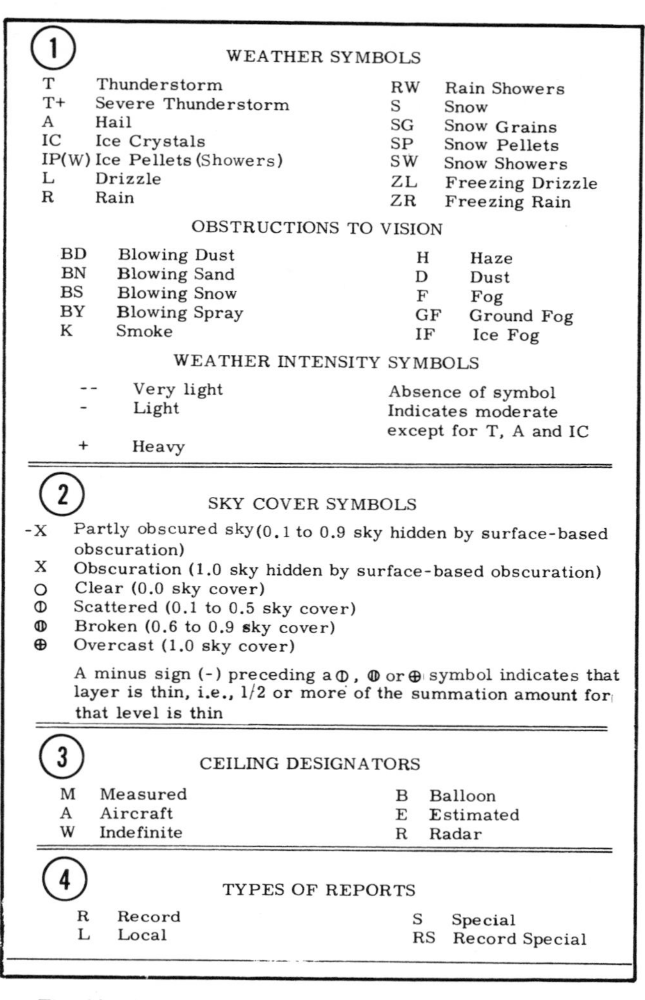

(1) WEATHER SYMBOLS

T	Thunderstorm	RW	Rain Showers
T+	Severe Thunderstorm	S	Snow
A	Hail	SG	Snow Grains
IC	Ice Crystals	SP	Snow Pellets
IP(W)	Ice Pellets (Showers)	SW	Snow Showers
L	Drizzle	ZL	Freezing Drizzle
R	Rain	ZR	Freezing Rain

OBSTRUCTIONS TO VISION

BD	Blowing Dust	H	Haze
BN	Blowing Sand	D	Dust
BS	Blowing Snow	F	Fog
BY	Blowing Spray	GF	Ground Fog
K	Smoke	IF	Ice Fog

WEATHER INTENSITY SYMBOLS

--	Very light	Absence of symbol Indicates moderate except for T, A and IC
-	Light	
+	Heavy	

(2) SKY COVER SYMBOLS

-X Partly obscured sky (0.1 to 0.9 sky hidden by surface-based obscuration)
X Obscuration (1.0 sky hidden by surface-based obscuration)
O Clear (0.0 sky cover)
⦶ Scattered (0.1 to 0.5 sky cover)
⦷ Broken (0.6 to 0.9 sky cover)
⊕ Overcast (1.0 sky cover)

A minus sign (-) preceding a ⦶, ⦷ or ⊕ symbol indicates that layer is thin, i.e., 1/2 or more of the summation amount for that level is thin

(3) CEILING DESIGNATORS

M	Measured	B	Balloon
A	Aircraft	E	Estimated
W	Indefinite	R	Radar

(4) TYPES OF REPORTS

R	Record	S	Special
L	Local	RS	Record Special

Fig. 20-14. Symbols used in sequence reports and terminal forecasts *(ESSA)*.

add 9. For instance, if the number were 733 on the sequence the pressure would be 973.3 mb.

The temperature is 15° F and the dew point is 5° F. If the temperature of the present air falls to the dew point visible moisture may be expected.

The wind is from 240° true at 13 knots with gusts to 20 knots. The altimeter setting is 30.18 inches of mercury.

NOTAM information follows and is separated from the sequence report by an arrow (→) with the name of the station affected. The name is separated from the following NOTAMS by the arrow (↘). The term 1/16 shows that the 16th NOTAM issued in the month of January is still in effect. You would check the NOTAM Summary for the full story (Fig. 20-16) for all of the current NOTAMS in plain (well, fairly plain) language. The other two NOTAMS for IPT concern runway conditions (UR).

Following the ABE (Allentown-Bethlehem-Easton) weather is a note that the NOTAMS follow the sequence for Phillipsburg (PSB), so you can check back.

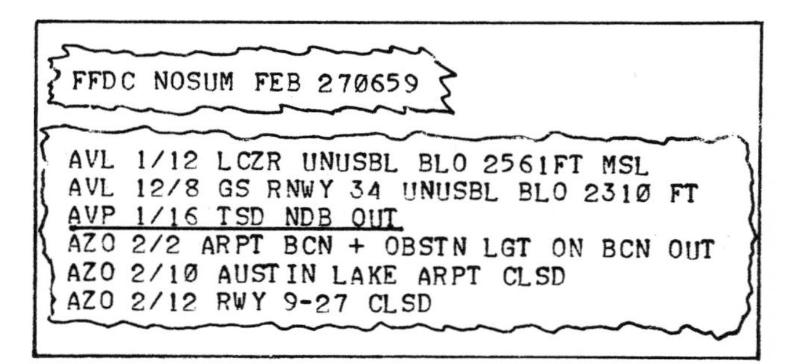

```
FFDC NOSUM FEB 27Ø659

AVL 1/12 LCZR UNUSBL BLO 2561FT MSL
AVL 12/8 GS RNWY 34 UNUSBL BLO 231Ø FT
AVP 1/16 TSD NDB OUT
AZO 2/2 ARPT BCN + OBSTN LGT ON BCN OUT
AZO 2/1Ø AUSTIN LAKE ARPT CLSD
AZO 2/12 RWY 9-27 CLSD
```

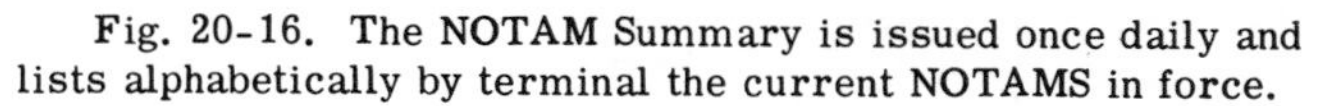

Fig. 20-16. The NOTAM Summary is issued once daily and lists alphabetically by terminal the current NOTAMS in force.

Other notes of interest: Dubois, Pa. (DUJ) has an estimated ceiling of 1800 broken (either broken or overcast constitute a ceiling) and the visibility is 2 miles variable in very light snow showers and blowing snow. The note at the end cites that the visibility is variable from 1 to 3 miles and the snow began at 29 minutes past the hour.

Bradford, Pa. (BFD) has issued a Special (S) indicating that changes of significance have occurred.

Once daily the summary of all NOTAMS is printed in alphabetical order. The NOSUM, as it is called, is sent and is given as valid at a certain *Zulu* time.

Fig. 20-16 is part of a NOSUM showing the NOTAM for AVP (Avoca or Wilkes-Barre, Pa.). It

```
Ø22 SA22Ø32ØØØ

BFD S E17⦷2VBS 215/5/1/2512G25E/ØØ1/ VSBY 1V3→BFD↘1/6
DUJ E18⦷2VSW--BS 241/5/-1/2918G24/Ø1Ø/VSBY 1V3 SB29
→DUJ↘1/2
PSB E2Ø⦷6BSH 231/8/Ø/2915G26/ØØ7
ABE →ABE↘1/15    1/17OA 1/22
IPT 4Ø⦶15 233/15/5/2413G2Ø/Ø18
→IPT↘1/16   1/36   2/3
AVP 35⦶15+ 214/13/-4/2514G21/ØØ9→AVP↘1/16
ABE O15+ 224/18/-4/2722G31/Ø16 NOTAMS FLW PSB
```

Fig. 20-15. Part of a sequence report.

notes that 1/16 AH concerns the fact that the Tobyhanna (identifier TSD) non-directional beacon (NDB) is out. With experience you'll remember the "Q" Code designation for various items such as AP for VOR, AH for radio beacon (homer) or UR (runway). Tobyhanna is about 16 miles southeast of Wilkes-Barre but falls within the area controlled by that FSS. Flight Service Stations will publish NOTAMS for nearby uncontrolled airports. Kalamazoo (AZO) sends the word that the Austin Lake Airport is closed (likely temporarily due to snow — this is quite common for smaller northern airports for a day or so at a time in the winter).

The NOSUM was released as of 0659 Zulu February 27th.

One tip on using the sequence report: as weather in the U.S. moves generally from west to east, the stations just *west* of the stations along your flight may give hints as to what may be expected later in your area of interest.

Sequences are *local* conditions. The weather may be fine at the stations reporting, but in the area between stations the crows may be walking.

Pilots sometimes complain that the last sequence report may have the airport below VFR minimums but in the meanwhile the early morning scud has busted out and it's now CAVU — but the official weather is still what the sequence says. Remember, they can't be running special observations every ten minutes. Specials will be coming out if the weather makes significant changes — but there may be a few minutes delay.

The visibility given by the sequence is *surface* visibility and you could have a different picture (for better or worse) at altitude.

Here are a couple of additional items concerning sequence reports:

1. Broken clouds or overcast clouds constitute a ceiling, so that if you have a report of "500 scattered, 700 broken, 2000 overcast," the ceiling is 700 feet. *Thin* broken clouds do *not* constitute a ceiling.

2. When the temperature and dewpoint are within 4° of each other, be alert for fog forming. The dew point is that temperature at which the present air would be saturated and visible moisture would form.

TELETYPE WEATHER FORECASTS

TERMINAL FORECAST

Terminal forecasts are issued every six hours for a 12-hour period. Principal airports also issue 24-hour terminal forecasts (these also are issued every six hours).

When you are looking at a terminal forecast, check the forecast against the sequences of the station. If the terminal forecast says the ceiling at a certain airport will be 3000 feet at 1400Z, and the 1400Z sequence gave the ceiling as 900 feet, you can expect worse conditions than forecast for the rest of the forecast period.

The cautious pilot is a pessimist as far as forecasts go. He tends to look on the dark side of the picture — and this is a lot better than looking at it through rose-colored glasses.

Fig. 20-17 is an actual terminal forecast. The terminal forecast is sent out for a period of 12 hours Zulu time (this one is for a period from 0500Z to 1700Z) but discusses the weather in terms of local standard time. (Don't ask *why* this is the case, just know it.)

Cleveland (CLE) will be clear with a visibility of more than 7 miles and the wind will be from 230 degrees *true* at 12 knots. *If the visibility is not given, it is expected to be more than 7 statute miles.*

<u>After 0800Z</u> — Ceiling 2000 broken (visibility 7 + miles), the wind will be from 230 degrees true at 12 knots with an occasional condition of 1000 feet obscuration and visibility 1 mile in light snow showers.

<u>After 1200Z</u> — Ceiling 1500 overcast, visibility 2 miles in light snow showers, wind from 230 degrees at 12 knots with an occasional ceiling 800 feet (obscuration), visibility 1 mile in light snow showers.

<u>After 1600Z</u> — Ceiling 2500 broken, visibility 7 + miles, wind from 250 degrees at 12 knots. This carries on through to the end of the forecast period (12 noon EST).

Note that after 1600Z Syracuse (SYR) will have an occasional ceiling of 400 feet obscured with 1 mile visibility in light snow showers (SW-) and blowing snow (BS).

A couple of last items: *If the wind is not given on a terminal forecast it is expected to be less than 10 knots.* All of the stations on this particular terminal

```
Ø5-17

CLE O 2312. Ø8ØØZ  C2Ø⦶ 2312 OCNL C1ØX1SW-.  12ØØZ C15⊕2SW-
2312 OCNL C8X1SW-. 16ØØZ C25⦶7 2512 OCNL SW-
TOL C25⦶ 2314 OCNL C1ØX2SW-. Ø9ØØZ 25⦶ 2715 OCNL C25⦶

ERI C25⊕7 222ØG OCNL C1ØX1SW-. 13ØØZ C15⊕2S- 2618G VRBL
C8X1S-
SYR C3Ø⦶ 2314 OCNL C15X2SW-. 16ØØZ C25⊕3SW-BS 2918G VRBL
C4X1SW-BS
```

Fig. 20-17. Part of an actual terminal forecast.

```
FA MSY Ø5ØØ45
ØlZ-13Z SUN

LA SRN MISS MOBILE AREA FLA W OF 85 DEG CSTL WTRS
HGTS ASL UNLESS NOTED

SYNS. WK TROF SRN LA NEWD TO GA WL MOVE SEWD OVR FLA CSTL WTRS A
RDG FM NW SPRDS OVR INTR LA SRN MISS BY 13Z

CLDS AND WX. NW LA MISS INTR 4ØØC12Ø⌽3ØØ⌽V⊕7 IN PTCHS LWRG C2⌽7
BY 13Z.
RMDR SE LA CSTL MISS MOBILE AREA 15⌽C6Ø-8Ø⌽V⊕7 TOPS TO 2ØØ WITH
SCTD EMBEDDED TSHWRS LCLY C1Ø-2Ø⊕2-4TRW TOPS TO 4ØØ DMNSHG
OVR LND AREAS AFT Ø6Z.
FLA W OF 85 DEG ADJT CSTL WTRS AND MISS SE LA CSTL WTRS
C8-15⌽4Ø-6Ø⊕1-3RW/TRW NMRS CLUSTERS OR LNS AND TOPS ABV 4ØØ.

ICG. LCLY MDT MXD IN BLDUPS ABV FRZLVL 11Ø-125

TURBC. SVR NEAR TSTMS

OTLK 13Z SUN-ØlZ MON. WK FRONTAL TROF LA CSTL WTRS NEWD TO SRN
GA WITH SHWRS AND TSTMS LCLY C1Ø⊕2TRW TOPS ABV 35Ø CONTG NEAR TROF.
ELSW PTCH C1Ø-2Ø⌽3-6HK DABRK IPVG 2Ø⌽C25Ø⌽ BY 18Z.
```

Fig. 20-18. An area forecast.

forecast are reporting winds and you'll notice that they are over 10 knots.

AREA FORECAST

The area forecast is issued every six hours for a 12-hour period and covers a large area such as states or portions of states. In addition, the outlook for an additional 12 hours is added to the end of the forecast.

Fig. 20-18 is an actual area forecast. The heading notes that it is an area forecast (FA) issued from New Orleans Moisant International Airport (MSY) on the 5th of the month (Ø5) at 0045 Zulu time. It is valid from 0100Z to 1300Z Sunday.

It covers Louisiana, Southern Mississippi, Mobile area and Florida west of 85 degrees west longitude (including coastal waters). Heights are given Above Sea Level unless noted.

Synopsis (SYNS). (This gives a general picture of the weather systems in the area.) — A weak trough from southern Louisiana northeastward to Georgia will move southeastward over Florida coastal waters. A ridge from the northwest spreads over the interior of Louisiana and southern Mississippi by 1300Z Sunday. (Note to English teachers and others interested in language use: The mix-up in tense, along with an original approach to punctuation, must be overlooked in the area forecast or you won't get the message.)

Clouds and Weather (CLDS AND WX) — Northwest Louisiana and the Mississippi interior (will have) 4000 scattered, ceiling 12,000 broken, 30,000 broken, variable to overcast, with visibilities 7 miles but in patches the conditions will be lowering to a ceiling of 200 feet broken, 7 miles visibility by 1300Z.

(For the) remainder of southeast Louisiana, coastal Mississippi, and Mobile area, conditions will be 1500 scattered, ceiling 6000-8000 broken, variable to overcast, visibility 7 miles, tops to 20,000 with scattered, embedded thundershowers. Locally, ceilings will be 1000 to 2000 overcast, visibilities 2 to 4 miles in thunderstorms and rain showers with tops to 40,000 feet. (These conditions will be) diminishing over the land areas after 0600Z.

Florida, west of 85 degrees, adjacent coastal waters and Mississippi and southeast Louisiana coastal waters, ceilings will be 800 to 1500 broken, 4000 to 6000 overcast with visibilities 1 to 3 miles in rain showers and/or thunderstorms with rain showers in numerous clusters or lines and tops will be above 40,000 feet.

Icing (ICC) — Locally moderate, mixed (clear and rime) in the buildups above (the) freezing level of 11,000 to 12,500 feet.

Turbulence (TURBC) — Severe near thunderstorms.

Outlook 1300Z Sunday — 0100Z Monday — There will be a weak frontal trough (in the) Louisiana coastal waters extending northeastward to southern Georgia with showers and thunderstorms with conditions locally of ceilings 1000 overcast, 2 miles visibility in thunderstorms and rain showers with tops above 35,000 feet continuing near the trof — er — trough. Elsewhere in the area there will be patches of ceilings 1000 to 2000 broken, visibilities 3-6 miles in haze and smoke at daybreak, improving (to conditions of) 2000 scattered, ceiling 25,000 broken by 1800Z.

Occasionally you'll find an abbreviated word that will stump you, but will be able to figure it out by getting the gist of the sentence. One problem is that the abbreviations are not always the same; for "sunrise" you may see SUNRS, SNRS, or SR.

Many small airports now have teletype machines

but no weather maps, and the area forecasts can take the place of maps in getting a general idea of the weather. The synopsis portion will geographically locate fronts and pressure area centers.

WIND INFORMATION

WINDS ALOFT REPORTS

Every six hours the actual winds aloft are observed and reported at major Weather Bureau Offices. The teletype reports are on a special meteorological circuit and are received by Weather Bureau Offices only. The code is complex, and you had better just ask the Weather Bureau personnel about the actual winds, rather than trying to memorize the code.

WINDS ALOFT CHARTS

The winds aloft charts are issued four times daily and a separate map of the U.S. is provided for each altitude reported on. (Fig. 20-19)

The wind arrow uses the "barb and pennant" presentation. Fig. 20-20 shows the symbols used.

A pennant represents 50 knots, a barb, 10 knots and a half barb equals 5 knots. The wind is blowing the way the arrow's going.

The number by the barb helps pin down the direction. The 5 shows that the wind is from 250° true (you might guess that it could be from 240° or 260°). The 9 indicates that the wind is from 290° true. Keep in mind that winds aloft directions are given to the nearest 10° *true* direction and velocities are in knots.

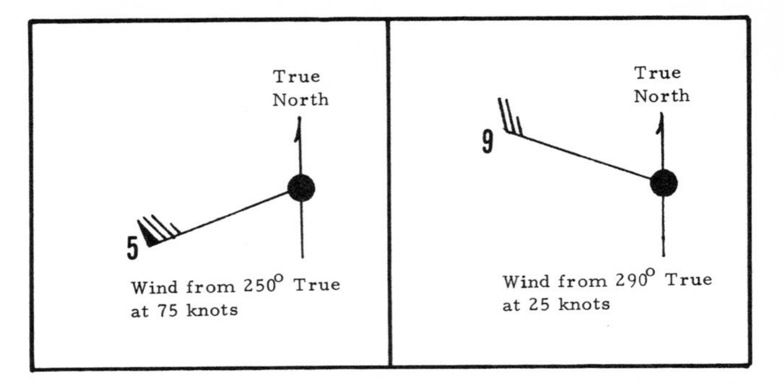

Fig. 20-20. Wind symbols.

The winds aloft chart represents actual winds at the time of observation and the information can be up to 9 hours old.

A winds chart is printed for the first standard altitude above the surface, then for 5, 8, 10, 14, 20, 25, 30, 35, 40, and 50 thousand feet. (It may be that a particular airport is above 5000 feet so that the first winds aloft reported for there would start at 8000 or even 10,000 feet, as would apply – notice in Fig. 20-19 that many of the high altitude western airports don't show up on the charts for lower altitudes.)

WINDS ALOFT FORECAST

The Winds Aloft Forecast or Forecast Winds is issued every six hours for a twelve-hour period in

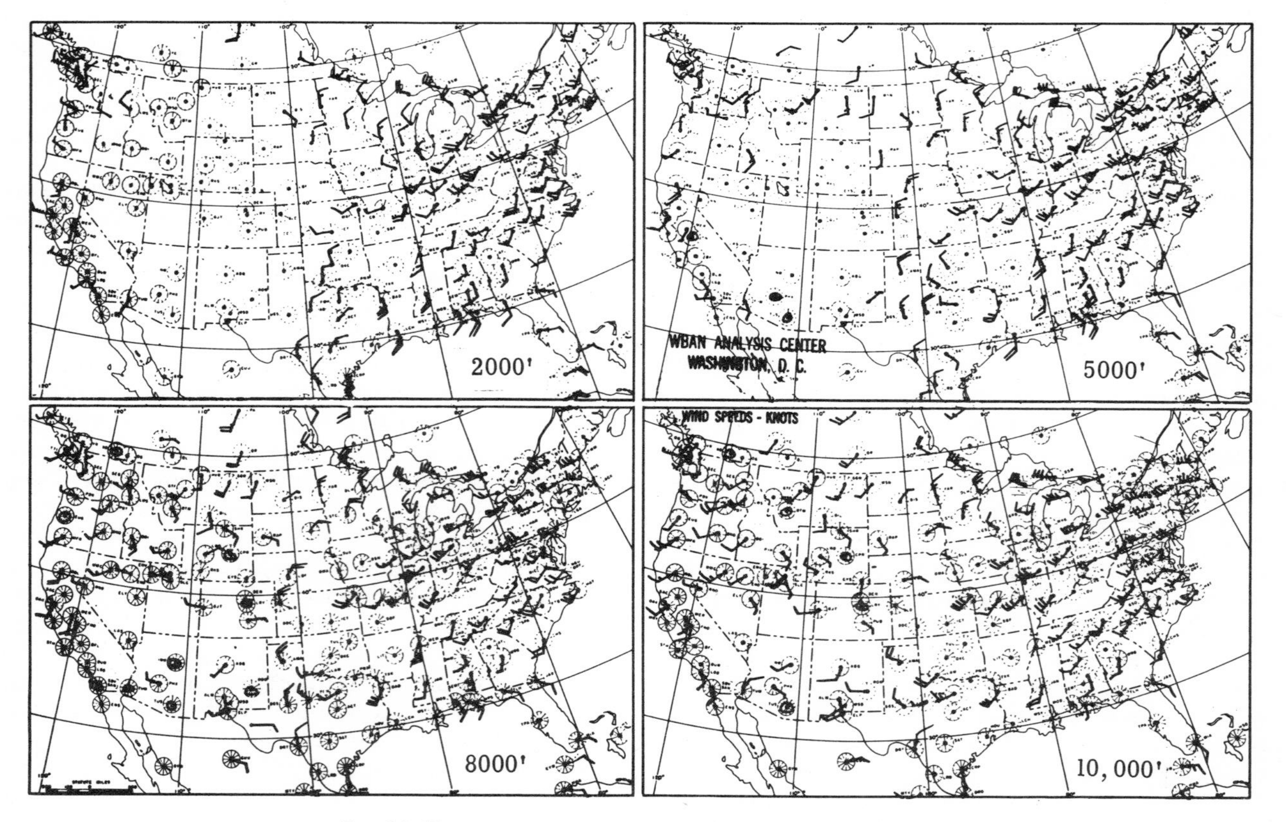

Fig. 20-19. Winds Aloft Charts *(Pilots' Weather Handbook)*.

```
FD WBC Ø41745
BASED ON Ø412ØØZ DATA
VALID Ø5ØØØØZ    FOR USE 21ØØ-Ø3ØØZ. TEMPS NEG ABV 24ØØØ

FT  3ØØØ     6ØØØ     9ØØØ     12ØØØ     18ØØØ     24ØØØ   3ØØØØ   34ØØØ   39ØØØ

CAE 2ØØ7 2411+Ø9 2515+Ø4 2618+ØØ 2726-13 2731-26 273942 284652 285Ø59
ATL 99ØØ 29Ø7+Ø9 291Ø+Ø5 2914+Ø1 2921-11 3Ø28-25 3Ø3742 3Ø4351 29466Ø
BHM 32Ø8 32Ø7+Ø8 311Ø+Ø5 3113+Ø2 3Ø22-1Ø 3Ø3Ø-24 3Ø4Ø41 3Ø4351 294661
JAN 32Ø8 3211+Ø8 3212+Ø6 3114+Ø3 3Ø2Ø-Ø9 3Ø27-23 29354Ø 29405Ø 284661
```

```
VALID Ø5Ø6ØØZ    FOR USE Ø3ØØ-Ø9ØØZ. TEMPS NEG ABV 24ØØØ

FT  3ØØØ     6ØØØ     9ØØØ     12ØØØ     18ØØØ     24ØØØ   3ØØØØ   34ØØØ   39ØØØ

CAE 21Ø9 24Ø9+Ø9 2612+Ø5 2716+Ø1 2823-12 2928-25 3Ø3642 3Ø4351 294759
ATL 99ØØ 99ØØ+Ø8 29Ø8+Ø5 2912+Ø2 2921-1Ø 3Ø26-24 3Ø3441 3Ø3951 29436Ø
BHM 31Ø7 31Ø5+Ø8 3ØØ8+Ø5 2911+Ø2 2919-Ø9 2925-23 293341 28385Ø 274461
JAN 3ØØ9 3Ø11+Ø8 2912+Ø5 2812+Ø2 2717-Ø9 2623-23 25304Ø 25395Ø 255Ø61
```

Fig. 20-21. The winds aloft forecast is issued every six hours for a twelve-hour period but is broken down into six-hour periods. The stations covered are Columbia, S.C., Atlanta, Ga., Birmingham, Ala. and Jackson, Miss.

Zulu time. Actually the information produced by computer is broken into two six-hour periods as shown by Fig. 20-21. The forecast winds are given for the entire list of stations for the first six hours of the forecast period and are then immediately repeated for the following six.

Figure 20-21 shows only four stations of the many covered by each winds aloft forecast, but they will do for a sample look at the coding system.

Looking at the heading for the *first* 6 hour period you can see that this is Forecast Winds (FD) issued by the Weather Bureau Center (WBC) on the 4th of the month (Ø4) at 1745 Zulu. The forecast is based on the 4th's 1200Z data and is (most) valid at 0000Z on the 5th. It's for use between the hours of 2100Z on the 4th and 0300Z on the 5th. It is noted that the temperatures are negative above 24,000 feet, and hence do not require repeated printing of a "minus" sign above that altitude.

The altitudes (3000 feet, 6000 feet, etc.) are listed across the top and are for all of the stations below, and are MSL values.

Note that the winds at 3000 feet at ATL (Atlanta) are forecast to be *light and variable* as indicated by the number (or symbol) "99ØØ." The winds at 12,000 feet at ATL are forecast to be from 290° (true) at 14 knots and the temperature will be +1°C (+Ø1) at that level (2914+Ø1). At 18,000 at ATL, the wind will be from 290° at 21 knots and the temperature will be -11°C (2921-11). At 30,000 feet above that station the wind will be from 300° (30) at 37 knots and the temperature will be -42°C (303742).

For a station like Denver, with its elevation of 5331 feet, the winds for 3000 feet MSL naturally would not exist and, since the forecast must begin at an altitude of *at least* 1000 feet above the surface, the 6000 feet level will not be given for Denver (9000 feet will be the first). The temperature is not given at the 3000 foot level at any place as you can see (Fig. 20-21).

The *second* 6 hours of the forecast indicates that it will be valid at 0600Z on the 5th (Ø5Ø6ØØZ) for use from 0300 to 0900Z. Again you are reminded that the temperatures listed are negative above 24,000 feet.

Atlanta's wind at 3000 is expected to *remain* light and variable (99ØØ) and the wind at 6000 is expected to *become* light and variable. All the winds are unusually light; for instance, during this period the wind at Jackson, Mississippi, at 12,000 feet will be from 280° at only 12 knots (2812). The temperature will be +2°C (+Ø2).

All in all it's pretty much coded like the surface winds covered earlier, but the method of reporting winds of 100 knots or more could catch you by surprise. As an example, you may see at one of the higher altitudes a figure group such as "74Ø15Ø." Obviously the wind is *not* from 740° (74) at 1 knot (Ø1), although the temperature *is* -50°C (5Ø). Since the Weather Bureau wants to stick to a six-digit code, winds above 100 knots will need a special treatment. Subtract 50 from the 74 to get 24 (240°) and add 100 to the Ø1 to get a wind from 240° true at 101 knots. The temperature is -50°C. So, in the impossible-looking cases, subtract 50 from the two digits for the direction and add 100 to the wind velocity. The temperature numbers are not affected.

ENROUTE WEATHER AIDS

SCHEDULED WEATHER BROADCAST

The scheduled weather broadcast, including Notice to Airmen information, is made at 15 minutes past the hour via navigational aids with voice facilities (VOR's and LF/MF non-directional beacons). The broadcast covers reporting points within approximately 150 miles of the broadcast station.

The broadcast will include the latest weather (as of 58 minutes past the last hour), alert notices (ALNOT), SIGMETS or AIRMETS, if applicable, pilot reports, radar reports, lost or overdue aircraft notices and NOTAMS or AIRADS (Airman Advisories) not published in AIM. (If you need winds aloft information you'll have to ask the FSS for it.)

TRANSCRIBED WEATHER BROADCASTS

Continuous, updated weather reports and forecasts are available at certain VOR and LF/MF facilities in the U.S. The tape is continually updated as new weather is available. Check the AIM for stations having this service.

The scheduled weather broadcasts at 15 minutes past the hour are a major aid. You can also get the benefit of two advisories issued by the Weather Bureau.

AIRMET

This is an advisory concerning weather of such a degree as to be potentially hazardous to inexperienced pilots but not necessarily hazardous to experienced pilots and covers such things as:

a. Moderate icing
b. Moderate turbulence
c. The initial onset of phenomena producing extensive areas of visibilities of less than two miles or ceilings less than 1000 feet, including mountain passes and
d. Winds of 40 knots or more within 2000 feet of the surface.

THE AIRMET is broadcast upon receipt and at 15 minutes past the hour thereafter as part of the regularly scheduled weather broadcast during the valid period.

SIGMET

The SIGMET is an advisory concerning *sig*nificant *met*eorological developments of such severity as to be potentially hazardous to transport category and other aircraft in flight. SIGMET advisories will cover:

a. Tornadoes
b. Lines of thunderstorms (squall lines)
c. Hail of 3/4-inch diameter or more
d. Severe and extreme turbulence
e. Heavy icing and
f. Widespread dust storms or sandstorms lowering visibilities to less than two miles.

The SIGMET is broadcast upon receipt and at quarter hour intervals (H + 00, H + 15, H + 30, H + 45) during the valid period. The H + 15 broadcasts will be a part of the scheduled weather broadcast.

SIGMETS and AIRMETS are also sent out on the teletype circuit. (Fig. 20-22)

Figure 20-22 is an actual AIRMET issued by the Washington (DCA) Flight Advisory Weather Service on the 5th of the month at 1025Z and is valid from that time until 1500Z on the 5th.

AIRMET ALPHA 1—This is the first AIRMET issued by Washington since 0000Z. If an AIRMET (or SIGMET) condition develops in a distinctly separate sector the advisory is identified as "Bravo 1," "Bravo 2," etc.

Over western and northern West Virginia visibility is generally less than 2 miles in ground fog with some local conditions near zero zero in dense fog. Conditions will be slowly improving at 1300Z and ending by 1500Z. See the area forecast from DCA at 1300Z for further information.

PILOT REPORT (PIREP)

The PIREP is one of the most valuable aids to the pilot because it gives actual enroute information "as it happens." Information on cloud tops, conditions of turbulence or precipitation or icing is useful.

If you run into unusual conditions let the nearest FSS know about it. The word will be passed on on the scheduled weather broadcast and by teletype. You would appreciate the same service.

```
FL DCA Ø5 1Ø25
Ø5 1Ø25-Ø5 15ØØ

AIRMET ALFA 1. OVR WRN AND NRN WVA VSBY GENLY LESS THAN 2 MI IN
GNDFG WITH SOME LCL CONDS NR ZERO ZERO IN DNS FOG. CONDS SLOLY
IPVG 13Z ENDG BY 15Z. SEE FA DCA 13Z FOR FRTHR INFO.
```

Fig. 20-22. An AIRMET.

SUMMARY

If you don't remember any of the rest of the points covered in this chapter, at least keep this one in mind: Always feel free to ask questions in the Flight Service Station or Weather Bureau Office. In fact, you should always make it a point to talk to the Flight Service Station or Weather Bureau personnel, even in a self-briefing type setup. They will always have more pertinent information about your trip than is in the printed matter. ("You're going to Rockfield? We just had a pilot call in and report severe turbulence right on your route," etc.) They might have local information that hasn't had time to be published or broadcast.

21. NIGHT FLYING

A certain amount of night flying experience is required for the ICAO commercial certificate (at least 5 hours of flight time at night, including at least 10 take-offs and landings as pilot in command). Even if you aren't planning on getting a commercial ticket, you may want to be able to use your airplane at night.

If you want to start a good argument, just place the subject of single-engine night flying in front of a group of pilots in an airport office. There's no doubt that the second engine increases the safety factor a great deal. If you had a choice, you would want to fly a multiengine airplane at night. *However, if your single-engine airplane is properly equipped and you are qualified, the final choice is left up to you.* The first several times you get beyond gliding distance of the airport at night, the engine will run very rough and will continue to do so until you again get within gliding distance of the airport. If you happen to know that it's mountainous or swampy down there, the engine will run even rougher. This is known as *Zilch's Law:* "The roughness of the engine is directly proportional to the *square* of the roughness of the terrain below and the *cube* of the pilot's imagination."

Actually, night flying is the most enjoyable type of flying if you can ignore Zilch's Law (and you will, after a while). The air is smooth, there is less traffic, and you will discover that you can see cities and other airplanes much more easily than in day flying under the same atmospheric conditions.

Before you start night flying you should be aware of the following points.

THE EYES HAVE IT

Owls and cats have it all over humans for night operation because their eyes are designed for it. The cat on the back fence is able to dodge those shoes because he sees them coming — he is a creature of nocturnal habits — but back to the human eye.

The retina, or layer of cells in the back of the eyeball that receives light and makes up the images you see, is made up of two distinct types of cells called cones and rods.

The cones (millions of them) are in the center area of the retina and are used to pick up color and detail and distant objects. These are the daytime cells, and you use them when you watch the girl next door sunbathing.

The rods (many more of them) are arranged around the cones and see color only as shades of gray. The rods aren't much good for detail or distance. At night, the rods are mostly used — this is why you can't make out colors in very dim light.

Because of the placement of the cones and rods in the retina, your picking out of objects will be slightly different for day and night.

Because the cones are in the center, the most detail will be found in the daytime at the point on which you are focusing and details to the side will be less distinct. In darkness, because the rods are working and the cones in the center of the retina are loafing, you can see an object better by not staring directly at it, but slightly to one side. (Fig. 21-1)

You'll never be able to see as well at night as in the day and Fig. 21-1 is merely given to show the comparative *areas* of coverage.

The shift from cones to rods is gradual; as the light decreases, more and more of the rods take over. The rods require a certain period of adaptation to the dark. That's why you sat on the fat lady's lap that time in the movie theatre — your eyes weren't dark-adapted.

It takes about 30 minutes to become night-adapted. The rods are not as sensitive to red lights as white, so that night fighter pilots are briefed in a room with red lights or wear red-lensed goggles before a flight. This is also why your airplane's instrument lights are red. Under red lights, you can use the cones for detailed cockpit work and still not lose dark adaptation.

If you're night-adapted, a flash of white light will ruin your careful work and the whole adaptation process has to begin again.

You'll probably have neither the time nor inclination to wear red goggles for 30 minutes before the flight, but you should have an idea of what to do for your night vision:

1. Avoid brilliant white lighting before flying, if possible.
2. If it appears that a search light, taxi light or landing light will be directed at you in the airplane, shut one eye. That eye will still be night-adapted when you open it again (after the bright light is gone).
3. Under normal flying conditions keep the instrument and cockpit lighting as low as practicable.
4. If you are over a brightly lighted city or other bright outside lights, turn the instruments up to full intensity.
5. When looking for objects at night: scan, don't stare. (Remember that center blind spot!)
6. *Always* carry a flash light when night flying *but* make sure that it has a red lens. You can cut a piece of red cellophane to fit the inside of the lens.

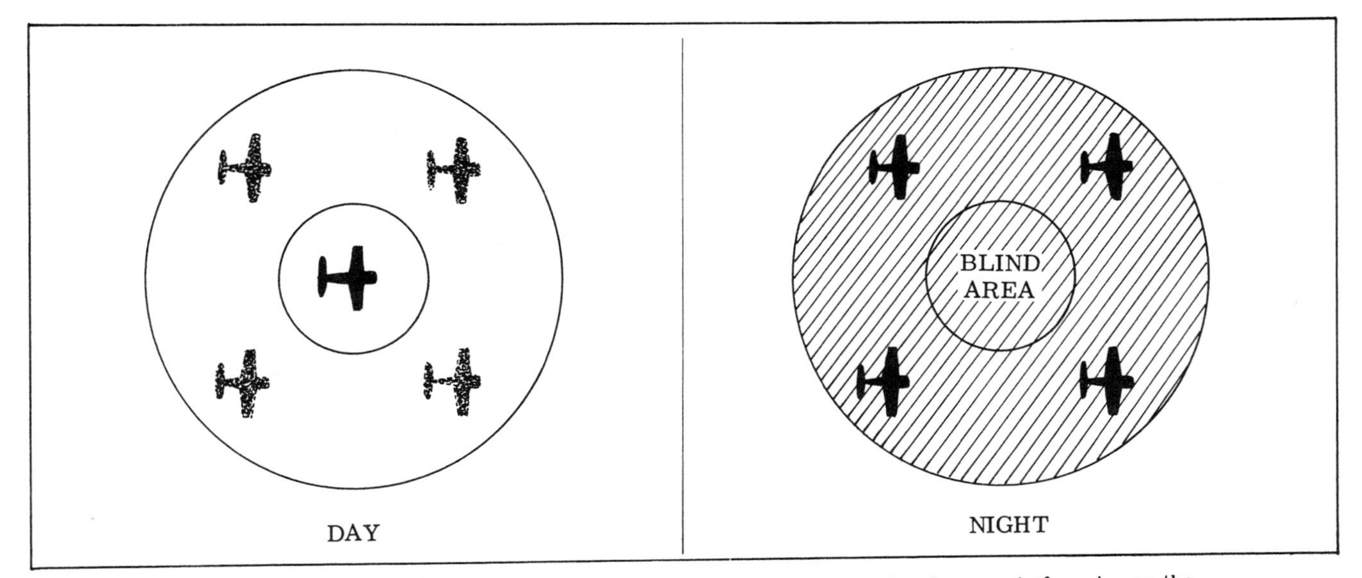

Fig. 21-1. A flight of five airplanes as seen by day and by night. The observer is focusing on the center of the group in both cases.

7. Don't stare at the instruments, it tires your eyes quickly.

Oxygen helps night vision but it's doubtful that you have it or want to use it. Alcohol and tobacco hurt night vision. If you were a night fighter, you could set up a daily regimen to help improve your night vision.

AIRPLANE LIGHTING

The airplane's navigation (or position) lights, like a lot of other aviation equipment and terminology, are arranged similar to those of ships. The lighting arrangement (red light on the left wing, green on the right and white light on the tail) can tell you approximately what that other airplane is doing.

Suppose that you see the red and green wing lights of an airplane dead ahead and on your altitude. Is he coming or going? One clue is the fact that the white tail light is very dim or not showing. The final touch would be if the red light is on the right and the green is on the left — he's headed your way!

The sailors used the term, "Red Right, Returning," meaning that when you see the red and green lights of another ship (or airplane) and the red light is on the right, the other vessel (airplane) is headed toward you — not necessarily on a collision course but at least in your general direction.

The airplane's position lights must be on *any time the airplane moves after sunset* — taxiing or flying. Also, the lights must be on if the airplane is parked or moved to within dangerous proximity of that portion of the airport used for, or available to, night operations.

The red rotating beacon, originally mandatory only for airplanes of 12,500 lbs. or more gross weight, is now being used for nearly all the airplanes flying at night, regardless of gross weight. The rotating beacon can be seen many miles away in good atmospheric conditions. Many pilots turn on the rotating beacons in the daytime also when visibility is down.

The landing light is a great aid but not as necessarily vital to night operation on a lighted field as might be considered. The idea is that if you find yourself at night with the landing light on the blink, you'll still have no trouble at all making a good landing by watching the runway lights and using a little power in the landing.

Don't fly without adequate cockpit lighting. If you fly into hazy conditions, it's possible that you may lose all visual references and have to go on instruments to get out of that area. You may also fly into clouds or fog before realizing it. You'd be in pretty sad shape without being able to see the instruments. With your present instrument training you'll have your hands full without having to hold a flashlight on the panel (if you remembered to bring one). The overhead cabin light can be used in this event, but it will wreak havoc with your night vision.

AIRPORT LIGHTING

One of the toughest parts of night flying into a strange, large airport is the taxiing. This is especially rough if you haven't been in there before in the daytime. Even a familiar airport is strange the first time you try to taxi at night. The only thing to do is to get help from ground control if it's available, otherwise just take it slow and easy until you find your destination on the field. The lighted taxiway signs will be a great help. One private pilot, on his first night taxi, was too proud to ask directions and managed to taxi out the airport gate and onto the freeway. He was last seen taxiing at 60 miles an hour, trying to keep from being run over by traffic (auto type). He didn't know that taxiway lights are blue.

The runway edge lights are white lights of controlled brightness and usually are at the lowest brightness compatible with atmospheric conditions.

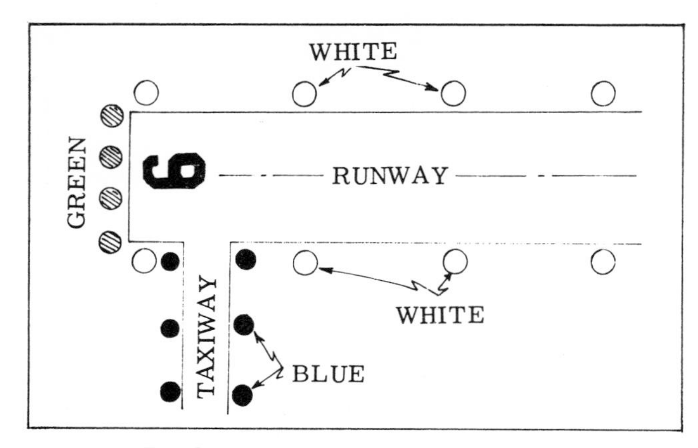

Fig. 21-2. Runway and taxiway lighting.

The runway lights in some airports may be nearly flush with the runway and others may be on stands a foot or two high. The lights marking the ends of the runway are green. (Fig. 21-2)

Airport beacons flash alternating green and white. This is because they are rotating and have the different-colored lights on opposite sides of the beacon head. The following should be noted:

1. Civilian airport beacons — green — white — green, etc.
2. Military airport beacons — green — white — white — green, etc.

The military airport beacon has a dual-peaked white flash between the green flashes to differentiate it from the civilian. Remember this because the beacons can be seen from a great distance and are used to help the pilot locate the airport at night. It would cause plenty of excitement if you were talking to a civilian tower and landed at a nearby SAC base by mistake.

If an airport beacon is on during the day, it means that the weather conditions at the airport are less than 3 miles visibility and/or 1000 feet of ceiling. It's no longer VFR.

The tetrahedron or wind tee on the airport will have steady lights outlining it during VFR night operations. If the weather gets below 1000 feet and/or 3 miles, the outlining lights are switched to flashing.

All dangerous obstructions near the airport are marked by red lights.

TIPS FOR NIGHT FLYING

The following suggestions are for the various procedures and phases of a flight as they affect night flying. *Get a night check-out.*

Fig. 21-3 shows the standard system of hand signals. For night operation, wands or flashlights are held for clarity.

STARTING

Be doubly careful in starting at night. The darkness could result in your starting the engine when someone was in the propeller arc. Turn on the position lights before starting. Granted, this won't help the battery, but it will serve to warn people on the ramp that you are going to start. If other airplanes' engines are running up close by, no one can hear your "CLEAR." Keep all other equipment off, particularly the taxi and landing lights and radios, while starting. If you can possibly work it, have a man standing outside as you start. If it's all clear, he'll give you a signal which consists of moving one of the wands or flashlights in a rotary motion and pointing to the engine to be started with the other wand.

TAXIING

After the engine or engines are started, turn on other needed equipment but leave the taxi or landing lights off until you are released by the taxi director.

A signal normally used at night but not shown in Fig. 21-3 is crossed wands or flashlights, meaning "stop" or "hold brakes."

Taxi slowly at all times and very slowly in congested areas at night. You'll find that judgment of distances is harder when you first start night operations. Your visibility is limited to the area covered by the taxi light and you don't want to have to suddenly apply brakes while moving at a high speed.

RUN-UP

Once you have reached the run-up position and have

Fig. 21-3. Standard system of hand signals *(Pilot Instruction Manual).*

stopped, turn the taxi or landing lights off to conserve electrical power. Use a red-lensed flashlight to check controls and switches not readily seen by normal instrument lighting. Make a pre-flight check carefully, as you should always do — day or night.

When you are ready to go, call the tower — or if you have no transmitter, turn on the landing light and turn toward the tower.

Airplanes with no transmitters may acknowledge tower instructions by blinking the position or landing lights.

Make sure the cockpit lighting is at the proper level.

TAKE-OFF

You will normally use the landing lights during the take-off roll, but the check pilot may have you make several take-offs and landings without it. You'll find that the runway edge lights and the landing light will be more than sufficient for helping you to keep the airplane straight. Don't just stare into the area illuminated by the landing lights. Scan ahead, using the runway edge lights as well. Your first feeling will be that of a great deal of speed because of your tendency to pick references somewhat closer than during day take-offs. Because of this, make sure that you have reached a safe take-off speed before lifting off — don't try any short or soft field take-off techniques for the first few times. Actually, only in an emergency would you attempt taking off from a very short or soft field at night.

Retract the gear when you are definitely airborne and can no longer land on the runway ahead. Establish a normal climb and turn off the landing lights. Turn off the boost pumps at a safe altitude.

AIRWORK

You'll find that you are referring more to the flight instruments at night than in the daytime because of the occasional lack of a definite horizon in sparsely populated areas. (Or any area for that matter.)

The worst situation will be on a hazy night in an area with a few scattered lights on the ground. The ground lights may appear as a continuation of the stars and you could talk yourself into a good case of vertigo. If you're in doubt, rely on the instruments.

You may have a tendency to fumble for controls in the cockpit at first. It's wise to spend extra time sitting on the ground in the cockpit until you could find any control blindfolded.

At night you may suddenly encounter instrument conditions. Make a 180° turn as discussed in Chapter 18.

APPROACH AND LANDING

The landing check list should be carefully followed, taking even more time than in daylight to insure no mistakes.

Make the downwind leg farther from the runway than in the daytime. All in all, plan on a wider, more leisurely pattern. Since judging distances also is more difficult, plan on a pattern that requires the use of power all the way around.

Turn the landing lights on after turning onto final. You will notice that if the night is hazy, your visibility will be less during the early part of the final approach with the landing lights on, but will improve as you get to the round-off height.

When you are using landing lights the landing technique should be very close to that used in day landings (power-off). If the landing lights are not used, make the touchdown still carrying some power.

The common tendency is to be a little fast on night landings; watch for it.

SUMMARY

You'll find that after the slight confusion of the first hop has passed you'll be sold on night flying. Summing up the chapter: Get a check pilot — and don't get rushed during night operations.

PART V / PREPARING FOR THE COMMERCIAL FLIGHT TEST

22. ORAL OPERATIONAL EXAMINATION

AIRCRAFT DOCUMENTS

On the flight test you'll be expected to know what papers are necessary for the airplane and to explain the purpose of each. You must know that the Registration Certificate (Certificate of Ownership), Airworthiness Certificate and Airplane Flight Manual (or equivalent form) must be in the airplane at all times. You must also be familiar with the documents that do not necessarily have to be carried in the airplane but must be readily available for inspection (airframe and power plant logbooks and inspection reports). If you fly to another field for the flight test, you'd better take the logbooks and inspection reports along.

DOCUMENTS THAT STAY IN THE AIRPLANE

AIRPLANE REGISTRATION

The Certificate of Registration contains the name and address of the owner, the aircraft manufacturer, model, registration number and the manufacturer's serial number and must be *displayed* in the airplane. The registration number is that number painted on the airplane. You can change the registration number of your airplane by applying to the FAA and paying a small fee. This is assuming that the number and letter combination you have chosen is not already in use. Many corporation planes use this system. For instance, the Jones Machinery Corporation may decide that the registration "N1234P" is too ordinary so they might apply for, and get, "N100JM" or some other number-letter combination more suited to its taste.

The manufacturer's serial number is permanent and is a means of identifying the airplane, even though the registration number has been changed several times. An airplane may have the serial number 6-1050, which means that the airplane is a Zephyr model 6, the 1050th airplane of that model made. The manufacturer's serial number is used for establishing the airplanes affected by a new service note or bulletin ("Zephyr Sixes, serials 379 through 624 must comply with this bulletin"). You could easily check the Registration Certificate for the manufacturer's serial number to see if your airplane is affected. If you own an airplane you'll know the serial number — or should.

When an airplane changes owners, or the registration number is changed, a new Registration Certificate must be obtained.

CERTIFICATE OF AIRWORTHINESS

The Certificate of Airworthiness is a document showing that the airplane has met the safety requirements of the Federal Aviation Agency and is "airworthy." It remains in effect indefinitely, or as long as the aircraft is maintained in accordance with the requirements of the FAA and, like the Certificate of Registration, must be *displayed* in such a manner that it can be readily seen by pilot or passenger. The Airworthiness Certificate itself will stay in the airplane indefinitely, but in order for it to be valid, the following must be complied with:

1. Privately Owned Aircraft (not operated for hire) — The aircraft shall have had a periodic (annual) inspection within the preceding 12 calendar months in accordance with the Regulations. The log-books and inspection forms will be a voucher for this.
2. Aircraft Used for Hire — In addition to the periodic (annual) inspection, the airplane used to carry passengers or for flight instruction for hire shall have had an inspection within the last 100 hours of flight time in accordance with the Regulations. This interval may be exceeded by not more than 10 hours when necessary to reach a point at which the inspection may be accomplished. In any event such time must be included in the next 100-hour interval. The periodic inspection will be accepted as a 100-hour inspection. Both the periodic and 100-hour inspections are complete inspections of the aircraft — identical in scope.
3. Progressive Inspection — The airplane you're flying may use the progressive inspection system. The owner or operator provides or makes arrangements for continuous inspection of the aircraft whereby the inspection work load may be adjusted or equalized to suit the operation of the aircraft or the need of the owner. Its purpose is to permit greater utilization of the aircraft. The owner using the progressive inspection must provide proper personnel, procedures and facilities before commencing such inspection. The use of the progressive inspection eliminates the need for periodic and 100-hour inspections during the period that the progressive inspection procedure is followed.

Check the logbooks and the inspection forms before taking the flight test to make sure that the airplane is airworthy. It would be plenty embarrassing if the flight examiner checks the airplane's papers and finds that the airplane has not been inspected as required and is not airworthy. No matter whether you begged, borrowed, hired or stole the airplane, the final responsibility will rest on you. The owner-operator will be in trouble too, but this won't make it any easier on you.

If there is no record of the required inspections, then the Airworthiness Certificate is null and void.

AIRPLANE FLIGHT MANUAL OR OPERATIONS LIMITATIONS

LIMITATIONS

To quote the FAA Regulations:

"An Airplane Flight Manual shall be furnished with each airplane having a maximum certificated weight of more than 6000 pounds.

"For airplanes having a maximum certificated Weight of 6000 pounds or less an Airplane Flight Manual is not required; instead, the information prescribed in this part for inclusion in the Airplane Flight Manual shall be made available to the operator by the manufacturer in the form of clearly stated placards, markings or manuals."

Some manufacturers' Owner's Manuals for airplanes of less than 6000 pounds maximum Weight contain all of the information required in an Airplane Flight Manual; for others the Owner's Manuals (or handbooks) only give specifications and descriptions of the systems plus normal and emergency procedures. They furnish "Operations Limitations" in the form of a separate sheet, or sheets, of information which sometimes is termed an "Airplane Flight Manual."

At any rate, become familiar with the following information for your airplane whether you can get it all from the Owner's Handbook or have to refer to a couple of sources. At least know where to look for the information if it is required.

1. Engine Limits — take-off, altitude (rpm, manifold pressure).
2. Fuel — minimum octane to be used and tank capacities.
3. Propeller
   (a) Type of propeller.
   (b) Propeller limitations — maximum and ranges of operations to be avoided.
4. Power Plant Instruments — engine instruments such as the oil temperature and oil pressure gages, cylinder head temperature, tachometer and the limits of operation (such as, "Do not exceed 260° F oil temperature").
5. Airspeed Operation Limits — calibrated airspeed (you remember that calibrated airspeed means that the instrument error is corrected).

   The airspeed operation limits section covers, in writing, the airspeed indication markings — white arc, green arc, red line. (You might review the section on Airspeed Indicators in Chapter 3 if the meaning of the markings has slipped your mind).
6. Flight Load Factors — positive and negative load factors, flaps up and flaps down.
7. Maximum Weight
8. Center of Gravity Range
   (a) Location of the datum.
   (b) Allowable center of gravity range in inches, measured from the datum.
   (c) Length of the mean aerodynamic chord — in inches.
9. Limitations Notes, such as:
   (a) "Intentional spins with flaps extended prohibited."
   (b) A list of approved acrobatics or maneuvers such as chandelles and lazy eights, with recommended entry speeds if such maneuvers can apply to the particular airplane.
10. Procedures of Operations — recommended procedures for starting, taxiing, take-off, climbs, cruise, approach and landing plus emergency procedures. (Note: For airplane of less than 6000 pounds gross weight this information is usually given in the form of Owner's Manuals or Owner's Handbooks.)
11. Performance Information: (Required for airplanes over 6000 pounds but is also included in most lighter airplanes' owner's manuals.)
    (a) Airspeed calibration table.
    (b) Take-off and landing distances at various weights, altitudes and wind conditions.
    (c) Rate of climb table or graph — rate of climb at various weights and altitudes.
    (d) Stalling speeds at various angles of bank at different flap settings.

WEIGHT AND BALANCE RECORD

This is usually a form especially printed for your model of airplane giving original empty weight, empty center of gravity and useful load. It will also show computations of the center of gravity for various loadings of your airplane, such as the C.G. with one pilot, full fuel, plus one, two or more passengers and baggage, if applicable. You can tell at a glance if you will remain in the C.G. envelope for these loadings. An entry is made in the aircraft logbook when equipment changes result in changes in empty C.G. and/or empty weight. The Weight and Balance Record is a part of the Airplane Flight Manual.

EQUIPMENT LIST

Normally the factory will include the original equipment list as part of the Weight and Balance Form. Significant changes will be noted in the logbooks, as well as on the Weight and Balance Form.

AIRCRAFT RADIO STATION LICENSE

If your airplane has any transmitting equipment make sure the radio station license is on board and in date. You must have a license for DME, if installed.

OTHER AIRCRAFT DOCUMENTS— AIRCRAFT AND ENGINE MAINTENANCE RECORDS

The Regulations state that the registered owner or operator shall maintain a maintenance record in a form and manner prescribed by the administrator

(FAA) which shall include a current and accurate record of the total time in service on the aircraft and on each engine, a record of inspections and the record of maintenance as required. Such records shall be:

(a) Presented for required entries each time inspection or maintenance is accomplished on the aircraft or engine.
(b) Transferred to the new registered owner or operator upon disposition of the aircraft or engine involved.
(c) Made available for inspection by authorized representatives of the administrator or board.

Paragraph (c) is the reason you should have the logbooks and inspection reports for the flight test.

THE LOGBOOKS

The logbook entry shall include the type and extent of maintenance, alterations, repair, overhaul, or inspection, and reflect the time in service and date when completed. The logbook should have entries when mandatory notes, service bulletins and airworthiness directives (sometimes called AD notes) are complied with. The regulations call for a separate, current and permanent record of maintenance accomplished on the aircraft and each engine, and the logbook *is* the record. Carve the information on a piece of granite if you so desire but this would be unhandy to haul around, so the usual procedure is to use a logbook.

The aircraft and each engine must have separate records.

Before you take the flight test, make sure that the logbooks you have are the right ones and they are up-to-date.

RECORD OF MAJOR REPAIRS AND ALTERATIONS

This is form ACA 337, a special form for major changes done after the plane leaves the manufacturer. The owner-operator will keep a copy and an entry is made in the logbook with a reference to the date or work order by number and approving agency. The ACA 337 will note the new empty Weight and empty C.G. and is normally attached to the Weight and Balance form in the airplane.

SUMMARY OF AIRCRAFT DOCUMENTS

You can accomplish a great deal by going over the documents with the owner-operator of the airplane you plan to use on the flight exam. Flight examiners can ask some embarrassing questions that would have been simple had the applicant just spent a few minutes checking the airplane's papers.

AIRPLANE RANGE, PERFORMANCE AND OPERATION

RANGE

The flight examiner may ask you the range of your airplane and it would behoove you to remember the best range figures for 65 and 75 per cent power (and maximum range if the information is available).

OPERATION

Know the airplane fuel, electrical and hydraulic system operations. (One commercial applicant didn't know where the circuit breaker panel was — things had been going so well that it hadn't been necessary for him to find it.) If you cannot come right out with an answer, at least be able to look it up in the Airplane Flight Manual or in the Owner's Manual. For instance, know about the alternate static pressure source, if the airplane has one.

PERFORMANCE

Be familiar with the take-off and landing distances (ground run *and* over a 50-foot obstacle).

He'll ask you the effects of temperature and altitude on performance. You might review Part 1 — "Airplane Performance for Pilots" again before taking the flight test.

AIRPLANE LOADING

FUEL AND OIL

You'll be asked the fuel and oil capacities of the airplane. Know what oil weight (viscosity) is presently being used and how much is normally carried, as well as the total oil capacity.

Be able to answer promptly when you are asked the minimum octane fuel used by your airplane.

AIRPLANE LOADING AND BAGGAGE CAPACITIES

You'll be expected to refer to the approved weight and balance data for the airplane. You might review "AIRPLANE WEIGHT AND BALANCE" in Chapter 10. You'll probably be asked for some practical computations of permissible fuel and payload (baggage and passenger) distributions. The FAA considers the following as standard for weight and balance computations:

Fuel — 6 pounds per gallon
Oil — $7\frac{1}{2}$ pounds per gallon

Use actual weights of persons.

Remember that the baggage compartment is placarded for *two* reasons: (1) structural and (2) C.G. considerations.

AIRPLANE LINE CHECK

Be prepared to answer *why* you are checking certain things on the airplane.

Start the line check in the cockpit. Make sure the ignition switch (or switches) and battery switch are off.

You will use the manufacturer's recommended procedure or will have your own line check procedure and no attempt will be made here to set one up. A pilot who finally sets up a line check in order to pass a flight check doesn't deserve to get a higher rating. Of course, *you* have used a thorough line check for every flight since the student pilot days, so the only difference on the flight test will be explaining why each item is inspected.

It's amazing how little some pilots know about the internal workings of their airplanes. This is more often true in cases where the pilot has rented airplanes all along. His faith in the operator is almost blind. It may be possible that the pilot has never bothered to drain the fuel strainers, always assuming that the "operator did it this morning." These pilots may not even know where to drain the fuel to check for water or dirt. It *has* happened.

When you become a commercial pilot or buy your own airplane you'll be carrying more of the preflight responsibility. As a commercial pilot you may operate away from the home base for days or weeks on a charter operation and good ol' Joe, the mechanic, won't be around to drain the fuel strainer and check the oil for you. It'll be up to you to decide whether repair work should be done and whom to contact at the strange field.

SUMMARY

As you know, the flight test is divided into four phases: I — Oral Operational Examination, II — Basic Piloting Technique, III — Precision Maneuvers and IV — Cross-country Flight. The failing of any one of these phases means that you don't get the certificate that day. If, through some ignorance of the airplane, you barely pass the oral examination, the nervousness engendered could affect the rest of the flight test. A good oral examination might help shift the balance in your favor if one of the flight maneuvers was a little weak. *Know your airplane.*

23. BASIC PILOTING TECHNIQUE

ACCURACY LANDINGS

The only maneuver required in this phase of the commercial flight test not previously covered in this book is the performance of three accuracy landings, touching down in normal landing attitude beyond and within 200 feet of an assigned line or mark. The approaches to the accuracy landings should be accomplished in a glide through a 180° change in direction. Flaps and moderate slips may be used in a normal manner.

As always in a power-off approach to hit a spot, it will be better to be slightly high. If there's any question, you'll be allowed to slip or use flaps. Things may work out so that neither slipping nor flap application is necessary. Nothing is more embarrassing on a flight test than to think you're high, apply full flaps or make a steep slip, and discover that enthusiasm caused you to *under*shoot. The line or mark that you are required to land beyond and within 200 feet of is considered to be a ditch. (Fig. 23-1)

If you are going to undershoot, recognize the fact and apply power. If the check pilot catches you trying to "stretch the glide," he may get a bad impression of your flying ability.

About correcting for overshooting: although the rate of sink will increase considerably by slowing up the airplane in the glide, this is not the place to get cocked up and dangerously slow. Violent maneuvering, excessive slips or dangerously low airspeeds will be disqualifying.

One aid in telling whether you'll hit the spot or not is to watch its apparent movement as you glide toward it on final. (Fig. 23-2)

There will be ground effect that will tend to carry you farther down the field than will be seen simply by the apparent movement of the spot. (Fig. 23-3)

Every approach will have some small amount of float. The amount depends on the approach speed. If you approach at 140 knots (normal approach speed 80 knots), the float distance will be extremely long. If you should approach at 61 knots (stall speed 60), the float distance will be quite short (the airplane kind of "squashes" onto the ground — Fig. 23-4).

Notice that the glide angle is the same for both airplanes. Review Fig. 8-7 if there are any questions.

You're more apt to pass if you overshoot slightly (touch down at 250 feet past the spot) than if you get violent and land 30 feet past the spot after doing everything short of spinning in.

If you clear the engine, don't clear it too long — even if it looks as though an undershoot is in the offing. The check pilot may not allow any clearing of

Fig. 23-1. The mark used for the accuracy landings is considered to be a deep ditch — don't undershoot.

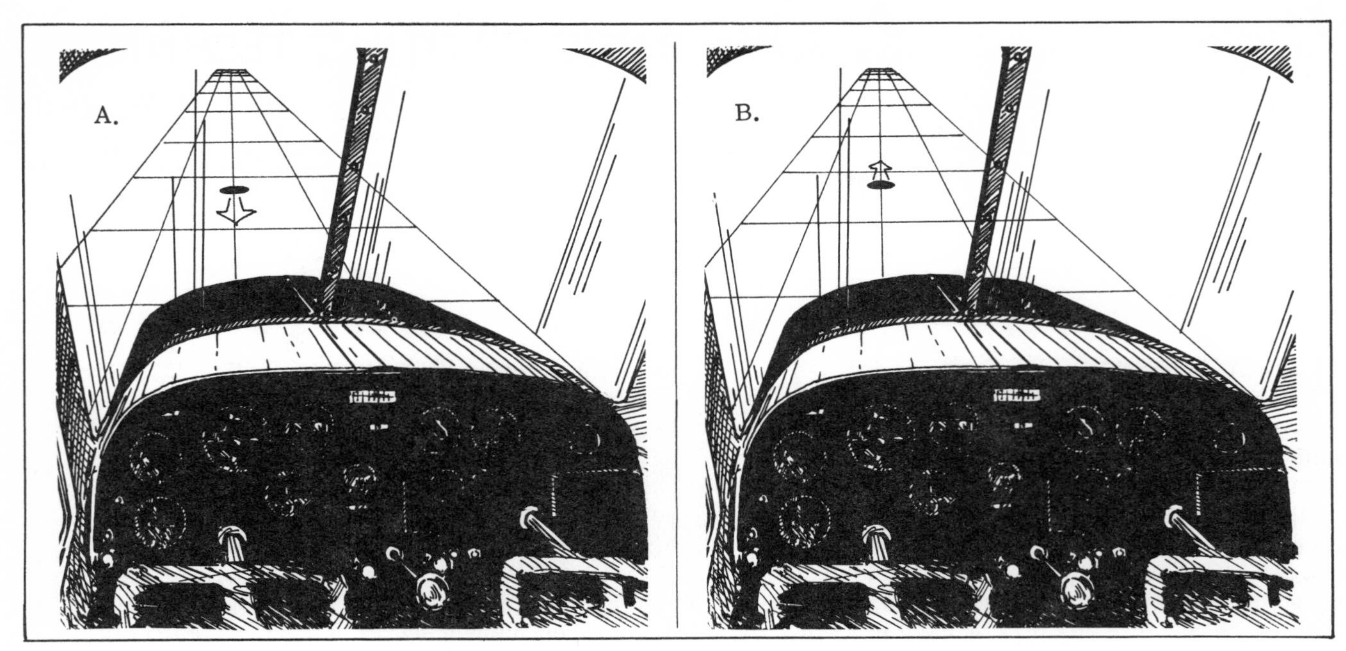

Fig. 23-2.

A. The spot is apparently moving toward the airplane. The airplane will overshoot — slip or use flaps in increments as needed.

B. The spot is apparently moving away from the airplane. The airplane will undershoot at the present airspeed and flap setting.

the engine on these three approaches. (Applicants have been known to "clear the engine" so well on an accuracy landing approach that they've dragged the airplane up to the runway from a sure undershoot of half a mile back — and the check pilot won't like this at all.)

Don't get so engrossed in the accuracy landing that you forget to put the wheels down, or skip other check list items.

OTHER BASIC PILOTING TECHNIQUE POINTS

Better review and sharpen up the following basic piloting techniques if necessary.

PRE-FLIGHT OPERATIONS

This includes engine starting, warm-up, run-up and pretake-off check. Use a check list if available. A good way to start off *wrong* on the flight test would be to (1) forget to holler "CLEAR!" before engaging the starter (2) blast somebody's airplane during the run-up or (3) use poor run-up techniques (heading downwind, for instance) that could possibly damage the engine.

TAXIING

You must maintain good control of the airplane, complying with local taxi rules and ground control instructions. Taxi slowly in congested areas and at a fast walk elsewhere.

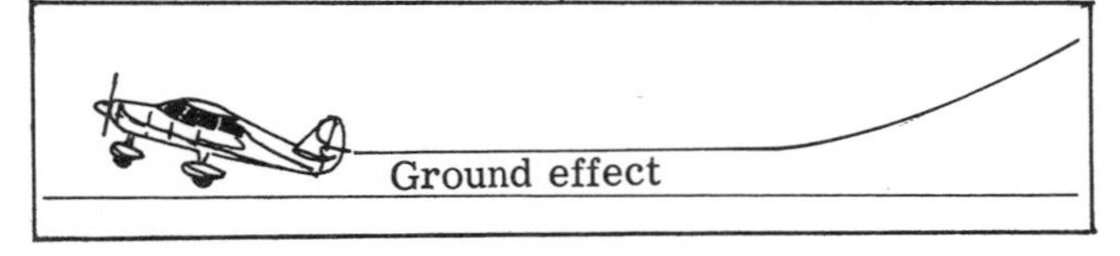

Fig. 23-3. Ground effect.

OTHER TAKE-OFFS AND LANDINGS

The following take-offs and landings should be demonstrated, either during the required accuracy landings or separately.

1. A slip to a landing (if three-control airplane used).
2. Crosswind take-off or landing.
3. Short field take-off and power approach to landing, as appropriate to a short field with surrounding obstructions.
4. Soft field take-off and landing.
5. Wheel landing in tailwheel-type airplane or stall landing in tricycle-type airplane.

These types of take-offs and landings are covered in Chapters 4 and 9 respectively.

TRAFFIC PATTERN

If you've been a little sloppy in the pattern, concentrate on smoothing out such things as traffic pattern altitude (better keep it within 100 feet of the prescribed altitude), wind drift correction, and voice communications.

FORCED LANDINGS

You might review Chapters 16 and 17 concerning forced landings. In the required 10 hours of dual on the commercial maneuvers the instructor will give you plenty of practice for emergencies at all altitudes.

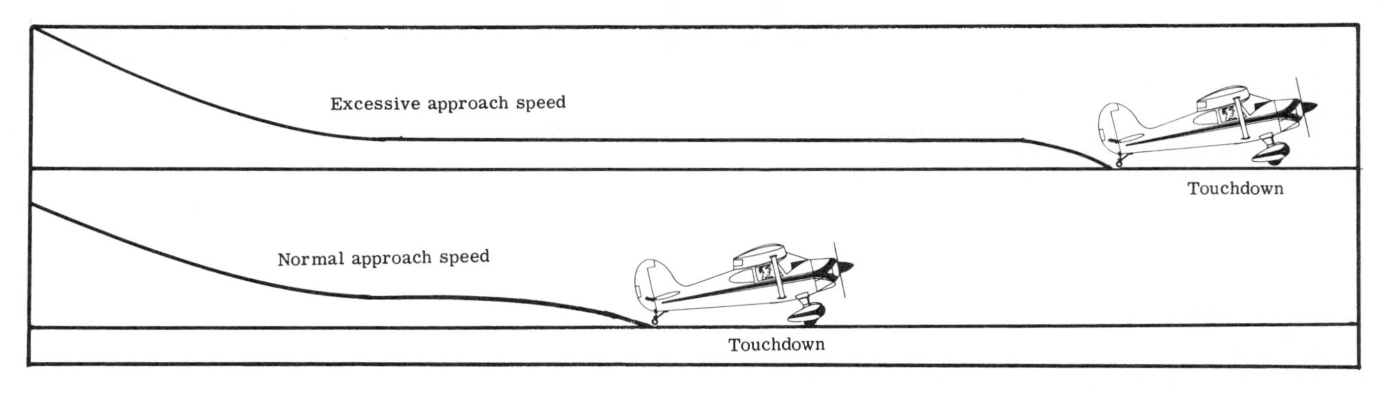

Fig. 23-4. Exaggerated comparison of approach speeds and touchdown points.

EMERGENCY OPERATION OF AIRPLANE EQUIPMENT

Check the Airplane Flight Manual for emergency operation of landing gear, flaps, electrical and fuel systems. Review Chapters 12 through 15.

ENGINE-OUT EMERGENCIES (MULTIENGINE)

If you use a multiengine airplane, be prepared to demonstrate the following operations:

1. Maneuvering with one engine out (feathering should be simulated if feathering-type propellers are used).
2. A demonstration of engine-out minimum control speed.
3. A comparison of the effectiveness of the engine-out best rate of climb speed as compared with faster and slower speeds.
4. Demonstrations of the effect on engine-out climb of the failure to feather a propeller, the extension of the gear and flaps, and combinations of these.
5. An approach and landing with slightly above idling power on one engine (zero thrust) to simulate a feathered propeller if feathering propellers are installed, or idling if the propellers aren't featherable.

SUMMARY

These Basic Piloting Techniques (Phase II) cover the maneuvers that might be used in everyday flying. The check pilot is interested in how you fly the airplane as normally required. Safety is the prime factor here, rather than precision flying.

24. PRECISION MANEUVERS

These maneuvers are required on the commercial flight test so that you may demonstrate your ability to fly in a precise manner. They are a measure of your airmanship and are specifically planned to have you fly the airplane through all speed ranges and in varying attitudes. The precision maneuvers are practice maneuvers and are seldom used in normal flying.

The maneuvers done at higher altitudes (1500 feet or higher) are considered as "high work" and are normally all done at the beginning of the flight test, although there is no written law about it. High work and low altitude maneuvers are grouped as such here to give a clearer picture of the probable sequence of the flight test. The stall series and slow flight will be covered in a separate chapter.

Here's a suggestion for any climbs or glides during the flight test (or any time). Don't climb or glide straight ahead for extended periods. Remember that there is a blind spot under the nose. Make all climbing turns *shallow* and keep a sharp lookout for other airplanes (a steeply banked climbing turn results in much turn and little climb).

HIGH WORK

THE 720° POWER TURN

You've done this maneuver before but now have closer limits to stay within. To review the maneuver briefly:

The 720° power turn required on the commercial flight test is a steep turn with a bank of at least 50° (a 60° bank is a good even number), carried through for two complete turns, or 720°. You will be allowed the following tolerances:

a. Altitude — ±100 feet
b. Bank Variation — ±5° from desired bank
c. Recovery Heading — 15° left or right of entry heading
d. No slip or skid will be maintained.

You will use approximately climb power in the turns (hence the name "power turns") to help maintain altitude at the steep angle of bank.

THE PROCEDURE

At a safe altitude and clear area pick a road or a prominent landmark on the horizon to use as a reference point. Don't just peer ahead looking for a point — there may be an outstanding one off the wing tip. Head toward the reference point and get settled on the chosen altitude.

You'll have to do a 720° turn in each direction and the check pilot may require that you roll directly from one into another.

Choose the direction of the first 720 and, after looking to make sure that you aren't turning into another airplane, start rolling into the turn and smoothly opening the throttle to climb power. You should use a 60° bank and have climb power established before you have turned the first 45°. Your job will be to maintain a near-constant bank.

A bank of 60° is pretty steep and the usual tendency is to lose altitude. If this occurs, you know that the bank must be shallowed in order to regain altitude. If you have a tendency to climb, a slight steepening of the bank may help you — *but* you only have about 5° of bank to vary on each side of the 60°.

In several places in earlier chapters it was mentioned that the load factor was 2 in a 60° bank and that the stall speed was increased by the factor of $\sqrt{2}$ or 1.414. The stall speed increases by 41 per cent in a 60-degree banked, constant-altitude turn. The stall speed increases because of the bank. The airplane slows because the angle of attack is increased to maintain altitude, and you are being squeezed in the middle. The power you are using helps to lower the stall speed as well as allowing you to maintain a constant altitude at a higher airspeed.

Check the nose, wing and altimeter as you turn. Keep a sharp eye out for the reference point and keep up with your turn.

The earlier you catch deviations, the fewer problems you'll have.

If the airplane is holding the bank and altitude, don't do anything. The most common problem is that the pilot spoils the ideal setup by thinking he should be doing *something* at all times.

Remember "torque" effects: the airplane is slower than cruise and you are using climb power. The tendency will be to skid slightly in the left turns and slip in the right turns. Slight right rudder may be needed to keep the ball centered.

There's no need of going into detail about the fallacy of trying to hold up the nose with top rudder during the turn: *don't.*

The ±100-foot altitude allowance means that at *no time* during the maneuver will you exceed those limits. Some guys figure that it doesn't matter how far they are off the original altitude during the turns as long as they are within 100 feet of it when they roll out. They find that the check pilot disagrees.

Start the roll-out about 45° before completing the second turn. As you roll out, throttle back to cruise power, even if you plan on rolling right back into a 720° turn in the opposite direction. The biggest problem most pilots have is keeping the nose from rising during the roll-out even if they have started throttling back. Imagine how tough it would be to keep from gaining altitude with excess power being used. If the check pilot wants an immediate turn in the opposite direction, you can smoothly reapply climb power as you establish the new bank.

COMMON ERRORS

1. Too much back pressure at the beginning of the roll-in, the nose rises and the airplane climbs.
2. Improper throttle handling — rough throttle operation at the beginning and end of the maneuver.
3. Attempting to use back pressure alone to bring the nose up, if it drops; forgetting that the bank must be shallowed.
4. Failure to keep up with the check point.
5. Letting the nose rise on roll-out, causing the airplane to climb.
6. Slipping or skidding throughout the turns.

THE CHANDELLE

The Chandelle is a maximum performance climbing turn with a 180° change in direction. It is a good training maneuver because of the speed changes and the requirement for careful planning.

THE PROCEDURE

Use the recommended entry speed as given in the Airplane Flight Manual or appropriate placard. Stay below the maneuvering speed, if you aren't sure about the entry speed. Cruise plus 10 per cent is a quick and dirty figure for Chandelle entry for many airplanes. As the proper speed is reached, set up a medium bank (25°-30°). The ailerons are neutralized. You may have to hold slight left rudder in the dive because of offset fin effect but check it for your airplane. Apply back pressure smoothly so that the airplane's bank does not actually change. It will *appear to steepen* as the nose moves up and around, but just make sure you keep the ailerons neutral — you'll be changing rudder pressure because of the torque effects as the speed changes.

The airplane will be turned slightly before you get the back pressure started — this is expected. The wrong thing to do during the initial dive and bank is

Fig. 24-1. A Chandelle to the right.

to try to keep the airplane headed for the reference point by holding top rudder. Expect the slight turn and don't worry about it.

As the climb is started and the airspeed drops, smoothly increase power to full throttle but don't cause the engine to overspeed. It's better to start the dive at cruise rpm (assuming a fixed pitch prop here) and try to maintain this by opening the throttle as the airspeed decreases in the climb.

At the 90° turn position the recovery is started and the airplane should be in a wings-level attitude with the airspeed just above the stall at the 180° position. (The nose should *not* be raised further after the 90° position; the first 90° is used to bring it up to the proper pitch attitude.) The nose is then lowered and the airplane returned to cruise attitude. The throttle is eased back to maintain cruise rpm as the speed picks up. Use whatever right rudder is necessary to take care of torque during this part of the maneuver.

Your main problem will be setting up the proper bank; too shallow and the airplane will stall before completing the turn; too steep and little climb is attained. You can visualize this by exaggeration — the effect of a 0° bank or a 90° bank in the dive portion.

The exact amount of bank depends on your airplane's characteristics, but for most trainers the initial bank should fall between 25° and 30° (the less power available, the steeper the initial bank).

It's best to do Chandelles into the wind so that you won't drift so far. The initial dive is done crosswind and the turn is made into the wind. Pilots practicing Chandelles have made the turns downwind and after several maneuvers have found themselves a considerable distance from the practice area. This is a very slow way to go cross country but is fast enough to get you to a new area before you realize it. Remember that the check pilot is interested in your *planning* as well as smoothness in the maneuver.

Figures 24-2A through 24-2F show a sequence of steps during a Chandelle to the *right*.

You'll be better off to stay to one side of the road used for reference. If you're right over it you will lose sight of it during part of the Chandelle and may over- or undershoot the 180° turn.

COMMON ERRORS

1. A too shallow initial bank resulting in the airplane's stalling before 180° of turn is reached.
2. A too steep initial bank resulting in all turn and little climb.
3. Poor coordination throughout, particularly failing to compensate for torque during the last 90° of turn.
4. Failing to roll out at 180° of turn; becoming so engrossed in the nose attitude and airspeed that the turn is neglected.
5. Excessive back pressure, stalling the airplane or too weak back pressure resulting in the airplane "dragging" itself around with little evidence of a high performance maneuver being shown.

With underpowered airplanes you may be lucky to finish the Chandelle at the same altitude you started, much less gain a great deal of altitude. This will be understood by the check pilot as he is more interested in your technique than a great deal of performance — which the underpowered airplane does not have. He knows that if you have the word, when you do get a chance to fly a more powerful airplane you'll get the extra performance.

You are to perform Chandelles consistently within 10° of the desired heading and recover within 5 knots of stall speed.

THE LAZY EIGHT

The lazy eight is one of the best maneuvers for finding out if a pilot has the feel of his airplane. It requires constantly changing airspeed and bank and because of this is more difficult than the Chandelle.

The lazy eight gets its name from the figure the nose apparently transcribes on the horizon: that of a figure 8 lying on its side (a "lazy" eight). For the sake of clarity at this point you can consider the lazy eight as being a series of wingovers. Unlike the wingovers, however, it would have no transition between maneuvers.

The lazy eights, like the Chandelle, should be done into the wind to avoid drifting too far from the original area.

PROCEDURE

Pick a well-defined reference point off the wing tip and preferably one that is into the wind. A point on the horizon is best so that you won't be moving in on it and distort your pattern.

Leaving the throttle at cruise setting, lower the nose and pick up an airspeed slightly above cruise (use the recommended entry speed, if available). Pull up smoothly and, as the nose crosses the horizon, start rolling into a bank toward the reference point. (For example's sake we'll say to the right.) The maximum bank should be 25°-30° and this at the 90° point of turn.

An extension of your line of sight over the nose should pass through the reference point at the 90° point of turn. The second 90° of turn consists of a shallow diving turn, rolling out until at the completion of 180° the wings are level and the point is now off the left wing. Ease the nose up smoothly and make a climbing turn to the left, following through as before, and you may continue the maneuver indefinitely. To sum up the lazy eight discussion as given before:

The maneuver is a climbing turn of ever steepening bank until the 90° point of turn is reached (25-30°) of bank) after which it becomes a diving turn of ever shallowing bank until the 180° point is reached, at which the wings are level. This is followed by a 180° combination climb and diving turn in the opposite direction. Figures 24-4 through 6 show how the reference point would appear as seen from the cockpit at different parts of the first 180° of the maneuver. The nose should have its highest pitch at 45° of turn; the lowest pitch will be at 135° of turn.

Keep the maneuver "lazy." One of the faults of

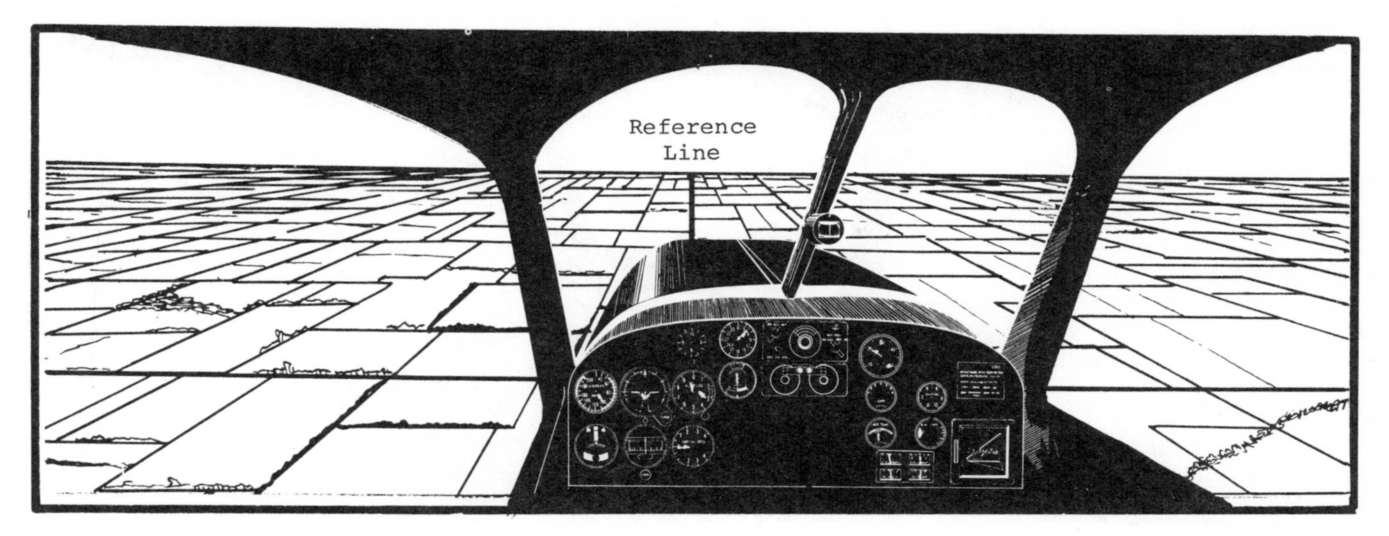

Fig. 24-2A. The Chandelle. Pick a reference, such as a road, and line up with it.

Fig. 24-2B. Ease the nose over to pick up the recommended entry speed. (Watch the rpm.) When the desired airspeed is reached, establish a 25-30° bank. Neutralize the ailerons and start applying steady back pressure.

Fig. 24-2C. As the airspeed starts decreasing, smoothly apply power to the climb value.

Fig. 24-2D. The roll-out is initiated at 90° of turn. The pitch attitude is held constant for the rest of the maneuver.

Fig. 24-2E. The airplane at 135° of turn. The airspeed is continuing to decrease as a constant nose pitch position is maintained. The roll-out is continuing.

Fig. 24-2F. The wings are level and the airspeed is just above the stall after 180° of turn.

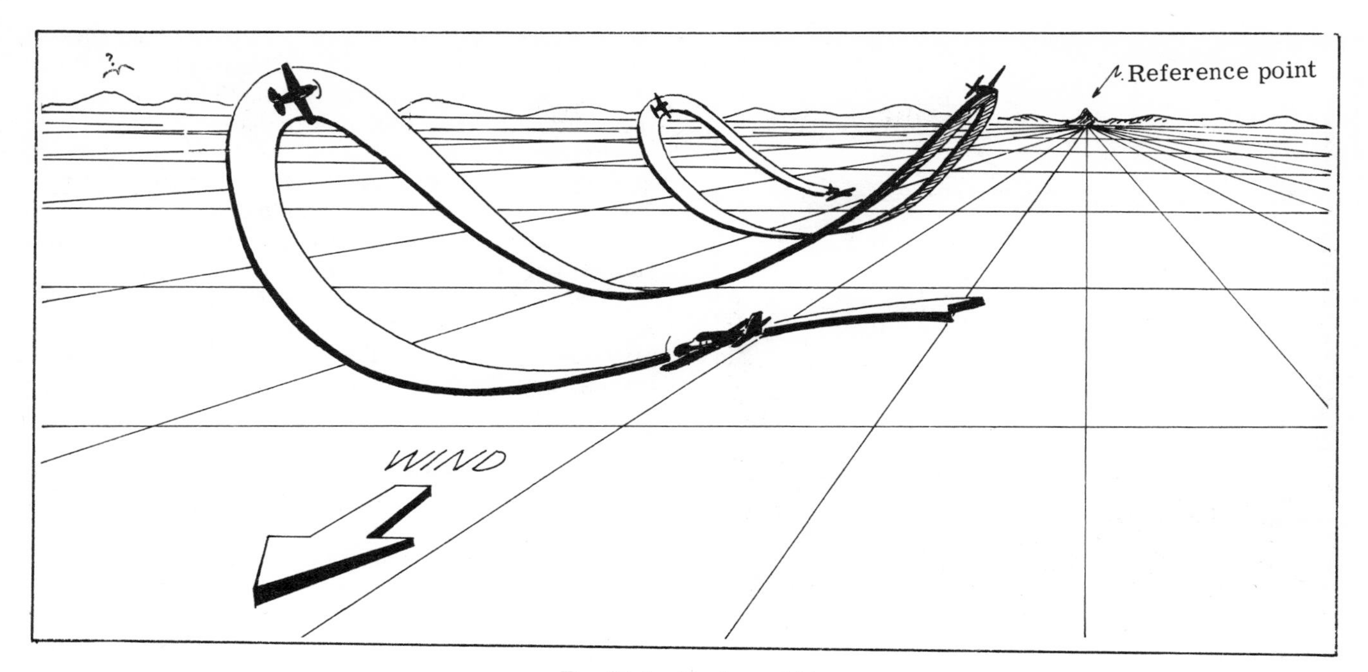

Fig. 24-3. The lazy eight.

most pilots is that as they get farther into the series of turns the faster and more frantic are their movements. You may have to make yourself relax as the series progresses.

You should always be at the same altitude at the bottom of the dive. If you tend to climb, decrease power slightly. If you tend to lose altitude, increase power as necessary. The maneuver should be symmetrical; that is, the nose should go the same distance below the horizon as above it.

You may have to consciously apply rudder to the top of the "loop" to keep the ball centered at all times. Back pressure usually is needed at the peak to make sure your line of sight goes through the point without the airplane slipping.

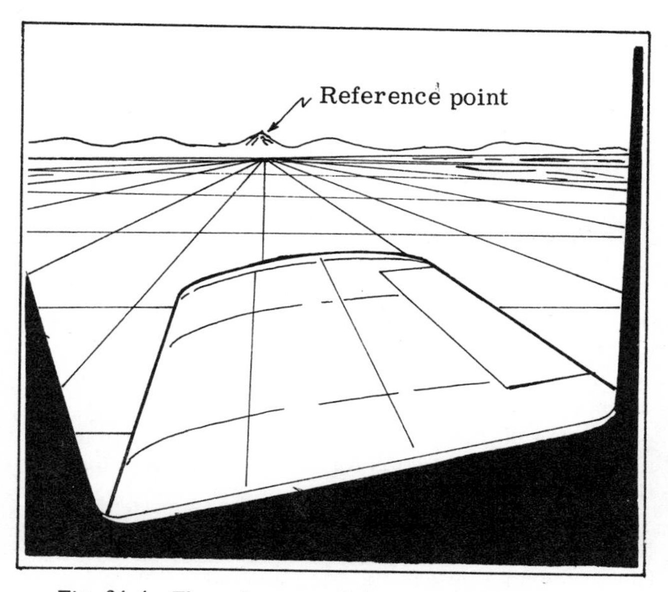

Fig. 24-4. The reference point as seen during the initial dive (lazy eight).

COMMON ERRORS

1. Poor coordination; slipping and skidding.
2. Too steep a bank at the peak of the maneuver.
3. Failing to maintain the same altitude at the bottom of the descents.
4. Losing the reference point.

You will be judged on planning, orientation, coordination, smoothness, altitude control and airspeed control.

1080° GLIDING SPIRALS ABOUT A POINT ON THE GROUND

This is a good maneuver for several reasons: (1) it checks the pilot's ability to stay oriented

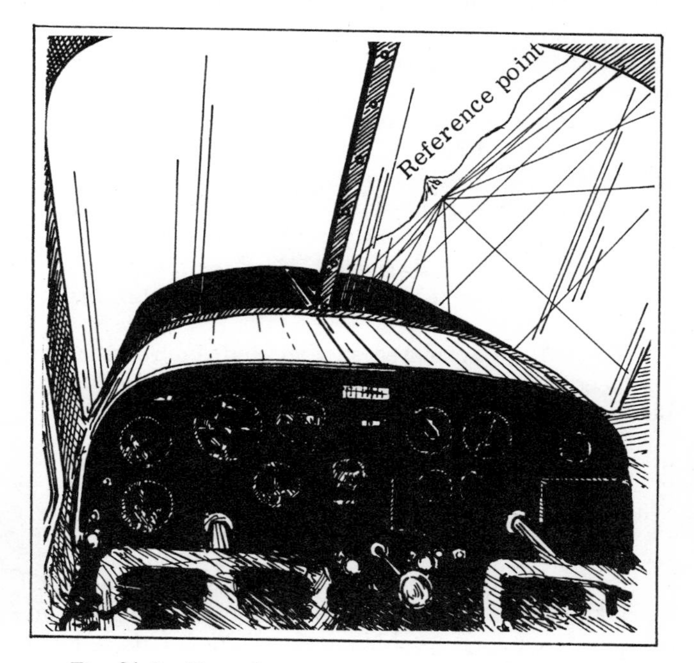

Fig. 24-5. The reference as seen at the 90° point of turn (25°-30° of bank) in the lazy eight.

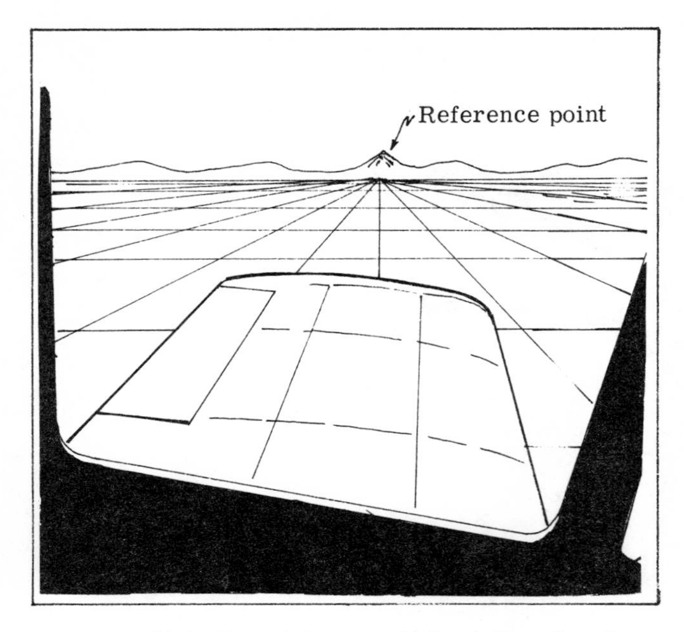

Fig. 24-6. The point as seen in the shallow dive after 180° of turn (lazy eight).

throughout a number of steep turns; (2) the pilot must correct for wind drift by varying the bank in order to stay over the same point; and (3) it requires the pilot's attention to be directed outside the cockpit while performing the maneuver, thereby requiring that he have good feel of the airplane.

Think back to S-turns across the road and the turns around a point: Remember how you varied the bank to take care of wind drift?

From S-turns and turns around the point you remember that *"the angle of bank is directly proportional to the groundspeed."* The greater the ground speed, the greater the angle of bank required. You varied the bank to fly a definite curved path over the ground. The principle is exactly the same for the gliding spiral about a point on the ground — except that you'll have the power at idle and will be descending.

The required maneuver is three full-turn spirals (1080°) in each direction with a bank of at least 50° at the steepest sector of each turn. *Following the S-turn and turns-around-a-point principles, you know that this point of steepest bank will be when you are headed directly downwind. The shallowest bank will be when you are headed directly upwind in the turn.*

You will be evaluated on your drift correction, airspeed control, coordination, orientation and vigilance for other traffic.

Your tolerances are:

Airspeed — ±8 knots

Bank — 45° to 55° maximum at steepest point.

Heading of recovery — 10° right or left of the entry heading.

Keep up with the number of turns and look around!

Make sure that you have sufficient altitude to complete the turns. It wouldn't look too good from a headwork standpoint if you expected to start a three-turn spiral at 1000 feet above the ground — the check pilot might wonder a little. If you aren't sure that there's enough altitude to do the job, tell the check pilot and climb to a better altitude. You should be recovered from the spirals at 1000 feet or better.

Figure 24-7 shows part of the 1080° spiral as seen by an observer directly above the point. The 360° turns are treated separately and perspective requires that the lowest turn appear to be of a smaller radius, but all turns should be of equal radius.

The check pilot normally will require the spirals as the last maneuver of the high work, since you can also use it to get down for the low work maneuvers. He may say, "forced landing" during one of the turns, requiring you to set up a simulated dead stick pattern to a nearby field. As you have been correcting for the wind you should know its direction but it's funny how blank your mind can go sometimes.

Back to the spiral itself — watch your coordination. Pilots who normally make extra smooth gliding turns sometimes get so engrossed in the reference point that they'll do anything to keep it where they want it, and the ball in the turn and slip instrument just barely stays in the cockpit.

The airplane will be in the clean configuration throughout the maneuver and an airspeed of about 50 per cent above the stall should be a rough figure to think of in terms of glide speed. Don't get too fast or the airplane may get out of hand for a few seconds (or at least appear that way to the check pilot). If you are too slow you could stall by pulling excess back pressure in the steep bank.

Practice three turns in each direction.

COMMON ERRORS

1. Poor correction for wind drift.
2. Poor coordination — slipping or skidding.
3. Too fast an airspeed in the spiral.
4. Failure to keep up with the number of turns.
5. Failure to look around for other traffic (staring at the point).
6. Poor planning, not being able to complete the maneuver without getting excessively low.

Remember to clear the engine.

LOW WORK

EIGHTS ON PYLONS

Eights on pylons, or on-pylon eights, are given on the flight test as a test of your ability to fly the airplane by sense and feel. You won't have much opportunity to be gazing around the cockpit during the performance of these maneuvers.

Eights on pylons consist of flying the airplane in a figure 8 around two reference points, or pylons, on the ground.

At cruising airspeed (assuming no wind) you'll find that at a certain altitude you will be able to lay the wing on an object and turn around it. At a lower altitude you'll find that the wing will tend to move ahead of the pylon; at a higher altitude, the wing will

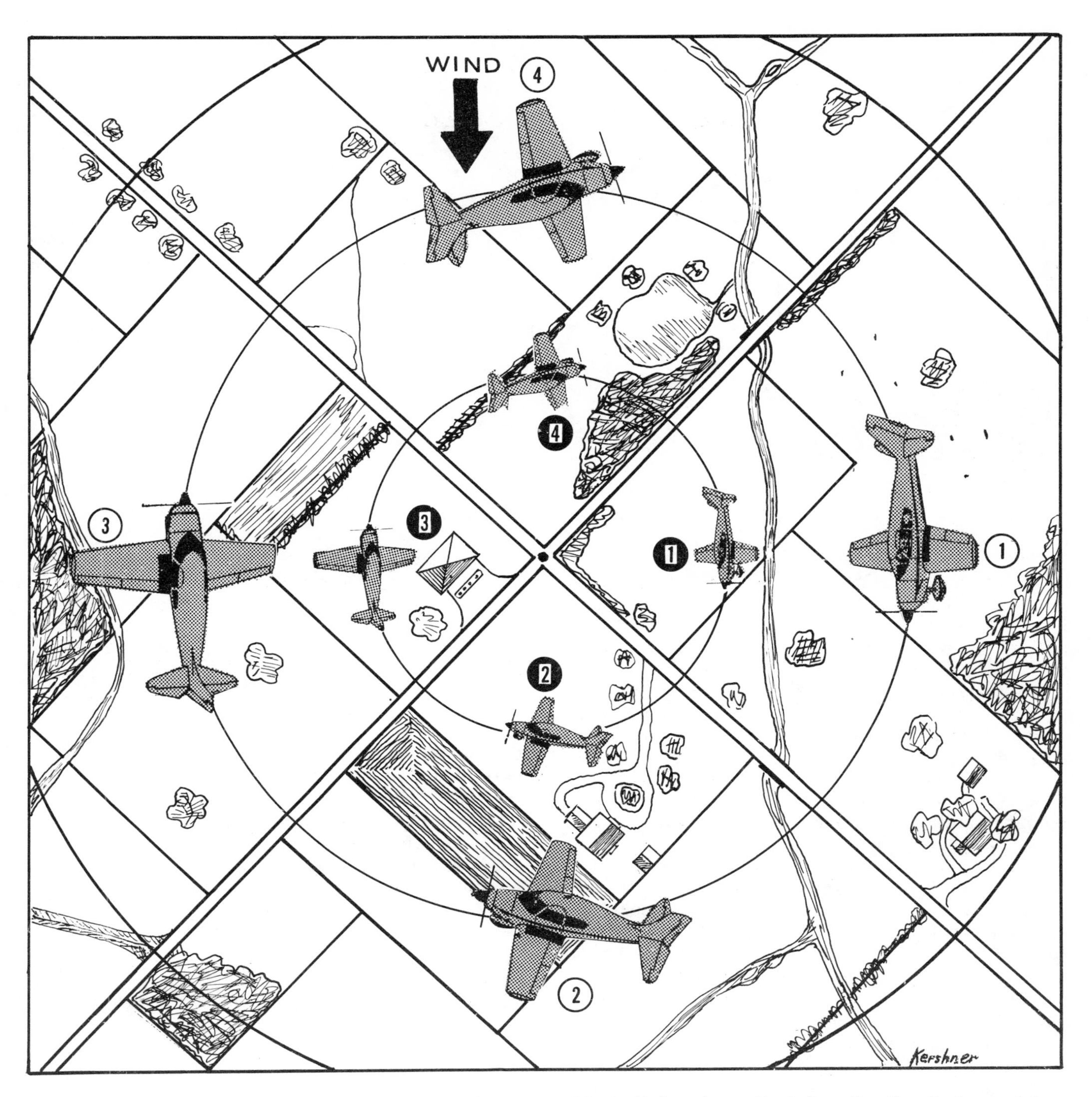

Fig. 24-7. The 1080° gliding spirals about a point. The turns would actually be a descending helix rather than the "separate" turns shown. Consider the need for varying the bank to maintain a constant radius, Point (1) on the circle will require the steepest bank (headed directly downwind) and (3) will require the shallowest bank in order to follow the circle. At Points (2) and (4) the airplane will have comparable banks but you will actually have a crab set up. (The turns will be balanced and the ball will be centered at all times.) The required bank and crab will vary smoothly around the circle.

Since the airplane must start at a fairly high altitude in order to complete the turns, the point is harder to see (and hold) at the beginning of the maneuver. In fact you'll find that things start really falling into place during that lowest, final, 360° turn. Because of the required higher initial altitude, this maneuver is more difficult than 720° turns around a point, which are done at a *constant* lower altitude.

apparently move behind the pylon. The correct altitude above the pylon to keep the wing on the pylon varies as the square of the airspeed and is called the "pivotal altitude."

The pivotal altitude actually depends on the relative velocity of the airplane to the pylon, so in actuality it will change with changes in ground speed.

(For a top view of the maneuver, take a look at Figure 24-8.)

You should have a rough idea of the pivotal altitude and can arrive at an approximation by the following: $\text{Pivotal altitude} = \frac{\text{airspeed squared (mph)}}{15}$

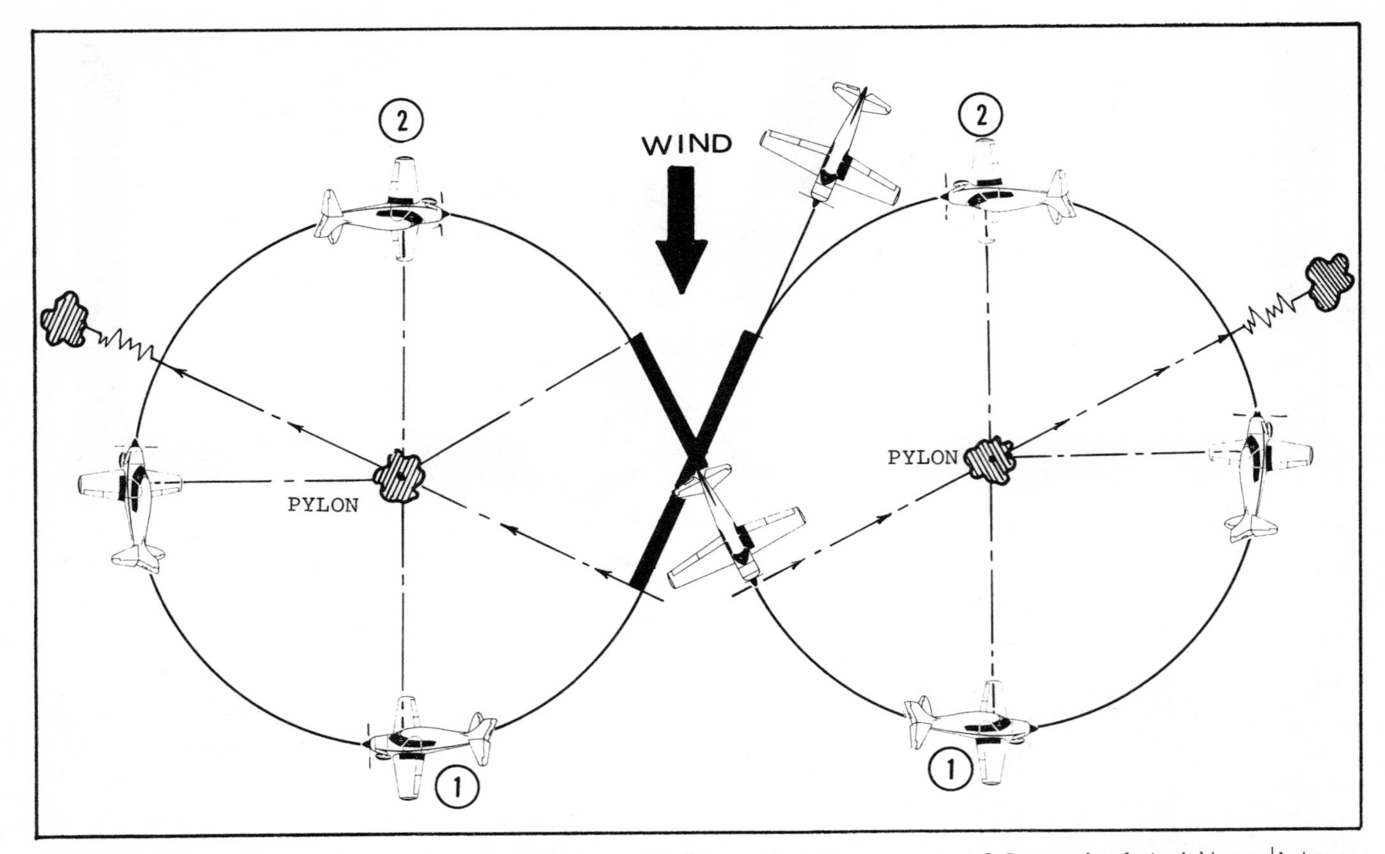

Fig. 24.8. On-pylon eights as seen from directly above. The strong black lines represent 3-5 seconds of straightaway between each pylon. The points "1" and "2" will be discussed later in this section.

or $\frac{\text{airspeed squared (knots)}}{11.3}$. If you plan on using an airspeed of 100 mph (87 knots) for the maneuver the pivotal altitude (altitude above the pylon) in feet would be $\frac{(100)^2}{15}$ or $\frac{(87)^2}{11.3}$ = *665* feet.

Then, in theory, assuming that your airspeed indicator is correct and there is no wind, you could fly up to the pylon, roll into a balanced turn, lay the wing tip on the pylon and be able to keep it there as long as you did not vary the bank or altitude.

The pivotal altitude remains the same for any angle of bank you choose to fly. (Fig. 24-9)

To be completely accurate, you should use the *true* airspeed in the equation for determining pivotal altitude because this (and the wind) establishes your speed relative to the pylon. If you are doing eights at a part of the country at an elevation of 10,000 feet you may find that using indicated airspeed can cause you to be flying too low to hold the pylon and you will wonder why. Of course, you would soon find the altitude by trial and error but would wonder what happened to the equation. For lower altitudes using indicated airspeed will give you something with which to start.

If you get such a steep bank that excessive back pressure is required, *slowing* the airplane, then you can expect a change in pivotal altitude.

Once you have an approximation of the pivotal altitude you can make minor adjustments to find the exact altitude required. You won't be able to read the altimeter that closely, but will know that you have the correct altitude when you're able to "hold the pylon."

About this idea of wingtip reference to the pylon -- you actually want to keep the pylon at a point perpendicular to the airplane's longitudinal axis as seen by you. If you are sitting directly over (or under)

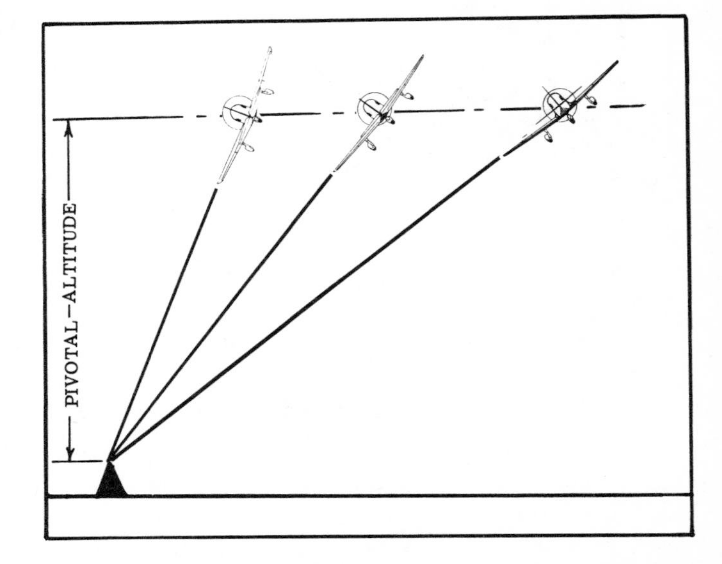

Fig. 24-9. The pivotal altitude is only affected by airspeed (and ground speed), not by the steepness of turn. Your indicated altitude will be the elevation of the terrain at the pylon plus the pivotal altitude.

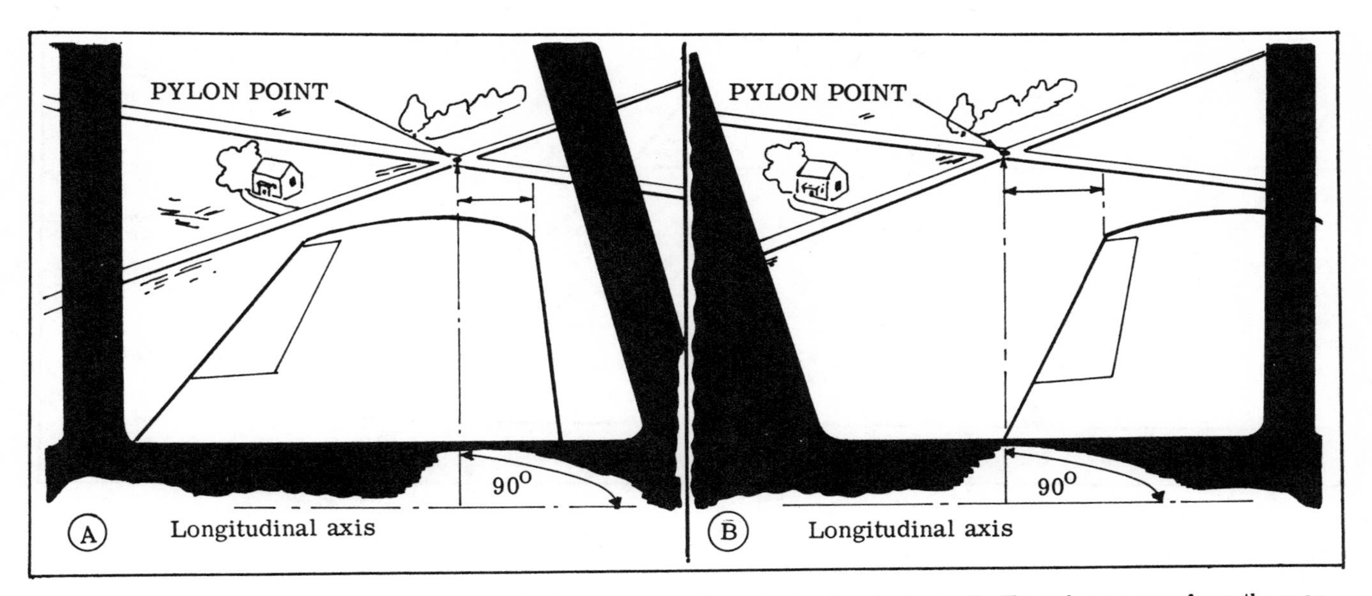

Fig. 24-10. **A. The pylon as seen from the front seat of a low wing tandem trainer. B. The pylon as seen from the rear seat of the same airplane.**

the wing, then the wing tip (or part of it) will be used for reference. If your position is well forward or aft of the wing, an imaginary point forward or aft of the wing tip must be used. (Fig. 24-10)

Fig. 24-11 shows the same idea as seen from *above* the airplane.

The correct reference may vary for the same airplane between pilots of different heights (and seat positions). You can see in Fig. 24-11 that if the pilot in the rear seat tries to use the wing tip as a reference he will not be able to "hold" the pylon, as the airplane will be turning into it which will result in a diving spiral. (On the other hand, if your reference is too far aft, you will tend to fly around the pylon in a nose high slip as you try to hold the pylon by fair means or foul.)

WIND EFFECTS

The effects of wind on the path around the pylon must be considered. On the downwind side the wind moves the airplane away from the pylon, requiring a

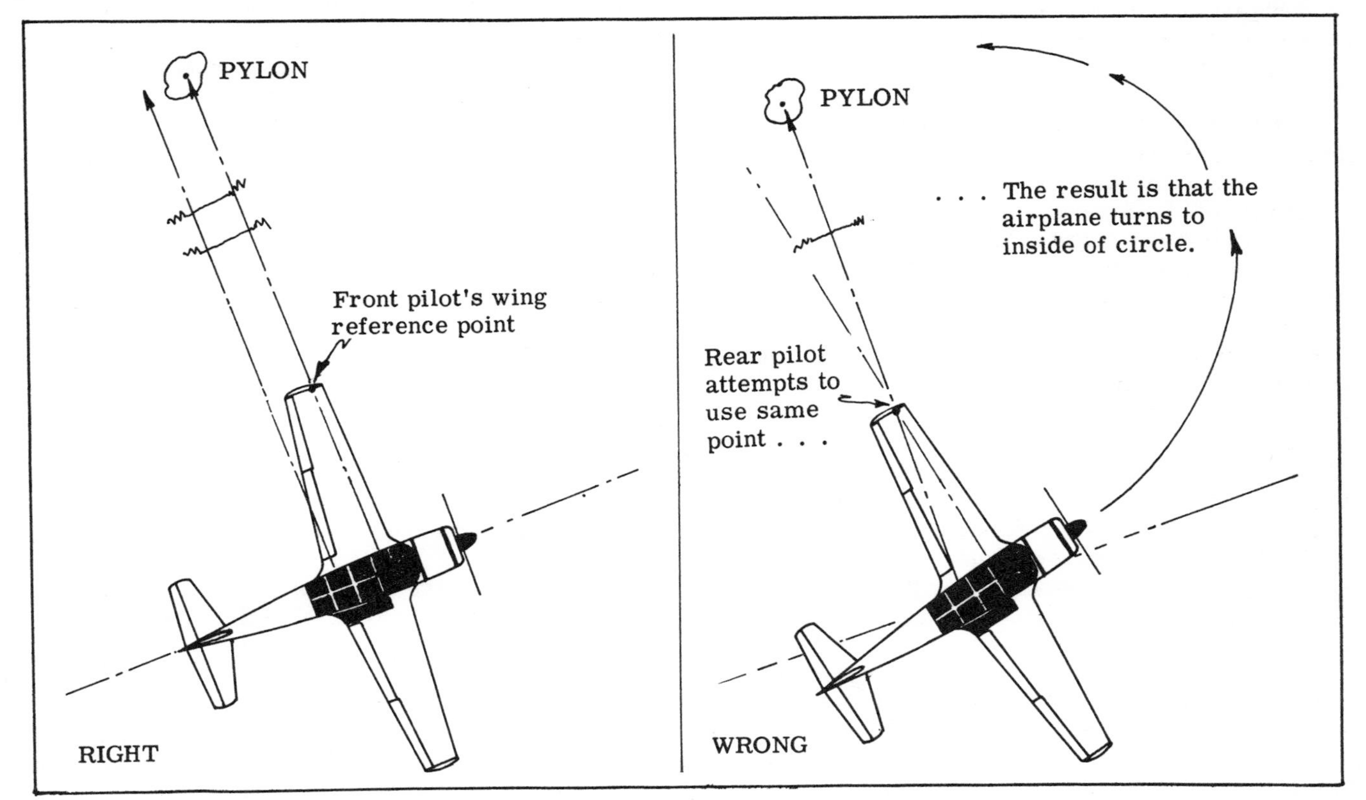

Fig. 24-11.

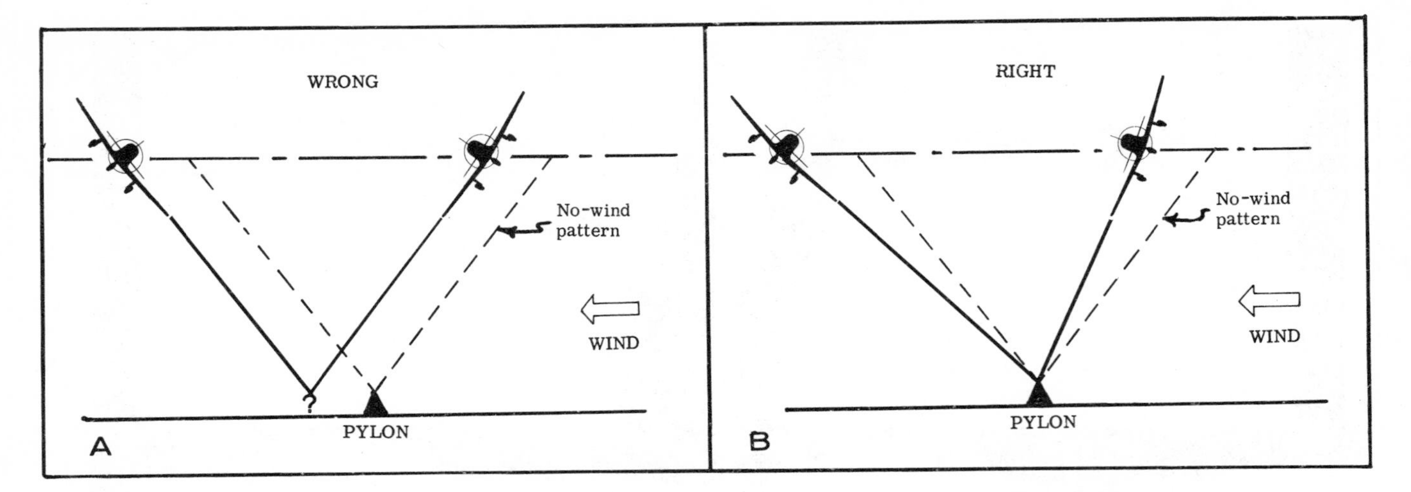

Fig. 24-12. Wind effects. (A) The pilot maintains the same degree of bank and is unable to hold the pylon. In (B) he varies the bank as needed to hold the pylon.

shallower bank to hold the pylon. On the upwind side the wind moves the airplane into the pylon, requiring a steeper bank to hold the pylon at the same relative position to the wing. (Fig. 24-12)

Remember, you are *not* trying to maintain a constant path around the pylon as you were for the (1) turns around a point, (2) *around* pylon eights or (3) spirals around a point. In this maneuver you are trying to keep the pylon at a constant relative position to *you.* Therefore, if the wind moves you close to the pylon, you must steepen the bank to keep the wing tip reference on it. A pilot doing this maneuver for the first time has a tendency to unconsciously slip back into the habit of correcting for wind as in the wind drift correction maneuvers.

As the pivotal altitude is a function of groundspeed, it will change as the airplane is heading upwind or downwind. When headed upwind the groundspeed is lower, therefore, the pivotal altitude is lower. Flying downwind, the reverse occurs.

You should fly around a single pylon several times before attempting your first pylon eight. You can control the steepness of the maneuver by entering the maneuver at varying distances from the pylon.

In effect, you will take care of required altitude changes by use of the elevators. You'll be getting a double effect: As an example, suppose that as you turn into the wind the pylon appears to move ahead of the wing, which shows that the airplane is too high for the groundspeed it's making — you are above the pivotal altitude for that groundspeed. Your correction will be forward pressure which will take care of two points at once: (1) it will lower the airplane's altitude and (2) also increase the airspeed (which in turn increases the groundspeed) so that a slight change in altitude by this method gets large results. Once you are set up in the pattern the altimeter will not be so important. You'll fly the airplane to keep the pylon at the correct relative position — and the pivotal altitude will take care of itself.

Some pilots' methods of holding pylons are unique to say the least. They skid or slip and go through all sorts of gyrations to keep the wing on the pylon. *The ball in the turn and slip should stay centered throughout.* You'll be graded on smoothness on the maneuver, too.

STEEP ON-PYLON EIGHTS

You'll be required to perform three consecutive steep and three consecutive shallow on-pylon eights on the flight test.

The steep eights should be performed with banks of at least 60° at the steepest sector of the turns. You are allowed to use one turn about a pylon to establish the pivotal altitude.

Enter the pylons as shown back in Fig. 24-8 and at the approximate pivotal altitude for your airplane. As you may not know the exact terrain elevation in the pylon area, be prepared to take advantage of that "free" turn around the pylon to get set up. During that first turn take a quick glance at the altimeter at points (1) and (2) (Fig. 24-8) if you are holding the pylon. The altitude should be the same at (1) and (2), as they are both crosswind. *This altitude will be the basic pivotal altitude and you will maintain this altitude in flying in the short straightaway.*

Notice that in entering downwind the turns are *into* the wind. This is the safest method because the pilot making a turn downwind would be misled by his groundspeed and in a strong wind might subconsciously tighten the turn and also pull the nose up — a dangerous procedure at steep banks and low altitudes!

You'll notice that for the steep eight you'll be turning around each pylon at a goodly clip and will be concentrating on what's happening *now.* You'll have to remember that you must roll out at a point that will allow you to fly to the other pylon to arrive at the proper distance from it. You can get so engrossed in the pylon in use that the turn is continued too far around.

If, in a strong wind, you head the airplane toward

the point of tangency of the turn around the other pylon, you'll have drifted so far from it that the maneuver may end up as a *shallow* eight for that pylon turn.

You'll find that the pylon's relative fore and aft position to the wing is controlled by fore or aft pressure (which in turn controls altitude and airspeed for this maneuver).

Make the roll-in on the pylons smooth and coordinated. The usual tendency is to roll in *too soon,* which requires shallowing the bank and then resteepening it. A smoother maneuver is one that requires a faster roll-in. So if in doubt, roll into the turn a little late and with a faster rate of roll. The roll-in should be timed so that it is completed just as the pylon reaches the correct relative position to the wing.

Set the power to cruise and leave it there. Check occasionally to make sure you still are carrying cruise power.

The pylon must stay within one foot (at the wing tip) of the reference point. You'll find that with proper planning this will be no problem.

SHALLOW ON-PYLON EIGHTS

The principles for the shallow eights are the same as for the steep ones. You will have the advantage of having more time in the turns to make corrections on the pylons, but on the other hand you'll also have more time to get fouled up. Plan ahead and *know* that as you turn into the wind the pylon will start to move ahead — and do something about it just *before* it starts to move. It's hard to get back on the pylon once you've lost it.

Once you have the reference pivotal altitude for the steep eights this applies to the shallow ones also (or vice versa). (Figs. 24-9 and 24-13)

You should not exceed a minimum bank (30^0) in the steepest sector of the turn.

Keep a sharp lookout for other traffic when performing either of these maneuvers.

EIGHTS — "BACK TO BACK"

After you've become proficient in the type of eights just discussed you should start doing them "back to back," that is, without any straightaway period. You'll find that this is more challenging and uses your time in the maneuver more effectively. You'll roll right into a turn around the other pylon without a straightaway, and a common mistake is to keep your attention on the one pylon too long. You will have to pick an extra pylon if you plan to do both steep and shallow on-pylon eights as shown in Fig. 24-13.

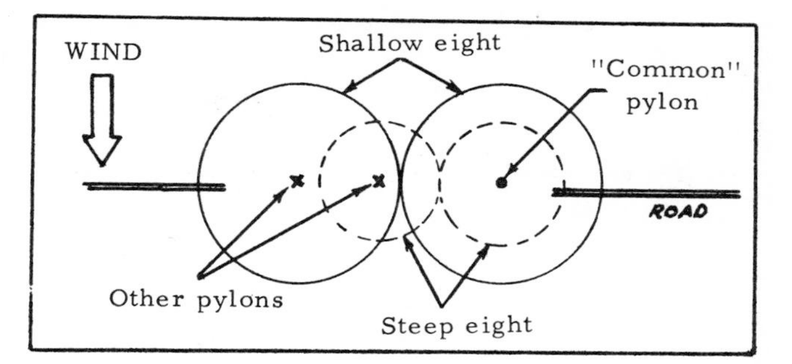

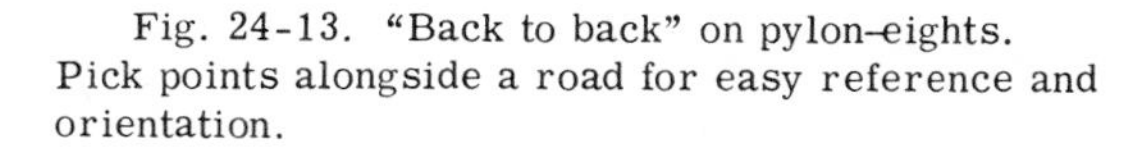
Fig. 24-13. "Back to back" on pylon-eights. Pick points alongside a road for easy reference and orientation.

PICKING THE PYLONS FOR THE EIGHTS

To do the eights on pylons, pick two well-defined points. The speed of your airplane has a lot to do with distance between pylons. Faster airplanes need more room, slower ones less. You'll want to pick pylons far enough apart so that the turns won't be crowded, yet close enough so that you won't have to file a flight plan going from one to the other After a little practice you'll be able to spot pylons of a suitable distance apart for your airplane.

It is suggested that road intersections make good pylons for low wing airplanes as you'll need some reference to find the pylons. If you are in a side-by-side airplane the pylon requiring a right turn will be hard to spot. You can pretty well tell the pylon position under the wing by the portions of the two roads you can see.

Look back to Figure 24-8. If you have to pick pylons where no crossroads are available, pick some distant object that will be in line with the pylon at the point of roll-in as indicated by the arrows in that Figure. It's pretty easy to "lose" your pylon if it happens to be a tree close to other trees.

Pick pylons so that a line drawn between them is perpendicular to the wind and, if possible, ones that are at the same elevation. This will be a good place to get a low altitude emergency given to you so always keep tabs on good fields nearby.

One good way of picking pylons is to fly crosswind and look down-wind. When you see pylons that can be used, make a turn and commence the maneuver. Figure 24-14 is a stylized view of the procedure.

COMMON ERRORS

1. Poor pylon picking — pylons too close or too far apart.
2. Skidding or slipping in an attempt to hold the pylons.
3. Poor planning in rolling out from the pylons.
4. Neglecting wind drift in the straightaway portions of the maneuver.
5. Poor altitude control in the straightaway.
6. Rolling into the pylon turn too soon.
7. Failure to check the area for other traffic.

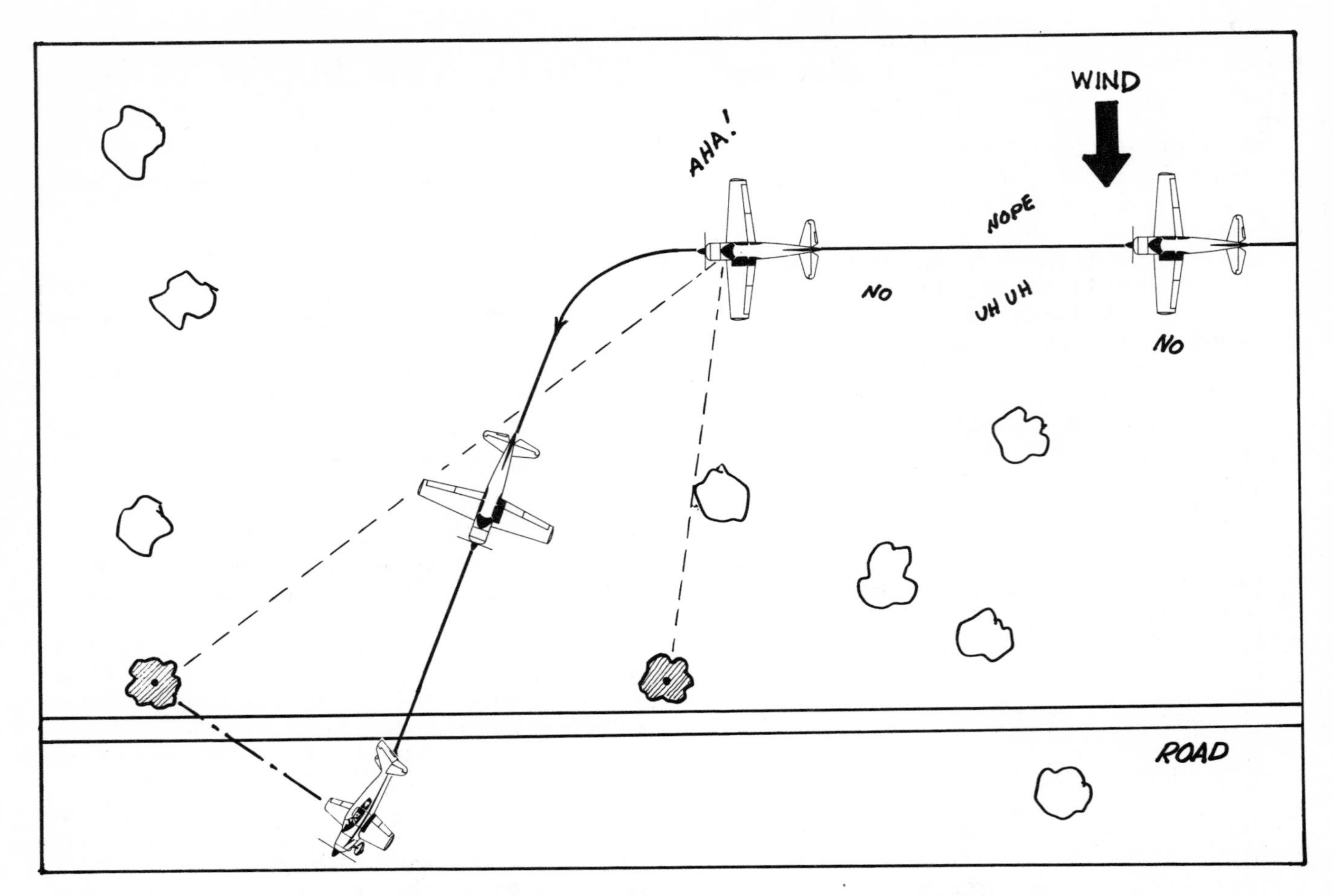

Fig. 24-14. Picking pylons precisely prevents the prolonging of practice periods.

25. STALLS AND SLOW FLIGHT

Although stalls are considered to be part of the precision maneuvers on the commercial flight test, they are numerous (and important) enough to warrant a separate chapter.

Stalls and stall recoveries will be required from straight climbing and gliding flight and from three other conditions: take-off and departure, landing approach, and accelerated maneuvering at reduced speeds. The required stalls are almost exactly like those on the private flight test.

In addition, this chapter will discuss cross control stalls and spins, not required on the flight test but a knowledge of which nevertheless will be useful to the private or commercial pilot.

As you were taught as a student, the stall is a function of angle of attack — not airspeed. The airplane can be stalled at any attitude and airspeed, but at higher airspeeds the airplane may come apart before it stalls.

The present FAA certification rules require that the airplane exhibit no unsafe rolling tendencies at the stall. The wings are leveled through coordinated use of ailerons *and* rudder — no more yawing the airplane with rudder to pick up a wing during the stall. The design methods used to aid lateral control during the stall were covered in Chapter 1 and you might review them now.

STRAIGHT-AHEAD STALLS

Always clear the area by making two 90° turns in opposite directions, or one 180° turn before doing stalls straight ahead. There will be a blind spot over the nose during the stall. Some pilots mechanically make the turns but don't look around. This is not only unsafe but a waste of time. If you don't look around during the turns it's as bad as no attempt to clear the area at all.

Be sure in all stalls that you have recovered at least 1500 feet above the ground.

APPROACH TO A STALL—POWER-OFF

The stall will not be allowed to progress to the break; recovery will be effected as soon as indications of the stall are felt.

PROCEDURE

1. Clear the area.
2. Carburetor heat on (if your airplane requires it) — throttle to idle.
3. Raise the nose to an angle of 10-15° above the horizon. Have some prominent object picked to help keep the nose lined up.
4. Pin the nose at that attitude by continued back pressure. Keep the wings level.
5. When indications of the stall are felt by sight, sound and feel, lower the nose and apply full power smoothly — don't ram the throttle open.

The check pilot may ask you to make a power-off recovery — which only means that the recovery is not quite as quick as is found by using power, but is still positive. The reason for a no-power recovery would be to simulate a dead stick glide during an emergency when you've let it get too slow and are approaching a stall (with no power available to help recover).

APPROACH TO A STALL—POWER-ON

The only difference found between this stall and the power-off version is that cruise power is used throughout, which makes it necessary for you to raise

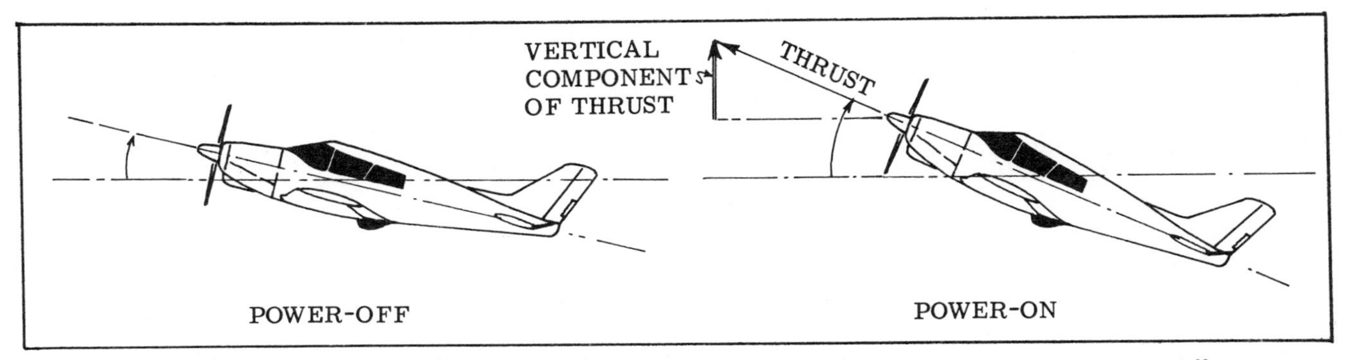

Fig. 25-1. A comparison of the attitudes of the power-off and power-on versions of the approach to a stall.

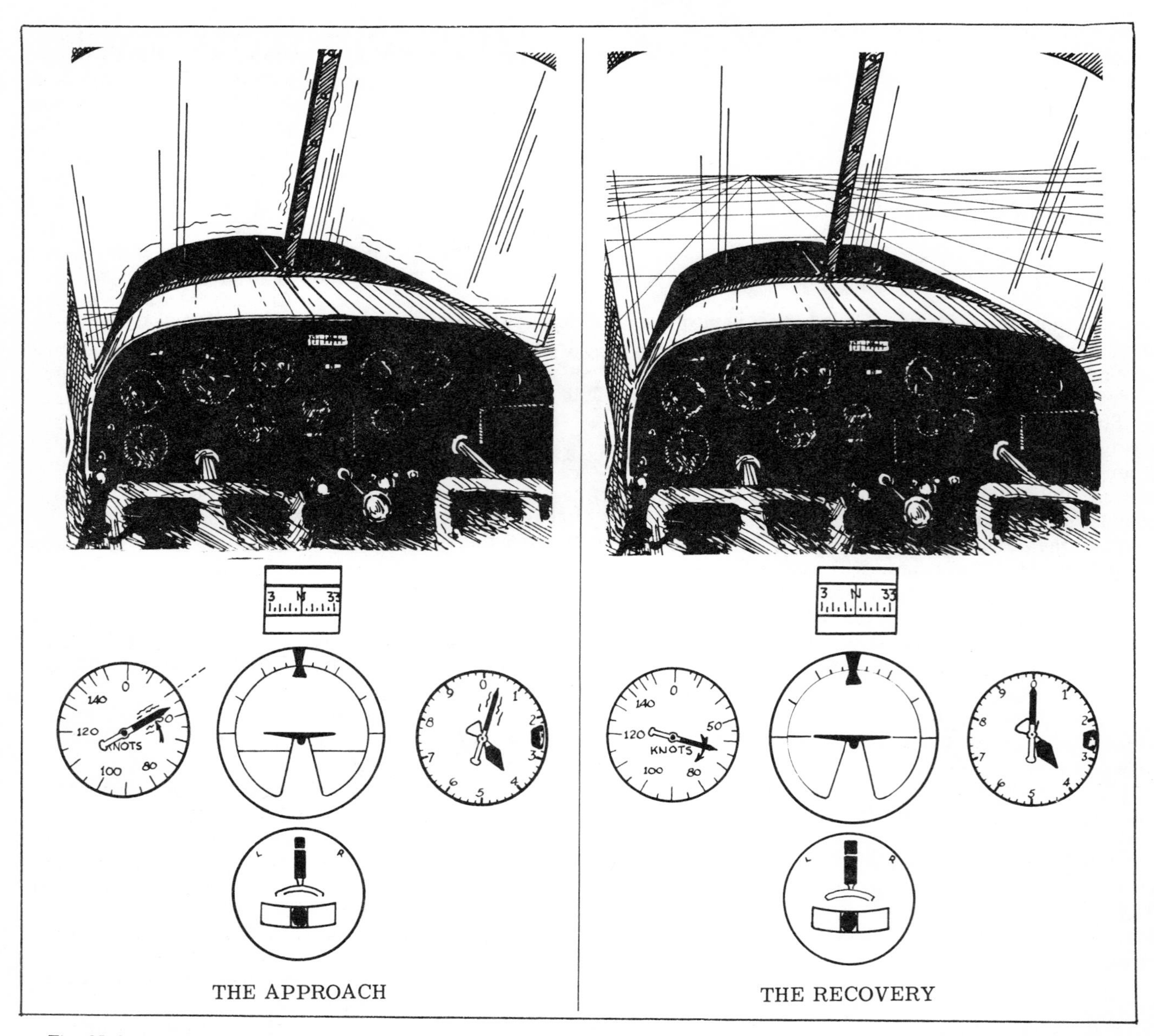

Fig. 25-2. The approach to a stall as seen from the cockpit and as indicated by the flight instruments. (This stall is sometimes called a partial stall.)

the nose slightly higher in order to get stall indications without too much delay. Power makes a difference in nose attitude and stall speed because of the added slipstream over the wings and the airplane's vertical component of thrust. Fig. 25-1 shows a comparison of the nose positions for the power-on and power-off approach to a stall. This also applies for differences in power-on or -off versions of other straight-ahead stalls.

The recovery for the approach to a stall also consists of applying full throttle. Get in the habit of automatically applying full power on all stalls unless you are told otherwise by the check pilot.

Fig. 25-2 shows an approach to a stall as it would be seen from the cockpit and as indicated by the instruments.

COMMON ERRORS — PARTIAL STALLS

1. Failure to clear the area.
2. Allowing the stall approach to progress to a break.
3. Lowering the nose before there are definite indications of the approaching stall.
4. Not keeping the wings level throughout.
5. Letting the nose wander.

NORMAL STALL

The major difference between the normal stall and the approach to a stall is that in this maneuver the back pressure is continued until a definite break occurs. Start the recovery as soon as the break occurs and apply full power. Have the nose slightly higher than in the approach to a stall to insure a clean break.

You may be required to do the maneuver power-on or power-off, and the principles and errors mentioned concerning approach to stalls still apply, so practice both versions of the normal stall.

COMPLETE OR FULL STALL

This stall is a good exercise for keeping the wings level -- and little else – but you should be familiar with it, if only for academic interest. (The flight examiner *could* ask you to delay recovery until the nose pitches down to the horizon – if you are in a light trainer.)

The maneuver may be done at cruise power or at idle as the first stall types mentioned.

In performing this stall pull the nose up higher (up to about 30° above the horizon) than was done for the normal stall and continue easing the control wheel back until the break occurs. In the complete stall you will not start a recovery until the nose has fallen to the horizon (keep that wheel all the way back!) at which point you release back pressure and apply full power.

You'll find that the power-on version may give you a little trouble in keeping the wings level and the nose tends to wander. As you can see, this is an exercise – you certainly wouldn't deliberately wait until the nose had fallen through to the horizon before starting recovery in an accidental stall at low altitudes!

SUMMARY OF THE STRAIGHT-AHEAD STALLS

The Approach to a Stall – recovery before the break.

The Normal Stall -- recover immediately after the break.

The Complete Stall – nose higher, recover after the break when the nose has moved down to the horizon.

Always clear the area.

Note that you have six possible combinations of the above stalls by doing each one with and without power.

Keep the wings level with coordinated controls and don't let the nose wander.

TAKE-OFF AND DEPARTURE STALLS

These stalls will simulate a situation that happens too often – that of a stall occurring during the take-off or climb-out. It may be caused by the pilot's distraction at the critical time or he may be just showing off.

The maneuver will be done at slightly above take-off speed and in take-off configuration (this means flaps if you normally use them for take-off and, of course, gear down) and at recommended take-off power. They will be done from straight climbs *and* climbing turns of 15°-20° constant bank.

If you have some semblance of a balanced turn the top wing will stall first in most airplanes and the airplane will roll in that direction. The reason being that as the stall is approached the wings start losing Lift and the airplane mushes and starts to slip. The highest wing gets interference from the fuselage and quits first. You can check this by watching the ball throughout the approach and stall. You'll notice that if you can no longer keep the ball centered – it will indicate a slip – the high wing will stall first. Don't blindly expect this to happen because if you stall at a higher airspeed and are skidding, the bottom wing may be slowed to a point where it might stall first. There is one model trainer on the market that has a tendency in a *balanced* climbing right turn to have the inside (right wing) drop first (the left turns are as expected). Also, your particular airplane may have been rigged laterally after leaving the manufacturer so that funny things happen in the stall. Of course, that's why you practice them anyway -- to see what *your* airplane does and to find out the best method of recovery. But normally, you can count on an "over the top" type of stall. (The airplane rolls away from the ball.)

To recover, get the nose down and then level the wings with coordinated controls. Unless the roll is particularly vicious you can recover by merely relaxing back pressure.

PROCEDURE

1. Slow the airplane to about 5 knots above the stall in the take-off configuration (lift off speed).
2. Apply climbing power and start a shallow, banked turn in either direction, pulling the nose up steeply. (Practice them from straight climbs, also.) Note the altitude at the "lift off."
3. When the stall breaks, relax back pressure, use opposite rudder if necessary to stop rotation, and then level the wings (if necessary) with coordinated controls.

In practicing these stalls you'll find that if the bank is steep you'll have trouble getting a clean break. The nose may drop, causing the airplane to descend in a tight circle, shuddering and buffeting – with no stall break.

Keep your head swiveling all during the approach and stall. (If you have a midair collision on the flight test, you will probably be failed.)

Practice the stall in both directions as well as straight ahead.

COMMON ERRORS

1. Too steep a bank in the turning stalls – no definite stall break. The bank angle must be between 15°-20°.
2. Too early a recovery -- recovering before a definite break.
3. Too late a recovery – the airplane is allowed to rotate too far before recovery is started.
4. Failure to make a steep climb and thereby delaying the stall as the airplane mushes.

5. Overeagerness to get the nose down; abrupt forward pressure on the wheel or stick, with the result that the nose is pushed too low and excessive altitude is lost during recovery.

APPROACH TO LANDING OR GLIDING-TURN STALL

These stalls might be considered as a power-off version of the departure stalls. The airplane will be in the landing configuration and the engine throttled back. This stall demonstrates what *could* happen if the airplane is allowed to get too slow during a landing approach. This type stall will be harder to get to break cleanly, particularly if you are practicing solo. Of course, you may have a great deal of trouble getting the airplane to stall cleanly during practice and then dope off someday and find that the airplane *can* stall if you get sloppy and distracted. Practicing these stalls is intended to help you recognize what could happen and learn how to recover as quickly and safely as possible.

PROCEDURE

Keep looking around throughout the approach and stall.

1. With the airplane in landing configuration, establish a normal gliding turn of 20° to 30° bank in either direction (after applying carburetor heat if called for). Be sure to practice the stalls in both directions -- not all approaches are made from left hand patterns.
2. Flatten the glide through continued back pressure until the stall occurs.
3. Stop any rotation as you simultaneously release back pressure and apply *full* power.
4. Clean up the airplane and establish a climb.

You'll demonstrate this stall from straight glides as well as from moderately banked (20°-30°) turns. After recovery, get altitude in climb power and configuration (clean up the airplane) to at least 300 feet above the altitude at which you regained full control effectiveness.

It requires no will power to lower the nose when you're practicing stalls at 3000 feet. It's very easy up there — you just get the nose down and don't pay any attention to it. *But,* at a couple of hundred feet altitude you'd have to force yourself to release back pressure to recover from a stall. Fatal accidents have occurred because pilots have gone back to their instinct and ignored training when under stress.

Practice recoveries without adding power — note the difference.

COMMON ERRORS

1. Too steep a bank in the turning stall.
2. Allowing the wings to become level through inattention during the stall approach. You wouldn't have this problem during an actual landing approach because you'd be watching the runway and would be turning as necessary to line up.
3. Too early a recovery -- not allowing the stall to break.
4. Overeagerness in getting the nose down — excessive forward pressure during the recovery.

ACCELERATED STALLS

Here is proof positive that a stall is a matter of angle of attack, not airspeed.

You'll not do any of the accelerated stalls at an air-speed of more than 1.25 times the unaccelerated stall speed because of the possibility of overstressing the airplane, particularly in gusty air. The flaps will be retracted, for this same reason.

Accelerated stalls will be done from a turn of at least 45° of bank. There are two reasons for this: (1) this is the condition under which most pilots actually encounter the accelerated stall in flight — when they try to tighten the turn by pulling back on the wheel or by trying to maintain altitude without sufficient power, and (2) to avoid inadvertently pulling the nose *straight up* abruptly in practice, getting a whip stall (the airplane stalls with the nose up so steeply that it tends to slide backwards).

Unless you are skidding, the high wing will stall first, as was found in the departure and approach stalls, but the roll will be faster. The recovery is standard; release back pressure and return to level flight through use of coordinated controls.

Plan on adding power during recovery unless the check pilot says otherwise.

The check pilot will require that you recover (1) immediately upon stall recognition and (2) after a full stall develops and the nose falls below level flight attitude.

In (1) the recovery is simple. Relaxing the back pressure at the right time can result in the airplane's recovering in straight and level flight, if the high wing stalled first.

In (2) the nose will be low and the airplane will have rolled over into a steep bank. Relax back pressure and then bring the airplane back to straight and level flight through use of coordinated controls.

Recover quickly with a minimum loss of altitude — but don't get overeager and re-stall it.

Don't get over cruise speed at any time during the recovery.

PROCEDURE — LEVEL FLIGHT ACCELERATED STALL

1. Using power as needed to maintain altitude, make a 45° (or more) banked turn. Slow the airplane to no more than 1.25 times the normal stall speed.
2. Increase the angle of attack in a moderate climb or constant altitude until a stall occurs. Power may be reduced below cruising to aid in producing the stall, but any decrease in the rate of climb or loss of altitude will relieve the load factor (and the stall is not "accelerated"). Remember the term

"accelerated" means that above-normal load factors are present at the stall.

3. Release back pressure, open the throttle and recover to straight and level flight using coordinated controls.

If you are flying an aerobatic category airplane the 1.25 factor does not apply, but keep it below the maneuvering speed or recommended speed for accelerated maneuvers.

Incidentally, some airplanes do *not* give a sharp stall break and, with these, when the elevator is full up, initiate your recovery.

COMMON ERRORS

1. Jerking the wheel back.
2. The opposite -- timid application of back pressure.
3. Too brisk forward pressure on recovery -- hanging occupants on seat belts and putting undue negative stress on the airplane.

MANEUVERING AT MINIMUM CONTROLLABLE SPEEDS

On the commercial flight test you'll be required to demonstrate straight flight and turns, with and without power, at such a slow speed that any further reduction of airspeed would produce loss of effective control. This will be done both in the cruising and landing configurations in aircraft equipped with retractable gear and/or flaps.

Remember to keep turns shallow because of the stall speed increase with angle of bank. Expect to use more power to maintain altitude in the turns.

The problem that most pilots have in the straight portions is that of maintaining heading. They concentrate so hard on airspeed that the airplane is allowed to turn at will.

The following tolerances will be maintained:

STRAIGHT AND LEVEL FLIGHT

Altitude	no consistent change of altitude in excess of 100 fpm
Airspeed	within 5 knots of the desired speed

CLIMBING AND GLIDING TURNS

Airspeed	within 5 knots of the desired speed
Bank	20^0 to 30^0

Make sure that you don't exceed 30^0 of bank.

A good exercise for you to practice is as follows:

AIRPLANE CONFIGURATION — CRUISE

1. Throttle back to the estimated power required to maintain altitude at a speed just above stall. Maintain altitude and heading as the airplane slows and adjust power as necessary. Fly straight and level for two minutes.
2. Increase power for climb as you raise the nose. Don't let the airspeed change. Maintain your heading -- torque will now be more of a problem. Climb straight ahead for one minute. Make shallow climbing turns in each direction (15^0-20^0 banks).
3. Throttle back and resume level slow flight.
4. Carburetor heat "ON," throttle closed. Set up a glide, maintain the slow flight airspeed. (Don't jerk the throttle closed but ease it back as the nose is lowered to the glide position. (Fig. 25-3)
5. Make shallow gliding turns (20^0-30^0 banks) in each direction.

You can repeat the exercise with the airplane in landing configuration.

The transition to the various attitudes will give you the most trouble.

If your plane has a stall warner, it should be sounding off throughout the exercise.

COMMON ERRORS

1. Poor altitude control during the transition from cruise to slow flight.
2. Heading problems during straight and level and during climb.
3. Stalling the airplane.
4. Excessive changes of speed during transition from straight and level to climb or glide.

MISCELLANEOUS STALLS, PLUS SPINS

CROSS CONTROL STALL

This stall is most likely to occur during the turn

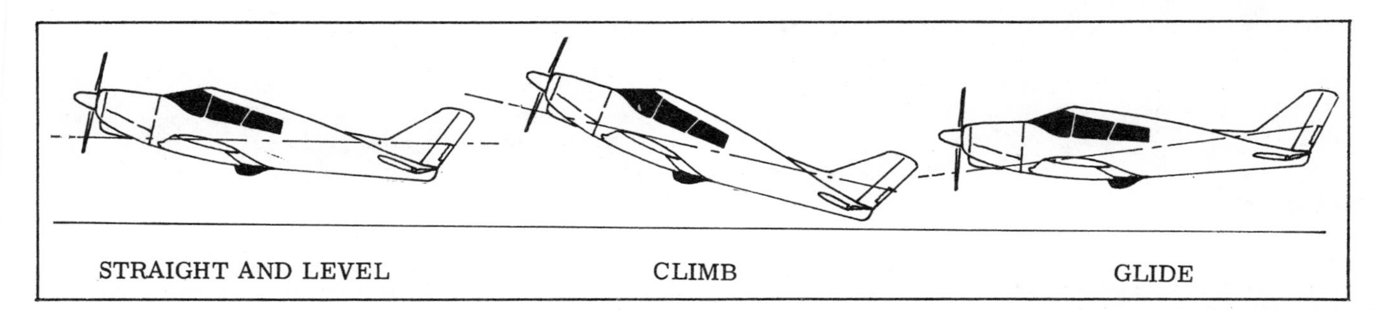

Fig. 25-3. Slow flight attitudes.

onto the final approach. A typical situation might be as follows:

The pilot sees that in the turn onto final he will go past the runway unless his turn rate is increased. Now, everybody knows that the stall speed increases with an increase in bank, so he figures that the best way to turn is by skidding around — and not increasing the bank. He starts applying inside rudder which increases the turn rate, but the outside wing is speeded up and the bank starts to increase. He takes care of this by using aileron against the turn, which actually helps drag the inside wing back farther. As the airplane is in a banked attitude, application of inside rudder (and opposite aileron) tends to make the nose drop, which is counteracted by increased back pressure. The situation? Perfect — for having one wing stall before the other, that is. True, the wing with the down aileron has more Lift (the down aileron acts as a flap, increasing the coefficient of Lift) but the Drag of that wing increases even more sharply which slows it more. So, the inside wing starts dropping which increases the angle of attack — and it stalls before the outside wing. Another term for this stall is "under the bottom stall" — an apt description.

You should be familiar with this stall and know that it *could* happen and note the best means of recovery.

Figures 25-4A through 25-4D show the probable sequence of such an actual stall at a low altitude. The series shows a *right* hand approach — an even more likely setup for such a stall, as you would be more apt to misjudge the turn.

PROCEDURE

1. Practice these stalls at a safe altitude and keep an eye out for other traffic. Use carburetor heat, if required. Make a shallow gliding turn in either direction.

2. Apply more and more inside rudder to cheat on the turn. Use opposite aileron as necessary to keep the bank from increasing.

3. Keep the nose up by increasing the back pressure.

4. When the stall break occurs, the roll will be rapid. Neutralize the ailerons, stop any further rotation with opposite rudder as you relax the back pressure.

5. Recover with coordinated controls and add power as the recovery progresses.

As the roll is fast, the bank may be vertical or past vertical. The nose will be low and speed will build up quickly. The usual error found in practice is that the airspeed is allowed to build up too high during recovery — which wastes altitude. Recover to straight and level flight with coordinated controls as soon as possible without overstressing the airplane or getting a secondary stall.

The bad thing about this stall is that it is most likely to occur at lower altitudes where recovery might be questionable. Practice it until you are able to recognize the conditions leading up to it — and then avoid those conditions.

COMMON ERRORS

1. Not neutralizing the ailerons at the start of the recovery.

2. Using too much opposite (top rudder to stop rotation, causing the airplane to slip badly.

Fig. 25-4A. The pilot sees on approach that unless the rate of turn is increased he'll fly past the runway and have to turn back to it.

Fig. 25-4B. This is sloppy flying and the other pilots will notice. He knows that the stall speed increases with a steeper bank so he decides to cheat by using inside rudder and skidding it around (and maybe use just a *touch* of opposite aileron).

Fig. 25-4C. Inside rudder means that the outside (left) wing will speed up, steepening the right bank. He applies more opposite (left) aileron. The cycle continues until he is in a shallow banked skid with well-crossed controls at a very low speed (he is using added back pressure to keep the nose up).

3. Hesitation in recovering — with a greater than necessary altitude loss.

SPINS

While spins are not required on the commercial flight test you should have some idea of the theory behind them.

A spin is an aggravated stall resulting in autorotation.

The airplanes that you've been flying are spin resistant; you have to make them spin. They should show no uncontrollable spin characteristics no matter how you use the controls. Most (though not all) light trainers will come out of a spin of their own accord if the pilot takes his hands and feet from the controls but this is not the most effective spin recovery method. You remember in the departure and approach stalls and the cross control stall that one wing stalled before the other and a rolling moment was produced. If you had held the wheel back and used rudder in the direction of roll a spin would have likely followed. The whole theory of the spin can be covered by realizing that one wing is stalled before the other and an imbalance of Lift is produced. If you continue to hold back pressure and hold rudder into the roll, this stalled condition remains and the roll continues as the nose "falls off" downward.

PROCEDURE — PRACTICE SPIN TO THE LEFT

1. Before doing any spins in an airplane you haven't spun before, it would be wise to have an instructor, experienced in spinning it, ride with you to demonstrate the entry and recovery procedures. Even if you've spun earlier models of this airplane, you'd better review the recovery procedure and

Fig. 25-4D. A view of the runway as seen from a near inverted position just after the stall occurs. Another statistic!

either talk to some of the local instructors or get them to ride with you. In newer models the manufacturer may have changed the geometry of the airplane with a more swept fin and rudder; or extended the propeller shaft; or added other factors that may have changed the airplane's recovery characteristics. Such changes can surprise you on that first spin.

Of course, you should make sure that the airplane is certificated for spinning and is properly loaded.

2. Get enough altitude so that you'll be recovered by 3000 feet above the ground (expect about 400 feet of altitude to be used for each turn of the spin in average, light two-place trainers).

3. Clear the area and start a normal power-off stall (use carburetor heat if recommended).

4. Just as the stall break occurs, apply — and hold — full left rudder; keep the wheel or stick full back. Some airplanes require a blast of power to get the spin started; the prop blast gives the rudder added effectiveness to yaw the airplane.

5. The nose drops as the airplane rolls but the full up-elevator does not allow the airplane to recover from the stall. The unequal Lift of the wings gives the airplane its rotational motion.

The spin is continued as long as the rudder and wheel are held as above.

The rotation of the airplane tends to continue the imbalance of Lift; the "down-moving" wing keeps its high angle of attack and remains stalled. The "up-moving" wing maintains a lower angle of attack.

If you should unconsciously relax back pressure during the spin, a spiral will result.

The properly executed spin is no harder on the airplane than a stall. A sloppy recovery puts more stress on the airplane than the spin itself.

Don't have the flaps down when practicing spins

because the airplane's spin characteristics may be changed, plus the fact that you might exceed the maximum flaps-down speed during the recovery.

THE RECOVERY

It would be well to first take the recovery step-by-step from a theoretical standpoint and then bring in some practical recovery procedures.

1. You used the rudder to induce the yaw in the spin entry, so opposite rudder should be applied to stop the yaw, equalize the Lift of the two wings, and stop autorotation. If the rudder effectively does this, it should be neutralized as soon as rotation stops or a spin in the opposite direction (a progressive spin) could be started.

2. At this instant of time in the theoretical look at recovery procedures, the rudder is neutral (the rotation has stopped) but you're still holding the wheel or stick full back and the airplane is still stalled, even though the nose appears to be pointed almost straight down. If you continued to hold the wheel full back the airplane would be buffeting and could, because of rigging, tend to whip off into a new spin in either direction.

The autorotation, or imbalance of Lift, has been broken, and now the stall recovery is initiated by (depending on the airplane) relaxing the back pressure *or* giving a brisk forward motion of the wheel or stick. For older and lighter trainers such as the J-3 *Cub*, Aeronca *Champion*, Taylorcrafts, or Cessna 120s, a slight relaxing of the back pressure was enough to assure that the stall (and spin) was broken. In some later trainers with higher wing loadings, the wheel or stick must be moved forward *briskly* well ahead of neutral to get the nose farther down and break the stall.

3. The third step is recovery from the dive after the autorotation and stall are broken. Sometimes the airspeed is allowed to get too high or too much back pressure is used on the pull-out (or both). Sometimes a pilot may get in a hurry to recover and relaxes back pressure (or uses a brisk forward motion as required) and immediately pulls back on the control wheel to "get it out of the dive," which results in a quick restalling of the airplane and a progressive (or new) spin could be set off. You've had this experience with plain everyday stalls; a too-quick recovery can put the airplane back into a stalled condition, and the process has to be started again.

Okay, the three steps mentioned were just that, a separation of required control movements for discussion purposes. You'll find that with some airplanes use of rudder alone won't stop the rotation, so that you wouldn't wait for it to take effect before applying forward pressure — which would help not only to break the stall but also aid the rudder in stopping the yaw and unbalanced Lift condition. In other words, you would apply full opposite rudder followed about a quarter turn later by brisk forward movement of the wheel and the controls held this way until rotation stops. Usually this procedure means that because of the almost simultaneous application of elevator with the rudder, the nose is down and the airspeed starts picking up as soon as the rotation stops; at this point the rudder is neutralized and back pressure is again applied to ease the airplane from the dive. (Some Owner's Handbooks suggest simultaneous use of rudder and elevator in the recovery.)

For some airplanes, ailerons against the spin speed it up, but for others the rate of rotation slows if ailerons are used opposite to the roll. A suggested all-around procedure, however, would be to leave the ailerons neutral throughout the spin and recovery to avoid possibly adding some unknown factors to the recovery.

SUMMARY OF SPINS

The Owner's Handbook or Airplane Flight Manual (and an instructor who has experience in spinning a particular airplane) will take precedence over the general look at spins as given here, but you might keep in mind the following notes about spin recoveries:

1. Most airplanes recover more promptly if the throttle is closed before using the aerodynamic controls.

2. A general recovery procedure would be to use full rudder opposite to the spin, followed almost immediately by a brisk forward movement, well ahead of neutral, of the wheel or stick. (Don't violently *jam* the wheel full forward — the airplane could be overstressed or an inverted spin entered.)

3. As soon as rotation has stopped, neutralize the rudder and use the elevator to help further break the stall or ease the airplane from the dive. One error found for pilots with comparative little spin recovery experience is that of continuing to hold opposite rudder after the rotation has stopped and airspeed builds up; this can cause heavy side loads on the vertical tail.

4. Normally docile airplanes can bite back if the center of gravity is near the aft limits of the envelope. Of course, you would not deliberately spin an airplane with people in the rear seats or with baggage back there but, if you get into a stall situation with this type of loading, don't let a rotation get started. For most airplanes the first two turns are an incipient spin condition and the spin can be stopped relatively easily. After that, moments of inertia can be such in the developed spin that recovery could be impossible. If it looks like the stall is getting out of hand, get that nose down farther with a brisk forward movement of the wheel (and you may decide that opposite rudder *and* wheel action is the best move to make). You can apologize to the sensitivities of the passengers later.

One good procedure for checking the spin characteristics (and your reaction) after getting the proper briefing and instruction and making sure the airplane is properly loaded is to set up on that solo practice less than one turn, recover, then do a turn and a half and recover, and so on until you've gotten the feel of the airplane in this area. (Climb back up after each spin.) If things start to feel funny to you that first spin entry or two, you can break it off before getting problems.

Altitude is money in the bank for spin practice.

26. CROSS-COUNTRY FLIGHT

Phase IV of the flight test is a cross-country flight and you may expect to be able to do the following:

CROSS-COUNTRY FLIGHT PLANNING

You'll plan for a trip of about three hours for your airplane and will include an intermediate stop for fuel.

In planning the flight you'll have to show that you can utilize available weather information (sequences, terminal and area forecasts plus actual and forecast wind data).

You'll plot the course, using a sectional or WAC chart and plan on using radio navigation aids as necessary to establish checkpoints and distances as well as estimating flying time, headings and fuel requirements. (The use of a computer or wind vector diagram for working out headings is desirable but not required.)

Don't fumble when it comes to finding out airport or other information such as NOTAMS in the *Airman's Information Manual* and also be able to get late teletype NOTAMS from Flight Service Stations and other sources.

You'll be allowed no more than 20 minutes to plan the trip, so don't dawdle. (Review Chapters 19 and 20.) *Don't* bring a pre-planned flight log. You'll have to do your planning for the flight examiner.

CROSS-COUNTRY FLYING

The examiner will give you the word when to get out on the planned trip. You'll fly enough of the trip to get established on course and have a good estimate of your groundspeed. He'll likely ask you to divert to an alternate airport. (He may select it or ask you to select it.) Establish the appropriate heading to the alternate within 10 minutes.

You'll be judged on your ability to make the desired track, correctly identify checkpoints, maintain heading and altitude and provide reasonable ETA's. It's also recommended that the flight include a landing at a strange airport.

You'll be expected to hold a heading within 10 degrees and track within one mile of the planned course. Your altitude tolerances enroute will be ±200 feet. Using the groundspeed established you should be able to compute an ETA to the point of first intended landing with an error of no more than 5 minutes per hour of cruising flight involved.

Be able to know the proper method of entering traffic at the strange field and make sure your communication procedures are sharp if you'll be dealing with a tower or FSS there. (You might review Chapter 10.)

CROSS-COUNTRY EMERGENCY

The examiner will simulate or ask you to simulate such things as engine overheating, partial or imminent power failure, being lost, weather problems (including loss of visual reference) icing, plus electric or hydraulic system failures.

In one emergency you'll be faced with partial power failure or imminent fuel exhaustion which will require an immediate landing. You'll be expected to use appropriate emergency procedures or simulate them if that is impracticable.

You might make sure that you have the idea of "dragging the area" down pat so that there will be no groping around if you have to pick an off-airport landing site. Watch your patterns and approach speeds. (Review Chapters 16 and 18.) Know your airplane systems.

RADIO AIDS TO VFR NAVIGATION

You'll demonstrate your ability to use radio aids such as following a particular VOR radial, following a low frequency (LF/MF) range course (if there are any left in your area) or using an ADF tuned to an appropriate facility. Make sure that the station is tuned properly and *identified.*

Know the procedures for obtaining direction finding (DF) steers or radar vectors. Review Chapter 19 and check the *Airman's Information Manual* for suggested procedures.

INSTRUMENT FLIGHT

You'll be required to demonstrate in simulated instrument flight your ability to control the aircraft manually by referring solely to the flight instruments. Chapter 18 covers the required maneuvers in detail.

SUMMARY OF THE CHAPTER

Part V of this book is based on the Flight Test Guide – Commercial Pilot – Airplane Single-Engine published by the FAA and it is suggested that you get a copy of this document at the local airport if available or order a copy from the U.S. Government Printing Office, Washington, D.C. 20402.

APPENDIX

STANDARD ATMOSPHERE CHART

Altitude (ft)	Pressure (in. Hg)	Pressure (psf)	Temp. (°C)	Temp. (°F)	Density-slugs per cubic foot
0	29.92	2116.22	15.0	59.0	.002378
1,000	28.86	2040.85	13.0	55.4	.002309
2,000	27.82	1967.68	11.0	51.9	.002242
3,000	26.82	1896.64	9.1	48.3	.002176
4,000	25.84	1827.69	7.1	44.7	.002112
5,000	24.89	1760.79	5.1	41.2	.002049
6,000	23.98	1695.89	3.1	37.6	.001988
7,000	23.09	1632.93	1.1	34.0	.001928
8,000	22.22	1571.88	-0.9	30.5	.001869
9,000	21.38	1512.70	-2.8	26.9	.001812
10,000	20.57	1455.33	-4.8	23.3	.001756
11,000	19.79	1399.73	-6.8	19.8	.001701
12,000	19.02	1345.87	-8.8	16.2	.001648
13,000	18.29	1293.70	-10.8	12.6	.001596
14,000	17.57	1243.18	-12.7	9.1	.001545
15,000	16.88	1194.27	-14.7	5.5	.001496
16,000	16.21	1146.92	-16.7	1.9	.001448
17,000	15.56	1101.11	-18.7	-1.6	.001401
18,000	14.94	1056.80	-20.7	-5.2	.001355
19,000	14.33	1013.93	-22.6	-8.8	.001310
20,000	13.74	972.49	-24.6	-12.3	.001267

CORRECTING ALTITUDE FOR NON-STANDARD TEMPERATURE

You may find the corrected altitude by using the following equations:

$$\text{Corrected altitude} = \text{Indicated Altitude} \times \frac{\text{OAT } (^{\circ}\text{C}) + 273}{\text{Standard Temp } (^{\circ}\text{C}) + 273}$$

The standard temperature used in the equation is the standard for the indicated altitude. To be absolutely correct, the calibrated altitude (corrected for instrument error) should be used, but indicated is usually close enough for most current altimeters.

As an example, assume an indicated altitude of 10,000 feet and an outside air temperature of +1°C (the standard temperature at 10,000 feet is -5°C). Using the equation:

$$\text{Corrected altitude} = 10{,}000 \times \frac{1 + 273}{-5 + 273} = 10{,}000 \times \frac{274}{268} = 10{,}224 \text{ feet.}$$

At 5000 feet the error would be half of that at 10,000 feet if the temperature were still 6°C higher than standard, since the altitude is a multiplier.

BIBLIOGRAPHY

Airman's Guide. U.S.G.P.O.

All-Weather Flight Manual (NAVAER 00-80T-60) issued by FAA Bureau of Flight Standards.

Ditching Sense. 1958. NAVAER 00-800-50.

Dommasch, Daniel O., *et al.* Airplane Aerodynamics. 2nd ed., 1957. Pitman Publishing Co. New York, N.Y.

Flight Test Guide – Commercial Pilot, Airplane. 1969. FAA Flight Standards Service. AC 61.117-1C.

G Sense. 1943. U.S. Navy.

Hurt, H. H., Aerodynamics for Naval Aviators (NAVWEPS 00-80T-80)

Kershner, W. K., *Instrument Flight Manual,* 2nd ed. Ames. Iowa, Iowa State University Press, 1970.

Kershner, W. K., *Student Pilot's Flight Manual,* 3rd ed. Ames, Iowa, Iowa State University Press, 1968.

Liston, Joseph, Power Plants for Aircraft. 1953. McGraw-Hill Book Co. New York, N.Y.

Perkins, C. D., and Hage, R. E. Airplane Performance, Stability and Control. 1949. John Wiley and Sons, New York, N.Y.

Pilot's Instruction Manual. 1959. U.S.G.P.O. Washington, D.C.

Pilot's Radio Handbook. 1962. U.S.G.P.O. Washington, D.C.

Pilot's Weather Handbook. 1955. U.S.G.P.O. Washington, D.C.

Shields, Bert. Air Pilot Training. 3rd ed., 1952. McGraw-Hill Book Co. New York, N.Y.

Wake Turbulence. AC 90-23A.

Airplane Owner's Manuals, Products referred to, or information received from companies:

Aero Commander, Inc., Bethany, Okla. – Commander 500.

Beech Aircraft Corp., Wichita, Kans. – Baron, Bonanza, Debonair, Queen Air 65, Twin Bonanza (E-50) and E18S.

Bendix Products, Aerospace Div., Bendix Aviation Corp., South Bend, Ind.

Cessna Aircraft Co., Wichita, Kans. – 150, 172, 180, 182, 210, 310.

Continental Development Corp., Ridgefield, Conn. – Instantaneous Vertical Speed Indicator (IVSI).

Hartzell Propeller, Inc., Piqua, Ohio.

Lycoming Division, AVCO, Williamsport, Pa. – Detail Engine Specifications.

Mooney Aircraft, Inc., Kerrville, Tex. – Mark 20.

Piper Aircraft Corp., Lock Haven, Pa. – Apache 160; Aztec B and C; Cherokee 140, 150, 160, 180 and 235; Colt; Comanche 180, 250 and 400; Pawnee; Super Cub; Twin Comanche.

INDEX

D

E

F

G

N

O

P

R

S

T

U

V

W